Electrochemical Strategies in Detection Science

RSC Detection Science Series

Titles in the Series:
1: Sensor Technology in Neuroscience
2: Detection Challenges in Clinical Diagnostics
3: Advanced Synthetic Materials in Detection Science
4: Principles and Practice of Analytical Techniques in Geosciences
5: Microfluidics in Detection Science: Lab-on-a-chip Technologies
6: Electrochemical Strategies in Detection Science

How to obtain future titles on publication:
A standing order plan is available for this series. A standing order will bring delivery of each new volume immediately on publication.

For further information please contact:
Book Sales Department, Royal Society of Chemistry, Thomas Graham House, Science Park, Milton Road, Cambridge, CB4 0WF, UK
Telephone: +44 (0)1223 420066, Fax: +44 (0)1223 420247
Email: booksales@rsc.org
Visit our website at www.rsc.org/books

Electrochemical Strategies in Detection Science

Edited by

Damien W. M. Arrigan
Curtin University, Perth, Australia
Email: d.arrigan@curtin.edu.au

ROYAL SOCIETY OF CHEMISTRY

THE QUEEN'S AWARDS FOR ENTERPRISE: INTERNATIONAL TRADE 2013

RSC Detection Science Series No. 6

Print ISBN: 978-1-84973-831-6
PDF eISBN: 978-1-78262-252-9
ISSN: 2052-3068

A catalogue record for this book is available from the British Library

Published by The Royal Society of Chemistry,
Thomas Graham House, Science Park, Milton Road,
Cambridge CB4 0WF, UK

Registered Charity Number 207890

For further information see our web site at www.rsc.org

Printed and bound by CPI Group (UK) Ltd, Croydon, CR0 4YY

Preface

Electrochemical methods of chemical analysis have been widely used for many years. This is seen most especially with the trusty pH electrode and conductivity meter, which are used widely in industrial and environmental applications. The everyday analyses of clinical samples for electrolytes, performed with ion-selective electrodes, and the mass-manufactured glucose test strips which place routine electrochemical measurements into the hands of non-scientists are further examples of the extensive use of electrochemical measurements. These examples illustrate that electrochemical methods and devices built on sound principles and evaluations serve vital functions outside of the research laboratory. Electroanalytical chemistry and electrochemical methods have the great advantage that they are sensitive and selective, can be easily made portable and miniaturized, and are often adaptable to new applications.

The purpose of this volume is to discuss recent advances in electrochemical methods and materials that may bring new strategies to bear on chemical and biochemical detection problems. The scope of the volume is to survey contemporary research and development advances within the areas of electrochemical detection based on new and re-vitalised methods, new materials with enhanced properties, and new devices that achieve better electroanalytical signal generation. As a result, the chapters collected here, written by leading researchers in the field, encompass advances in electrodes and devices, from microscale to nanoscale, electrochemical nanomaterials, and electrochemical behaviour and applications of soft interfaces and materials.

Chapter 1 deals with advances in stripping voltammetric detection of metals and addresses the replacement of mercury electrodes for this application. Chapter 2 continues the discussion of electrode devices, dealing specifically with microfabricated electrodes and their use in biomedical

RSC Detection Science Series No. 6
Electrochemical Strategies in Detection Science
Edited by Damien W. M. Arrigan

Published by the Royal Society of Chemistry, www.rsc.org

detection, while Chapter 3 discusses the use of electrodes in conjunction with microfabricated chips for chemical/biochemical separations and detection. Microelectrodes also feature in Chapter 4, in terms of scanning electrochemical microscopy, especially for biological systems characterization. Chapters 5–8 bring the discussions to the nanoscale, dealing with detection of nanoparticles (Chapter 5), nanofabricated electrode devices (Chapter 6), carbon nanomaterials (Chapter 7), and dispersible nanoparticle electrodes (Chapter 8). The final two chapters discuss electroanalytical opportunities derived from soft (liquid–liquid) interfaces (Chapter 9) and from room temperature ionic liquids (Chapter 10).

It is hoped that this collection of chapters will provide the interested reader with an introduction to some of the recent hot topics in electrochemical detection research and provide a platform for the design and development of further improvements in this area of detection science.

Damien W. M. Arrigan
Perth, Australia

Contents

RSC Detection Science Series No. 6
Electrochemical Strategies in Detection Science
Edited by Damien W. M. Arrigan

Published by the Royal Society of Chemistry, www.rsc.org

CHAPTER 1

Advances in Stripping Analysis of Metals

ANASTASIOS ECONOMOU*[a] AND CHRISTOS KOKKINOS[a,b]

[a] Department of Chemistry, National and Kapodistrian University of Athens, Panepistimiopolis Zografou, Athens 157 71, Greece; [b] Department of Chemistry, University of Ioannina, 45110 Ioannina, Greece
*Email: aeconomo@chem.uoa.gr

1.1 Introduction

Electrochemical stripping analysis (SA) has been established as a successful trace analysis technique for more than three decades. Stripping analysis is a general term that encompasses a wide family of electroanalytical techniques that make use of a "preconcentration" step of the analyte on the surface of the working electrode. In SA, the preconcentration step is followed by the detection step in which the accumulated analyte is detected and quantified by means of a voltammetric or chronopotentiometric scan. While in solution-phase electrochemical detection the redox signal is determined by the rate of mass transport of the electroactive analyte to the electrode surface, in SA the redox signal is determined by the amount of the accumulated analyte. Therefore, the accumulation step is responsible for the high sensitivity of SA whereas the different potential methods of accumulation of the analyte on the working electrode (discussed in Section 1.2) and the multi-parametric nature of SA provide versatility, wide applicability and enhanced selectivity. Typical metals and semi-metals that can be determined with SA are listed in Table 1.1. The theory and practice of SA are described in specialized monographs[1,2] and general review articles.[3–5]

RSC Detection Science Series No. 6
Electrochemical Strategies in Detection Science
Edited by Damien W. M. Arrigan

Published by the Royal Society of Chemistry, www.rsc.org

Table 1.1 Metals and semi-metals that can be determined by SA (in italics are the species that are normally determined only after electrolytic accumulation, underlined are the species that are normally determined after non-electrolytic accumulation and in bold are the species that can be determined after either electrolytic or non-electrolytic accumulation).

												Al					
		Sc	Ti	V	Cr	Mn	Fe	Co	Ni	**Cu**	**Zn**	Ga	Ge	**As**	*Se*		
		Y	Zr		Mo	Tc			Pd	**Ag**	**Cd**	**In**	**Sn**	**Sb**			
		La	Hf						Pt	**Au**	**Hg**	*Tl*	**Pb**	*Bi*			

Ce	Pr				Eu			Dy	Ho			Yb	Lu
Th		U											

A brief account of the birth and gradual evolution of SA is provided in earlier work.[1,5,6] SA was first conceptualized and applied in 1931 by Zbinden to determine low concentrations of copper at a platinum electrode.[7] However, not until the 1950s was SA revisited when the hanging mercury drop electrode was developed by Kemula. In the same decade, the remarkable sensitivity of SA was stressed by many workers and new types of thin mercury-film electrodes were introduced. In the 1960s, the theory for SA and pulsed potential–time waveforms was developed. The commercial availability of low cost multi-purpose apparatus and the expanding use of SA to many new analytical applications took place in the 1970s. The development of non-electrolytic preconcentration approaches and of automated flow-through manifolds, together with the introduction of microelectrodes for SA, dominated the 1980s and 1990s. Recent advances in SA are concentrated on research on new eco-friendly electrode materials, the extended use of microfabrication technologies and miniaturization and the development of integrated devices for on-site monitoring. Therefore, this chapter will not focus on specific applications but will try to highlight developments in sensor types, fabrication and materials.

1.2 The Principle of Stripping Analysis

The distinctive feature of SA in comparison to other electroanalytical techniques is the accumulation/preconcentration step. In most applications involving analysis of metals, preconcentration is carried out with electrolysis of the target metal cations to the respective metals (and some special cases to metal oxides) and deposition on the surface of the working electrode[1,2] (Figure 1.1A). Following accumulation, the accumulated metals are oxidized (or the metal oxides reduced) to the respective cations and stripped back into the solution. Depending on the nature of the working electrode surface, the

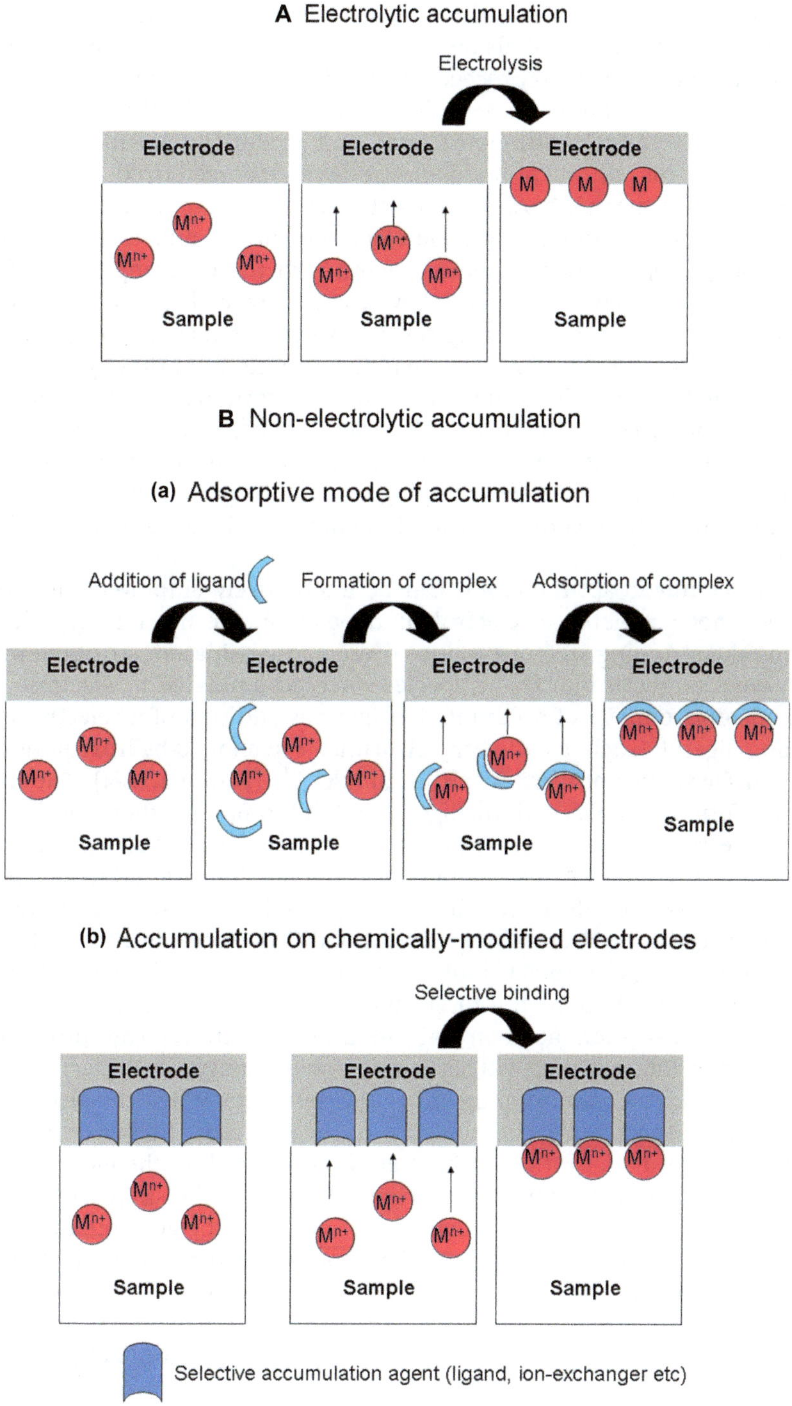

Figure 1.1 Modes of accumulation in SA of metals.

deposited metal can form a thin film, an amalgam or an alloy with the electrode. Reduction of metals on inert electrodes (such as carbon, platinum or boron-doped diamond) proceeds *via* the formation of a monolayer of the target metal (underpotential deposition, UPD) followed by the deposition of the bulk-phase metal. Although simple and potentially useful, electrolytic preconcentration on inert solid electrodes is characterized by multiple stripping peaks that are hard to interpret.[8,9] This is due to the easy formation of intermetallic compounds on the electrode surface and to the different oxidation (stripping) potentials associated with the different states of the metal deposits. Amalgams are typically generated when heavy metals deposit on mercury electrodes or during the deposition of mercury on gold electrodes. Finally, alloys are formed when bismuth, tin or antimony electrodes are used for electrolytic accumulation of alloy-forming metals. Amalgamation and alloying facilitate the electrolytic deposition and stripping of metals and lead to stripping signals that are easier to interpret. Following electrolytic accumulation, the stripping step can be performed by an anodic voltammetric scan, a constant oxidation current or by a chemical oxidizing agent.[1,2]

Non-electrolytic accumulation can be alternatively employed for metals that are not amenable to electrolytic deposition, are not readily oxidized during the stripping step or produce overlapping oxidation stripping peaks with other target or interfering species when accumulated by electrolysis.[10] The adsorptive mode of accumulation involves addition of a selective complexing ligand (such as dimethylglyoxime, catechol, 8-hydroxyquinoline) with surface-active properties in the sample[10–13] (Figure 1.1B-a). The metal cations form a complex with the ligand and the complex is then physisorbed on the electrode surface. The stripping step is usually based on the reduction of the metal cation in the accumulated complex but the reduction of the ligand or catalytic effects can be also exploited.[13,14] A second strategy is based on the utilization of chemically-modified electrodes.[10,15,16] In this case, the working electrode is bulk- or surface-modified with a suitable agent that can selectively bind the target metal cation. In the context of SA of metals, accumulation at chemically-modified electrodes can proceed by electrostatic interactions (with a clay or ion-exchanger modifier) or complexation (with a complexing ligand as the modifier). When the electrode is immersed in the sample, the modifier on the electrode interacts and binds with the target metal cations, which are accumulated on the electrode surface in their native oxidation state (Figure 1.1B-b). In the simplest scenario, the stripping step can be a reductive scan in which the accumulated metal cations are reduced.[9] A more complex protocol involves potentiostatic reduction of the accumulated metal cations followed by an anodic stripping step in which the target metals are oxidized.[9] Again, stripping can be performed by either a voltammetric scan or a (reductive or oxidizing) constant current.

The nomenclature in SA is not standardized and usually more or less established empirical abbreviations define a technique by reference to the accumulation mode, the stripping step or both. To avoid the confusion caused by this practice, the guidelines proposed earlier are adopted here.[17] When voltammetry is used in the stripping step, the terms anodic stripping voltammetry (ASV, anodic scan) and cathodic stripping voltammetry (CSV, cathodic scan) are used. In ASV and CSV, the peak-shaped current–potential (i–E) response is recorded and the oxidation peak current is measured and related to the respective metal concentration in the sample. The voltammetric scan is normally performed in the differential-pulse (DP) or the square-wave (SW) mode, which offer better rejection of the background current and lower limits of detection.[1] When the stripping step is performed by an oxidant in solution or a constant current (irrespective of its polarity), the term stripping chronopotentiometry (SCP) is used. In SCP, the potential of the working electrode is recorded with respect to time as the target analytes are oxidized (by the application of current or chemically) and the residence time (time during which the potential of the working electrode remains stable) is related to the respective metal concentration in the sample. In modern instruments, the response is converted into a peak-shape by plotting the differential dt/dE with respect to potential. The relative strengths of the potentiometric and voltammetric stripping modes have been reviewed.[18] In both cases, the mode of accumulation of the target analyte(s) should be explicitly defined as electrolytic or non-electrolytic.[17]

Some more specialized variations of SA, particularly useful for metal speciation, include the "competitive ligand equilibration–adsorptive cathodic stripping voltammetry" and the "absence of gradients and Nernstian equilibrium stripping (ANGES)" approaches.[19] Abrasive SA is widely used for the elemental analysis of solid samples of archeological[20] or forensic[21] interest.

In SA, the sensitivity critically depends on the amount of the analyte accumulated onto the surface of the electrode during the preconcentration step and, therefore, on the rate of mass transport. Efficient mass transfer of the analyte to the surface of the working electrodes during the preconcentration step is normally carried out under conditions of forced convection by mechanical stirring of the solution or rotation of the electrode. However, in addition to these conventional methods of enhancing the mass transport efficiency, some unconventional techniques to enhance mass transfer (ultrasound, microwaves or heating of the working electrode) have been developed. The principles, enhancing mechanisms, instrumentation and applications of these approaches have been reviewed.[22] In "sonoelectrochemistry", ultrasound is applied with frequencies between 20 and 100 kHz and power densities from *ca.* 8 to 500 $W\,cm^{-2}$. The high energy density transferred by the sound wave promotes turbulent solution flow and cavitational effects that lead to reduction of the diffusion layer thickness and *in situ* cleaning/activation of the electrode. Heating of electrodes causes heat to stream out of the heated surface and warm up the solution near the electrode. Density gradients between hot and cold regions induce laminar flow. This "micro-stirring" effect is more efficient than mechanical stirring and can be exploited during the accumulation step in SA. Electrode heating can be achieved by means of a high frequency AC current or microwave heating. Finally, application of constant or high-frequency magnetic fields have been reported for SA.

1.3 Advances in Electrodes, Sensors and Devices

1.3.1 Carbon Electrodes

Carbon still remains one of the most important and widely used electrode materials in SA. Applications of carbon-based electrodes in SA of metals have been reviewed previously.[23,24] Glassy carbon, impregnated graphite, carbon paste, carbon fibres, carbon-containing composite materials, carbon nanomaterials or screen-printed inks can be used; graphite ink screen-printed electrodes will be covered in Section 1.3.2.

Glassy carbon (GC) is still a widely used material in SA. Since the sensitivity of bare GC in the SA of metals is low, the GC surface is usually subject to preliminary (*ex situ*) surface modification and/or *in situ* modification to achieve the desired surface properties. Thin films of metals, conducting polymers, selective organic reagents, clay minerals and ion-exchangers have been reported for surface modification of GC.[23] Occasionally, more than one modifying layers is used. Metal films (mercury, bismuth, gold, antimony, copper) facilitate the accumulation of metal cations and the stripping procedure of deposited metals. Conducting polymers (polyaniline, cellulose acetate, Nafion) can provide selectivity against macromolecules in complex samples and may serve as cation exchangers, thus facilitating the accumulation of metals. Organic ligands such as crown ethers and EDTA (ethylenediaminetetraacetic acid) (sometimes immobilized in polymers) are used to fabricate chemically-modified electrodes suitable for non-electrolytic accumulation of metal cations.

Carbon paste electrodes (CPEs) are made by homogenizing graphite and a water-immiscible binding liquid (paraffin, silicon oil, tricresyl phosphate). These CPEs are suitable for *in situ* or *ex situ* surface modification with metal films (mercury, bismuth, antimony) or metal oxides (which are converted into the respective metals by reduction).[23,25] CPEs are also very convenient for bulk modification by incorporating organic ligands (crown compounds, calixarenes), clays (vermiculite and montmorillonite) and ion exchangers (resins) within the paste material in order to facilitate metal preconcentration.[23,25]

Carbon composite electrodes are made of graphite, carbon, glassy, carbon or diamond powders mixed with polymeric binders such as epoxy resins, methacrylate, silicon, styrene–acrylonitrile and polyester. Although there are some reports of unmodified composite electrodes in SA of metals, most applications involve bulk or surface modification of composite materials by metal films or organic ligands that lead to higher sensitivity and better selectivity.[24,26]

Nanoscale carbon-based materials (single-walled carbon nanotubes (SWCNTs), multi-walled carbon nanotubes (MWCNTs), fullerenes and graphene) exhibit many characteristics that make them ideal as electrode materials. Due to their strong adsorption ability, nanoscale carbon-based materials can be used as sorbents for metal ions.[27] Therefore, there are many recent reports of the detection of multiple heavy metal ions by SA following a sorption/preconcentration step on nanoscale carbon-based electrodes either unmodified or modified with chelators, metal films or clays.[28–36]

Boron-doped diamond (BDD) consists of a diamond lattice *p*-type doped with boron typically to $10^{20}\,cm^{-3}$ (which corresponds to approximately 1 boron atom per 1000 carbon atoms) ensuring good conductivity and electrochemical properties. Over the last 15 years diamond electrodes have been used for a wide range of electrochemical applications and one of the main areas on which research has focused is the use of unmodified or metal film-modified boron-doped diamond as a mercury-free alternative for analysis using SA.[37] Due to their inertness, BDD electrodes are compatible with ultrasound-, microwave-, heating- and laser-enhanced deposition in SA.

1.3.2 Screen-Printed Electrodes

The fabrication of screen-printed electrodes (SPEs) involves squeezing the printing material (an ink paste) through a screen mesh (which defines the electrode's geometrical pattern) onto an electrically insulating substrate. The advantages of screen printing as a method for electrode fabrication are the low cost (in terms of the capital cost of the printing apparatus and the cost of consumables), the flexibility in selecting and modifying the electrode material, the potential for miniaturization and the scope for mass production of disposable sensors. SPEs have been widely used in SA of metals and their fabrication and applications have been covered in comprehensive reviews.[9,38,39] The printing material is normally some type of graphite-containing ink but gold, platinum and silver SPEs can be produced by using ink loaded with these metals. Additionally, screen-printed microelectrodes or microelectrode arrays with various configurations can also be prepared by different technologies.[24] Bare SPEs have been reported with some success but these are normally of limited utility due to the complicated voltammograms obtained. SPEs covered with mercury films and, more recently, with bismuth and antimony films are the most popular such sensors combined with electrolytic ASV and SCP.[38,39] SPEs made by using gold-containing ink or coated with gold are also frequently used for the determination of mercury and arsenic while silver and platinum SPEs have been occasionally used. Numerous applications employing chemically-modified SPEs have been reported for non-electrolytic accumulation of metals.[38,39] A chelating ligand can be incorporated on the surface of the SPE by drop-coating or into the bulk of the SPE by mixing in with the carbon ink prior to printing.

1.3.3 Solid Amalgam Electrodes

Recent research has indicated that solid amalgam electrodes can, in many cases, successfully substitute mercury electrodes in SA. Amalgam electrodes are essentially solid electrodes consisting of an amalgam of mercury with another metal (such as silver) or of an amalgam dispersed in a conductive polymer or pasting liquid. The classification, fabrication methods and analytical applications of amalgam electrodes in SA of metals have been extensively reviewed.[40] Amalgams are claimed to be non-toxic since the mercury contained in the amalgam is not readily released.[41] The main challenges in developing solid amalgam electrodes approaching the performance of their mercury counterparts are: (a) to obtain a sufficiently wide cathodic potential window (since the presence of various metals in the amalgam normally causes a reduction of the cathodic overpotential of mercury) and (b) to achieve a reproducible and stable surface.[42] In practical terms, silver amalgam electrodes in conjunction with ASV and electrolytic deposition are the most widely used for SA of metals.[40,42]

1.3.4 "Green" Metal and Metal-film Electrodes

In 2000, a new type of "green" metallic electrode, the bismuth-film electrode (BiFE), was introduced, consisting of a thin layer of bismuth electroplated on a conductive substrate.[43] The BiFE was the bismuth counterpart of the existing mercury film electrode (MFE), itself widely used in SA of heavy metals.[44] This pioneering work initiated an intense research effort for the development of different types of eco-friendly electrodes for SA. Bismuth-based electrodes are the most promising "green" sensors for SA of heavy metals using both electrolytic and non-electrolytic accumulation as demonstrated in some earlier reviews.[45–48] The attractions of these transducers are their low toxicity combined with the ability to form low-temperature alloys with many metals similar to the amalgams formed with mercury. In its simplest form, the electrode consists of a bismuth rod or bismuth-containing composite or carbon paste. BiFEs are the most popular bismuth electrodes and can be fabricated by *in situ* or *ex situ* electroplating of a thin bismuth film on a conductive substrate (noble metals, different forms of carbon) from a Bi(III) solution or by deposition on glass or silicon *via* thin-layer microelectronic techniques.[49–51] The bismuth film can also be coated with polymeric materials (such as Nafion) to improve the selectivity and sensitivity. Recently, there is an increasing number of applications involving the use of carbon-based nanomaterials (nanotubes, graphene oxide) in conjunction with BiFEs.[52,53] Another interesting possibility is to incorporate a bismuth precursor compound (bismuth oxide or an insoluble bismuth salt) in carbon paste or screen-printed graphite inks;[54,55] when used for cathodic electrolytic accumulation, the bismuth precursor compound is converted into metallic bismuth that forms a film on the surface of the electrode.

Following the success of bismuth electrodes, antimony-film electrodes (SbFEs) were introduced in 2007.[56] Antimony is also a less toxic electrodic material than mercury and has been shown to exhibit excellent performance in more acidic media.[57] The usual fabrication method of antimony (SbFEs) electrodes is again electroplating but microfabrication approaches have also been used.[58,59] Finally, tin-film,[60,61] lead-film,[62,63] gallium-film,[64] selenium-film[65] and tantalum-film[66] electrodes have been reported as mercury-free sensors for SA of metals. The main drawback of all these alternative electrodes compared to mercury is their limited anodic range since these metals are oxidized at a more negative potential than mercury. The comparative performance of electroplated and precursor-modified bismuth, tin and antimony electrodes with the benchmark MFE for the detection of Cd(II) and Pb(II) by DPASV is illustrated in the Figure 1.2.

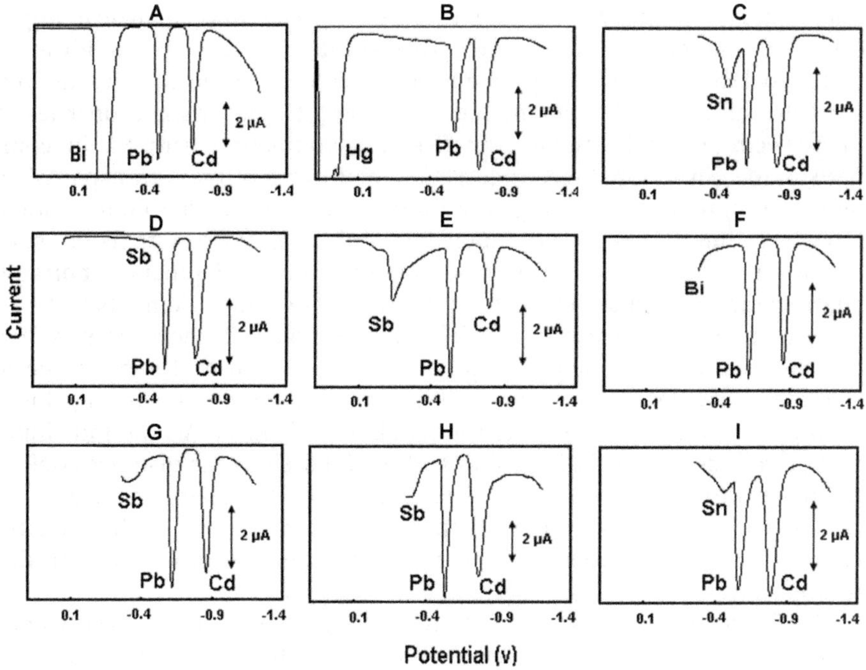

Figure 1.2 Comparative DPASV responses in a solution containing 25 μg L^{-1} Pb(II) and 50 μg L^{-1} Cd(II) at *in situ* electroplated screen-printed sensors: (A) BiFE, (B) MFE, (C) SnFE, (D) SbFE (in acetate buffer), (E) SbFE (in 0.01 mol L^{-1} HCl) and at screen-printed sensors modified with 4% (w/w): (F) bismuth oxide; (G) antimony oxide; (H) antimony oxalate hydroxide; (I) antimony tin oxide. Buffer, 0.1 mol L^{-1} acetate (pH 4.5) (except (D); preconcentration time, 240 s; preconcentration potential, −1.4 V (except (C) and (I), −1.5 V); plating metal concentrations, 1 mg L^{-1}.
(Reproduced with permission from ref. 60.)

Gold electrodes have been widely used mainly for the detection of mercury[67] and aresenic[68] and to a lesser extent for the detection of some other heavy metals. A drawback of bare gold electrodes is that the simultaneous determination of Cd(II) and Pb(II) is not possible.[69]

1.3.5 Microelectrodes and Microelectrode Arrays

IUPAC defines as a microelectrode any electrode whose characteristic dimension is, under the given experimental conditions, comparable to or smaller than the diffusion layer thickness (δ)[70] (typically ranging from a few μm to 50 μm). Microelectrodes exist in different geometries and can be arranged as single microelectrodes or as arrays (ensembles) of microelectrodes. The most popular arrangements of single microelectrodes are those of disk, hemisphere, sphere-caps and cylinder while arrays (or ensembles) of microelectrodes can be configured as microdisks, microchemispheres, microbands and interdigitated structures.[6,71] The first analytical applications of microelectrodes were recorded during the 1980s when advances in microelectronic technology made it possible to fabricate micron-sized structures in a reproducible manner.[6,72] Several characteristic properties of microelectrodes make them attractive in combination with SA in comparison to conventional electrodes:[6,71–73] high rates of mass transport, low ohmic drop, reduced charging current and enhanced signal-to-noise ratios. Moreover, due to their small dimensions, microelectrode systems are suitable for miniaturization, which makes possible the development of portable analytical instruments for on-site measurements of trace elements.[72,73]

A drawback in the use of single microelectrodes is the extremely low currents that are obtained. A solution to this is the use of regular microelectrode arrays (MEAs) whereby multiple microelectrodes are arranged in a geometrical pattern and operated in parallel.[6,71–73] In MEAs, the individual microelectrodes' critical dimension and the inter-electrode distances define the mass transport regime and are critical for the MEA performance.[6,71–74] Under conditions of non-overlapping diffusion regimes, the electrochemical signal increases in proportion to the number of microelectrodes in the array.

The fabrication methods and analytical applications of microelectrodes and MEAs in SA of metals have been comprehensively reviewed.[6,71–73] Construction of MEAs is usually based on thin-film microelectronic technologies as well as screen-printing. Single microelectrodes may consist of carbon fibres, carbon microdisks, platinum, gold, silver, iridium and alloy microdisks and microwires.[6,71] MEAs can consist of carbon, platinum, gold, iridium, rhodium, indium tin oxide (ITO), BDD and alloy microelectrodes. These can be used directly in SA of metals using UPD but multi-element analysis is usually complicated by intermetallic compound formation, multiple stripping peaks and narrow cathodic potential range. More common is the modification of these substrates with mercury, bismuth and gold films as well as with antifouling coatings (such as agar and Nafion).[6,71–73]

1.3.6 Microfabricated, Cell-on-a-chip, Paper-based and Wearable Devices

Advances in, and wider access to, microfabrication technologies have enabled the realization of different cell-on-a-chip and integrated miniaturized devices for SA of metals. Single "green" bismuth, antimony and tin sensors have been fabricated by thin layer microengineering approaches[50,59,75] and this work was extended to complete cell-on-a-chip devices.[76–78] A series of publications has reported the implementation of thin-film integrated three-electrode devices with working electrodes made of copper, palladium and bismuth and their application to SA of metals.[79–81]

Microfluidic devices combine the advantages of microfabrication with convenient delivery of low volume of sample using flow in microchannels. The earliest example of such a device, consisting of a microchannel in glass with electrokinetic flow and a mercury-coated SPE electrode for Ni(II) detection by SA, was reported in 2000.[82] Plastic fluidic chips with integrated electrodes made of carbon-loaded polystyrene have been fabricated and used for metal detection by SA.[83–85] Several other designs of micromachined fluidic chips with varying degrees of complexity have been reported for trace metal detection using bismuth[86–89] and gold or silver[90–92] working electrodes. An integrated electronic tongue device with gold working electrode suitable for the detection of heavy metals by CSV and ASV in wastewater or seawater is illustrated in Figure 1.3. The monitor employs a gold electrode for Mn and As CSV detection (cell 1) and a mercury-covered silver electrode for Zn, Cd, Pb and Cu detection by ASV (cell 2).

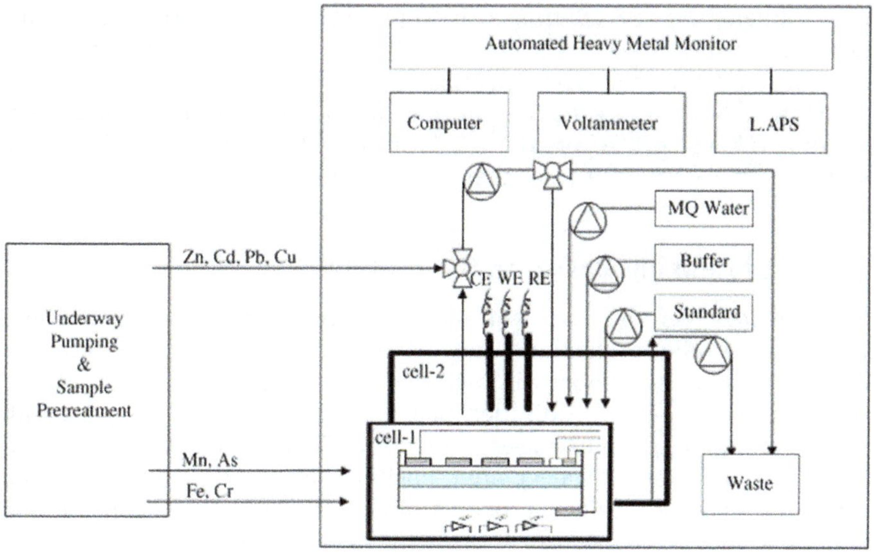

Figure 1.3 Scheme of the electronic tongue instrument system for detection of Mn and As by CSV and Zn, Cd, Cu and Pb by ASV. (Reproduced with permission from ref. 92.)

Paper-based devices (or μPADs) constitute a new exciting field of research.[93,94] In μPADs, an hydrophilic paper substrate is patterned with hydrophobic areas using different technologies to manipulate sample flow. In the context of electrochemical detection, electrodes are printed on the paper layer enabling SA detection of different metals.[95–99]

Finally, wearable sensors (integrated on either a textile material or directly on the epidermis) can be used for multi-analyte monitoring in the user's environment.[100] Wearable electrochemical sensors based on screen-printing have been used for heavy metals detection by ASV after electrolytic preconcentration in marine environments[101] or for the detection of gunshot residue by abrasive stripping voltammetry.[102]

1.3.7 Flow Systems and Probes for Environmental and Personal Exposure Monitoring

Advantages of coupling SA to flow manifolds include better control of the preconcentration efficiency, the scope for increased automation, faster optimization, sample pretreatment, calibration and analysis, higher precision and accuracy and lower consumption of sample and reagents.[103] Flow injection analysis (FIA), sequential injection analysis (SIA) and multicomutation manifolds combined with wall-jet, thin layer or tubular cells have been reported for SA of metals.[103] A relatively recent development is the lab-on-a-valve (LOV) arrangement which allows integration of on-line miniaturized sample manipulation and detection modules within a specially-designed multi-port selection valve.[104–106] Flow systems also allow easy hyphenation of SA with spectroscopic techniques (mass spectroscopy (MS) inductively coupled plasma-MS (ICP-MS) and atomic emission spectrometry (ICP-AES)) for metal analysis.[107–111]

Different flow probes for monitoring of trace metals by SA in the environment have been reported.[5,112–116] On the other hand, portable on-line systems can be used as personal exposure monitors to trace metals.[117]

1.3.8 Biosensing Based on SA of Metals

SA of metals can be exploited for the indirect detection of different biomolecules (oligonucleotides, proteins) or whole cells.[118–122] The principle of these assays is to label the target biomolecules with a reported biomolecular probe (*via* hybridization or bioaffinity interactions) tagged with metal (silver, gold) nanoparticles or quantum dots (nanocrystals composed of metal sulfides or selenides). After labelling, the metal nanoparticles and quantum dots are oxidized or dissolved, respectively, to release the respective metal cations, which are then detected and quantified by SA. Specific strategies to implement detection of several biomolecules and cells through SA of metal-containing labels are reported in recent reviews.[118–122] An example is illustrated in Figure 1.4 that demonstrates a method for the

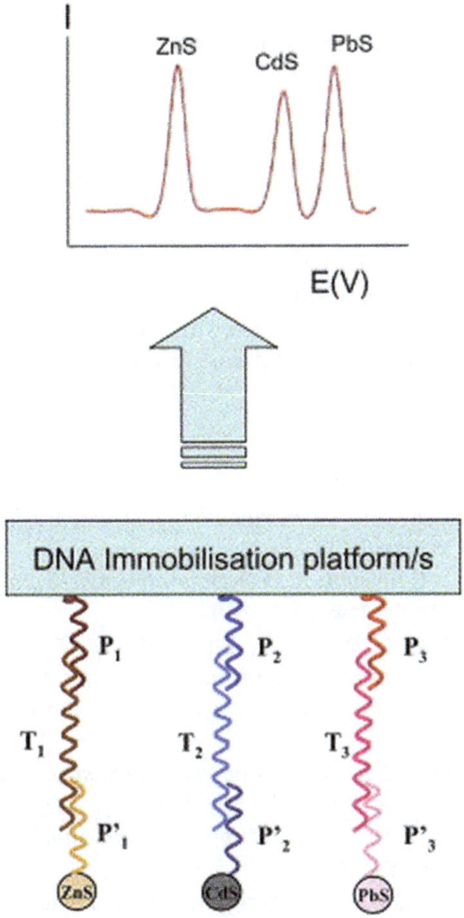

Figure 1.4 Schematic of multiple detection of DNA. DNA probes P'_1, P'_2 and P'_3 bearing different DNA sequences with different nanoparticles (ZnS, CdS and PbS, respectively) that enable the simultaneous detection of three DNA targets T_1, T_2 and T_3 hybridized with corresponding DNA-capturing probes P_1, P_2 and P_3 immobilized onto a direct or an indirect (magnetic particles) transducing platform(s). (Reproduced with permission from ref. 123.)

simultaneous assay of three oligonucleotide targets after labelling DNA probes with different nanoparticles (ZnS, CdS and PbS quantum dots); the quantum dots produce well-defined and resolved stripping peaks for Zn, Cd and Pb at a mercury-coated glassy carbon electrode.[123]

1.4 Conclusions

SA, more than 80 years after its first application, is still an active field of research. While the basic principles and instrumentation remain unchanged, the recent research trends revolve around new materials, microfabrication technologies and biosensing. Undoubtedly, the introduction of bismuth electrodes (as well as antimony and tin electrodes) as substitutes of mercury electrodes has brought about a revival of SA as a viable and environment-friendly trace metal analysis technique. Screen-printing electrodes and microelectrodes continue to receive attention as sensors in SA using both electrolytic and non-electrolytic accumulation. Microfabricated lab-on-a-chip and microfluidic devices as well as paper-based and wearable sensors constitute emerging fields in which SA plays a predominant role as detection technique. Finally, the applicability of SA is further extended by its ability to detect metallic labels in order to indirectly monitor biomacromolecules in clinical samples.

References

1. J. Wang, *Stripping Analysis: Principles, Instrumentation and Applications*, VCH, Deerfield Beach, FL, 1985.
2. Kh. Brainina, E. Neyman, *Electroanalytical Stripping Methods*, John Wiley & Sons, Inc., New York, 1993.
3. J. Wang, B. Tian, J. Wang, J. Lu, C. Olsen, C. Yarnitzky, K. Olsen, D. Hammerstrom and W. Bennett, *Anal. Chim. Acta*, 1999, **385**, 429.
4. Kh. Z. Brainina, N. A. Malakhova and N. Y. Stojko, *Fresenius' J. Anal. Chem.*, 2000, **368**, 307.
5. J. Wang, *Comprehensive Analytical Chemistry*, ed. S. Alegret and A. Merkoçi, Elsevier, 2007, ch. 6, vol. 49, pp. 131–141.
6. S. Daniele, M. A. Baldo and C. Bragato, *Curr. Anal. Chem.*, 2008, **13**, 215.
7. C. Zbinden, *Bull. Soc. Chim. Biol.*, 1931, **13**, 35.
8. G. Herzog and D. W. M. Arrigan, *TrAC, Trends Anal. Chem.*, 2005, **24**, 208.
9. K. C. Honeychurch and J. P. Hart, *TrAC, Trends Anal. Chem.*, 2003, **22**, 456.
10. J. Wang, *Electroanalytical Chemistry*, ed. A. J. Bard, Marcel Dekker, New York, 1989, ch. 1, vol. 16, pp. 1–81.
11. R. Kalvoda and M. Kopanica, *Pure Appl. Chem.*, 1989, **61**, 97.
12. M. G. Paneli and A. Voulgaropoulos, *Electroanalysis*, 1993, **5**, 355.
13. C. M. G. van den Berg, *Anal. Chim. Acta*, 1991, **50**, 265.
14. A. Bobrowski and J. Zarebski, *Curr. Anal. Chem.*, 2008, **13**, 191.
15. D. W. M. Arrigan, *Analyst*, 1994, **119**, 1953.
16. W. Kutner, J. Wang, M. L'her and R. P. Buck, *Pure Appl. Chem.*, 1998, **70**, 1301.
17. A. G. Fogg and J. Wang, *Pure Appl. Chem.*, 1999, **71**, 891.
18. N. Serrano, J. M. Díaz-Cruz, C. Ariño and M. Esteban, *Electroanalysis*, 2007, **19**, 2039.
19. M. Pesavento, G. Alberti and R. Biesuz, *Anal. Chim. Acta*, 2009, **631**, 129.

20. A. Doménech-Carbó, M. T. Doménech-Carbó, L. Osete-Cortina, J. V. Gimeno-Adelantado, S. Sánchez-Ramos and F. Bosch-Reig, *Electroanalysis*, 2003, **15**, 1465.
21. A. M. O'Mahony and J. Wang, *Electroanalysis*, 2013, **25**, 1341.
22. P. Gründler, *Curr. Anal. Chem.*, 2008, **13**, 263.
23. N. Y. Stozhko, N. A. Malakhova, M. V. Fyodorov and Kh. Z. Brainina, *J. Solid State Electrochem.*, 2008, **12**, 1185.
24. N. Y. Stozhko, N. A. Malakhova, M. V. Fyodorov and Kh. Z. Brainina, *J. Solid State Electrochem.*, 2008, **12**, 1219.
25. K. Kalcher, I. Švancara, R. Metelka, K. Vytras, A. Walcarius, *Encyclopedia of Sensors*, ed. C. A. Grimes, E. C. Dickey and M. V. Pishko, American Scientific Publishers, 2006, vol. X, pp. 1–147.
26. A. Merkoçi, S. Alegret, *Comprehensive Analytical Chemistry*, ed. A. Merkoçi and S. Alegret, Elsevier, 2007, vol. 49, pp. 143–161.
27. J. G. Yu, X. H. Zhao, L. Y. Yu, F. P. Jiao, J. H. Jiang and X. Q. Chen, *J. Radioanal. Nucl. Chem.*, 2014, **299**, 1155.
28. A. Safavi, M. Sorouri and P. Khanipou, *Electroanalysis*, 2014, **26**, 359.
29. A. Afkhami, H. Khoshsafar, H. Bagheri and T. Madrakian, *Mater. Sci. Eng., C*, 014, **35**, 8.
30. A. M. Ashrafi, S. Cerovac, S. Mudric, V. Guzsvány, L. Husáková, I. Urbanová and K. Vytras, *Sens. Actuators, B*, 2014, **191**, 320.
31. J. Wei, D. Yang, H. Chen, Y. Gao and H. Li, *Sens. Actuators, B*, 2014, **190**, 968.
32. J. T. Zhang, Z. Y. Jin, W. C. Li, W. Dong and A. H. Lu, *J. Mater. Chem. A*, 2013, **1**, 13139.
33. W. Wonsawat, S. Chuanuwatanakul, W. Dungchai, E. Punrat, S. Motomizu and O. Chailapakul, *Talanta*, 2012, **100**, 282.
34. C. Hong, A. Wong and M. Pumera, *RSC Adv.*, 2012, **2**, 6068.
35. C. M. Willemse, K. Tlhomelang, N. Jahed, P. G. Baker and E. I. Iwuoha, *Sensors*, 2011, **11**, 3970.
36. A. K. Wanekaya, *Analyst*, 2011, **136**, 4383.
37. S. E. Ward Jones and R. G. Compton, *Curr. Anal. Chem.*, 2008, **13**, 170.
38. K. C. Honeychurch, *Insci. J.*, 2012, **2**, 1.
39. X. Niu, M. Lan, H. Zhao, C. Chen, Y. Li and X. Zhu, *Anal. Lett.*, 2013, **46**, 2479.
40. B. Yosypchuk and J. Barek, *Crit. Rev. Anal. Chem.*, 2009, **39**, 189–203.
41. T. Navrátil, I. Svancara, K. Mrázová, K. Nováková, I. Sestáková, M. Heyrovsky, D. Pelclová, *Sensing in Electroanalysis*, ed. K. Kalcher, R. Metelka, I. Svancara, K. Vytras, University of Pardubice, Pardubice, 2011, vol. 6, pp. 23–53.
42. E. Mikkelsen and K. H. Schrøder, *Electroanalysis*, 2003, **15**, 679.
43. J. Wang, J. M. Lu, S. B. Hocevar, P. A. M. Farias and B. Ogorevc, *Anal. Chem.*, 2000, **72**, 3218.
44. A. Economou and P. R. Fielden, *Analyst*, 2003, **128**, 205.
45. C. Kokkinos and A. Economou, *Curr. Anal. Chem.*, 2008, **4**, 183.

46. I. Svancara, C. Prior, S. B. Hocevar and J. Wang, *Electroanalysis*, 2010, **22**, 1405.
47. A. Economou, *TrAC, Trends Anal. Chem.*, 2005, **24**, 334.
48. J. Barón-Jaimez, M. R. Joya and J. Barba-Ortega, *J. Phys.: Conf. Ser.*, 2013, **466**, 012025.
49. C. Kokkinos, I. Raptis, A. Economou and T. Speliotis, *Electroanalysis*, 2010, **22**, 2359.
50. C. Kokkinos, A. Economou, I. Raptis and C. E. Efstathiou, *Electrochim. Acta*, 2008, **53**, 5294.
51. I. Yanagimachi, N. Nashida, K. Iwasa and H. Suzuki, *Sci. Technol. Adv. Mater.*, 2005, **6**, 671.
52. J. Ping, Y. Wang, J. Wu and Y. Ying, *Food Chem.*, 2014, **151**, 65.
53. R. Ouyang, W. Zhang, S. Zhou, Z. L. Xue, L. Xu, Y. Gu and Y. Miao, *Electrochim. Acta*, 2013, **113**, 686.
54. N. Lezi, C. Kokkinos, A. Economou and M. I. Prodromidis, *Sens. Actuators, B*, 2013, **182**, 718.
55. M. Khairy, R. O. Kadara, D. K. Kampouris and C. E. Banks, *Electroanalysis*, 2010, **22**, 1455.
56. S. B. Hocevar, I. Svancara, B. Ogorevc and K. Vytras, *Anal. Chem.*, 2007, **79**, 8639.
57. E. Tesarova, L. Baldrianova, S. B. Hocevar, I. Svancara, K. Vytras and B. Ogorevc, *Electrochim. Acta*, 2009, **54**, 1506–1510.
58. C. Kokkinos and A. Economou, *Sens. Actuators, B*, 2014, **192**, 572.
59. C. Kokkinos, A. Economou, I. Raptis and T. Speliotis, *Electrochem. Commun.*, 2009, **11**, 250.
60. M. Maczuga, A. Economou, A. Bobrowski and M. I. Prodromidis, *Electrochim. Acta*, 2013, **114**, 758.
61. E. Czop, A. Economou and A. Bobrowski, *Electrochim. Acta*, 2011, **56**, 2206.
62. A. Bobrowski, A. Królicka, M. Maczuga and J. Zarebski, *Sens. Actuators, B*, 2014, **191**, 291.
63. M. Korolczuk, A. Stepniowska and K. Tyszczuk, *Int. J. Environ. Anal. Chem.*, 2009, **89**, 727.
64. K. Tyszczuk, M. Korolczuk and M. Grabarczyk, *Talanta*, 2007, **71**, 2098.
65. Y. Nagaosa, P. Zong and A. Kamio, *Microchim. Acta*, 2009, **167**, 241.
66. W. Zhang, Z. Liu, S. Zhu, J. Chen and G. Xu, *Electrochem. Commun.*, 2010, **12**, 1291.
67. C. Gao and X. J. Huang, *TrAC, Trends Anal. Chem.*, 2013, **51**, 1.
68. D. E. Mays and A. Hussam, *Anal. Chim. Acta*, 2009, **646**, 6.
69. G. M. S. Alves, J. M. C. S. Magalhães, R. Tauler and H. M. V. M. Soares, *Electroanalysis*, 2013, **25**, 1895.
70. K. Stulik, C. Amatore, K. Holub, V. Marecek and W. Kutner, *Pure Appl. Chem.*, 2000, **72**, 1483.
71. X. Xie, D. Stueben and Z. Berner, *Anal. Lett.*, 2007, **38**, 2281.
72. G. Herzog and V. Beni, *Anal. Chim. Acta*, 2013, **769**, 10.
73. V. Beni and D. W. M. Arrigan, *Curr. Anal. Chem.*, 2008, **13**, 229.

74. P. Tomčík, *Sensors*, 2013, **13**, 13659.
75. C. Kokkinos, A. Economou and T. Speliotis, *Electrochem. Commun.*, 2014, **38**, 96.
76. C. Kokkinos, A. Economou and M. Koupparis, *Talanta*, 2009, **77**, 1137.
77. C. Kokkinos and A. Economou, *Talanta*, 2014, **125**, 215.
78. C. Kokkinos, A. Economou and I. Raptis, *Anal. Chim. Acta*, 2012, **710**, 1.
79. W. Kang, X. Pei, A. Bange, E. Haynes, W. R. Heineman, I. Papautsky, *Book of Abstracts: 17th International Conference on Miniaturized Systems for Chemistry and Life Sciences*, 27–31 October 2013, Freiburg, Germany, pp. 901–903.
80. P. Jothimuthu, R. A. Wilson, J. Herren, X. Pei, W. Kang, R. Daniels, H. Wong, F. Beyette, W. R. Heineman and I. Papautsky, *Electroanalysis*, 2013, **25**, 401.
81. W. Kang, X. Pei, W. Yue, A. Bange, W. R. Heineman and I. Papautsky, *Electroanalysis*, 2013, **25**, 2586.
82. J. Wang, R. Polsky, B. Tian and M. P. Chatrathi, *Anal. Chem.*, 2000, **72**, 5285.
83. N. G. Naseri, S. J. Baldock, A. Economou, N. J. Goddard and P. R. Fielden, *Anal. Bioanal. Chem.*, 2008, **391**, 1283.
84. N. G. Naseri, S. J. Baldock, A. Economou, N. J. Goddard and P. R. Fielden, *Electroanalysis*, 2008, **2008**, 448.
85. Z. M. Redha, S. J. Baldock, P. R. Fielden, N. J. Goddard, B. J. Treves Brown, B. G. D. Haggett, R. Andres and B. J. Birch, *Electroanalysis*, 2009, **21**, 422.
86. P. Jothimuthu, R. A. Wilson, J. Herren, E. N. Haynes, W. R. Heineman and I. Papautsky, *Biomed. Microdevices*, 2011, **13**, 695.
87. Z. Zou, A. Jang, E. T. MacKnight, P. M. Wu, J. Do, J. Sub Shim, P. L. Bishop and C. H. Ahn, *IEEE Sens. J.*, 2009, **9**, 586–594.
88. P. Jothimuthu, E. N. Haynes, I. Papautsky, *Book of Abstracts: Transducers 2009*, Denver, CO, USA, 21–25 June 2009, pp. 1846–1848.
89. Z. Zou, A. Jang, E. MacKnight, P. M. Wu, J. Do, P. L. Bishop and C. H. Ahn, *Sens. Actuators, B*, 2008, **134**, 18.
90. R. A. B. da Silva, E. G. N. de Almeida, A. C. Rabelo, A. T. C. da Silva, L. F. Ferreira and E. M. Richter, *J. Braz. Chem. Soc.*, 2009, **20**, 1235.
91. M. R. Gongora-Rubio, M. B. A. Fontes, Z. M. da Rocha, E. M. Richter and L. Angnes, *Sens. Actuators, B*, 2004, **103**, 468.
92. H. Men, S. Zou, Y. Li, Y. Wang, X. Ye and P. Wang, *Sens. Actuators, B*, 2005, **110**, 350.
93. E. W. Nery and L. T. Kubota, *Anal. Bioanal. Chem.*, 2013, **405**, 7573.
94. M. Benhabib, X. J. Li, *Microfluidic Devices for Biomedical Applications*, ed. X. J. Li and Y. Zhu, Woodhead Publishing Ltd, Cambridge, 2013, pp. 492–522.
95. Z. Nie, C. A. Nijhuis, J. Gong, X. Chen, A. Kumachev, A. W. Martinez, M. Narovlyansky and G. M. Whitesides, *Lab Chip*, 2010, **10**, 477.
96. Q. M. Feng, Q. Zhang, C. G. Shi, J. J. Xu, N. Bao and H. Y. Gu, *Talanta*, 2013, **115**, 235.

97. T. Nurak, N. Praphairaksit and O. Chailapakul, *Talanta*, 2013, **114**, 291.
98. H. L. Fang, H. X. Zheng, M. Y. Ou, Q. Meng, D. H. Fan and W. Wang, *Sen. Actuators, B*, 2011, **153**, 369.
99. S. N. Tan, L. Ge and W. Wang, *Anal. Chem.*, 2010, **82**, 8844.
100. J. R. Windmiller and J. Wang, *Electroanalysis*, 2013, **25**, 29.
101. K. Malzahn, J. R. Windmiller, G. Valdés-Ramírez, M. J. Schöning and J. Wang, *Analyst*, 2011, **136**, 2912.
102. A. J. Bandodkar, A. M. O'Mahony, J. Ramírez, I. A. Samek, S. M. Anderson, J. R. Windmiller and J. Wang, *Analyst*, 2013, **138**, 5288.
103. A. Economou, *Anal. Chim. Acta*, 2010, **683**, 38–51.
104. M. Miró and E. H. Hansen, *Anal. Chim. Acta*, 2012, **750**, 3.
105. Y. Wang, Z. Liu, J. Tang, G. Yao and X. Hu, *Anal. Methods*, 2011, **3**, 731.
106. Y. Wang, Y. Wu, J. Xie, H. Ge and X. Hu, *Analyst*, 2013, **138**, 5113.
107. A. J. Baca, A. B. De La Ree, F. Zhou and A. Z. Mason, *Anal. Chem.*, 2003, **75**, 2507.
108. N. Parthasarathy, M. Pelletier, M. L. Tercier-Waeber and J. Buffle, *Electroanalysis*, 2001, **13**, 1305.
109. G. X. Cao, O. Jimenez, F. Zhou and M. Xu, *J. Am. Soc. Mass Spectrom.*, 2006, **17**, 945.
110. J. R. Pretty, G. J. Van Berkel and D. C. Duckworth, *Int. J. Mass Spectrom.*, 1998, **178**, 51.
111. A. J. Baca, Y. Garcia, A. L. Briseno and F. Zhou, *J. Electroanal. Chem.*, 2011, **513**, 25.
112. J. Wang, B. Tian, J. Wang, J. Lu, C. Olsen, C. Yarnitzky, K. Olsen, D. Hammerstrom and W. Bennett, *Anal. Chim. Acta*, 1999, **385**, 429.
113. J. Wang, J. Wang, J. Lu, B. Tian, D. MacDonald and K. Olsen, *Analyst*, 1999, **124**, 349.
114. C. B. Braungardt, E. P. Achterberg, B. Axelsson, J. Buffle, F. Graziottin, K. A. Howell, S. Illuminati, G. Scarponi, A. D. Tappin, M. L. Tercier-Waeber and D. Turner, *Mar. Chem.*, 2009, **114**, 47.
115. J. Buffle and M. L. Tercier-Waeber, *TrAC, Trends Anal. Chem.*, 2005, **24**, 172.
116. M. L. Tercier-Waeber, F. Confalonieri, M. Koudelka-Hep, J. Dessureault-Rompré, F. Graziottin and J. Buffle, *Electroanalysis*, 2008, **20**, 240.
117. W. Yantasee, Y. Lin, K. Hongsirikarn, G. E. Fryxell, R. Addleman and C. Timchalk, *Environ. Health Perspect.*, 2007, **115**, 1683.
118. A. Merkoçi, *Electroanalysis*, 2013, **25**, 15.
119. L. Ding, A. M. Bond, J. Zhai and J. Zhang, *Anal. Chim. Acta*, 2013, **797**, 1.
120. M. T. Castañeda, S. Alegret and A. Merkoçi, *Electroanalysis*, 2007, **19**, 743.
121. S. Marín and A. Merkoçi, *Electroanalysis*, 2012, **24**, 459.
122. A. Merkoçi, *Biosens. Bioelectron.*, 2010, **26**, 1164.
123. A. Merkoçi, M. Aldavert, S. Marín and S. Alegret, *TrAC, Trends Anal. Chem.*, 2005, **24**, 341.

CHAPTER 2

Development of Microelectrode-based Biosensors for Biomedical Analysis

ANTON GUIMERÀ, ELISABET PRATS-ALFONSO, ROSA VILLA AND F. JAVIER DEL CAMPO*

Instituto de Microelectrónica de Barcelona, IMB-CNM (CSIC), Campus de la Universidad Autónoma de Barcelona, 08193 - Bellaterra, Barcelona, Spain
*Email: javier.delcampo@csic.es

2.1 Introduction to Microelectrode Arrays

Electroanalytical devices have a long tradition in the biomedical area, with the use of pH electrodes and the amperometric measurement of blood oxygen dating back to the 1950s,[1] and the early introduction of the glucose biosensor in the 1960s.[2] These first devices were bulky and measurements had to be carried out *ex vivo,* even during surgical procedures.

The limitations inherent to those large and cumbersome instruments were seen as opportunities in terms of miniaturization, which eventually led to the introduction of devices that relied on microelectrodes and microelectrode arrays. In this context, diagnostics and the monitoring of chemical and biochemical parameters have not been the only areas to take advantage of microfabricated devices in biomedicine. In fact, electrophysiology and drug discovery have been the main beneficiaries of these advances in miniaturization and electrochemical science. Figure 2.1 shows different

RSC Detection Science Series No. 6
Electrochemical Strategies in Detection Science
Edited by Damien W. M. Arrigan

Published by the Royal Society of Chemistry, www.rsc.org

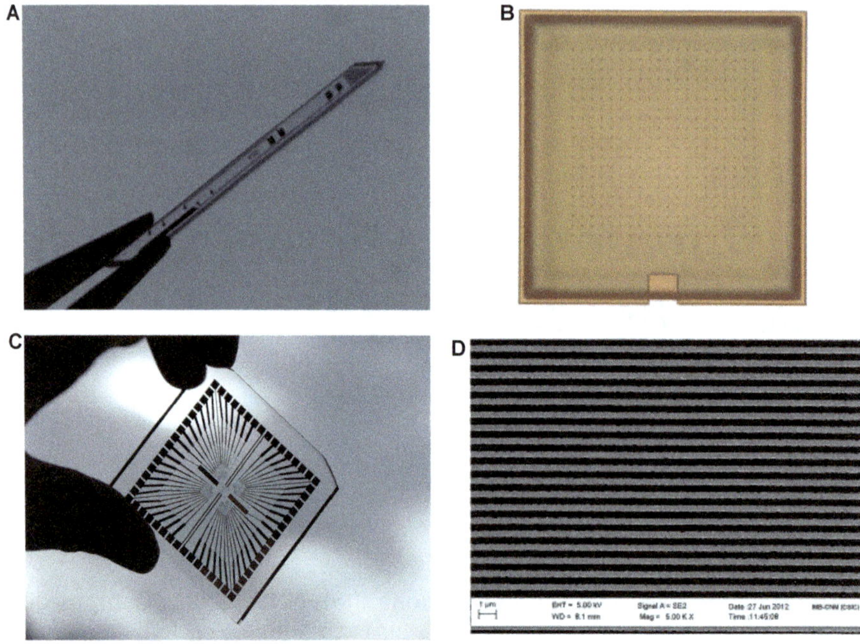

Figure 2.1 Microelectrode arrays: (A) microelectrodes on a needle; (B) pyrolyzed photoresist film (carbon) microelectrode array; (C) individually addressable microelectrode array for cell studies; (D) SEM detail of Au interdigitated microband electrodes.

microelectrode arrays used in electroanalysis. The first microfabricated microelectrode arrays, similar to that shown in Figure 2.1(C), were used back in 1972 for the *in vitro* study of heart cells.[3] Since then, electrophysiology and drug discovery have been the most active fields for (individually addressable) microelectrode arrays and impedance analysis.[4–7]

This chapter is about the development of miniaturized electroanalytical devices for diagnostic applications, with an emphasis on microfabricated biosensor systems.

2.1.1 Why Microelectrodes?

Microelectrodes, broadly defined as electrodes having at least one dimension in the micrometre range, offer many advantages compared to macroelectrodes.[8–11] Microelectrodes allow the possibility to measure in small volumes and/or with a very high spatial resolution *without perturbing* their surroundings. Second, their small size enables them to observe much higher mass transport rates, which means enhanced sensitivity and rapidity of response. Third, they display very low capacitive currents and experience very low *iR*-drop levels, which allow them to work under physiological conditions without the need for added supporting electrolyte. These advantages are particularly relevant in biomedical applications, where accurate *in vivo* measurements may be required, or available sample volumes are in the low μL range.

On the other hand, their small size also results in low-current signals (pA–nA) that are hard to measure and call for particularly sensitive equipment and adequate shielding from background electrical noise. One way to overcome these drawbacks and improve the signal-to-noise ratio of individual microelectrodes is to group them in arrays with individual microelectrodes connected in parallel. The additive nature of the current results in signal enhancements, so that it is possible to keep some of the advantages of individual microelectrodes, particularly greater sensitivity, while recording currents in the μA range that are much easier to measure and more unlikely to be masked by noise. However, microelectrode arrays experience larger *iR* drops than macroelectrodes of the same area. This is because electrode resistance increases with decreasing microelectrode size. As the number of microelectrodes in an array increases, the current passing through the array is enhanced by hemispherical diffusion, but the overall array resistance is much higher than that of an equivalent macroelectrode because the actual active area is still very small. To illustrate this point, imagine a disk microelectrode array composed of 10 μm radius microelectrodes distributed in a square lattice with a centre-to-centre separation of 100 μm. This array features 100 microelectrodes per mm^2. However, the combined area of these microelectrodes is only 0.03 mm^2. This means that the electrical resistance of such an array is approximately 30 times greater per mm^2 than a macroelectrode occupying the same surface area. The conclusion is that one should be careful when deciding to use microelectrode arrays in low conductivity media.

Regarding microelectrode geometries, the most commonly used in arrays are disks and bands (Figure 2.2, left- and right-hand side, respectively). Disks are normally used in batch measurements such as patch-clamp experiments or in stagnant solutions. Band microelectrodes, on the other hand, are better suited for so-called generator–collector experiments,[12] conductivity

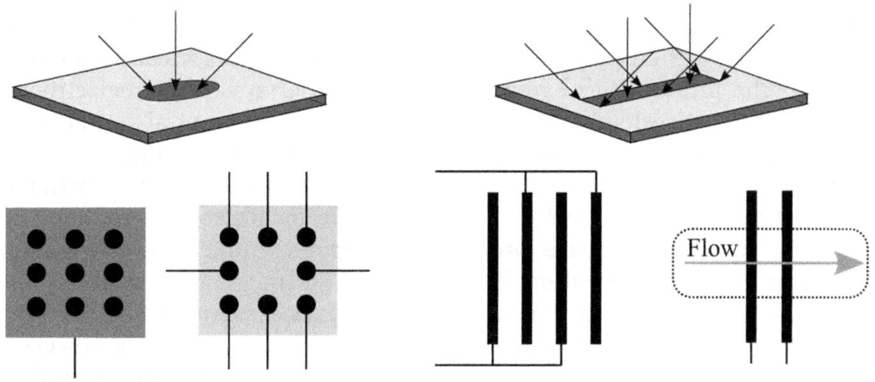

Figure 2.2 Schematic representation of commonly used electrode geometries and electrode arrangements.

measurements,[13] and in flow applications.[14–17] Microband electrode arrays usually contain at least two individually addressable bands or sets of microbands, as in the case of interdigitated electrodes (IDEs).[12,18,19]

Other microelectrode geometries used in arrays are rings[20–23] and pillars.[24,25] Although these geometries also benefit from enhanced mass transport rates, their fabrication is more complicated and costly than that of more prosaic disk and band-based devices.

The arrangement of the microelectrodes within an array also deserves consideration. Microelectrodes may be regularly or randomly distributed in an array, and they may be connected in parallel or be individually addressable.[26–29] Most of the microelectrode arrays used in electroanalytical applications are regular arrays of microelectrodes polarized in parallel.[8,10] However, the importance and tradition of electrophysiology dictate that individually addressable disk microelectrodes dominate in biomedical applications. Microelectrode arrays can be used in electrophysiology applications to study the interactions between electrogenic cells with a higher spatial resolution than conventional electrodes.[30–32] The main application fields in electrophysiology are the study of cardiac cells and tissues, the response of the retina to light impulses, and of course the activity of neurons.

Microelectrode arrays take advantage of their small size to probe the electrode–solution interface, and enable the study of phenomena that are otherwise inaccessible to macroelectrodes.[33–36]

2.1.2 Microfabrication Opens New Horizons

Voltamperometric detection at microelectrode arrays was not reported until 1985–1986 by Anderson *et al.*, who produced a flow cell using an array of microband electrodes.[37,38] Since then, electroanalytical applications of microelectrode arrays span across several medical specialties and include the determination of electrolytes, gases, hormones, proteins and pathogens, demonstrating the importance and utility of microelectrode arrays.[5,6,10,39–45]

The availability of microfabrication techniques outside the electronics industry led to the development of new microelectrode-based analytical tools able to tackle hitherto unapproachable problems. Microfabrication not only enables the production of microelectrodes and arrays of microelectrodes, such as in the development of multi-analyte platforms, but also their integration with other functions, *e.g.* microfluidics, on a single chip.

Microfabrication involves the creation of structures on planar substrates, mainly silicon wafers, using a combination of photolithographic and additive and/or subtractive processes. These techniques were originally developed for the fabrication of transistors and integrated circuits by the microelectronics industry.[46] Eventually, the excess capacity arising from the industry's rapid development resulted in the application of these processes to other areas of science and technology, which led to an entire new field known as micro-electro-mechanical-systems (MEMS). This includes lab-on-a-chip devices, also known as μ-total analysis systems (μ-TAS), and in turn electrochemical devices.[47–49] In a broad sense, the lab-on-a-chip concept involves the integration of as many laboratory operations as possible on a single miniaturized device, so as to minimize waste, improve safety and favour portability and automation of instrumentation. Despite this, examples of microelectrode arrays integrated with electronics on the same chip are relatively uncommon,[50–55] and the most usual case is to find them separated from the instrumentation.

One reason is that while the microfabrication of electrodes is relatively straightforward, fabricating an application specific integrated circuit (ASIC) involves tens of lithographic levels and is much more costly. Add this to the fact that microfabricated electrodes do not last nearly as long as an integrated circuit (IC) and it normally makes more sense to keep microelectrodes and ICs physically separated. Microfabrication techniques have evolved from the silicon industry and it is nowadays quite common to find microelectrode-based devices fabricated on glass and polymeric substrates alike.[56–59] This transition in materials from silicon to plastic represents gains in terms of costs but also of manufacturability and functionality of new devices.

Another critical aspect in the fabrication of microsystems, including microelectrode arrays, is their packaging. Packaging, or encapsulation, refers to how a chip is interfaced with the macro-world, how the chip is connected and how these connections are protected from the application environment so that the chip remains fully functional longer. Chip packaging has very important cost implications, and often is the *main* cost driver behind microsystems. This chapter explores the most common approaches to the fabrication of microelectrode-based devices and their packaging, and will discuss some of their advantages and limitations.

2.1.3 Biomedical Applications of Microelectrodes

Electrochemical devices for biomedical applications follow three main streams: electrophysiology,[60–62] basic (medical) research tools[40,49,63] and diagnostic devices.[63–65] The first two areas aim at the discovery of new knowledge, for which they seek higher levels of performance while development cost becomes a second-order issue. The diagnostic area, on the other hand, is more concerned with analytical reliability, ease of use, scalability and low-cost manufacturing processes. This is clearly illustrated by the difference in cost and functionality found between a typical glucose test strip and a microelectrode array used in electrophysiology. Different fabrication techniques discussed in this chapter are applied in the construction of all of these devices, but this chapter focus will on diagnostic applications.

Nowadays there is a clear trend towards personalized healthcare where point-of-care devices are seen as crucial to provide timely and reliable information.[66,67] Perhaps the best-known example is the glucometer, a small hand-held device that is daily used by millions of diabetic patients for the determination of their blood glucose concentration.[68] Here development emphasizes decreasing production costs and increasing functionality and "intelligence" of the devices by providing more data management capabilities. Thus, some of today's glucometers come with important information management functions, such as storing and analysing data series *in the cloud* that helps patients manage their health condition more effectively and improve their overall life-quality.

Personal care devices for the diagnosis of cardiovascular diseases may not enjoy the same degree of maturity, but significant efforts are nevertheless being dedicated to the rapid diagnostic of common acute conditions such as stroke or heart failure. It is these less mature areas that provide more opportunities to explore new and more daring detection strategies.

Therefore, we will focus on the detection of cardiovascular disease markers to illustrate the benefits of microfabricated electrochemical devices.

2.2 Design of Microelectrode Array-based Systems

Target analytes and application scenarios dictate most of the design aspects behind detection systems, including electrochemical devices. Here we will focus on those design aspects underpinning the development of electrochemical biosensors. Figure 2.3 represents the different aspects surrounding biosensor design, and how they influence each other. Starting from the target analyte, biosensor design implies choosing the right bioreceptor and deciding on the most suitable detection technique, while transducer materials and bioreceptors define the functionalization approach. Last, sensitivity and other performance requirements dictate transducer geometry and measurement technique. This section provides an overview of these biosensor construction elements.

2.2.1 Analytes and Bioreceptors

The presence of certain biomarkers in a body fluid (tear, blood, sweat and so on) above a certain concentration indicates illness or a health disorder. Biosensors can be devised to detect, monitor and quantify the concentration of such biomarkers and help medics make better decisions. Although

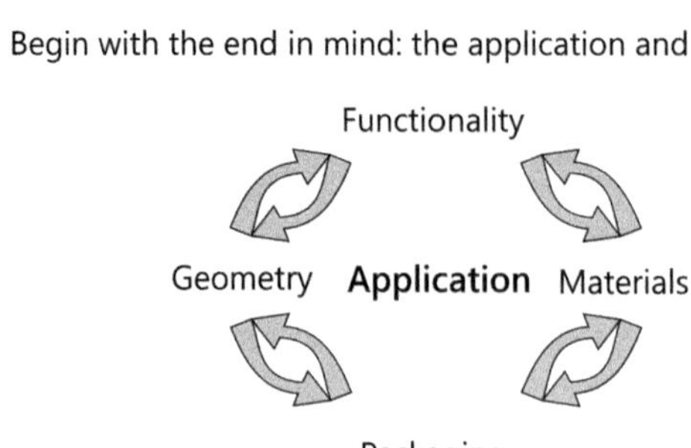

Figure 2.3 Main aspects of the biosensor design process.

Table 2.1 Typical detection strategies found for the main analyte types in biomedicine.

Analyte	Suitable (bio)receptors	Electrochemical techniques	Sensor type	Ref.
Electrolytes	Ionophore	Potentiometry	Ion-selective electrodes (ISE)	Bakker and Pretsch[69]
Gases	Semipermeable membrane/ none	Voltammetry, amperometry	Chemical sensors	Floate and Hahn,[70] Griveau and Bedioui[71]
Metabolites	Enzymes	Amperometry, voltammetry	Biosensors	Baumann and Karst[72]
Toxins	Enzymes/ antibodies	Amperometry, voltammetry	Chemical sensor, biosensor	Ray *et al.*[63]
Hormones	Antibodies/ aptamers	Amperometry, voltammetry, EIS (electrochemical impedance spectroscopy)	Biosensor	Mazzei *et al.*[73]
Proteins	Antibodies	Amperometry, EIS	Biosensor	Batchelor-McAuley *et al.*[74]
Pathogens	Antibodies/ nucleic acids	Amperometry, EIS	Biosensor	Kiilerich-Pedersen *et al.*[75]

different biorecognition elements may be used in the detection of any given target analyte, one is usually more adequate than the others in a given application environment.

Table 2.1 summarizes the most common biorecognition elements used as a function of analyte type, and outlines some possible detection strategies.

The most common (bio)recognition elements in biosensing are enzymes and antibodies. Nucleic acids, aptamers and other biomimetic receptors are also described in the literature, but are much less common in practice.

2.2.1.1 *Enzymes*

Enzymes are highly specific molecular machines that catalyse numerous (bio)chemical transformations thanks to their active site, which is buried within a protein shell. Access to the active site is restricted not only sterically, but there are also strongly pH-dependent coulombic effects arising from the charges of the amino acid residues on these access regions. This, together with the enzyme size and structure, is what makes enzymes so specific, and is also why enzymes are better suited to work on low molecular weight species such as sugars, lipids and certain metabolites in very narrow pH and medium conductivity ranges. Electrochemical enzyme based biosensors monitor enzyme activity as a function of target analyte concentration.[76]

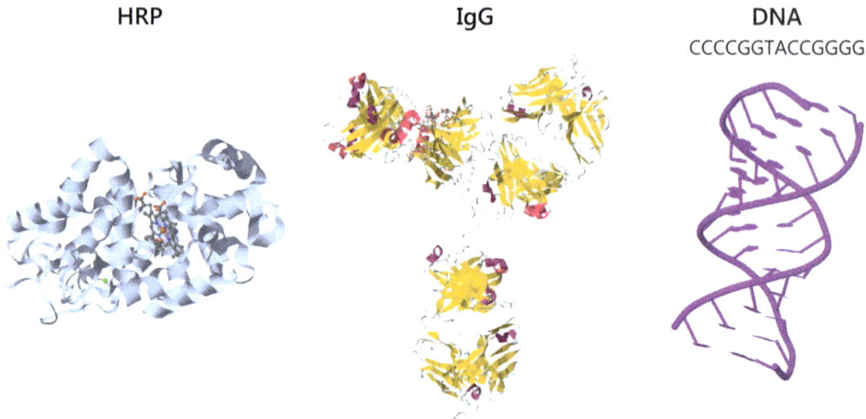

Figure 2.4 Representation of the enzyme horseradish peroxidase, IgG, and a double strand DNA fragment.

Since direct electron transfer to the active site is very difficult, electrochemical enzyme-based biosensors rely on indirect measurements, such as monitoring changes in the concentration of a co-substrate or of species produced by the enzymatic reaction. If either these co-substrates or the products of the enzymatic reaction are electroactive, they can be monitored amperometrically. On the other hand, if their concentration changes result in local conductivity and/or pH changes, then potentiometric or conductimetric probes may be used. Figure 2.4 shows the enzyme horseradish peroxidase, perhaps the most common enzyme used in biosensing applications.

2.2.1.2 Antibodies

Antibodies (Ab), on the other hand, are produced by the immunologic system in response to "foreign" bodies such as viruses, bacteria, and large bodies such as implants. Antibodies are immunoglobulin proteins consisting of two identical heavy chains and two identical light chains all assembled by disulfide bonds that support the antibody structure itself, that present one crystallisable fragment region (Fc), and two antigen-binding fragments (Fab). The binding sites are found in the N-terminal regions, where the antibody will recognize the antigen (Ag). Antibodies are perhaps the most commonly employed biorecognition systems in rapid tests including biosensors (for analytes at concentrations in the 10^{-6}–10^{-10} M range). This recognition is due to the perfect assembly between the lateral chain of the amino acids that perform the antibody sequence and the antigen or protein target. The great variety of functional groups present in these lateral chains allows the entrapment between the two biomolecules, through weak forces, such as van der Waals forces, hydrogen bonds and hydrophobic interactions, and also strong forces such as ionic pairing; this last force is clearly affected by the pH of the solution in which the Ab is immersed.

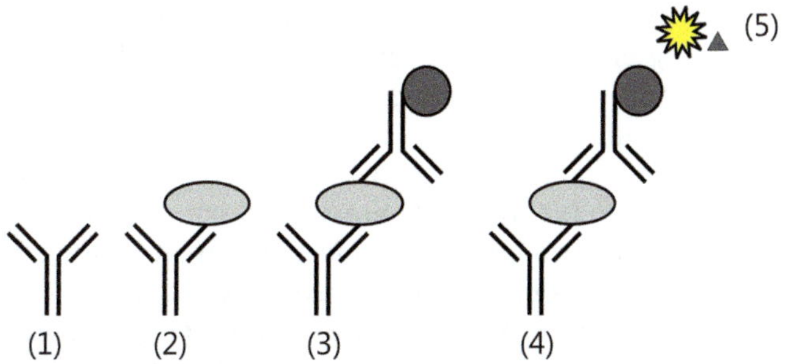

Figure 2.5 Five step "Sandwich" ELISA: (1) coating surface with the capture antibody; (2) antigen binding to the capture antibody; (3) binding of the detecting antibody to the antigen; (4) labelled (enzyme-linked) secondary antibody binding to detecting antibody; (5) addition of an enzyme substrate to monitor enzyme activity.

Antibodies are immobilized on an electrode surface mostly through lysine, their predominant amino acid and which contains a free ε-amino group.[77]

Their high affinity constants towards their antigens (target analytes in the case of biosensors) make regeneration of the transducer surface difficult and limits their application in practice to single-use devices.

Electrochemical immunosensors are based on corresponding enzyme-linked immunosorbent assays (ELISA),[78,79] and both direct (label-less) and indirect (using redox mediators) detection strategies can be used. Figure 2.5 depicts the steps involved in a typical sandwich ELISA assay.

Label-less immunosensors most typically rely on electrochemical impedance spectroscopy (EIS), which enables the measurement of both faradaic and non-faradaic processes simultaneously and with a high sensitivity. However, enzyme-labelled antibodies coupled to amperometric detection are much more frequent in electrochemical immunosensors. Horseradish peroxidase is the most common enzyme used, although others such as alkaline phosphatase and β-galactosidase can also be used.[80]

Assay sensitivity may be further enhanced using microelectrodes of the right geometry; for instance, interdigitated microband electrodes enable the performance of generator–collector experiments, which can boost the output signal by an order of magnitude, and improve the detection limit.[81]

2.2.1.3 Oligonucleotides, Aptamers, and Other Bioreceptors

Nucleic acids and molecularly imprinted polymers can also be used as recognition elements in the construction of biosensors. DNA biosensors are mostly used in pathogen detection and identification, as single-strand DNA chain fragments can also be immobilized on transducer surfaces. Detection is enabled by the specificity of the base-pairing interactions between complementary sequences; although a minimum sequence of 20 base pairs are recommended for these biosensors to be reliable. Different detection techniques can be applied, including impedance spectroscopy, and amperometry using electroactive species and quantum dots.[82,83]

Aptamers are small synthetic single-stranded DNA or RNA fragments that can bind with high affinity and specificity to non-nucleic acid target molecules. They are synthesized in a process known as SELEX (systematic evolution of ligands by exponential enrichment), which was first reported back in 1990.[84,85] Aptamers can bind to molecular targets with affinities and specificities that match and sometimes even exceed those of antibodies, which they expect to replace. Allegedly, they can be designed and synthesized to recognize many small molecules. However, the best known and seemingly one of the few working aptamers known to date is that for the detection of thrombin.[86]

Lastly, molecularly imprinted polymers (MIPs) are synthetic receptors that offer similar recognition abilities to natural bioreceptors.[87] MIPs are templated structures produced by co-polymerization of functional and cross-linking monomers in the presence of the target analyte (the imprint molecule), and act as a molecular template. Imprint molecule and functional monomers initially form a complex that is held in position by the polymeric structure after polymerization. Next, the molecule is removed and the analyte binding sites are revealed. The use of MIPs has recently been reported in the amperometric detection of cardiac troponin T, where *o*-phenylenediamine (*o*-PD) film was electrodeposited in the presence of cTnT as a template in acetate buffer (0.5 M, pH 5.2) on gold substrates.[88]

The next step once a biorecognition element has been chosen is to find the most suitable route to link it to the transducer. The following section covers the most common surface functionalization strategies used in biosensor development.

2.2.2 Transducer Functionalization Strategies

Figure 2.6 summarizes the main surface functionalization approaches and chemistries that can be adopted, which are: (i) adsorption, (ii) avidin–biotin interaction, (iii) covalent attachment and (iv) matrix entrapment.

In addition to understanding the structure and functionality of the chosen biorecognition element, the surface composition and structure also need to be known for successful attachment of the bioreceptor molecules. Successful attachment means that the bioreceptor keeps its functionality as far as possible, that it is easily accessible to target molecules in the sample, and that the transducer is able to pick up the recognition event.

The materials most commonly used in the microfabrication of electrochemical devices are metals such as gold[89] and platinum, but others such as iridium[90] and indium tin oxide[91,92] can also be used. In addition, passivation materials are mostly silicon oxide, silicon nitride, and polymers such as PDMS[93] and PMMA. More recently, the use of beads has become increasingly important in the development of biosensors. The surface of these particles is typically latex or polystyrene, and may present different

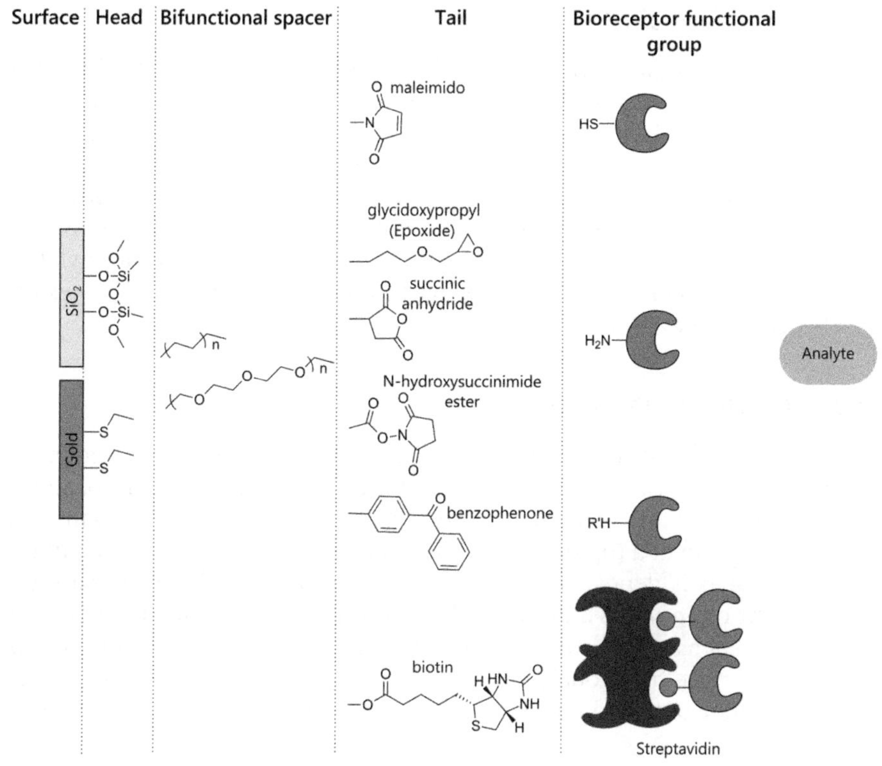

Figure 2.6 Main surface functionalization approaches.

functional groups ready for subsequent attachment of biomolecules. These functional groups include hydroxyl, carboxylate, amine, and thiol. On the other hand, avidin-coated beads are also very common due to the convenience of using the strong affinity between avidin and biotin.[80]

2.2.2.1 Adsorption

Adsorption is the non-covalent, spontaneous attachment of biomolecules to a surface.[94] Adsorption falls within one of two categories, depending on the strength of the interaction between adsorbate (the substance that adsorbates) and adsorbent (the substrate). *Chemisorption* is governed by *strong, short-range* interactions, usually (but not necessarily) of covalent nature. In contrast, *physisorption* is governed by *weak, long-range* interactions, usually Van der Waals forces. Examples of chemisorption are the non-covalent formation of thiol-based self-assembled monolayers on gold surfaces, and the adsorption of aromatic systems on metal surfaces, which may result in electrode passivation, as in the case of 3,3′,5,5′-tetramethylbenzidine (TMB) on gold electrodes.[95] On the other hand, examples of physisorbed systems are proteins such as enzymes, antibodies (IgG) and bovine serum albumin (BSA) used to block surfaces.

2.2.2.2 Avidin–Biotin Interaction

Although non-covalent, this interaction is very strong and specific. The interaction of biotin with the proteins avidin and streptavidin is among the strongest non-covalent affinities known ($K_a = 10^{15}$ M^{-1}). The binding occurs between the bicyclic ring of biotin and a pocket within each of the four protein sub-units. This system is typically used in combination with magnetic beads, as in many of the instances tabulated in Table 2.5. To allow this interaction, the protein or the bioreceptor should be previously biotinylated. This reaction is commonly used for bioconjugation purposes and is extensively described by Hermanson.[80]

2.2.2.3 Matrix Entrapment

This strategy consists of the encapsulation of bioreceptors within a polymer/gel matrix. There are two ways to achieve this. In a bottom-up approach the polymer/gel matrix is generated *in situ* in the presence of the relevant biomolecules, while in a top-down approach a mixture of biomolecules and the polymer matrix is deposited on a surface.[96] Examples of the former are the entrapment of enzymes and antibodies within electrodeposited conducting polymers[97] such as polyaniline, PANI, PEDOT, or Prussian blue. Examples of top-down surface functionalization are spraying and screen printing of inks containing not only carbon or metal particles but also enzymes and redox mediators.

2.2.2.4 Covalent Attachment: The Importance of Bifunctional Spacers

Linking biomolecules to a given surface usually requires the participation of so-called spacers, as steric hindrances can complicate the direct attachment of biomolecules onto surfaces. Spacers usually present a hydrocarbonated chain with a different functional group at each end: a head group and a tail group (Figure 2.6).

Moreover, the two ends should ideally be different to avoid crosslinking during the functionalization procedure. The spacer mid-section can have different lengths, the choice of which is justified by the nature of the biomolecule and the surface, and the transduction method.

Table 2.2 summarizes the most relevant functionalities that are used in the covalent attachment of biomolecules to surfaces, and the different reactions involved.

In addition, it is worth mentioning that other approaches can be used to attach biomolecules. Click chemistry[106] and native chemical ligation[107] are both examples of smart reaction strategies that are occasionally used.[80]

Some of the most common crosslinkers are glutaraldehyde, *N*-hydroxysuccinimide (NHS), maleimide, or epoxide. These reactions used in surface chemistry are almost quantitative and their mechanism avoids the necessity

Table 2.2 Functional groups involved in bioconjugation reactions.

Biomolecule functional group	Tail group bifunctional spacer	Reaction conditions	Ref.
Amines ($-NH_2$)	(i) *N*-Hydroxysuccinimide esters; (ii) epoxide (glycidoxypropyl...); (iii) anhydride; (iv) glutaraldehyde	(i) EDC (ii) direct (iii) direct (iv) NaCNBH3/ direct	(i) Tuteja *et al.*[98] (ii) Valera *et al.*[99] (iii) Kiilerich *et al.*[75] (iv) Ge *et al.*[100]
Carboxylates (-COOH)	Amine	Previous activation EDC +sulfo-NHS	Grabarek and Gergely[101]
Thiols (-SH)	Maleimide acid Disulfide exchange	Direct Direct	Houseman *et al.*[102] Pavlovic *et al.*[103]
Reactive hydrogen compound (H-R′)	Benzophenone	UV light	Jung *et al.*[104]
Biotin	Biotin	Streptavidin	Hou *et al.*[105]

of purification, only several washes on the surface are needed to eliminate any by-product obtained, which allows an adequate bioconjugation of the analytes.[108]

In the case of gold surfaces, although the Au–SH interaction is not covalent itself, it is sufficiently strong as to be considered covalent by some authors,[109] and thiols with different spacer lengths are extensively used to covalently attach biomolecules to their tail groups.[110] The importance of choosing the right thiol spacer cannot be overstressed. Among other things, long alkane chains result in highly ordered and packed layers, but they hinder electron transfer in amperometric applications, which compromises sensitivity. Shorter chains, on the other hand, provide higher sensitivities but are more soluble and tend to present pinholes that facilitate non-specific interactions. In summary, the most suitable spacer for each application can only be found by experiment.

In the case of silicon surfaces, covalent attachment of functional spacers by silanization is straightforward. It is performed by vapour deposition of the silane over the surface in a vacuum chamber to avoid polymerization and multilayer deposition of the silanes. Commonly used silanes are 3-aminopropyltriethoxysilane (APTES) and 3-glycidoxypropyltrimethoxysilane.

Bioreceptors are typically linked through –COOH or $-NH_2$ groups of the lateral chains of their amino acids, but the functionalization reaction conditions need to be controlled, especially pH and salinity, to ensure that a favourable orientation is achieved, and that the biomolecule keeps its functionality after the process.

For instance, antibodies consist of amino acid chains supported by thiol bonds. When a derivatization needs to be performed, the shape and the function of the antibody have to be preserved. Then the best option is usually to use the lateral chains of their outer residues. The most exposed are the epsilon amino group of the lysines, carboxylates from the arginine residues, hydroxyls, and the hydrophobic groups of other amino acids. In addition, amine is the preferred group in bioconjugation reactions because its use does not normally imply protein structure changes.

As a final note, after a surface has been functionalized it is important to prevent its reaction with other biomolecules. Different blocking agents can be used for this, the most common of which are bovine serum albumin (BSA), cysteine and poly(ethylene glycol).

2.2.3 Electrochemical Transduction

Electrochemical systems involve faradaic and non-faradaic phenomena. Non-faradaic phenomena are related to the electrical properties of the medium and the electrical double layer (EDL) structure. Faradaic processes, on the other hand, involve electron transfer events leading to oxidation or reduction of chemical species at the electrode surface.[112]

Microelectrode arrays can be used in combination both with DC or AC electrochemical techniques. DC techniques are suitable for the study of faradaic processes whereas AC techniques, on the other hand, are more adequate for the study of non-faradaic processes such as interfacial capacitance and medium conductivity.

The following sections briefly describe the main electrical, charge transfer and mass transport phenomena that need to be considered in the design of miniaturized devices.

2.2.3.1 The Electrical Double Layer (EDL)

When a metal electrode is immersed in an electrolyte solution, an electrical potential develops across the electrode–electrolyte interphase. As a result of this electric field, charge spreads non-uniformly in the direction normal to the electrode surface. Intuitively, the metal surface charges attract counter ions from the electrolyte solution, giving rise to an electrical double layer (EDL) in the liquid phase.

The EDL can be thought of as a molecular capacitor, where one plate is represented by the charges in the metal and the other plate by the ions in the solution. The first EDL model was introduced by Helmholtz in 1879, defining the EDL as layer depleted from charges within 5 nm from the electrode surface. However, this model fails to predict the dependency of double layer with potential and electrolyte concentration, and was subsequently refined by Gouy and Chapman, and later by Stern.[112] The theory considers both the coulombic forces and thermal motion to better define the counter ion distribution across this region. The Debye–Hückel approximation simplifies the Gouy–Chapman mathematical expression when small surface potential values (<100 mV) are considered. On the other hand, the Stern theory overcomes the limitation of the Gouy–Chapman theory for large surface potentials, considering counter ion size and binding properties.

Surface conductivity is another important phenomenon that needs to be considered when dealing with low conductivity media. In this case, the double layer is more conducting than the bulk solution due to the comparatively higher ion concentration at the solid–solution interface. This results in an increase of the current density in the EDL region. Moreover, this effect is enhanced in the presence of an electrical potential gradient parallel to the solid surface. Interestingly, the surface conductivity phenomenon affects all solid surfaces, and biosensors can take advantage of the impact of the surface modification on this phenomenon.[113] Based on this, IDE structures can be built to detect variations on the surface conductivity in the inter-electrode space, taking advantage of the low geometrical dimensions achieved with standard planar microtechnologies.

2.2.3.2 Electron Transfer

When designing microelectrode-based devices, it can be assumed that electron transfer is infinitely fast[114,115] and that mass transport is limiting. However, we believe that the importance of electron transfer justifies the following short introduction. Electrochemical (faradaic) processes are typically represented as:

$$\mathrm{O} + ne^- \underset{k_a}{\overset{k_c}{\rightleftharpoons}} \mathrm{R} \tag{2.1}$$

where O is an oxidized species, n is the number of electrons involved (e^-) and R is the reduced species of the redox couple. Assuming the case of a metal electrode, with a high density of electronic states forming its conductiobands, the average energy of its electrons is located around the Fermi level, E_F.[116] This energy level can be modified by applying an external potential and allowing the injection or subtraction of electrons from the conductor.[115] This adjustment of the Fermi level at the electrode results in the tunnelling of electrons between the conductor (E_F) and the corresponding molecular orbital of the species in solution across the EDL. This can be an acceptor orbital, called LUMO (lowest unoccupied molecular orbital), or a donor orbital, called HOMO (highest occupied molecular orbital). The main electron transfer models are the Marcus–Hush model and the Butler–Volmer model. The Butler–Volmer model is a macroscopic model deriving from classical thermodynamics, and therefore it cannot be used in molecular studies. This is in contrast to the Marcus–Hush model, which considers two terms in the standard free energy of activation. One is related to bond length and angle changes in the molecule due to electron transfer and the other is related to the reorganization of the solvent molecules around the molecule.[117] Due to its relative simplicity compared to Marcus' model,

Butler–Volmer kinetics are mostly used in electrochemical simulations.[115,118] The Butler–Volmer expression is:

$$j = k_c^0 \exp\left[-\alpha n F \frac{(E - E^{0'})}{RT}\right][\mathrm{R}] - k_a^0 \exp\left[(1-\alpha) n F \frac{(E - E^{0'})}{RT}\right][\mathrm{O}] \quad (2.2)$$

where j is the current density, [O] and [R] are the concentrations of the oxidized and of the reduced species, respectively, k_c^0 and k_a^0 are the heterogeneous electron transfer rate constants for the reduction and oxidation reactions, α is the charge transfer coefficient, generally assumed to be 0.5, E is the electrode potential, and $E^{0'}$ the formal potential of the O/R couple; R is the gas constant and T the absolute temperature.

2.2.3.3 *Mass Transport*

For electron transfer to occur at an electrode, electroactive species need to approach it within tunnelling distance. Mass transport may be governed by three distinct phenomena, as summarized in the following flux equation:[112]

$$j_{\text{mass transport}} = j_{\text{diffusion}} + j_{\text{migration}} + j_{\text{convection}} \quad (2.3)$$

or:

$$j_i = -D\nabla C_i - \frac{z_i F}{RT} D_i C_i \nabla \varphi + C_i \vec{v} \quad (2.4)$$

Diffusion is the microscopic and random movement of species in a concentration gradient, from regions of higher concentration to regions of lower concentration, as denoted by the negative sign in eqn (2.5). Fick's laws mathematically describe the mass flux and the concentration of material as a function of position and time. The flux of a species $c(x,y,z)$, defined as the number of moles passing through a certain position per unit time and area, is:

$$j_{\text{diffusion}} = -D_i \nabla C_i \quad (2.5)$$

Eqn (2.5) is known as Fick's first law, and the proportionality factor D is known as diffusion coefficient. Fick's second law describes the concentration changes over time, and is mathematically described as:

$$\frac{\partial C_i}{\partial t} = C\nabla^2 C_i \quad (2.6)$$

Due to viscous forces the velocity of the fluid is always zero next to a solid surface, and diffusion is always the transport mode governing the final approach of electroactive species to an electrode surface. This is of paramount importance in all electroanalytical measurements where mass transport is the limiting process, as the current density is simply the diffusional flux normal to the electrode surface:

$$i = nFj_{\text{diffusion}} = -nFD\frac{\partial[C_i(x,t)]}{\partial t}\bigg|_{x=\text{electrode}} \tag{2.7}$$

where n, F, j, D and c take their usual meanings. (Note: in this case, x coordinate means the perpendicular coordinate to the electrode surface.)

Migration is the microscopic movement of charged particles under the influence of an electric field.[112] The application of an electric potential (ϕ) between two electrodes establishes an electric field that drives the movement of ions or other charged species in the solution. Migrational flux is mathematically described as:

$$j_{\text{migration}} = -\frac{z_i F}{RT} DC_i \cdot \nabla\phi \tag{2.8}$$

where z is the charge of the ions $c(x,y,z)$, and $\nabla\phi$ is the electric field gradient. In most electroanalytical experiments migration can be neglected if a moderately high concentration of inert electrolyte (support electrolyte) is used.[112] It is good practice to use a supporting electrolyte concentration at least 100 times greater than that of the electroactive species under study. Then it can be safely assumed that the charge in the solution is transported by the supporting electrolyte components,[119] and the contribution of migration to the overall current can be neglected.

Last, *convection* refers to the macroscopic relative movement of solution in the vicinity of an electrode. There are two forms of convection: natural convection and forced convection. Natural or free convection is the macroscopic movement of layers of fluid due to temperature, pressure or density gradients, and there are only a few studies describing how to account for it theoretically.[120,121] Forced convection, on the other hand, occurs when the solution movement is due to the application of external forces. Mathematically, convective mass transfer is described by:

$$j_{\text{convection}} = C_i \cdot \vec{v} \tag{2.9}$$

where $\vec{v}$ represents to the fluid velocity. There are different ways to induce such external forces in microsystems, based on different physical phenomena. The most common affecting miniaturized flow devices are mechanical[15,17] or electroosmotic pumps,[122,123] capillarity forces,[124–126] and even centripetal forces.[127–129]

2.2.3.4 Considerations in DC Detection at Microelectrode Arrays

The most common DC detection techniques used in electrochemical biosensors are amperometry and cyclic voltammetry. Microelectrode arrays are typically used to amplify electroanalytical currents taking advantage of the favourable mass transport rates experienced by single microelectrodes.

The design of microelectrode arrays focuses on optimizing their output current which is ultimately controlled by diffusional mass transport. However, this is a non-trivial task and requires the use of numerical simulations to fully and faithfully describe diffusion to an ensemble of microelectrodes. The current flowing through a microelectrode array made up of N identical microelectrodes is indeed the sum of the currents going through them. However, assuming that this is going to be equal to N times the steady state current of each microelectrode, as per eqn (2.10) for a disk microelectrode array, is often mistaken:

$$I_{\text{array}} = N \cdot I_{\text{microelectrodes}} = N4nFDCr \tag{2.10}$$

There are two main reasons why this approximation fails. First, this expression assumes that the microelectrodes are inlaid on the substrate. However, most microfabricated microelectrodes are either recessed or slightly protruding.[130] In the case of recessed microelectrodes, the current expression developed by Bond *et al.*[131] provide a good description of the current, as it considers the additional resistance to diffusion posed by the recess height:

$$I_{\text{recessed}} = \frac{4nFDCr}{\left(\dfrac{4L}{\pi r} + 1\right)} \tag{2.11}$$

The second reason why the current at microelectrode arrays does not follow eqn (2.10) is because microelectrodes in an array compete for the depletion of material, and their diffusion layers eventually overlap, leading to the observation of gradually decreasing currents.[27,132] Figure 2.7 shows the four distinct transport regimes that microelectrode arrays observe, and which depend on the relationship between the size of individual diffusion layer thicknesses, d, the microelectrode size, r, and the separation between adjacent microelectrodes. These four regimes are as follows (from top to bottom in Figure 2.7): (i) planar diffusion to each microelectrode as $\delta \ll r$, (ii) radial diffusion to each microelectrode as $\delta \approx r < d$, (iii) a transition zone when $r < \delta \approx d$, and (iv) planar diffusion to the microelectrode array when $\delta > d$. Regime (ii) is the optimum one and it corresponds to the case when the response of the whole array is equivalent to that of a single microdisk, times the number of microelectrodes in the array. Regime (iv), on the other hand, consists of a complete overlap of individual diffusion layers that results in the array behaving as an electrode of the size of the entire micro- or nanoelectrode array. Hence for most microelectrode arrays steady state currents are observed in regime (ii), and then as the diffusion layer grows the current tails off in regime (iv). Note that nanoelectrode arrays are a special case because their size is typically a few micrometres, so they show steady state currents corresponding to microelectrode behaviour, regardless of the extent of overlap between adjacent diffusion layers.[133]

The Einstein–Van Smoluchowski's equation $\delta = \sqrt{(2Dt)}$ can be used to estimate the time that it takes for adjacent diffusion layers to overlap, where

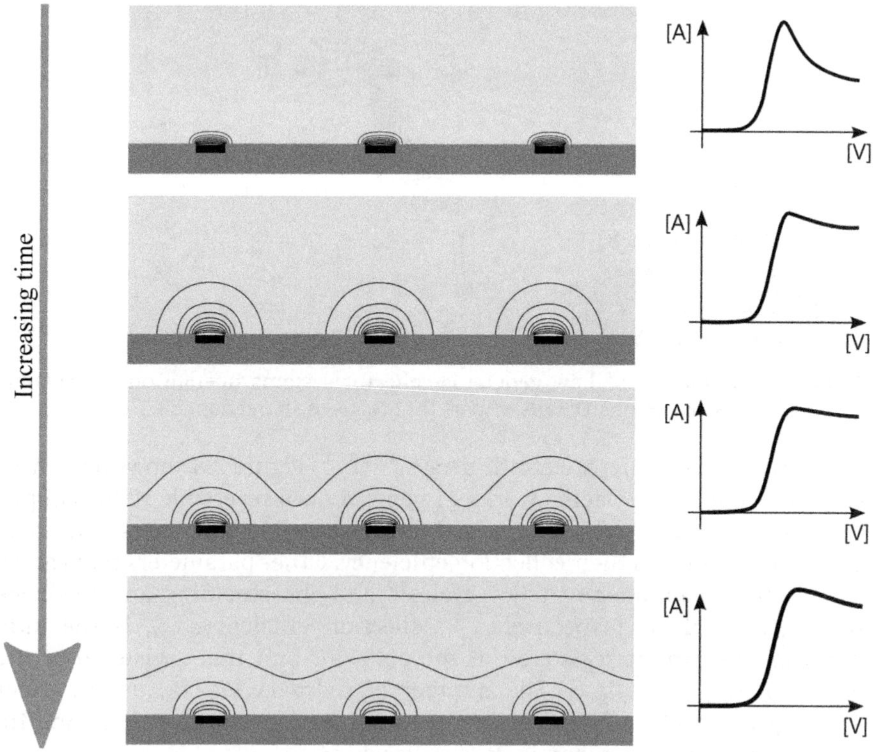

Figure 2.7 The four transport regimes observed at microelectrode arrays. Note that, in the case of recessed electrodes, a true steady state current is not attained (see text for details).

in this case it is better to consider d as the edge-to-edge distance between adjacent microelectrodes, as using the centre-to-centre distance may lead to overestimations that get worse for larger microelectrodes. However, the most accurate means to making such predictions is to use simulations,[115] and eqn (2.10), (2.11), and $\delta = \sqrt{(2Dt)}$ should be used with caution.

2.2.3.4.1 Generator–Collector Detection at Microelectrodes. One-way to use this shielding effect of neighbouring microelectrodes to an analytical advantage is to work in generator–collector or "redox cycling" mode. This operation mode consists of polarizing adjacent microelectrodes at potentials such that species generated at one of them (the generator) are converted back into their original state at the other (the collector). This species will then diffuse away from the "collector" towards the "generator" electrode, and the cycle repeats itself for as long as the electrodes are polarized. This is commonly known as "redox cycling", and it offers important sensitivity and current enhancements.[81]

Generator–collector systems mostly take the form of interdigitated microband arrays, although other constructions are also possible, such as

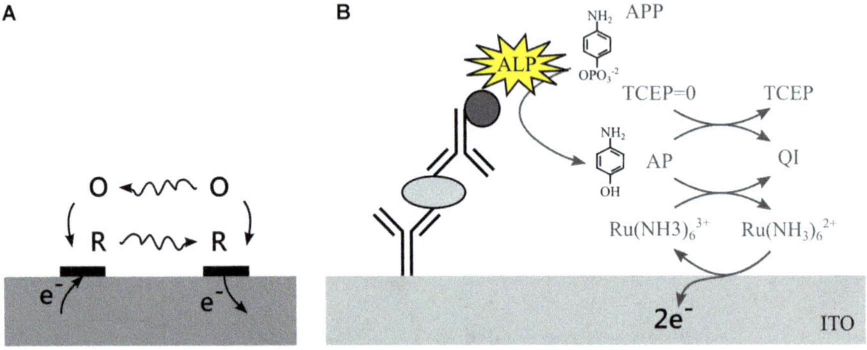

Figure 2.8 Depiction of how generator–collector systems operate on a reversible redox couple. Details of part (b) are given in Section 2.4.2.

ring-recessed disk microelectrode arrays.[20–22,114] Figure 2.8 shows how these generator–collector systems works in general on a reversible redox couple. Generator–collector systems are defined by several parameters, the most common of which is their collection efficiency. Other parameters that can be found are the number of redox cycles[19] and the shielding and feedback factors, S_F and F_F, respectively.[134] Collection efficiency, i_{eff}, is the ratio between the current registered at the collector and that registered at the generator ($i_{eff} = -i_{col}/i_{gen}$). The number of redox cycles, R_c, on the other hand, gives an idea of the number of times that species shuttle to and fro between adjacent microbands. It is defined as:

$$R_c = \left(1 - i_{eff}^2\right)^{-1} \tag{2.12}$$

The shielding and feedback factors also quantify the effect of neighbouring microbands when polarized at the same potential (shielding) or in generator–collector mode (feedback). As noted by Niwa *et al.*, the accuracy of R_c (and by extension of S_F and F_F) is strongly dependent on the experimental error associated to the determination of i_{eff}.[19] Although collection efficiency depends largely on the distance between adjacent microbands, and smaller gaps result in greater efficiencies,[19,134] the aspect ratio of the microbands also has an important effect on the observed collection efficiency. Closely spaced, protruding-microbands display the highest efficiencies.[135,136]

The actual geometry of the microbands has a significant impact on the diffusion-controlled currents observed at interdigitated microelectrode arrays. The most important theoretical work reported to date on the behaviour of interdigitated microband electrodes considers planar systems.[28,134,137,138] However, as the gap between microbands narrows, especially in the case of nanoelectrodes, the microband "z" dimension becomes increasingly important. In the case of nanobands it is possible to find aspect ratios of 1 or more.[135,136] These devices can no longer be considered planar, and numerical simulations are required to understand their behaviour fully.[135]

An important advantage of these systems is that the current is only amplified for electrochemically reversible systems. This means that species undergoing fast coupled chemical reactions, such as ascorbic and uric acids, which are typical interferences in biological samples, may not be so problematic if the detection strategy for a certain analyte, as in some of the immunosensors described below, is based on redox cycling. As long as the lifetime of the electrogenerated species is shorter than their time-of-flight towards the collector electrode, collection efficiency will tend to be zero.[139] Again, simulation is the only way to reliably account for both diffusion and coupled chemical kinetics simultaneously in the design of generator–collector devices.[12]

Typical interdigitated electrode arrays used for redox-cycling detection present microband widths and gaps ranging between 1 and 10 μm, although optical projection methods enable the fabrication of sub-micrometric dimensions.[12] On the other hand, nanofabrication techniques[140] enable the construction of interdigitated nanoband electrodes with pitch distances down to 100 nm.

2.2.3.5 Considerations in AC Detection at Microelectrode Arrays

Electrochemical impedance spectroscopy (EIS) can be used to study the dynamic response of microelectrode based systems over a very wide frequency range.[141] Ohm's law defines the impedance as the relation between voltage and current ($Z = V/I$) (Figure 2.9). This can be obtained by applying a small perturbation in either current or potential, and analysing the response of the other. Impedance value is expressed as a complex magnitude, as it contains information both on the relation between current and potential and on their phase shift. The dielectric properties of tissues and solutions,

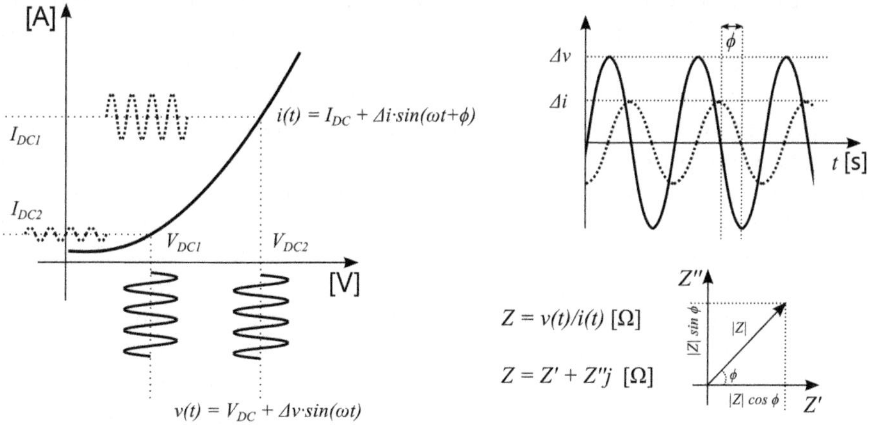

Figure 2.9 Intuitive representation of the relation between voltage and current for DC (left) and AC (right) measurements. Graphical representation of the impedance as a complex value (bottom right).

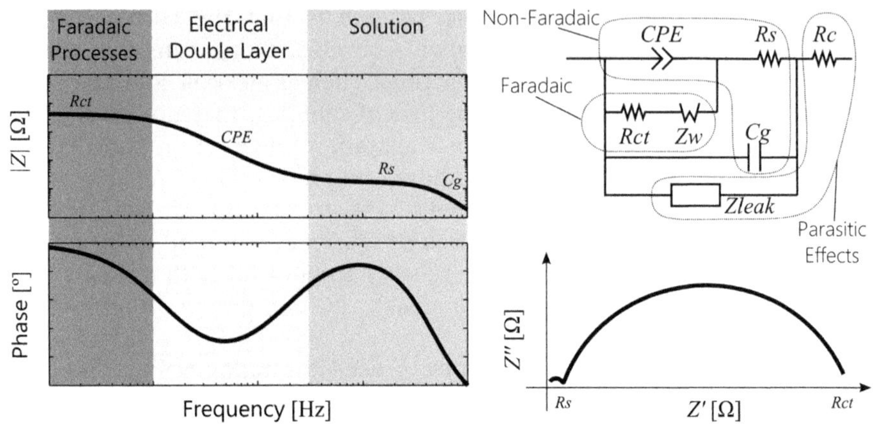

Figure 2.10 Most accepted equivalent circuit used to describe the impedance measured in electrochemical systems. The typical spectra described by this equivalent circuit are depicted in a Bode plot (left) and in a Nyquist plot (right). In addition, the main parameter of the equivalent circuit are indicated together with the phenomena that controls each frequency range in the graphical representation.

together with capacitive processes associated to the EDL, can be studied in detail by analysing the impedance spectrum.[111,142,143]

Figure 2.10 shows the two usual representations found in the literature to present the impedance spectra. Bode plots depict the impedance magnitude and phase shift as a function of frequency, usually on a logarithmic scale to facilitate the observation of different events across the whole frequency spectrum. Bode plots are commonly associated with the representation and analysis of non-faradaic phenomena. Nyquist plots, on the other hand, represent the impedance imaginary part *versus* its real part. In this case, both axes are linear in scale and displayed in 1 : 1 aspect ratio. Without the compression effect of the logarithmic scales, the large magnitude values obtained at the low frequency range can be easily observed. This makes Nyquist plots ideal for the study of faradaic phenomena.[141,144]

Equivalent circuits are commonly used to facilitate the description and interpretation of impedance spectra. Physical phenomena can be expressed in terms of electronic components, which facilitate their interpretation and quantification. This approximation becomes especially useful when several phenomena coexist in the same frequency range. Figure 2.10 shows the Randles circuit,[111,141] which is accepted for the description of electrochemical systems. Equivalent circuits are a powerful tool when they are backed by a physico-chemical model describing the system under study. However, using equivalent circuits to analyse experimental results can also be very risky.[145] The main issue is that the same impedance spectra can be produced by different circuits topologies and values. This is because complex equivalent circuits have too many degrees of freedom, which leads to many possible valid solutions. Defining the narrower possible value ranges for each parameter appears to be a good approach to isolate the different phenomena, as this course of action reduces the number of possible solutions. However, choosing too narrow ranges might predetermine the final solution, which may not necessarily be a correct one.

The following provides some important considerations for each of the branches in this circuit, and which refer to parasitic, non-faradaic, and faradaic processes.

2.2.3.5.1 Parasitic Effects. Beyond the physico-chemical phenomena detailed above, other events may decrease the quality of the obtained results. Note that when an electrochemical system is miniaturized, technological issues and material properties must also be considered. Undesired phenomena need to be isolated to avoid data interpretation mistakes.

Parasitic effects may come from the contacts and wires used in the experimental set-up.

Miniaturization and integration requires small connection tracks, which may introduce considerable electrical resistance. As shown in Figure 2.10, such *contact resistance* (R_c) must be considered in series with the electrochemical cell.[146,147] This aspect should be specially borne in mind when high currents and low impedances are expected. In addition, the use of non-metallic materials such as polymers might require connection tracks with larger section due to the low conductivity values of those materials. The contact resistance can be indirectly assessed through an accurate EIS characterization using solutions of different conductivity. The solution resistance, R_s, decreases with increasing medium conductivity while the contact resistance remains constant, and it can be obtained from the impedance spectra at the high frequency range.

In contrast, when large impedance values are expected, capacitive coupling and leak currents through insulating materials should be considered. This requires special attention when thin passivation layers are used to improve electrode access by diffusion. This might result in a large capacitor between the connection track and the medium solution. As shown in Figure 2.10, this *leakage impedance* (Z_{leak}) should be considered in parallel with the electrochemical cell. Accurate assessment of this impedance is complex and depends on the measurement setup and connections routing on the chip. On one hand, current leak through wires and connections can be assessed by performing an open circuit measurement. In this case, an infinite impedance value is expected and, therefore, the obtained impedance is only related to the limits of the electronic equipment and with the parasitic effects of wires and connections. This issue requires special attention when all the electrodes of the electrochemical cell are on the same device and their connections are close together. This could produce a capacitive coupling between connection tracks and, therefore, should be taken into account in the design of both the experimental setup and any sensing devices.

On the other hand, evaluation of the current leaks requires special attention when large insulated connection areas are in contact with the solution. In some applications, such as the electrodes placed in a microneedle (Figure 2.1a), this cannot be avoided. Therefore, the direct evaluation of these current leaks cannot be separated from the phenomena at electrode surface. To overcome this limitation, fabrication of a test device without the insulation aperture at the electrode is suggested as an internal control.

2.2.3.5.2 Non-Faradaic Processes. Next in line with the above, the electrical structure of the medium and the interface bring about a series of non-faradaic processes that are linked to the solution resistance, R_s, and the double layer capacitance, C_d, connected in series to it.

From an electrical point of view, the *electrical double layer (EDL)* may be considered as a capacitor, the value of which is directly proportional to the electrode area. For the commonly used metals this value is about 20 μF cm^{-2}. However, non-idealities such as surface roughness or molecular adsorption at the electrode surface change the frequency response and then a capacitor is no longer able to reproduce the obtained impedance behavior.[148,149] In particular, the phase shift is usually less than $-90°$ for real electrodes and the slope of modulus decrease is smoother than the behaviour of an ideal capacitor (Figure 2.11). The constant phase element (CPE) is a mathematical

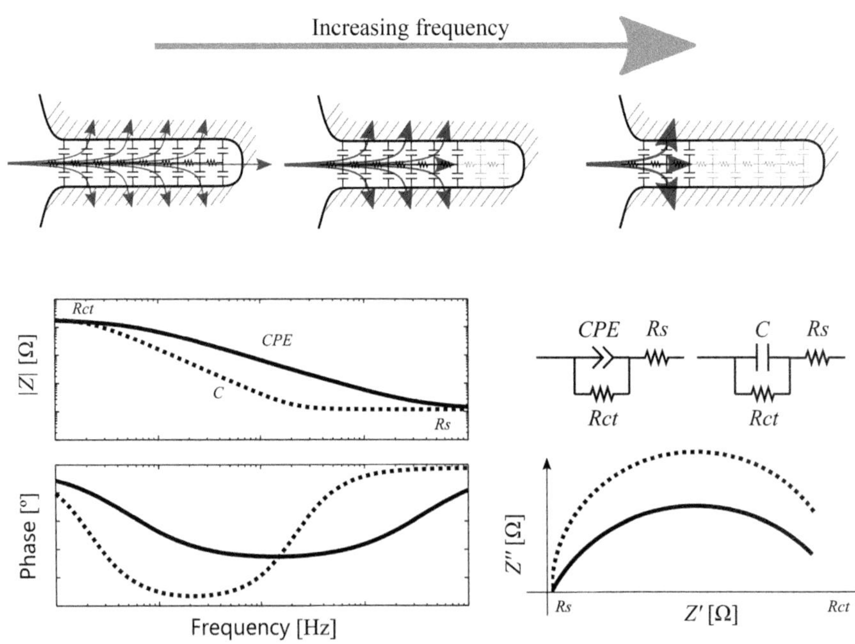

Figure 2.11 Intuitive representation of electrode surface micropore behaviour, and its comparison with transmission line structures: as frequency increases, the current does not reach the deeper areas of the groove, which results in a decrease of the capacitive behaviour and in the addition of a resistive behaviour. The variation observed in the graphical representation of the impedance due to the CPE behaviour are also indicated, solid line for CPE behaviour and dashed line for the equivalent capacitive behaviour.

entity devised to account for these non-idealities, and is consistently used in the literature[7] as a substitute of the classical capacitor. The impedance of a single CPE component is defined as following:

$$Z_{\mathrm{CPE}} = \frac{1}{K_{\mathrm{CPE}}(j\varpi)^{a}} \tag{2.13}$$

where K_{CPE} can be regarded as the equivalent capacitance value. Therefore, CPE is also considered as a sort of pseudo-capacitance. Moreover, a ranges from 0 to 1; when $a=0$ the CPE behaves as a simple resistance, and when $a=1$ the CPE behaves as a pure capacitor.

The physico-chemical explanation for parameter a considers several things, the most important of which is surface roughness: the microscopic pores and grooves of the electrode surface behave as TL-like structures.[150] An intuitive explanation for such behaviour is displayed in Figure 2.11. As frequency increases, the current does not reach the deeper areas of the surface grooves. This results in a reduction of the "effective" electrode area with increasing frequency, and in the appearance of certain resistive behaviour. This is enhanced when the electrode is modified and additional surface roughness is introduced. These kinds of modification such as metal electrodeposition are used to increase the electrode area by increasing the surface roughness. Another source for CPE behaviour can be found in the molecular adsorption of charged molecules on the electrode surface, which produces irregularities and inhomogeneities in counter ions concentration within the EDL.

The modification of electrodes using porous material such as membranes and polymers that act as selective membranes or protect the electrode against medium aggressions also results in the appearance of constant phase elements. In these cases, more complex models are needed to interpret impedance spectra in the EDL control region.[151,152]

At higher frequencies the EDL impedance becomes lower than that of the solution, and the spectrum is mainly controlled by the electrical properties of the latter, mainly the solution resistance, R_s. As the EDL capacitor value is related to the electrode area, the transition between the frequency range controlled by the EDL and the frequency range controlled by the medium solution also depends on the electrode area. Consequently, for large area electrodes the solution electrical properties control the spectrum over a wide frequency range. Conversely, for micro- and nano-electrodes, the EDL controls most of the frequency spectrum, making it possible to mask the bulk solution contribution, but the two parameters are strongly interrelated.

The measured solution resistance strongly depends on the cell geometry. The *cell constant* is the proportionality factor relating the electrical properties of the medium solution with the measured impedance when it is controlled by the medium solution:

$$Z = K(p + \varepsilon\varpi j) \tag{2.14}$$

where p is the medium resistivity, E is the medium permittivity, ϖ is the angular frequency and K is the *cell constant.* As shown in Figure 2.12, the

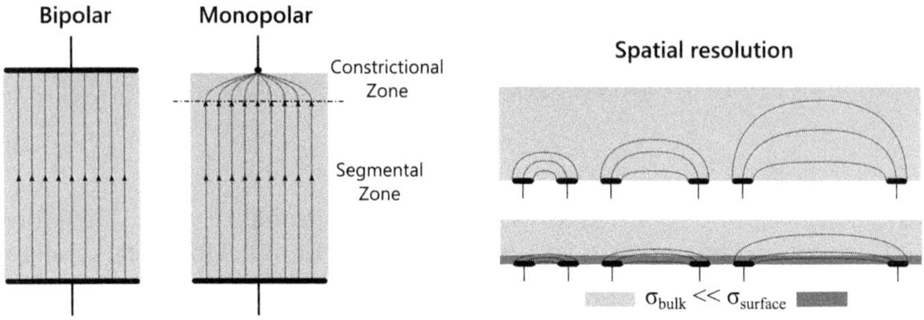

Figure 2.12 Representation of the electric field gradient for different electrode arrangements. In the case of two opposite electrodes of the same size (bipolar), the electric field is uniformly distributed in the medium. However, when the electrodes differ significantly in size, or the electrodes are placed on the same surface, the resulting electric field is not uniformly distributed. As a result, the measured impedance is mainly controlled by electrical properties of regions with higher current density.

geometry and arrangement of the electrodes define the direction of the electric field gradient and, therefore, the direction of the current density vector. When the electric field is uniformly distributed between two opposite electrodes, the cell *constant* can be obtained using simple geometrical relations. Conversely, when different electrode sizes are used, a constriction in the electric field appears close to the smaller electrode. As a result of this constriction, the measured impedance is mainly controlled by electrical properties of this region. Moreover, in the analytical systems based on microtechnology, the electrodes are usually distributed over a planar surface. This results in elliptical current distributions and cell constants that are hard to describe using analytical expressions. Some approximations such as the assumption of point electrodes have been suggested to find theoretical approximations of the cell constant.[13]

For closer electrode configurations, as in the case of interdigitated microelectrodes (IDE), the current does not penetrate into the bulk solution beyond a certain distance that is close in magnitude to the size of the microelectrodes in the array. This, combined with the fact that surface conductivity can be significantly greater than bulk medium conductivity, results in an increase in the current density near the device surface. In this case, the linear relation (K) between the solution electrical properties and the measured impedance is broken for lower solution conductivity values.[153]

2.2.3.5.3 Faradaic Processes. Lastly, in the presence of electroactive species, mass transport and electron transfer processes also show in impedance spectra. Due to the nonlinearity of these phenomena, particularly of mass transport, the obtained results are strongly dependent on variations of the chosen DC operation point. This is why it is typically recommended to choose DC operating points as close as possible to the formal potential of the redox couple under study, and also to use small AC signal amplitudes. In addition, reference electrodes are required to control the working electrode processes.[119]

In the case of microfabricated devices, the use of symmetric electrode configurations, such as the interdigitated microband structure, are an attractive alternative to avoid the use of reference electrodes. When the geometry and material of both electrodes are identical, and both electrodes are affected by the same phenomena to the same extent they observe the same electrode potential, and the resulting open circuit potential becomes null ($V_{oc} = 0$ V).

2.2.3.6 Other Issues Affecting Microfabricated Devices

To improve the performance of the electrochemical analysis systems, the integration of electrodes inside *micro-channels* to create *lab-on-a-chip* devices has seen important growth during the last decade.[14,48,154] In these systems, measurements are carried out inside a micro-channel where the reduced channel dimensions produce current density constrictions. In smaller channels, these constrictions together with the use of medium–low conductivity solutions can lead to considerable resistances that can affect the system behaviour at different levels. First, the counter electrode potential may increase up to values that are inaccessible to the common electronic equipment. Second, such high counter electrode potentials can destroy the electrode itself by oxidation. Last, depending on the electrode distribution and geometry, the potential may not be not homogeneously distributed over the entire electrode surface, resulting in unexpected behaviours. Numerical simulations, including finite difference[115] and finite element methods (FEM),[155] are a powerful tool to analyse the *iR*-drop and potential distribution in these systems.

Having considered the main aspects affecting the design of any miniaturized electrochemical (bio)sensing device, the following sections deal with the fabrication of such systems.

2.3 Overview of Microfabrication and Rapid Prototyping Techniques

Microelectrodes and their arrays can be fabricated in many different ways.[156] Thus, the encapsulation of micro- and nanofibres leads to the production of random arrays,[157] and screen printing techniques[158] enables the fabrication of microelectrode arrays in carbon-based materials. However, these techniques are very limited in scope. Microelectrodes in a random array are not individually addressable. Microfabrication techniques, on the other hand, enable the fabrication of structures with lateral sizes less than 1 μm, and the deposition of layers with controlled thicknesses down to a few atoms.[159] Figure 2.13 summarizes the main fabrication techniques involved in the

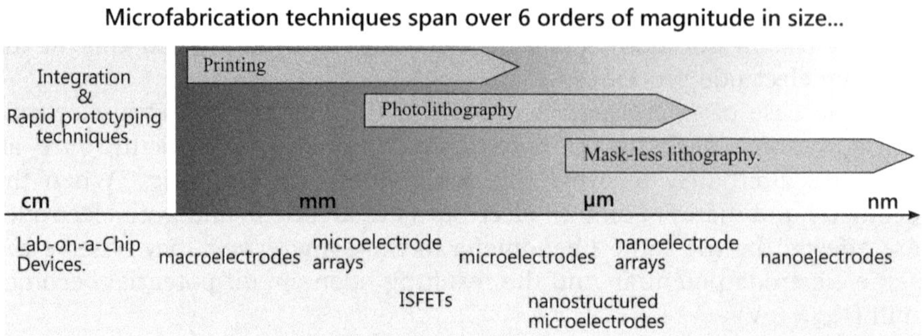

Figure 2.13 Summary of the main fabrication techniques involved in the miniaturization of electrochemical devices.

miniaturization of electrochemical devices, including microelectrode arrays. In addition to microfabrication techniques, standard manufacturing techniques may be required for the fabrication of fluidic components and interfaces that connect the microelectrodes with the macro-world. These techniques involve mostly cutting and bonding operations, and extend the range of materials beyond silicon and glass, and are most relevant in the development of flow devices, where several design–fabricate–test iterations may be required before reaching an adequate performance level.[14] So-called rapid prototyping techniques speed up the development cycle by enabling the introduction of design modifications in very little time and at very low cost.[160] Following the introduction of microfabrication techniques, we will present the most relevant prototyping and fabrication techniques applied to the development of miniaturized electroanalytical devices.

2.3.1 Microfabrication Processes

Microfabrication involves the combination of a series of lithographic, and additive and subtractive processes to create structures of micrometric and/or nanometric features on planar substrates.[46] To achieve these goals successfully, the entire microfabrication environment needs to be strictly controlled from the point of view of temperature, humidity, particle concentration and airflow. A cleanroom is a facility with such strict environmental control, and is where microfabrication processes take place. Temperature and humidity changes can easily affect photoresist curing processes, leading to reproducibility issues in lithography, which is at the core of any microfabrication procedure. Particle contamination control is a keystone because the size of air particles (namely, dust) is already of the same order of magnitude as that of the structures that are being fabricated, and the presence of a single speck of dust on a wafer can ruin an entire device.

Airflow in a cleanroom is controlled to avoid turbulences and to ensure that particles are pushed downwards, and in general a positive pressure is kept inside to avoid the presence of particles. In addition to environment control, certain areas, such as photolithography and nanofabrication, and certain characterizations need to be protected against vibrations to ensure that features in the micron range and below can be produced.

Microfabrication techniques can be classified according four different kinds of processes:

1. Lithography
2. additive processes
3. subtractive processes
4. thermal processes.

Although lithography may be considered an additive technique, as it involves the deposition of photoresist layers, here we have given it consideration as a class on its own because it is generally used to transfer patterns and structures *in combination* with either additive or subtractive processes.

2.3.1.1 Lithography: Photolithography and Maskless Techniques

Lithography is at the core of any microfabricated device, as it is the set of operations involved in the transfer of patterns from a computer design onto a physical substrate.[161,162] Photolithography involves the use of photocurable resists that are selectively activated using UV light through suitable photo masks, or are exposed to electron beams in the case of nanolithography.[163] There are two classes of photoresist: positive-tone resists and negative-tone resists. This[164] classification stems from whether the light-exposed areas are removed or not in the development stage. Figure 2.14

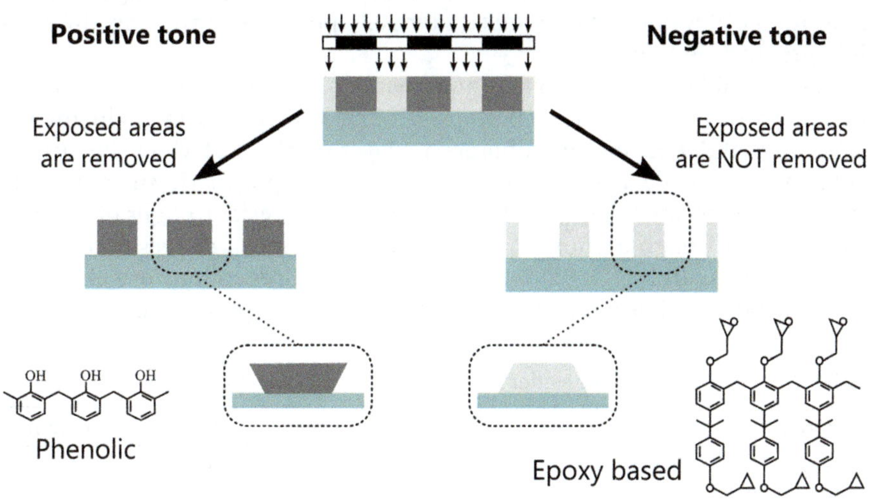

Figure 2.14 Positive *versus* negative photoresists.

shows the different behaviour of positive and negative resists, as well as the chemical structures of the most common resist types. One important difference between positive and negative tone resists is the shape of their resulting profiles after development, which can be used to advantage in subsequent processes. Thus, the "T" profile or undercut resulting in negative tone resists is very useful in lift-off procedures used in the definition of metallic parts (see below). Positive resists, on the other hand, may be more suitable masks in wet-etching operations. Photoresists are mixtures of a resin and a photoactive compound in a suitable organic solvent; they work on the basis of the different solubilities of this mixture before and after exposure to UV light. Thus, in a positive photoresist, resist solubility increases upon photoactivation,[165,166] whereas in a negative resist photoactivation results in crosslinking of resist components that decreases solubility.[167]

The choice of resist comes hand in hand with the choice of photo mask, which depending on its polarity can be either bright field or dark field. In addition to mask polarity, which is controlled by the following process and the resist tone, the size of the features to be transferred also conditions the type of photo mask that is used. Table 2.3 summarizes the main characteristics of commercial photo masks.

Quartz photo masks represent the highest quality, which in turn is reflected in their price. Because of their high transparency to UV light, quartz photo masks are used to define structures smaller than 1–3 μm, or when deep UV resists (active at 365 nm or less) are used, as soda lime is opaque in that range. For slightly larger structures and for most photoresists it makes more sense to use soda-lime glass masks or even high-density acetate films. The dark areas in soda-lime glass and quartz photo masks are defined by a thin chromium layer, whereas ink is used in high-density acetate films. Due to their low cost, acetate films can be used for prototyping purposes or as aids in the development of mix-and-match processes involving different lithographic methods.[12]

Once suitable photoresist and photo masks have been chosen, the process of lithography involves several steps, among which we can highlight the following: (i) deposition of a photoresist layer on the surface of the desired substrate—typically a silicon or a glass wafer, but polymeric substrates are gaining ground more recently; (ii) exposure of the resist to light of suitable wavelength through a photo mask; (iii) development of the resist to define the structures; and (iv) baking of the resist structures to improve chemical, mechanical and thermal resistance during follow-up processes.

Table 2.3 Parameters to consider during photo-mask selection.

Material	CD (μm)	Defect size (μm)	Defect density (cm^{-2})
Quartz	1	0.25–1	0.1–0.2
Soda-lime glass	3	0.25–1	0.1–0.2
Film	8–30	2	Unknown

Photoresists are deposited on thoroughly dried substrates by spin-coating, which results in controlled and homogeneous resist layers. Typically, photoresist layers are between 1 and 2 μm thick, although some applications demand nanometre-thin layers (such as e-beam lithography) or much thicker layers (as in the production of templates for electrodeposition as in the LIGA process,[168,169] or in the direct fabrication of microfluidic structures using SU-8, a negative tone epoxy resist).[170] After spin coating, a "soft-bake" and planarization step is recommended to remove the solvent from the photoresist, which improves photoresist behaviour and prevents the photoresist from sticking to the photo mask in contact-mode exposure, and ensures feature homogeneity across the entire wafer area. This is when the chromium side of the photo mask is placed directly in contact with the resist. Contact mode is used to improve the definition of the transferred structures and provides the highest resolution. Other modes of operation, shown in Figure 2.15, are the "proximity mode" and the "projection mode". In proximity mode, the photo mask is brought to a few microns from the photoresist, and it is used to protect the photo mask from photoresist contamination. The drawback in this case is a lower resolution than in contact mode, and features smaller than 3–5 μm are difficult to attain. Lastly, "projection mode" involves the use of lenses between the photo mask and the photoresist substrate, and provide much higher resolution than contact-mode. Current projection equipment provides resolution of less than 1 μm. However, to do this, only a small portion of the photo mask is imaged and stepped many times over the substrate surface. This is why these systems are known as step-and-repeat systems or steppers.[171] Following exposure of the photoresist, the transferred structures are developed, after which the substrates are dried and, in some cases, a hard-bake step is required to stabilise the structures further.

2.3.1.2 Additive Techniques

Microfabrication nearly always involves the deposition of new material layers on the substrate. In the case of microelectrode arrays and other

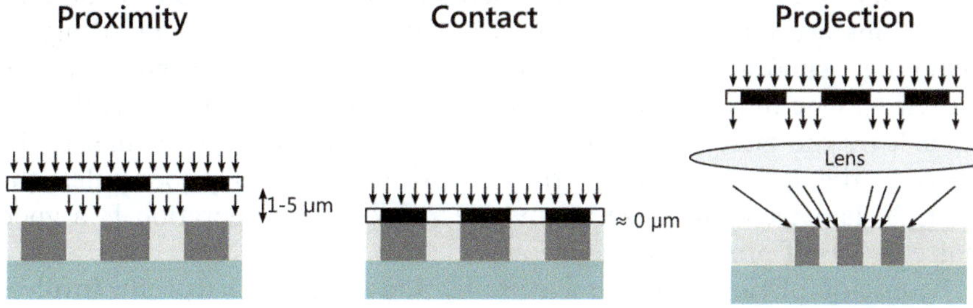

Figure 2.15 Diagrammatic representation of the three operation modes in optical lithography. Note that projection lithography only exposes a small substrate area, and hence the exposure of an entire wafer requires a series of step-and-repeat operations.

electrochemical devices, the most common additive processes used are the deposition of metals and of dielectric materials. Metals are used to create electrodes conducting tracks and contact pads. Although they are commonly deposited by physical vapour deposition, PVD, processes,[172] sometimes metal structures can be selectively grown by electrodeposition using photoresist templates.

2.3.1.2.1 Physical Vapour Deposition. Among the many PVD techniques available, here we shall only discuss sputtering[173] and evaporation,[174] since they are the most commonly used in electrode microfabrication. The deposition of material using PVD techniques involves three steps: (i) the formation of the coating vapour from a suitable target material, (ii) transport of vapour molecules to the substrate surface, and (iii) condensation of vapour onto the substrate.[46]

In sputtering, a target disc of the material to be deposited is set at a very negative potential *versus* the substrate platform while bombarded with argon cations. These argon cations are created in a plasma by different means (DC or RF), and as they hit the target surface metal ions are sputtered away, giving rise to a "cloud" of material, which is then deposited on the substrate that is resting on the anode. The advantages of sputtering compared to evaporation are better adhesion to the substrate and better lateral coverage (*i.e.* in the metallization of pillars and other vertical structures). One downside is that the substrate is easily heated because sputter deposition rates can be low, which prevents the metallization of certain polymeric substrates unless suitable heat dissipation measures are taken.

Evaporation is another important deposition technique. It differs from sputtering in that the trajectory of evaporated material is much more directional than in sputtering. There are two ways to evaporate material. One way is known as thermal evaporation,[175] and the other is known as e-beam evaporation.[46] In thermal evaporation the material is evaporated by heating a tungsten crucible in a high vacuum. The material is boiled off; this can present problems such as contamination by formation of alloys with the crucible material, or spurting of large particles on the substrate. On the other hand, e-beam evaporation, which works by evaporation of material after atoms are impinged by an electron beam, results in films of higher quality deposited at faster deposition rates than thermal evaporation and sputtering. One drawback of evaporation compared to sputtering is that, as atoms leave the material following a straight line path perpendicular to the substrate, the deposition of material on vertical structures is poor unless the substrate is rotated. Because of this, deposition by evaporation is a good choice in lift-off processes (see below).

The microfabrication of microelectrodes and their arrays typically involves the deposition of several metal layers adding up to between 100 and 200 nm. Normally a thin (10–30 nm) layer of Cr or Ti is deposited to improve the adhesion of other metals. In some cases an additional diffusion barrier consisting of Ni (*ca.* 10–30 nm) is deposited over the Ti layer to prevent the formation of Ti–Au intermetallic compounds during subsequent thermal processes.[20,176,177] Last, a slightly thicker layer (30–150 nm) of the electrode material is deposited. Following metal deposition, the fabrication process continues either by performing a lift-off operation or with a lithographic step prior to a wet etch.

2.3.1.2.2 Chemical Vapour Deposition (CVD) and Other Thermal Processes. The most common thermal process used in the microfabrication of electrodes is the formation of silicon oxide when silicon wafers are used as substrate, the hardening of photoresists used as mask in the dry etching of silicon nitride and silicon oxide layers used as passivation, and the CVD deposition of silicon oxide and silicon nitride layers for passivation purposes. In addition, CVD can also be used to produce carbon-based materials such as nanotubes and graphene, and glassy-carbon-like materials can be synthesised by pyrolysis of thick photoresist films.[170]

The passivation of silicon wafers is achieved in a thermal process above 1000 °C in the presence of oxygen or water. While thicknesses around 1 μm suffice to provide electric insulation from the silicon substrate and avoid leakage currents, layers as thick as 5 μm can be achieved. However, due to the process kinetics, producing such thick layers is difficult and lengthy.

On the other hand, the passivation of microelectrodes using SiO_2 and Si_3N_4 is typically deposited by CVD. CVD consists of the deposition of material from reacting gas precursors that flow towards the substrate. CVD processes are considered thermal processes because the deposition of material on the substrate occurs at high temperatures. An advantage of plasma-enhanced CVD (PECVD) is that it enables the deposition of layers at lower temperatures, which can of great help when processing delicate substrates, or when the gas constituents or the adsorbed products react with the substrate. One example is the formation of a Si–Au eutectic above 351 °C,[178] which causes certain microelectrodes to require continuous electrochemical activation.[177] SiO_2 is a very common dielectric material employed in microfabrication; it can be used as passivation material in the fabrication of microelectrode arrays. SiO_2 is formed by PECVD at temperatures above 270 °C using SiH_4 and O_2 to yield SiO_2 and H_2O. To prevent water leakage through the passivation layer, SiO_2 layers 500 μm thick, typical deposition temperatures are nearly always above 350 °C. The results are very stable, well adhered and nearly impermeable layers, but when they are deposited over gold, the latter is contaminated by silicon down to a few nanometres. Unfortunately, silicon atoms directly on the metal surface are easily oxidised, severely affecting the electron transfer ability of the electrodes. This can be overcome by following different activation protocols.[179] Alternatives to SiO_2 as passivation layer for microelectrode array fabrication are negative photoresists, such as SU-8, parylene,[180] polyimides, and more advanced materials such as benzocyclobutenes, BCB.[181] It is important to use as thin as possible a passivation layer, but ensure that its dielectric constant is sufficiently low so that the device does not show significant background capacitive currents.

2.3.1.3 Subtractive Techniques

Subtractive techniques are about the selective removal of material from the substrate. Here we will focus on etching techniques, both wet and dry, and will also mention lift-off processes.

When used in combination with lithography, etching processes enable the definition of structures in exposed areas. The most important parameters in any etch technique are etch rate, selectivity, directional control, uniformity, and throughput.

The selectivity of an etchant enables the choice of a suitable mask for the process. The mask should withstand the etch process conditions but be easy to remove afterwards in a process that does not damage the substrate, *i.e.* stripping a resist layer in an acetone bath. Typical mask materials in microfabrication are photoresists, silicon oxide, and metal layers.[182]

2.3.1.3.1 Wet Etching. Wet etching involves the chemical removal of material by immersion of a substrate in a bath with a suitable etchant.[183] Wet etching can be used in electrode microfabrication to pattern metals such as aluminium, titanium, nickel, chromium and gold, among others; in addition, wet etching of silicon and glass can be used in the construction of structures for microfluidic devices.[182,184] Barring the case of silicon, wet etching is typically isotropic, which means that the same amount of material is removed from the substrate in the horizontal and vertical directions. This is due to diffusion controlling the etch reaction, and has important implications in mask design and the process critical dimensions, as structures with aspect ratio greater than 1 : 1 are difficult to achieve by wet etching. On the other hand, wet etching can be highly selective and uniform and it is normally used in the removal of thin layers and for cleaning purposes.

2.3.1.3.2 Dry Etching. Dry etching consists of the removal of material in a reactive ion plasma, and the technique is commonly known as reactive ion etching, RIE.[185] The high directionality of the electric field inside the reaction chamber makes dry etching anisotropic, as the etch rate in the direction perpendicular to the substrate surface is significantly higher than etching in the lateral direction. Structures with very high aspect ratio can be achieved by controlling the plasma conditions, or by more advanced processes such as the Bosch process, where surface passivation and etch steps are alternated many times to achieve structures hundreds of microns high.[25]

2.3.1.3.3 Lift-off. Most materials can be etched either by wet or dry processes. However, some materials are extremely hard to etch, *e.g.* platinum.

In this case, microstructures are more easily defined by a process known as lift-off, which involves the deposition of the material to pattern on top of the mask. This technique involves a lithographic step followed by the deposition of the desired material throughout the entire substrate surface. Thus, the material to pattern is equally deposited on the mask and on the areas where it is needed. Next, the mask, which is usually a photoresist, is removed in a wet bath composed of the appropriate solvent. Acetone is normally used when the mask is a simple photoresist layer, but in some cases strong etchants may be required. Removing the mask removes the material deposited on top of it, neatly leaving the desired structures behind. To facilitate mask removal, masks with negative profiles or undercuts are sought. Sacrificial layers, negative-tone resists and image reversal resists such as Clariant's AZ-5214E are the most common strategies used to achieve this.[12]

2.3.2 Rapid Prototyping Techniques

Another class of fabrication techniques that complement microfabrication are so-called rapid prototyping methods. Prototyping techniques are used to reduce the cost of microfabrication and certain technical risks. Microfabrication techniques are costly and can be very time consuming, so it is essential to optimize the design of microfabricated parts. It is good practice to design new systems from the outside in because ultimately any miniaturized device needs to be interfaced and used in a "macro" world. Thus, it is important to first consider the application environment, with a special focus on electrical and fluidic connections. Chip packaging often drives the final cost of microfabricated devices, so introducing it during the development stages saves microfabrication time and cost. These savings are achieved by enabling numerous iterations in a short time and using low cost materials. Once a functional packaging has been attained, it is possible to proceed with the microfabrication of microelectrodes and other chip-based parts. Giving a full account of prototyping techniques is beyond the scope of the present chapter, but the following sections will summarise the most common and accessible prototyping techniques used in the development of electrochemical devices, and particularly flow systems. In terms of materials, poly(methyl methacrylate), PMMA, and polydimethylsiloxane, PDMS, are without doubt the most common, but laminates, *i.e.* polymeric films and adhesive tapes, are becoming increasingly important in the development of flow devices as they enable the construction of fully-functional devices.[186]

As in the case of microfabrication processes, prototyping techniques can also be classified according to their "additive" or "subtractive" character.

2.3.2.1 Subtractive Techniques

Cutting and milling techniques start with a piece of material and carve their way through it until all the desired features are produced.

2.3.2.1.1 Milling Techniques. Milling is among the oldest techniques used in the manufacture of objects, including electrochemical flow cells.[187] Milling uses rotary cutting tools to carve structures from a block of material. The introduction of micrometric size tools in the 1990s has enabled the fabrication of microfluidic devices.[188] Milling can be used to produce very high-quality moulds for casting of PDMS or other suitable polymers,[189,190] or it can be used to produce microfluidic channels directly on a suitable substrate—such as PMMA, PTFE or aluminium. However, milling finds its most common application in the fabrication of chip holders and flow cells. Important parameters in milling are tool rotation and material feed rates. Although most tool manufacturers provide guidelines for these parameters depending on the type of material to mill, such guidelines should be taken only as a starting point and adjusted depending on experience. Material quality varies from one manufacturer to another, and tools wear out with use. Milling is used to produce mm–cm-sized objects with features down to a few hundred microns. Although it is possible to find end-mill tools with diameters down to 35–50 μm,[191] these can only be used by extremely qualified staff running high-end equipment where spindle vibration is strictly controlled.

2.3.2.1.2 Blade Cutting. Blade cutting has been around for a long time in the graphic arts industry, where it is routinely used to cut out vinyl and adhesive films. However,[192,193] it is a relatively new technique in the development of electrochemical devices. Cutter plotters are classified according to how they hold the blade. In the most common and affordable ones, the blade can rotate freely in the head, and the offset between the blade tip and its rotation axis facilitates the orientation of the cutting edge in the cutting direction. These are known as "drag knife" plotters, and enable the production of structures down to 200 μm for materials 50–150 μm thick. At the other end there are so-called "tangential" plotters where the blade is fixed and the head is rotated and oriented in the cutting direction by a motor. These enable the fabrication of smaller features than drag-knife plotters with much higher precision, but they are also significantly more expensive. Drag knife tools are normally fine for prototyping purposes. Blade cutting has been used in the construction of fully functional microfluidic devices made in pressure sensitive adhesive tapes, and also in the development of not very complex paper microfluidic devices.[58]

2.3.2.1.3 Laser Cutting. Laser cutting works by melting and vaporizing material using a laser beam.[194] An advantage of laser cutting compared to milling or blade cutting is that it is a "contactless" technique, which means that the material does not suffer mechanical stress during the cutting operation. Different laser types are used for manufacturing and prototyping. Desktop and low-power CO_2 lasers (10.6 μm wavelength, in the long-infrared range) are an excellent choice for prototyping purposes, as they enable working on a wide range of materials barring carbon and metals, and some other high-density materials.[195,196] Laser cutting is faster than milling, but the high local temperatures achieved around the cut line can lead to problems such as the formation of burr in some polymeric materials, or to cracking of brittle materials such as glass and ceramics. In addition, as in the case of blade cutting, feature sizes below a few hundred microns are hard to achieve with CO_2 lasers. Excimer lasers allow for features down to *ca.* 5 μm, but they are much more costly than CO_2 systems.[197,198] On the other hand, as in the case of milling, operating conditions such as laser power and frequency and head speed need to be calibrated for each material along with the desired cutting depth.

2.3.2.1.4 Sand Blasting. Sand blasting consists of the bombardment of a substrate with abrasive particles of micrometric size through a suitable mask. Particles are carried by Venturi effect[199] in a stream of compressed air at several bars of pressure, and cast against a masked substrate. The mask may be made of material resistant to the sand grain impacts, such as metals or special vinyls. Vinyl masks can be cut out by blade or laser cutting, and the pattern sizes are typically limited to the resolution of those techniques (around 300 μm).

Sand blasting can be used to carve channels, although because controlling the depth can be a daunting task with most manual equipment, sand blasting is more suitable as a means to open ports in thick (>300 μm) glass substrates that are difficult to process by milling or laser techniques.[200]

2.3.2.2 Additive Techniques

2.3.2.2.1 Laser Sintering and 3D Printing. Although their operating principles are radically different, both techniques enable the bottom-up construction of complex 3D parts with minimum feature size down to 100–200 μm. In laser sintering a powdered material is sintered by action of a laser beam and parts are built layer by layer.[201] In 3D printing, on the other hand, material is extruded through a nozzle and deposited on a substrate, also layer by layer. These additive manufacturing techniques can produce objects in a very wide range of materials such as plastics, metal and ceramics, although plastics are the most suitable for prototyping of electrochemical devices.

These techniques are very suitable for the production of chip holders and flow cells. 3D printing technology is advancing rapidly, and printer resolution is improving significantly; it is possible to find machines able to print to features down to 30 μm,[202] and the most affordable ones already reach the 100 μm resolution.

2.3.2.2.2 Replica Moulding. Replica moulding consists of casting or injecting a material in a mould that has previously been fabricated by other methods, including lithography, milling, cutting and more recently even 3D printing.[203,204] The most common application of replica moulding in the development of electrochemical devices is the fabrication of PDMS parts.

PDMS is an extremely easy to use silicone elastomer: it is a two-component mixture that can cure under room conditions using a Pt-based catalyst. This means that PDMS shows nearly no shrinkage upon curing, in contrast to common silicones that use tin-based catalysts. PDMS is non-toxic and biocompatible; it is transparent, it can be covalently bonded to itself and to silicon and glass substrates,[205] which allows for multi-level structures including on-chip valves and pumps,[49] and it is highly permeable to gases, which, depending on the final application, can be either an advantage or a disadvantage. Moreover, it is possible to modulate its mechanical and surface properties by chemical and thermal means. These properties make PDMS and its derivatives excellent materials for prototyping and in the construction of research tools, including electrochemical devices. An important step in the fabrication of PDMS parts is the adequate mixing of its components due to their viscosity. In addition to thorough stirring, one can add a small amount of pigment to the mixture. The mixture can be considered complete once the colour of the mixture becomes homogeneous. Bubbles formed during mixing can be extracted by vacuum, or by leaving the PDMS to rest in a refrigerator for an hour or two before casting. PDMS takes about 24 h to cure at room temperature, but if parts are cured in an oven at 60–80 °C, the curing process may be completed within 1–2 h. Lastly, although they may not be an issue in most cases, unreacted PDMS components can be extracted using hexane. On the other hand, PDMS is unsuitable for work with organic solvents, as most of them will swell the PDMS significantly.[206] PDMS can be used to make very high-fidelity replicas with nanometric features, as in stamps for nanoimprint lithography.[207] This means that it will stick quite strongly to mould surfaces, and to facilitate the release of PDMS parts demoulding agents such as fluorosilanes are recommended. Moulds and master parts may be silanized by spraying, or in a desiccator from ethanolic fluorosilane solutions.

The following section will illustrate the above techniques applied to the development of devices for use in diagnostic applications.

2.4 Miniaturized Biomedical Diagnostic Devices

The broad term "point of care device" refers not only to portable or disposable devices, but in general to any method or device that operates outside the clinical laboratory. Point-of-care detection systems aim to simplify the analytical procedures and yet offer rapid and reliable diagnostic information that can help the patient or a first-aid team make the right decision. Thus, important requirements asked of point-of-care devices are portability, ease of use, and cost-effectiveness.[67,208] Microfabrication allows the development of novel analytical systems that usually meet the requirements of small size and rapid results. However, these are seldom easy to use, and require extensive development to meet the demanding cost, performance and regulations.[209]

Electrochemical detection methods are ideally suited for their integration in miniaturized systems, as clearly pointed out by Nyholm.[210] Microelectrodes can be easily fabricated and the instrumentation required to carry out potentiometric and amperometric measurements is relatively simple and cheap to fabricate in comparison with optical methods, which are currently the main detection method in miniaturized systems, including biosensors.

The following presents different approaches for the electrochemical detection of different biomarkers, mostly based on immunosensors. We will begin by describing the development of immunosensors on 3D interdigitated microband electrodes for label-less detection using an impedance spectroscopy-based technique. Next, two approaches for the amperometric diagnosis of acute heart failure using two different markers will be presented. The cases discussed cover a very wide range of functionalization, fabrication and transduction approaches and will illustrate the potential of miniaturized electrochemical devices in the biomedical area.

2.4.1 Development of Microfabricated Immunosensors for Label-free Detection

Immunoassays have become the gold-standard in clinical diagnostics to detect biomarkers associated to several diseases, using specific antibodies or immunoglobulins. Immunoassays can be applied to a great variety of biosensor formats, leading to the corresponding immunosensors, a powerful subclass of which rely on EIS detection.

Impedance biosensors are a promising tool for point of care devices. Their high sensitivity and their ability to perform label-free detection facilitate manipulation while avoiding several incubation steps. In addition, EIS can be easily implemented in portable devices. However, impedimetric sensors are also subject to limitations. For instance, EIS biosensors require exhaustive controls to ensure their selectivity and to discriminate the contribution of any interfering agents.

Impedimetric biosensors offer many advantages such as the capacity to reduce in size the electrical equipment needed and also, as described above, the possibility to study both faradaic and non-faradaic processes. This section presents the development of an impedimetric immunosensor based on the detection of surface conductivity changes.[153] This device relies on interdigitated microband electrodes and is validated with the most common reaction used as a model in biosensors, the antibody interactions of the immunoglobulin G (IgG) antigen–antibody pair, one of the most standard reactions in immunochemistry.

As mentioned above, interdigitated structures allow the detection of variations in the surface conductivity arising from surface concentration changes of the biomolecules attached to the surface. These variations are obtained by extracting the R_s value of the equivalent circuit described in Section 2.2.3.5 (Figure 2.10).

As shown in Figure 2.16, the device was fabricated using conventional microelectronic techniques as follows.[113,211] A 2500 nm thick silicon dioxide layer was thermally grown on a 4 inch silicon wafer. Next, a 230 nm thick layer of tantalum silicide ($TaSi_2$), a highly conductive material, was deposited by magnetron sputtering. The first photolithographic step defined collector

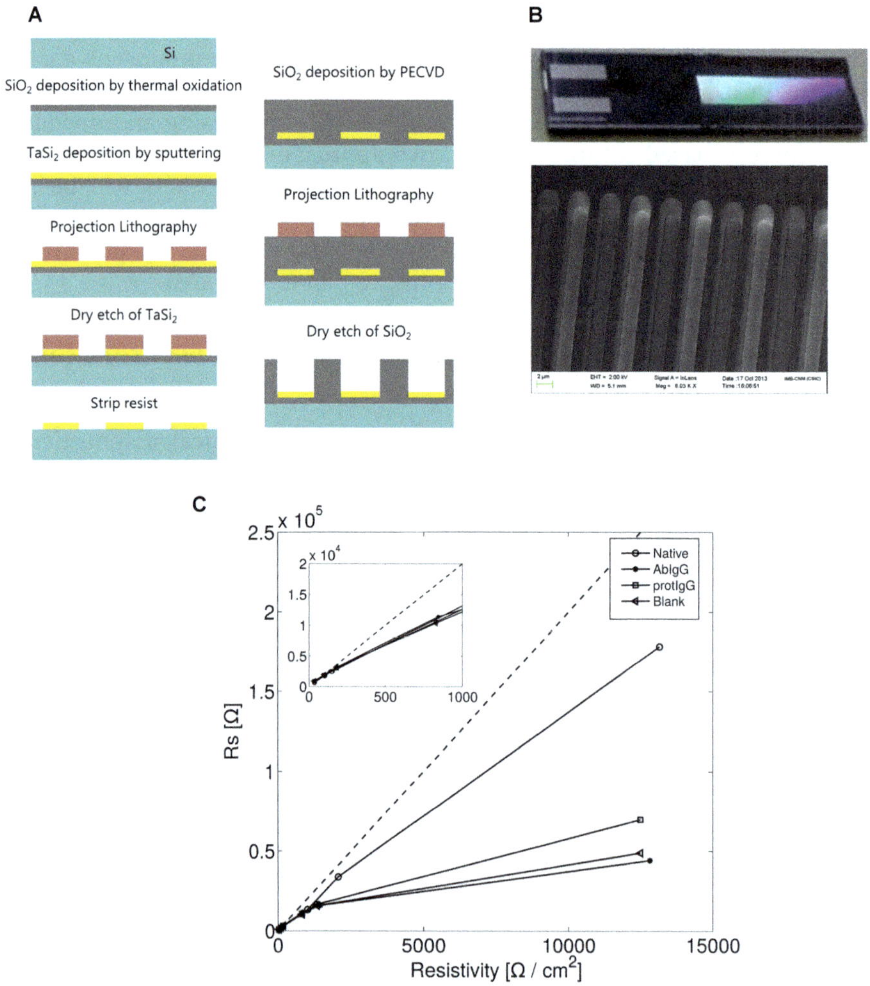

Figure 2.16 (A) Fabrication process for the 3D interdigitated microband array and (B) photographs of the device. (C) Resistance, R_s, determined from the EIS spectra *versus* the solution resistivity for studied sensors: native device (SiO_2 layer), AbIgG, ProtIgG and Blank. Inset: the linear part of the sensor response in highly conductive solutions. A dashed line shows the expected response of a sensor without the surface conductivity phenomenon. The slope of this line represents the cell constant.

bars and the digits of each electrode. The patterning was performed in a reactive ion etching step. This resulted in an interdigitated electrode array with 216 digits of 3 μm width and 3 μm gap between the adjacent electrode digits. The electrode digits are 1.5 mm long. To form the contact pads 1 μm of aluminium was deposited and patterned using standard photolithographic and wet etching steps leaving metal only at the end of the two collector bars. In the final step the wafer with formed interdigitated devices is covered with a 4 μm thick silicon dioxide layer deposited by a low pressure chemical vapour deposition (LPCVD). LPCVD SiO_2 is less dense than thermally grown SiO_2, but it is equally insulating. Photolithography was then used to define the trenches to be opened in the oxide layer over the electrodes digits and over contact pads. These zones were opened by deep reactive ion etching (DRIE), which resulted in nearly vertical walls. In this way capillaries were formed in silicon dioxide over the electrode digits. The capillaries over each electrode digit are 3 μm wide, 4 μm high, 1.5 mm long and are open at their top. After being cut from the wafer the sensors were glued to a printed circuit board substrate with copper leads and were wire bonded for electrical connections. Contact pads and wires were encapsulated using epoxy resin.

IDE sensors were cleaned with ethanol and deionized water and were immersed in 1 M NaOH for 30 min to enrich the surface with silanol groups. After the activation step the chips were rinsed with DI (deionised) water and ethanol and dried again with nitrogen. The chips were exposed to 2.5% v/v 3-glycidoxypropyltrimethoxysilane in ethanol for 16 h at room temperature. X-Ray photoelectron spectroscopy (XPS) experiments showed that the tantalum silicide surface was also covered with a thin (0.2–0.5 nm) layer of a native oxide formed at room temperature. Therefore, chemical modification with silane, and subsequent antibody immobilization, takes place not only on the SiO_2 barrier surface but also on the surface of the electrode digits.

The terminal epoxy group of the silanes attached to the surface of the SiO_2 readily reacts with amino groups of biomolecules, fixing them onto the surface. Thus, immobilization of the IgG antibody (1.0 $\mu g\, mL^{-1}$ in carbonate buffer, pH 9.6) proceeded overnight at 4 °C. The sensors were rinsed with PBS (phosphate-buffered silane) buffer and dried with a N_2 stream. Reaction with IgG protein for the final detection was performed by immersing the chips into a 5 $\mu g\, mL^{-1}$ protein IgG solution in PBS for 30 min at room temperature.

The blank control experiment (silanized electrode with the antibody immobilized) consisted of exposing a sensor to a PBS solution without IgG protein. Figure 2.16(C) shows the response of these sensors at different stages, and the control. These are referred to as Native: SiO_2 surface without any modification, AbIgG: modified with antibodies, protIgG: after reaction with protein, and Blank: not exposed to protein.

The sensitivity of interdigitated microband electrodes to variations of the surface conductivity can be defined as the R_s variation (ΔR_s) when the contribution of the surface conductivity is almost zero. The sensitivity enhancement is shown in Figure 2.16(C), where lower conductivity media result in higher sensitivities. The different layers of functionalization over the surface of the biosensor measured in solution with 1 μS cm^{-1} conductivity (the lowest achievable) show a measurable variation in the R_S of ΔR_S Native: 40%, ΔR_S ProtIgG: 75%, ΔR_S AbIgG: 84%, giving a wide detection range.

We have described the fabrication of a sensitive EIS immunosensor, the novelty of which lies in the monitoring of surface conductivity rather than capacitance or electron transfer rate changes. This sensitivity enhancement stems from an optimized design, allowing the development of a novel generation of surface conductivity biosensors for the detection of protein IgG, although this platform can be used for the detection of other proteins. In the following section we will explore more specific cases applied to the detection of cardiovascular disease biomarkers.

2.4.2 Microfabricated Devices for the Electrochemical Detection of Cardiovascular Disease Markers

Cardiovascular diseases, CVD, are among the main causes of death worldwide, followed by infectious and parasitic diseases, cancer and respiratory infections. This motivates the development of new diagnostic methodologies based on screening cardiac biomarkers. Correct risk stratification and treatment of acute myocardial infarction (AMI), and acute coronary syndrome (ACS) rely heavily on the availability of rapid diagnosis methods. Table 2.4 summarises the main parameters surrounding current CVD markers.

Table 2.4 Main clinical parameters for some of the most important cardiovascular disease markers.[64]

Biomarker	Disease[a]	Cut-off levels (ng mL^{-1})	Time to rise (h)	Peak time (h)	Back to normal levels (d)
Troponin I	AMI	0.01–0.1	4–6	12–24	6–8
Troponin T	AMI	0.05–0.1	4–6	12–24	7–10
Myeloperoxidase	Inflammation	>350 (>11.9 U mL^{-1})	ND[b]	ND	ND
CRP[c]	AMI	Low risk: <10^{3}; medium risk: 1–3×10^{3}: high risk: 3–15×10^{3}	ND	ND	ND
Interleukin-6	Inflammation/ risk factor	Low risk: <10^{-3}; medium risk: 1–2×10^{-3}; high risk: >2×10^{-3}	ND	ND	ND

[a]AMI = acute myocardial infarction.
[b]ND = not determined.
[c]CRP = C-reactive protein.

This section reviews the recent literature on the detection of some of the above markers using microfabricated devices, with an emphasis on the detection of troponin (cTn), myeloperoxidase (MPO) and C-reactive protein (CRP). Cardiac troponins, cTn, are currently considered the biochemical "gold standard" for the diagnosis of AMI, owing to their excellent specificity and sensitivity.[212–214] cTnI and cTnT are highly specific to the heart muscle and are normally not present in the blood stream. Troponin blood concentrations above a certain cut-off level provide 100% diagnostic sensitivity for AMI with serial sampling. Troponin is key in the regulation of skeletal and cardiac muscle contraction. As Figure 2.17 (left-hand side) shows, troponin is formed by three different subunits—troponin I (TnI), troponin C (TnC) and tropoinin T (TnT). In the figure, TnI, TnC, and TnT are depicted in red, yellow and green, respectively. Cardiac TnT and TnI are presented in different forms from the skeletal muscle and TnC does not have cardiac isoforms but forms binary complexes with TnI and ternary complexes with TnI and TnT. That is why the detection of cTnT is one of the main goals for the diagnosis of MI. Several clinical trials and patients' stratification studies include the monitoring of cTnT[212–214] showing the correlation between the presence of cTnT in patients arriving in emergency rooms after chest pain. Various studies have demonstrated that cardiac troponin levels increase dramatically during the 24 h following an AMI episode up to levels of around 50 $\mu g\ L^{-1}$.

Myeloperoxidase is a heme oxidase that catalyses the oxidation of H_2O_2 and Cl^- to yield hypochlorous acid, a highly oxidizing species. The concentration of myeloperoxidase in serum increases as a result of neutrophil and monocyte activity triggered by inflammation caused by eruption or ruptured arteries. Increased myeloperoxidase concentrations correlate with high risks of subsequent cardiovascular damage. This correlation enables the early diagnosis of cardiovascular disorders, particularly long-term acute myocardial infarction (AMI). As an enzyme, the concentration of myeloperoxidase found may be expressed in activity units per mL of sample, or in units of weight per volume (mL). The normal activity levels of myeloperoxidase are around 6.6 $U\ mL^{-1}$, and concentrations above 12 $U\ mL^{-1}$ or 350 $ng\ mL^{-1}$ are considered anomalously high.[64]

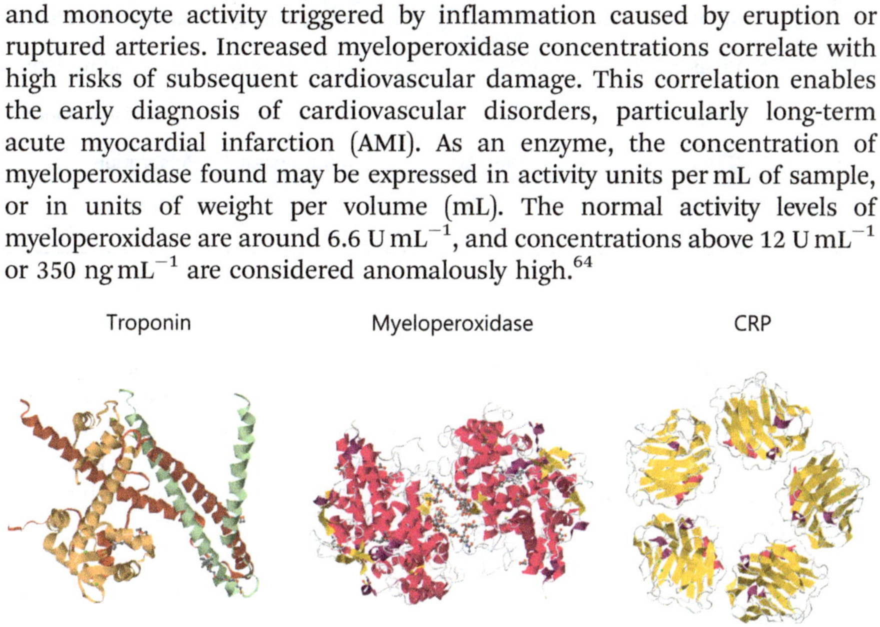

Figure 2.17 Ribbon representation of the molecular structures of troponin, myeloperoxidase and C-reactive protein.

Table 2.5 Summary of electrochemical detection methods for CVD biomarkers.

Biomarker	Electrode material	Substrate	Assay	Functionalization strategy
cTnI	Au	PDMS	Immunoassay	Silanization + EDC/NHS
cTnI	Au	Au	Immunoassay	Peptide SAM
cTnI	ITO[a]	ITO	Immunoassay	Avidin–biotin
cTnI	GC	Graphene oxide	Immunoassay	EDC/NHS
cTnI	VACNF	C	Immunoassay	EDC/NHS
cTnI	ITO	Nitrocellulose	Immunoassay	Sulfo-SMCC
cTnI	Au	Carboxylated graphene	Immunoassay	EDC/NHS
cTnT	Al	Al	Immunoassay	Glutaraldehyde–SAM
cTnI	Si nanowire	SiO_2	Immunoassay	Silanization + glutaraldehyde
cTnT	Ag paste	Latex beads	Immunoassay	Streptavidin–biotin
cTnT	Au	Au	Immunoassay	SAM/glutaraldehyde
cTnT	Au	Latex beads	Immunoassay	Streptavidin–biotin
cTnT	Au SPE	Latex beads	Immunoassay	Streptavidin–biotin
cTnT	Au macroelectrode	Au	MIP	Unclear
MPO	Au	Au	Immunoassay	SAM/EDC/NHS
MPO	C	C (SPE)	Enzymatic	Not a biosensor
MPO	Au, Pt	Latex (magnetic beads)	Immunoassay	Streptavidin–biotin
MPO	Unspecified	IrO_x	Immunoassay	Streptavidin–biotin
CRP	Unspecified	IrO_x	Immunoassay	Streptavidin–biotin
CRP	C fibre	PDMS-Au	Immunoassay	Adsorption
CRP	Au	Au	Immunoassay	SAM/EDC/NHS

[a]ITO, indium tin oxide.

C-reactive protein, CRP, is also used as a risk indicator of coronary events, and in the optimization of the statin therapy in primary and secondary prevention settings of cardiovascular disease. CRP is not as specific an indicator of CVD as myeloperoxidase or cTn, as it is produced in the event of inflammation throughout the body. However, it is believed that CRP blood concentrations less than 1 $\mu g\,mL^{-1}$ are normal, concentrations in the 1–3 $\mu g\,mL^{-1}$ range represent an average risk of developing a cardiovascular disease, but that the risk becomes high if the CRP concentration is greater than 3 $\mu g\,mL^{-1}$.[64]

Table 2.5 presents a summary of some of the most recent studies reporting on the electrochemical detection of troponin, myeloperoxidase

Detection technique	Detection limit (ng mL^{-1})	Dynamic range (ng mL^{-1})	Sample	Controls	Ref.
Amperometry	0.15	0.2–10 000	PBS	Not reported	Ko *et al.*[93]
QCM	110	0–3000	PBS	Not reported	Wu *et al.*[215]
Amperometry	0.01–0.1	Up to 10	Serum	Not reported	Akanda *et al.*[92]
SWV	0.05×10^{-3}	Up to 50	PBS	Not reported	Feng *et al.*[216]
EIS	0.2	0.25–100	PBS	Yes	Periyakaruppan *et al*[217]
Amperometry	0.1×10^{-3}	Up to 100	Serum	Not reported	Akanda *et al.*[218]
Conductance	10^{-3}	10^{-3}–1	PBS/air?	Not reported	Tuteja *et al.*[98]
EIS-capacitive	Not stated	0.07–7	Serum/PBS	unclear	Vasconcelos *et al.*[219]
MOSFET	10^{-6} (PBS) 3×10^{-5} (serum)	Up to 10	PBS/serum	Not reported	Chua *et al.*[220]
CV	0.2	0.2–10	Serum	Not reported	Silva *et al.*[221]
QCM	0.0015	0.1–0.5	Serum	Not reported	Fonseca *et al.*[222]
Amperometry	0.02	0.05–1	Serum/PBS	Yes	Abad *et al.*[59]
Amperometry	0.02	0.05–1	Serum/PBS	Yes	deÁvila *et al.*[223]
CV	10^{-3}	Unclear	PBS	Not reported	Karimian *et al.*[88]
EIS	10^{-3} on diatom membranes; 0.5 on AlO_2 membranes; 5 directly on Au	Up to 1000	Serum	Yes	Lin *et al.*[224]
Amperometry	0.6 U mL^{-1}	0–18 U mL^{-1}	Acetate buffer/serum	Yes	Windmiller *et al.*[225]
Amperometry	0.4	0.9–60	Plasma	Yes	Barallat *et al.*[226]
EIS	0.5	1–1000	Serum	Yes	Venkatraman *et al.*[90]
EIS	1	10–10^5	Serum	Yes	Venkatraman *et al.*[90]
SWV	220	500–2000	Acetate buffer/serum	Not reported	Zhou *et al.*[227]
EIS	10^{-3}	10^{-3}–10^3	Serum	Yes	Lin *et al.*[224]

and C-reactive protein (CRP), three of the main cardiovascular disease biomarkers. Table 2.5 includes some studies that do not use microfabricated electrodes but which are nonetheless relevant. Other biomarkers such as interleukin 6, creatine-kinase and B-type natriuretic peptide have been omitted altogether; first, because they are less reliable than the others, and second because the cases described here cover the main possible detection strategies mostly based on microfabricated devices.

Most of the studies reviewed during the preparation of this chapter are immunosensors, although in the case of the enzyme myeloperoxidase direct detection is also possible. Thus, Windmiller *et al.* reported on the detection of myeloperoxidase activity.[225] That work used a carbon screen printed electrode using 3,3′,5,5′-tetramethylbenzidine (TMB) as redox mediator to monitor the activity of myeloperoxidase in acetate buffer and in human serum samples. Unfortunately, the much lower activity of myeloperoxidase compared to other peroxidases, combined with its low concentration levels and the presence of interfering agents such as ascorbic and uric acid, make the direct detection approach rather difficult. Barallat *et al.*[226] reported an amperometric immunoassay using magnetic bead capture for sample pre-concentration—this increased the amount of active myeloperoxidase in the vicinity of an electrode, and made the measurement of the enzyme activity more accessible. The work consists of the microfabrication of gold micro-electrodes on silicon chips that are subsequently housed in a holder made by laser sintering. The holder enables the capture and release of magnetic beads directly on the working electrode surface. The assay relies on the capture of myeloperoxidase on streptavidin coated magnetic beads. Biotinylated anti-myeloperoxidase monoclonal antibodies were linked to them, and any remaining streptavidin was blocked with biotin to avoid non-specific adsorption of other proteins. These functionalized beads were added to diluted serum samples and, following an incubation period, they were recovered with a magnet and washed with PBS. After that they were transferred onto the surface of a gold electrode, where TMB was used to monitor the activity of the captured myeloperoxidase. The linear range of the reported methodology was 0.9–60 $ng\,mL^{-1}$, and a detection limit of 0.4 $ng\,mL^{-1}$. These values correspond to a range of 45–3000 $ng\,mL^{-1}$ and a detection limit of 20 $ng\,mL^{-1}$ in the undiluted sample. The total time of analysis was 30 min, excluding the preparation of the immune functionalized beads. Another example of myeloperoxidase detection worth mentioning is that reported by Venkatraman *et al.*,[90] who used iridium oxide nanowire electrodes to quantify myeloperoxidase and CRP by electrochemical impedance spectroscopy. Vertically aligned IrO_x nanowires were grown on TiN film electrodes by chemical vapour deposition using oxygen and an iridium organometallic precursor. These nanowires were 10–50 nm in diameter and 0.3–0.5 μm long, which results in a massive surface that can be covered with antibodies and, thereby, increase sensitivity. The surface of the nanowires was modified by non-specific adsorption of streptavidin, followed by incubation with biotin-conjugated antibodies. Electrochemical impedance spectroscopy was used to monitor myeloperoxidase and CRP capture by their corresponding antibodies on the electrode surface, and a simplified Randles circuit without the faradaic branch was used for data analysis purposes. The reported results include a linear range spanning three orders of magnitude in a logarithmic plot, both for myeloperoxidase and CRP. However, these results should be taken with care, as the authors do not report any controls to demonstrate that other proteins do not interfere with their assay to a significant extent. This note of caution also applies to many of the studies summarized in Table 2.5, as only a few of them provide control data.

The work of Lin *et al.* on the detection of myeloperoxidase and CRP is worth noting.[224] This study reports on the use of silica and alumina nanotemplates deposited on gold microelectrodes as a means to increase sensitivity. Antibodies are immobilized on the gold microelectrode under the nanopore structure on a 3,3′-dithiodipropionic acid di(*N*-hydroxysuccinimide ester) self-assembled monolayer. The NHS group was used to bind to streptavidin, to which biotinylated antibodies could bind. The capture of myeloperoxidase and CRP at the corresponding antibodies modified the interfacial capacitance, which was monitored by EIS. The use of nanopore templates was demonstrated to improve the linear range and detection limit for both myeloperoxidase and CRP, reaching detection limits down to 1 pg mL^{-1} and linear ranges spanning over five orders of magnitude, up to 1 μg mL^{-1}.

Besides myeloperoxidase and CRP, troponin is an increasingly important biomarker used in the early diagnosis of acute myocardial infarction.[228] The studies reported in the literature address mainly the detection of cTnC and cTnI. The following applications have been chosen not only because they are good examples of how miniaturized electroanalytical devices have evolved both in terms of materials and fabrication technologies but also show how this evolution is clearly driven towards marketable developments.

Most of the following studies rely on amperometric detection at gold electrodes. An exception is the work of Tuteja *et al.*, which looks at conductance changes induced by the capture of cTnI over the surface of a graphene sheet between two gold bands.[98] Graphite flakes were exfoliated by a method of lithium intercalation, and the resulting graphene platelets were oxidized in acid. This oxidation resulted in the formation of –COOH groups on the graphene, and these groups were in turn used to bind anti-cTnI antibodies in a reaction with EDC/NHS. The functionalized graphene sheets were deposited over the surface of a silicon chip featuring several individually addressable gold microbands. The resulting biosensor shows a detection limit close to 0.1 pg mL^{-1}, and a logarithmic dynamic range between 1 pg mL^{-1} and 1 ng mL^{-1}.

Another example where microbands are used was reported by Akanda *et al.*[91,92] In this case, an amperometric measurement is performed at indium tin oxide (ITO) electrodes to detect the activity of an enzyme-labelled anti cTnI antibody. What makes this work particularly interesting is not only the use of ITO electrodes, but the introduction of a homogeneous redox-cycling approach. Usually, redox cycling is carried out at systems with two sets of electrodes termed "generator" and "collector" electrode, respectively, because the products of one electrode (the generator) reaction are detected at the other (the collector). In this other case, the authors used an electroactive species in solution instead of a second set of electrodes, and used that to shuttle electrons back and forth, as depicted in Figure 2.8(b).

ITO electrodes were coated by adsorbed avidin, and biotinylated anti-cTnI were subsequently bound to them. The assay was a full sandwich ELISA, where the secondary antibody was labelled by alkaline phosphatase (ALP). Aminophenyl phosphate (APP) was chosen as enzyme substrate because it leads to the production of 4-aminophenol (AP), a reversible redox species.

At the electrode, 4-aminophenol is oxidized to 4-quinone imine (QI). Note that APP is not electroactive at the oxidation potential of AP, so it is possible to detect the product without interference from its parent molecule. To achieve the redox cycling effect, the authors added an excess of tris(2-carboxyethyl)phosphine (TCEP) to the solution. TCEP readily reduces QI back to AP, but it does not react with APP and was only very slowly oxidized at the ITO electrode. Thus, the oxidation current from the AP generated by the enzyme was enhanced by the presence of TCEP in solution. The authors went one step further and added a third redox couple to further enhance the detection current, in what they called an ECC redox cycling approach, but which may have also increased their background currents. In any case, the result is a very low detection limit for cTnI, at 10 fg mL^{-1}, and a very wide dynamic range spanning up to 10 ng mL^{-1}.

Flow devices have also been developed for the detection of cardiovascular disease markers. Flow devices can bring additional advantages such as accurate handling of small reagent and sample volumes, amenability for automation, and convective mass transport enhancements.

One of the first studies addressing the electrochemical detection of cTnI described a flow device consisting of a PDMS channel block that was bonded to a glass chip featuring interdigitated 10 μm microband electrodes (separated by 5 μm gaps) operated in generator–collector mode,[93] as schematized in Figure 2.18.

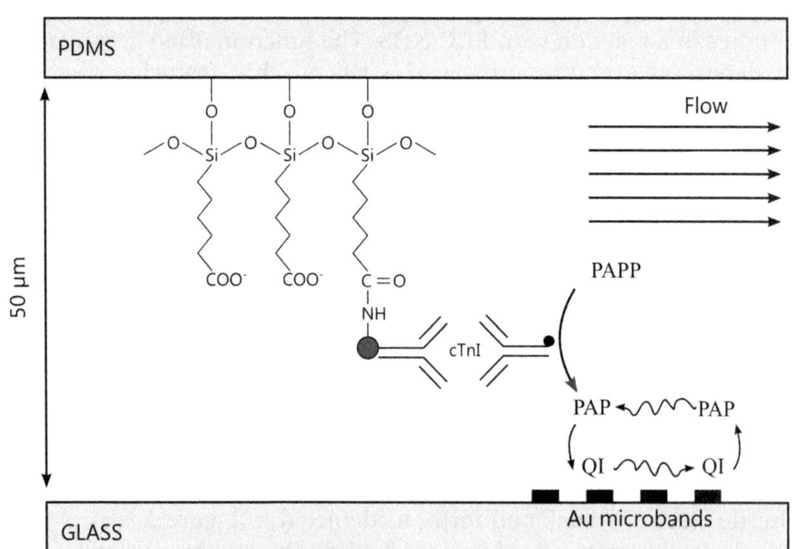

Figure 2.18 Ko *et al.* immobilized anti-cTnI antibodies on a PDMS microchannel, and detected aminophenol produced by alkaline phosphatase-labelled secondary antibodies. Redox cycling at gold microband electrodes was used to enhance the signal.[93]

A PDMS block containing a micro channel, defined by replica moulding from a SU-8 master, was bonded to the Pyrex chips following a 30 s activation in an air plasma. Once in place, it was functionalized by silanization with 7-octenyltrichlorosilane in toluene. Permanganate and periodate were then used to oxidize the vinyl groups in the silane chains into carboxylates that were next used to link to the amines in the antibodies in a reaction with EDC/NHS. The assay consisted of flowing a sample that could be pre-incubated with alkaline phosphatase-labelled anti-cTnI secondary antibodies in a one-step ELISA-type assay. After this, aminophenyl phosphate was passed through the device and the resulting aminophenol was detected at the microband array, which enhanced the signal by redox-cycling. Although it is not clear in their report, the authors must have performed some kind of electrochemical electrode activation. Despite exposing the electrodes to the silanization, and PDMS functionalization reagents, including blocking with bovine serum albumin (BSA), which must have very adversely affected the electron transfer capability of the gold microbands, the authors nevertheless observed a detection limit *ca.* 0.2 ng mL^{-1}, and a dynamic range going up to 10 μg mL^{-1}. This article highlights the main drawback of covalently attaching PDMS to microchips, which is the limited lifetime of these systems. Immunosensors are mostly single use devices, and from that point of view it might make sense to functionalize a PDMS part, as PDMS is relatively cheap, and replica moulding enables the fabrication of multiple identical pieces with relative ease. On the other hand, microelectrodes could be used many more times before they had to be discarded. However, once the PDMS is bonded to the chip, the entire set needs to be thrown away after a single detection event has been carried out. While it may be acceptable in a real application setting, this approach becomes extremely wasteful in a research and development environment. An approach relying on magnetic beads could have extended the useful lifetime of a device like this, and the following example will help to illustrate this point.

In 2012, Abad and co-workers[229] described the design and fabrication of an all cyclo-olefin polymer (COP)-based microfluidic device capable of capturing magnetic beads and performing electrochemical detection of cTnT at a series of gold microband electrodes. Sample preparation steps were to be carried out outside the device, according to the procedure outlined in Figure 2.19, and less than 1 ng mL^{-1} in a 1 : 50 human serum dilution could be quantified by amperometry under flow conditions.

The chip, shown in Figure 2.19, was designed for single use and to be amenable for mass production using low-cost materials. The chip was the size of a microscope slide and contained six independent measuring units, which allowed the simultaneous or sequential performance of multi-analyte measurements on the same chip. The chip holder was fabricated using a 3D printer.

The chip consisted of a microfluidic cartridge, containing the sensing microelectrodes, and an external holder for fluidic and electrical connections. The microfluidic cartridge was fabricated from 188 μm-thick

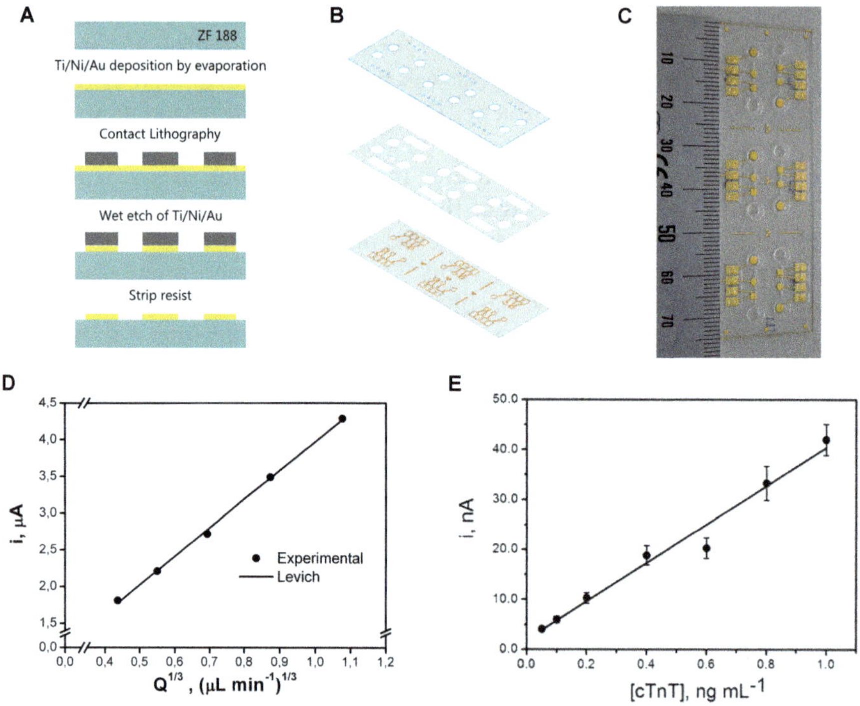

Figure 2.19 (A) Microelectrode fabrication process on COP substrates; (B) blow-out representation of the polymeric device, shown in (C). (D) Levich calibration plot and (E) system response to cTnT in serum samples. (Parts (D) and (E) reproduced from ref. 161 with permission from Springer.)

Zeonor® and two-side adhesive sheets, both of which are commercially available. As shown in Figure 2.19, layer 3 contains the patterned gold microelectrodes. The middle and top layers, which defined channel and microelectrode sizes, and the fluidic inlets and outlets, were cut in a drag-knife plotter.

The gold microelectrodes were fabricated using standard photolithographic techniques as depicted in Figure 2.19(A). First, a Zeonor wafer (100 mm diameter) was metallized by e-beam evaporation (15 nm Ti + 15 nm Ni + 150 nm Au). The metallized Zeonor wafer was then coated with positive photoresist S1813 (Fujifilm), and subsequently exposed to UV light through a bright-field high-resolution patterned transparency. After developing the resist, the metal layer was removed from the undesired areas in different etching baths. Thus, gold was etched in an I_2/I^- solution, and Ni was etched in a 4 : 1 dilution of concentrated HNO_3 (70%). Finally, the titanium layer was etched in a propylene glycol : HF mixture. Excess resist was removed in an acetone bath, and the wafer was rinsed with isopropyl alcohol and water. Finally, a drop of Ag/AgCl paste was deposited on the corresponding reference electrode.

The assay relied on the formation of an anti-cTnT antibody/cTnT/HRP-labelled antibody sandwich on the surface of magnetic beads outside the device. The magnetic beads were subsequently then injected into the device, where they were captured by detachable permanent magnets. The system worked by monitoring the activity of the HRP label attached to the secondary antibody closing the sandwich, using H_2O_2 and 3,3′,5,5′-tetramethylbenzidine (TMB), a redox mediator. In the presence of H_2O_2, HRP oxidized TMB, which could be reduced back at the working electrode. This reduction current was directly proportional to the concentration of HRP, which itself is proportional to the concentration of cTnT. Because TMB is known to adsorb and passivate gold electrodes, the working electrodes were calibrated before and after each cTnT determination. This was done using equimolar ferro/ferricyanide solutions, and comparing the resulting currents between them and with the theoretical prediction using Levich's equation for the operating flow conditions. Figure 2.19(D) shows typical calibration voltammograms recorded in this system, and Figure 2.19(E) shows calibration plots for the determination of cTnT in the range 0.05–1 ng L^{-1}.

Measurements repeated over a five-day period on different batches of magnetic beads yielded an RSD value of 10.4 % for a cTnT concentration of 0.4 ng L^{-1}. While this deviation may be acceptable, it is on the high side, and is probably due to the propagation of errors during bead preparation, sample extraction, and magnetic capture in the system. Due to the complexity and fat and protein content of blood and serum samples, these were diluted 50 times before analysis. This dilution level had also given good results in the allosteric detection of HIV antibodies using microelectrode arrays.[230]

Although this work has demonstrated that rapid prototyping methods can be used successfully in the fabrication of fully working devices, this approach presents two important weaknesses, which are that sample preparation and immunocapture need to be performed separately, and that syringe pumps and a certain degree of automation are needed. The consequence is that such a system would still require a certain bench space and trained staff to operate it. However, the two weaknesses inherent to this system may be overcome using lateral flow devices, as demonstrated by Akanda *et al.* in a very recent work.[218] On the one hand, external pumps are not required as liquids naturally flow across lateral flow membranes by capillary action. On the other hand, lateral flow membranes are routinely used to perform simple immunoassays, and so the immobilization of reagents and the performance of elementary sample preparation steps is also known and well established.

Lateral flow tests are ubiquitous and routinely used for the detection of hormones as in the ubiquitous pregnancy tests, drug metabolites in roadside testing, and pathogens in veterinary and food safety environments. These devices are made by a combination of different porous membranes that meet different functions and which are assembled on an adhesive backing card. Different reagents, mainly surfactants and antibodies, are immobilized at different points along the flow direction, so that they can be dragged and/or react with the sample as it flows through the device. Detection in these devices normally occurs at two coloured bands: the first signals whether an analyte was present and the second ensures that the device was working properly. The bands are normally formed by the concentration of particles (latex, polystyrene, or colloidal gold) conjugated to antibodies that are captured over a line in the detection zone. Although most devices can be read by the naked eye, recently the introduction of optical readers has improved detection limits and enabled assay quantification. Again, the most popular example is Alere's Clearblue ovulation test, which optically measures the concentration of two female hormones and correlates their concentration with the likelihood of being pregnant. Although most detection systems are optical, the integration of electrodes in lateral flow devices has also been attempted for at least ten years.[231] More recently, advances in lateral flow membrane materials has generated the exciting topic known as paper microfluidics, which promises to revolutionize lab on a chip devices and bring low cost diagnostic devices to the third world and to resource limited environments in general.[67]

Going back to the case of troponin detection, Akanda *et al.* transferred their assay described above at ITO electrodes to a lateral flow format with amperometric detection, as schematized in Figure 2.20. In this case a substrate pad consisted of an asymmetric polysulfone membrane that contained the electroactive reagents TCEP, the enzyme substrate 4-amino-1-naphthyl-galactopyranoside, and hexaammineruthenium(III). β-Galactosidase conjugated anti-troponin I IgG was immobilized on the main nitrocellulose membrane a few millimetres downstream from the substrate path, and anti-troponin IgG was immobilized further downstream, near the location where the working electrode was assembled. The asymmetric pore membrane provides additional time for the sample to dissolve the conjugated antibodies and allow them to react with the sample troponin and be subsequently captured in the detection zone. The operation of this device is schematized in Figure 2.20, and the detection reaction is shown in Figure 2.8(b). Interference from ascorbic and uric acid, which are present in human serum, was avoided by working at the low potentials required for the oxidation of the Ru(II) species generated in the ECC reaction between Ru(III) and aminonaphthol. This approach enabled the detection of down to 1 pg mL^{-1} cTnI, and a dynamic range up to 100 ng mL^{-1} in controlled solutions.

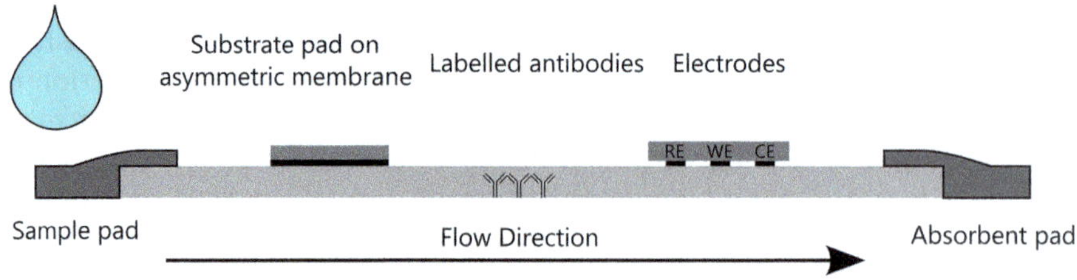

Figure 2.20 Lateral flow test with electrochemical detection.[151]

The authors also reported that they tried their electrochemical lateral flow test with real samples and found that its performance compared favourably compared to a commercial laboratory instrument. Although they used a commercial laboratory potentiostat in their measurements, the integration of a small and low cost potentiostat in a portable and disposable device is perfectly feasible, and small power sources are also commercially available at competitive costs, pointing to the spread of disposable quantitative analytical devices in the not so distant future.

2.4.3 Further Biomedical Applications of Miniaturized Electrochemical Devices

When a new biosensor needs to be developed, multiple choices to make the device viable arise: geometries, measurement techniques, surfaces, *etc.* The type of analyte determines some of these parameters, but even with some of these factors settled the possibilities are innumerable, as evidenced by the number of reviews available on the subject. Lei[65] focused on diagnostics application using microfluidics, while Xu *et al.*[232] addressed integration in microsystems. Mohammed and Desmulliez,[233] Varshney and Li[234] and Griveau and Bedioui[71] focus on specific analytes, as we did in the previous section with the cardiac diseases. Table 2.6 summarizes a series of different electrochemical biosensors in diagnostic applications.

2.5 Concluding Remarks and Outlook

The evolution of electrochemical devices is closely linked to advances in fabrication techniques and construction materials. Microfabrication techniques began to be used for the construction of microelectrode arrays for electroanalytical applications in the 1980s, and have contributed enormously to biomedical science. Although the most significant contributions of microfabricated devices are in the area of electrophysiology, where individually addressable disk microelectrode arrays thrive, their role in the development of novel diagnostic tools cannot be overemphasized.

This chapter has focused on microfabricated biosensors for the detection of cardiovascular disease markers to illustrate a development path that was already followed by glucose meters. Such a path involves a transition from the research laboratory towards marketable applications, and is clearly represented by the adoption of more affordable construction materials and device architectures compatible with mass production. This has been illustrated in the case of troponin immunosensors, where analytical methods were first developed using silicon-based microelectrode arrays, and the eventual transfer of these methods to flow devices, which allowed the simplification of sample treatment procedures. These flow devices gradually evolved too, from relatively cumbersome architectures based on traditional materials such as silicon, glass and PDMS towards more functional constructions based on polymeric materials and lateral flow membranes.

Table 2.6 Different electrochemical biosensors employed in diagnostic applications.

Geometry (band or disk)	Clinical application	Detection system: enzymatic, immuno, DNA	Measurement technique: amperometry, EIS	Signal amplification	Surface	Functionalization	Fluidic system integrated	Ref.
Interdigitated ultramicro-electrodes	Virus infection: herpes simplex; cytomegalovirus; Epstein–Barr (virus)	DNA-SH	Amperometry (redox recycling-ELISA)		Gold	Thiol	—	Albers *et al.*[4]
Planar electrodes	Cardiac diseases: atrial; fibrillation and atrial flutter: drug digoxin detection (glucoside)	Immuno-antibody	Impedance AC (electrochemical species)		Screen printed carbon electrodes + conductive polyaniline	Biotin-sulfo-NHS	—	Barton *et al.*[235]
Planar electrodes	Cancer: AFP,CA125, CA153,CEA (proteins)	Immuno-antibody	Impedance AC (electrochemical species)	AuNPs–MWCNTs bioconjugate	Poly(ethylene terephthalate) substrates	Chitosan glutaraldehyde	—	Ge *et al.*[100]
Band electrodes	Diabetes, obesity: cortisol (hormone)	Immuno-antibody	Square wave voltammetry	Au nanowires	Au	Thioctic acid NHS-biotin	Microfluidics	Kumar *et al.*[236]

Disk electrodes	Diabetes, obesity: cortisol (hormone)	Immuno-antibody	Amperometry		Pt	5-Carboxy-pentanethiol	Fluidic system	Yamaguchi *et al.*[237]
Interdigitated bands	Diabetes, obesity: cortisol (hormone)	Immuno-antibody	Cyclic voltammetry		Au	DTSP	Microfluidic system	Vasudev *et al.*[238]
Diskelectrodes	Breast, colon and ovarian cancer: Anti-p53 (antibody)	Peptide fragment	Impedance AC (electrochemical species)		Au	Peptide-thiol	No flow	Prats-Alfonso *et al.*[239]
Interdigitated bands	Influenza infections (virus)	Aptamers	Impedance PBS		PEDOT:TsO	Succinic acid	No flow	Kiilerich-Pedersen *et al.*[75]
Disk electrodes	Cancer (micro RNA)	Aptamer	Impedance AC (electrochemical species)	Gold nanoparticle/protein	Screen printed carbon electrodes	Aptamer-SH linker	No flow	Labib *et al.*[240]
Disk electrodes	Hormone disorders: estradiol (hormone)	Aptamer	Cyclic voltammetry		Gold	Thiol modified ssDNA	No flow	Kim *et al.*[89]
Interdigitated bands	Parkinson's disease: dopamine (small molecule)	Immuno-antibody	Capacitance		SiO_2	APTES + boronate	No flow	Lu *et al.*[241]

Similar trends can be found in other areas of diagnostics, where the development of point-of-care devices is booming, largely facilitated not only by miniaturization *per se* but also by advances in electronics and information technologies.

Electrochemical techniques are ideally suited as detection methods in point of care devices. Microelectrodes of different materials can be easily integrated on a wide range of substrates, including paper, and their associated instrumentation (the potentiostat) is simpler and less costly than optical detectors. Electroanalytical techniques are quantitative and provide access to low detection limits in a short time, as required by most biomedical applications. Although calibration and the importance of electrode passivation need to be considered and overcome when developing electroanalytical diagnostic devices, there is wide scope for developing successful electrochemical diagnostic applications.

References

1. L. C. Clark, R. Wolf, D. Granger and Z. Taylor, *J. Appl. Physiol.*, 1953, **6**, 189–193.
2. L. C. Clark and C. Lyons, *Ann. N. Y. Acad. Sci.*, 1962, **102**, 29–45.
3. C. A. Thomas, P. A. Springer, L. M. Okun, Y. Berwaldn and G. E. Loeb, *Exp. Cell Res.*, 1972, **74**, 61–66.
4. J. Albers, T. Grunwald, E. Nebling, G. Piechotta and R. Hintsche, *Anal. Bioanal. Chem.*, 2003, **377**, 521–527.
5. A. F. M. Johnstone, G. W. Gross, D. G. Weiss, O. H. U. Schroeder, A. Gramowski and T. J. Shafer, *Neurotoxicology*, 2010, **31**, 331–350.
6. I. L. Jones, P. Livi, M. K. Lewandowska, M. Fiscella, B. Roscic and A. Hierlemann, *Anal. Bioanal. Chem.*, 2010, **399**, 2313–2329.
7. S. Grimnes and G. Martinsen, *Bioimpedance and Bioelectricity Basics*, Academic Press, 2nd edn, 2014.
8. C. Amatore, *Physical Electrochemistry: Science and Technology*, 1st edn, Marcel Dekker, Inc., 1995.
9. S. Szunerits and L. Thouin, in *Handbook of Electrochemistry*, ed. C. G. Zoski, Elsevier, Amsterdam, 2007, p. 880.
10. O. Ordeig, J. del Campo, F. X. Munoz, C. E. Banks and R. G. Compton, *Electroanalysis*, 2007, **19**, 1973–1986.
11. R. G. Compton and C. E. Banks, *Understanding Voltammetry*, Imperial College Press, 2nd edn, 2010.
12. F. J. Del Campo, L. Abad, X. Illa, E. Prats-Alfonso, X. Borrisé, J. M. Cirera, H.-Y. Bai and Y.-C. Tsai, *Sens. Actuators, B*, 2014, **194**, 86–95.
13. W. Olthuis, W. Streekstra and P. Bergveld, *Sens. Actuators, B*, 1995, **24**, 252–256.
14. F. J. Del Campo, *Electrochem. Commun.*, 2014, **45**, 91–94.
15. R. G. Compton and P. R. Unwin, *J. Electroanal. Chem. Interfacial Electrochem.*, 1986, **205**, 1–20.

16. R. G. Compton, M. B. G. Pilkington and G. M. Stearn, *J. Chem. Soc., Faraday Trans. 1*, 1988, **84**, 2155.
17. J. A. Cooper and R. G. Compton, *Electroanalysis*, 1998, **10**, 141–155.
18. M. C. Zaretsky, M. Lama and J. R. Melcher, *IEEE Trans. Electr. Insul.*, 1988, **23**, 897–917.
19. O. Niwa, M. Masao and T. Hisao, *Anal. Chem.*, 1990, **62**, 447–452.
20. D. Menshykau, A. M. O'Mahony, F. Javier del Campo, F. Xavier Munoz and R. G. Compton, *Anal. Chem.*, 2009, **81**, 9372–9382.
21. D. Menshykau, A. M. O'Mahony, M. Cortina-Puig, F. Javier del Campo, F. Xavier Munoz and R. G. Compton, *J. Electroanal. Chem.*, 2010, **647**, 20–28.
22. D. Menshykau, M. Cortina-Puig, F. Javier del Campo, F. Xavier Munoz and R. G. Compton, *J. Electroanal. Chem.*, 2010, **648**, 28–35.
23. E. O. Barnes, A. Fernandez-la-Villa, D. F. Pozo-Ayuso, M. Castano-Alvarez, G. E. M. Lewis, S. E. C. Dale, F. Marken and R. G. Compton, *J. Electroanal. Chem.*, 2013, **709**, 57–64.
24. R. Prehn, M. Cortina Puig, F. J. Munoz-Pascual, F. J. Del Campo and L. Abad, *J. Electrochem. Soc.*, 2012, **159**, X1.
25. D. Sanchez-Molas, J. Pablo Esquivel, N. Sabate, F. Xavier Munoz and F. Javier del Campo, *J. Phys. Chem. C*, 2012, **116**, 18831–18846.
26. T. J. Davies and R. G. Compton, *J. Electroanal. Chem.*, 2005, **585**, 63–82.
27. T. J. Davies, W.-J. Sarah, C. E. Banks, d. C. Javier, M. Roser, M. F. Xavier and R. G. Compton, *J. Electroanal. Chem.*, 2005, **585**, 51–62.
28. I. Streeter, N. Fietkau, J. delcampo, R. Mas, F. X. Muñoz and R. G. Compton, *J. Phys. Chem. C*, 2007, **111**, 12058–12066.
29. P. M. Lewis, S. L. Bullard, R. E. Gawley and F. Ingrid, *Anal. Chem.*, 2010, **82**, 1659–1668.
30. U. A. Aregueta-Robles, A. J. Woolley, L. A. Poole-Warren, N. H. Lovell and R. A. Green, *Front. Neuroeng.*, 2014, **7**, 15.
31. M. P. Ward, P. Rajdev, C. Ellison and P. P. Irazoqui, *Brain Res.*, 2009, **1282**, 183–200.
32. S. F. Cogan, *Annu. Rev. Biomed. Eng.*, 2008, **10**, 275–309.
33. C. Amatore, C. Lefrou and F. Pflüger, *J. Electroanal. Chem.*, 1989, **270**, 43–59.
34. C. Amatore and C. Lefrou, *J. Electroanal. Chem.*, 1990, **296**, 335–358.
35. C. Amatore, E. Maisonhaute and G. Simonneau, *J. Electroanal. Chem.*, 2000, **486**, 141–155.
36. P. Fortgang, C. Amatore, E. Maisonhaute and B. Schöllhorn, *Electrochem. Commun.*, 2010, **12**, 897–900.
37. J. L. Anderson, *Abstr. Pap. Am. Chem. Soc.*, 1985, **190**, 63-ANL.
38. L. E. Fosdick, J. L. Anderson, T. A. Baginski and R. C. Jaeger, *Anal. Chem.*, 1986, **58**, 2750–2756.
39. R. Hintsche, M. Paeschke, U. Wollenberger, U. Schnakenberg and B. A. L. T. Wagner, *Biosens. Bioelectron.*, 1994, **9**, 697–705.
40. A. Stett, U. Egert, E. Guenther, F. Hofmann, T. Meyer, W. Nisch and H. Haemmerle, *Anal. Bioanal. Chem.*, 2003, **377**, 486–495.

41. C. K. Yeung, F. Sommerhage, G. Wrobel, A. Offenhusser, M. Chan and S. Ingebrandt, *Anal. Bioanal. Chem.*, 2007, **387**, 2673–2680.
42. F. A. Aguiar, M. C. Rosamond, D. Wood and R. Kataky, *Analyst*, 2008, **133**, 1060.
43. M. D. Johnson, R. K. Franklin, M. D. Gibson, R. B. Brown and D. R. Kipke, *J. Neurosci. Methods*, 2008, **174**, 62–70.
44. J. Orozco, F.-S. Cesar and J.-J. Cecilia, *Sensors*, 2010, **10**, 475–490.
45. R. Chen, Y. Li, K. Huo and P. K. Chu, *RSC Adv.*, 2013, **3**, 18698.
46. M. J. Madou, *Fundamentals of Microfabrication and Nanotechnology*, CRC Press, Boca Raton, 2012.
47. A. Manz, N. Graber and H. M. Widmer, *Sens. Actuators, B*, 1990, **1**, 244–248.
48. A. Arora, G. Simone, G. B. Salieb-Beugelaar, J. T. Kim and A. Manz, *Anal. Chem.*, 2010, **82**, 4830–4847.
49. P. Neužil, S. Giselbrecht, K. Länge, T. J. Huang and A. Manz, *Nat. Rev. Drug Discovery*, 2012, **11**, 620–632.
50. P. Caillat, D. David, M. Belleville, F. Clerc, C. Massit, F. Revol-Cavalier, P. Peltie, T. Livache, G. Bidan and A. Roget, *Sens. Actuators, B*, 1999, **61**, 154–162.
51. P. M. Levine, P. Gong, R. Levicky and K. L. Shepard, *IEEE, J. Solid State Circ.*, 2008, **43**, 1859–1871.
52. B. Y. Lee, S. M. Seo, D. J. Lee, M. Lee, J. Lee, J.-H. Cheon, E. Cho, H. Lee, I.-Y. Chung, Y. J. Park, S. Kim and S. Hong, *Lab Chip*, 2010, **10**, 894.
53. Q. Xue, C. Bian, J. Tong, J. Sun, H. Zhang and S. Xia, *Sens. Actuators, A*, 2011, **169**, 282–287.
54. T. Kong, R. Su, B. Zhang, Q. Zhang and G. Cheng, *Biosens. Bioelectron.*, 2012, **34**, 267–272.
55. Y. Huang and A. J. Mason, *Lab Chip*, 2013, **13**, 3929–3934.
56. X. Illa, O. Ordeig, D. Snakenborg, A. Romano-Rodrguez, R. G. Compton and J. P. Kutter, *Lab Chip*, 2010, **10**, 1254–1261.
57. P. S. Nunes, P. D. Ohlsson, O. Ordeig and J. P. Kutter, *Microfluid. Nanofluid.*, 2010, **9**, 145–161.
58. M. Focke, D. Kosse, C. Mller, H. Reinecke, R. Zengerle and F. von Stetten, *Lab Chip*, 2010, **10**, 1365–1386.
59. L. Abad, F. J. Del Campo, F. J. Munoz-Pascual, L. J. Fernández, D. Calavia, G. Colom, J. P. Salvador, M. P. Marco, V. Escamilla-Gómez, B. Esteban-Fernández de Ávila, S. Campuzano, M. Pedrero, J. M. Pingarrón, N. Godino, R. Gorkin III and J. Ducrée, *Electrophoresis*, 2012, **33**, 1–8.
60. P. G. Gross, E. P. Kartalov, A. Scherer and L. P. Weiner, *J. Neurol. Sci.*, 2007, **252**, 135–143.
61. M. Kwiat, D. Stein and F. Patolsky, *Curr. Opin. Biotechnol.*, 2013, **24**, 654–663.
62. C. A. Brunello, V. Jokinen, P. Sakha, H. Terazono, F. Nomura, T. Kaneko, S. E. Lauri, S. Franssila, C. Rivera, K. Yasuda and H. J. Huttunen, *J. Nanobiotechnol.*, 2013, **11:11**.

63. S. Ray, P. J. Reddy, S. Choudhary, D. Raghu and S. Srivastava, *J. Proteomics*, 2011, **74**, 2660–2681.
64. A. Qureshi, Y. Gurbuz and J. H. Niazi, *Sens. Actuators, B*, 2012, **171–172**, 62–76.
65. K. F. Lei, *J. Lab. Autom.*, 2012, **17**, 330–347.
66. P. B. Luppa, C. Mller, A. Schlichtiger and H. Schlebusch, *TrAC, Trends Anal. Chem.*, 2011, **30**, 887–898.
67. P. K. Drain, E. P. Hyle, F. Noubary, K. A. Freedberg, D. Wilson, W. R. Bishai, W. Rodriguez and I. V. Bassett, *Lancet Infect. Dis.*, 2014, **14**, 239–249.
68. S. E. Clarke and J. R. Foster, *Br. J. Biomed. Sci.*, 2012, **69**, 83–93.
69. E. Bakker and E. Pretsch, *Angew. Chem., Int. Ed.*, 2007, **46**, 5660–5668.
70. S. Floate and C. E. W. Hahn, *J. Electroanal. Chem.*, 2005, **583**, 203–211.
71. S. Griveau and F. Bedioui, *Anal. Bioanal. Chem.*, 2013, **405**, 3475–3488.
72. A. Baumann and U. Karst, *Exp. Opin. Drug Metabol. Toxicol.*, 2010, **6**, 715–731.
73. F. Mazzei, R. Antiochia, F. Botre, G. Favero and C. Tortolini, *Bioanalysis*, 2014, **6**, 225–245.
74. C. Batchelor-McAuley, G. G. Wildgoose and R. G. Compton, *Biosens. Bioelectron.*, 2009, **24**, 3183–3190.
75. K. Kiilerich-Pedersen, J. Daprà, S. Cherré and N. Rozlosnik, *Biosens. Bioelectron.*, 2013, **49**, 374–379.
76. B. R. Eggins, *Chemical Sensors and Biosensors*, John Wiley & Sons Ltd, Chichester, 2002.
77. M. C. Garnett, *Adv. Drug Deliver. Rev.*, 2001, **53**, 171–216.
78. G. B. Wisdom, *Clin. Chem.*, 1976, **22**, 1243–1255.
79. R. M. Lequin, *Clin. Chem.*, 2005, **51**, 2415–2418.
80. G. T. Hermanson, *Bioconjugate Techniques*, Academic Press, London, 2013.
81. P. Tomčík, *Sensors*, 2013, **13**, 13659–13684.
82. T. G. Drummond, M. G. Hill and J. K. Barton, *Nat. Biotechnol.*, 2003, **21**, 1192–1199.
83. A. Sassolas, B. D. Leca-Bouvier and L. J. Blum, *Chem. Rev.*, 2008, **108**, 109–139.
84. A. D. Ellington and J. W. Szostak, *Nature*, 1990, **346**, 818–822.
85. C. Tuerk and L. Gold, *Science*, 1990, **249**, 505–510.
86. L. C. Bock, L. C. Griffin, J. A. Latham, E. H. Vermaas and J. J. Toole, *Nature*, 1992, **355**, 564–566.
87. K. Haupt and K. Mosbach, *Chem. Rev.*, 2000, **100**, 2495–2504.
88. N. Karimian, A. P. F. Turner and A. Tiwari, *Biosens. Bioelectron.*, 2014, **59**, 160–165.
89. Y. S. Kim, H. S. Jung, T. Matsuura, H. Y. Lee, T. Kawai and M. B. Gu, *Biosens. Bioelectron.*, 2007, **22**, 2525–2531.
90. V. L. Venkatraman, R. K. Reddy, F. Zhang, D. Evans, B. Ulrich and S. Prasad, *Biosens. Bioelectron.*, 2009, **24**, 3078–3083.

91. M. R. Akanda, M. A. Aziz, K. Jo, V. Tamilavan, M. H. Hyun, S. Kim and H. Yang, *Anal. Chem.*, 2011, **83**, 3926–3933.
92. M. R. Akanda, Y. L. Choe and H. Yang, *Anal. Chem.*, 2012, **84**, 1049–1055.
93. S. Ko, B. Kim, S. S. Jo, S. Y. Oh and J. K. Park, *Biosens. Bioelectron.*, 2007, **23**, 51–59.
94. Peter Atkins and J. de Paula, *Atkins' Physical Chemistry*, Oxford University Press, 2014.
95. E. Baldrich, F. Javier del Campo and F. Xavier Munoz, *Biosens. Bioelectron.*, 2009, **25**, 920–926.
96. M. Trojanowicz, O. Geschke, T. K. V. Krawczyk and K. Cammann, *Sens. Actuators, B*, 1995, **28**, 191–199.
97. M. Trojanowicz and T. K. V. Krawczyk, *Mikrochim. Acta*, 1995, **121**, 167–181.
98. S. K. Tuteja, V. Priyanka, A. Bhalla, A. K. Deep, Paul and C. R. Suri, *Anal. Chim. Acta*, 2014, **809**, 148–154.
99. E. Valera, J. Ramón-Azcón, F.-J. Sanchez, M.-P. Marco and Á. Rodríguez, *Sens. Actuators, B*, 2008, **134**, 95–103.
100. S. Ge, L. Ge, M. Yan, X. Song, J. Yu and J. Huang, *Chem. Commun.*, 2012, **48**, 9397–9399.
101. Z. Grabarek and J. Gergely, *Anal. Biochem.*, 1990, **185**, 131–135.
102. B. T. Houseman, E. S. Gawalt and M. Mrksich, *Langmuir*, 2002, **19**, 1522–1531.
103. E. Pavlovic, S. Oscarsson and A. P. Quist, *Nano Lett.*, 2003, **3**, 779–781.
104. Y. Jung, J. M. Lee, J.-w. Kim, J. Yoon, H. Cho and B. H. Chung, *Anal. Chem.*, 2009, **81**, 936–942.
105. Y. Hou, S. Helali, A. Zhang, N. Jaffrezic-Renault, C. Martelet, J. Minic, T. Gorojankina, M.-A. Persuy, E. Pajot-Augy, R. Salesse, F. Bessueille, J. Samitier, A. Errachid, V. Akimov, L. Reggiani, C. Pennetta and E. Alfinito, *Biosens. Bioelectron.*, 2006, **21**, 1393–1402.
106. F. C. Schenk, H. Boehm, J. P. Spatz and S. V. Wegner, *Langmuir*, 2014, **30**, 6897–6905.
107. J. A. Camarero, Y. Kwon and M. A. Coleman, *J. Am. Chem. Soc.*, 2004, **126**, 14730–14731.
108. J. M. Goddard and J. H. Hotchkiss, *Prog. Polym. Sci.*, 2007, **32**, 698–725.
109. J. C. Love, L. A. Estroff, J. K. Kriebel, R. G. Nuzzo and G. M. Whitesides, *Chem. Rev.*, 2005, **105**, 1103–1170.
110. E. Baldrich, O. Laczka, F. Javier Del Campo and F. Xavier Munoz, *J. Immunol. Methods*, 2008, **336**, 203–212.
111. *Impedance Spectroscopy: Theory, Experiment, and Applications*, ed. E. Barsoukov and J. R. Macdonald, John Wiley & Sons, Hoboken, 2005.
112. A. J. Bard and L. R. Faulkner, *Electrochemical Methods: Fundamentals and Applications*, 2nd edn, John Wiley & Sons Inc, New York, 2001.
113. A. Bratov, N. Abramova, J. Ramón-Azcón, A. Merlos, F. Sánchez-Baeza, M.-P. Marco and C. Domínguez, *Electrochem. Commun.*, 2008, **10**, 1621–1624.

114. D. Menshykau, F. Javier del Campo, F. Xavier Munoz and R. G. Compton, *Sens. Actuators, B*, 2009, **138**, 362–367.
115. R. G. Compton, E. Laborda and K. R. Ward, *Understanding Voltammetry: Simulation of Electrode Processes*, Imperial College Press, London, 2014, p. 249.
116. P. Atkins and J. de Paula, *Atkins' Physical Chemistry*, Oxford University Press, Oxford, 2014.
117. R. A. Marcus, *Rev. Mod. Phys.*, 1993, **65**, 599–610.
118. D. Britz, *Digital Simulation in Electrochemistry*, Lecture Notes in Physics, Springer, Berlin, 2005.
119. R. G. Compton and G. H. W. Sanders, *Electrode Potentials*, Oxford University Press, Oxford, 1996.
120. C. Amatore, C. Pebay, L. Thouin, A. Wang and J. S. Warkocz, *Anal. Chem.*, 2010, **82**, 6933–6939.
121. C. Amatore, O. V. Klymenko and I. Svir, *Anal. Chem.*, 2012, **84**, 2792–2798.
122. H. Yu, X. Xu, J. Sun and T. You, *Cent. Eur. J. Chem.*, 2012, **10**, 639–651.
123. A. C. Moser, C. W. Willicott and D. S. Hage, *Electrophoresis*, 2014, **35**, 937–955.
124. W. Dungchai, O. Chailapakul and C. S. Henry, *Anal. Chem.*, 2009, **81**, 5821–5826.
125. J. L. Osborn, B. Lutz, E. Fu, P. Kauffman, D. Y. Stevens and P. Yager, *Lab Chip*, 2010, **10**, 2659–2665.
126. X. Li, D. R. Ballerini and W. Shen, *Biomicrofluidics*, 2006, **8**(1), 011301.
127. M. Madou, J. Zoval, G. Jia, H. Kido, J. Kim and N. Kim, *Annu. Rev. Biomed. Eng.*, 2006, **8**, 601–628.
128. N. Godino, E. Vereshchagina, R. Gorkin, III and J. Ducree, *Microfluid. Nanofluid.*, 2014, **16**, 895–905.
129. T. Li, Y. Fan, Y. Cheng and J. Yang, *Lab Chip*, 2013, **13**, 2634–2640.
130. X.-J. Huang, A. M. O'Mahony and R. G. Compton, *Small*, 2009, **5**, 776–788.
131. A. M. Bond, D. Luscombe, K. B. Oldham and C. G. Zoski, *J. Electroanal. Chem.*, 1988, **249**, 1–14.
132. M. C. Henstridge and R. G. Compton, *Chem. Rec.*, 2012, **12**, 63–71.
133. N. Godino, X. Borrise, F. Xavier Munoz, F. Javier del Campo and R. G. Compton, *J. Phys. Chem. C*, 2009, **113**, 11119–11125.
134. A. J. Bard, J. A. Crayston, G. P. Kittlesen, T. V. Shea and M. S. Wrighton, *Anal. Chem.*, 1986, **58**, 2321–2331.
135. M. Odijk, W. Olthuis, V. A. T. Dam and A. van den Berg, *Electroanalysis*, 2008, **20**, 463–468.
136. J. I. Heo, D. S. Shim, G. T. Teixidor, S. Oh, M. J. Madou and H. Shin, *J. Electrochem. Soc.*, 2011, **158**, J76–J80.
137. K. Aoki, *J. Electroanal. Chem.*, 1989, **270**, 35–41.
138. C. Amatore, C. Pebay, C. Sella and L. Thouin, *ChemPhysChem*, 2012, **13**, 1562–1568.
139. B. J. Feldman, S. W. Feldberg and R. W. Murray, *J. Phys. Chem.*, 1987, **91**, 6558–6560.

140. N. Alayo-Bueno, Universidad Autónoma de Barcelona, 2013.
141. M. E. Orazem and B. Tribollet, *Electrochemical Impedance Spectroscopy*, Wiley-Interscience, 2011.
142. A. Guimerà, X. Illa, E. Traver, M. Plata-Cordero, J. Yeste, C. Herrero, C. Lagunas, M. J. Maldonado and R. Villa, *Biomed. Microdevices*, 2013, **15**, 849–858.
143. A. Ivorra, R. Gómez, N. Noguera, R. Villa, A. Sola, L. Palacios, G. Hotter and J. Aguiló, *Biosens. Bioelectron.*, 2003, **19**, 391–399.
144. C. Gabrielli, M. Keddam, N. Portail, P. Rousseau, H. Takenouti and V. Vivier, *J. Phys. Chem. B*, 2006, **110**, 20478–20485.
145. E. T. McAdams and J. Jossinet, *Bioelectrochem. Bioenerg.*, 1996, **40**, 147–152.
146. A. Bratov, N. Abramova, A. Ipatov and A. Merlos, *Talanta*, 2013, **106**, 286–292.
147. S. Grimnes and Ø. G. Martinsen, *J. Phys. D: Appl. Phys.*, 2007, **40**, 9–14.
148. R. De Levie, *Electrochim. Acta*, 1965, **10**, 113–130.
149. J. C. Wang and J. B. Bates, *Solid State Ionics*, 1986, **18–19**(Part 1), 224–228.
150. A. Lasia, *J. Electroanal. Chem.*, 1995, **397**, 27–33.
151. R. de Levie, *Electrochim. Acta*, 1963, **8**, 751–780.
152. R. de Levie, *Electrochim. Acta*, 1964, **9**, 1231–1245.
153. A. Guimerà, G. Gabriel, E. Prats-Alfonso, N. Abramova, A. Bratov and R. Villa, *Sens. Actuators, B*, 2015, **207**, 1010–1018.
154. P. Neuzil, C. D. M. Campos, C. C. Wong, J. B. W. Soon, J. Reboud and A. Manz, *Lab Chip*, 2014, **14**, 2168–2176.
155. I. J. Cutress, E. J. F. Dickinson and R. G. Compton, *J. Electroanal. Chem.*, 2010, **638**, 76–83.
156. C. Zoski, *Handbook of Electrochemistry*, Elsevier, 2014.
157. S. Fletcher and M. D. Horne, *Electrochem. Commun.*, 1999, **1**, 502–512.
158. F. Tan, J. P. Metters and C. E. Banks, *Sens. Actuators, B*, 2013, **181**, 454–462.
159. S. M. George, *Chem. Rev.*, 2010, **110**, 111–131.
160. C. K. Chua, J. F. Leong and C. S. Lim, *Rapid Prototyping, Principles and Applications*, 3rd edn, World Scientific, Singapore, 2010.
161. T. Ito and S. Okazaki, *Nature*, 2000, **406**, 1027–1031.
162. G. L. T. Chiu and J. M. Shaw, *IBM J. Res. Dev.*, 1997, **41**, 3–6.
163. R. P. Seisyan, *Tech. Phys.*, 2011, **56**, 1061–1073.
164. J. M. Shaw, J. D. Gelorme, N. C. LaBianca, W. E. Conley and S. J. Holmes, *IBM J. Res. Dev.*, 1997, **41**, 81–94.
165. B. Hunek and E. L. Cussler, *AIChE J.*, 2002, **48**, 661–672.
166. P. C. Tsiartas, L. W. Flanagin, C. L. Henderson, W. D. Hinsberg, I. C. Sanchez, R. T. Bonnecaze and C. G. Willson, *Macromolecules*, 1997, **30**, 4656–4664.
167. H. Lorenz, M. Despont, N. Fahrni, N. LaBianca, P. Renaud and P. Vettiger, *J. Micromech. Microeng.*, 1997, 7, 121–124.
168. E. W. Becker, W. Ehrfeld, P. Hagmann and A. Maner, *Microelectron. Eng.*, 1986, **4**, 35–56.

169. A. Rogner, J. Eicher, D. Munchmeyer, R. P. Peters and J. Mohr, *J. Micromech. Microeng.*, 1992, **2**, 133–140.
170. F. Javier del Campo, P. Godignon, L. Aldous, E. Pausas, M. Sarrion, M. Zabala, R. Prehn and R. G. Compton, *J. Electrochem. Soc.*, 2011, **158**, H63–H68.
171. B. J. Lin, *Microelectron. Eng.*, 1987, **6**, 31–51.
172. D. M. Mattox, *Handbook of Physical Vapor Deposition (PVD) Processing*, Elsevier, 2010.
173. D. Depla, S. Mahieu and J. E. Greene, in *Handbook of Deposition Technologies for Films and Coatings*, ed. P. M. Martin, Elsevier Inc., 2010, pp. 253–296.
174. S. I. Shah, G. H. Jaffari, E. Yassitepe and B. Ali, in *Handbook of Deposition Technologies for Films and Coatings*, ed. P. M. Martin, Elsevier Inc., 2010, pp. 135–252.
175. Z. R. Dai, Z. W. Pan and Z. L. Wang, *Adv. Funct. Mater.*, 2003, **13**, 9–24.
176. O. Laczka, E. Baldrich, F. X. Muñoz and F. J. del Campo, *Anal. Chem.*, 2008, **80**, 7239–7247.
177. O. Ordeig, C. E. Banks, T. J. Davies, J. d. Campo, F. X. Muñoz and R. G. Compton, *J. Electroanal. Chem.*, 2006, **592**, 126–130.
178. R. F. Wolffenbuttel, *Sens. Actuators, A*, 1997, **62**, 680–686.
179. L. M. Fischer, M. Tenje, A. R. Heiskanen, N. Masuda, J. Castillo, A. Bentien, J. Emneus, M. H. Jakobsen and A. Boisen, *Microelectron. Eng.*, 2009, **86**, 1282–1285.
180. Y. Temiz, A. Ferretti, Y. Leblebici and C. Guiducci, *Lab Chip*, 2012, **12**, 4920–4928.
181. D. Burdeaux, P. Townsend, J. Carr and P. Garrou, *J. Electron. Mater.*, 1990, **19**, 1357–1364.
182. M. Bu, T. Melvin, G. J. Ensell, J. S. Wilkinson and A. G. R. Evans, *Sens. Actuators, A*, 2004, **115**, 476–482.
183. J. M. Bustillo, R. T. Howe and R. S. Muller, *Proc. IEEE*, 1998, **86**, 1552–1574.
184. C. Iliescu, B. Chen and J. Miao, *Sens. Actuators, A*, 2008, **143**, 154–161.
185. F. Laermer, S. Franssila, L. Sainiemi and K. Kolari, in *Handbook of Silicon Based MEMS Materials and Technologies*, ed. V. Lindroos, S. Frannssila, M. Tilli, M. Paulasto-Krocke, A. Lehto, T. Motooka and V.-M. Airaksinen, Elsevier Inc., 2010, pp. 349–374.
186. J. Kim, Y. Shin, S. Song, J. Lee and J. Kim, *Sens. Actuators, B*, 2014, **202**, 60–66.
187. J. Yamada and H. Matsuda, *J. Electroanal. Chem.*, 1971, **30**, 271–278.
188. C. R. Friedrich and M. J. Vasile, *J. Microelectromech. Syst.*, 1996, **5**, 33–38.
189. N. Godino, F. J. del Campo, F. X. Munoz, M. F. Hansen, J. P. Kutter and D. Snakenborg, *Lab Chip*, 2010, **10**, 1841–1847.
190. N. Godino, D. Snakenborg, J. P. Kutter, J. Emnéus, M. F. Hansen, F. X. Muñoz and F. J. d. Campo, *Microfluid. Nanofluid.*, 2010, **8**, 393–402.

191. http://www.kyoceraprecisiontools.com/.
192. D. A. Bartholomeusz, R. W. Boutte and J. D. Andrade, *J. Microelectromech. Syst.*, 2005, **14**, 1364–1374.
193. J. Do, J. Y. Zhang and C. M. Klapperich, *Robotics and Computer-Integrated Manufacturing*, 2011, **27**, 245–248.
194. H. Klank, J. P. Kutter and O. Geschke, *Lab Chip*, 2002, **2**, 242–246.
195. D. Snakenborg, H. Klank and J. P. Kutter, *J. Micromech. Microeng.*, 2004, **14**, 182–189.
196. M. Bowden, O. Geschke, J. P. Kutter and D. Diamond, *Lab Chip*, 2003, **3**, 221–223.
197. E. C. Harvey and P. T. Rumsby, Fabrication techniques and their application to produce novel micromachined structures and devices using excimer laser projection, in *Micromaching and Microfabrication Process Technology III*, ed. S. C. Chang and S. W. Pang, vol. 3223, SPIE, 1997, p. 26.
198. E. C. Harvey, P. T. Rumsby, M. C. Gower and J. L. Remnant, Conference on Micromachining and Microfabrication Process Technology, Austin, Tx, 1995.
199. P. A. Tipler and G. Mosca, *Physics for Scientists and Engineers*, W. H. Freeman, 2004.
200. L. Szekely and A. Guttman, *Curr. Anal. Chem.*, 2006, **2**, 195–201.
201. http://www.eos.info/en.
202. http://cubify.com/en/ProJet.
203. D. C. Duffy, J. C. McDonald, O. J. A. Schueller and G. M. Whitesides, *Anal. Chem.*, 1998, **70**, 4974–4984.
204. J. C. McDonald, D. C. Duffy, J. R. Anderson, D. T. Chiu, H. Wu, O. J. Schueller and G. M. Whitesides, *Electrophoresis*, 2000, **21**, 27–40.
205. M. A. Eddings, M. A. Johnson and B. K. Gale, *J. Micromech. Microeng.*, 2008, **18**(6), 067001.
206. L.-H. Hung, R. Lin and A. P. Lee, *Lab Chip*, 2008, **8**, 983–987.
207. D. Qin, Y. Xia and G. M. Whitesides, *Nat. Protoc.*, 2010, **5**, 491–502.
208. V. Gubala, L. F. Harris, A. J. Ricco, M. X. Tan and D. E. Williams, *Anal. Chem.*, 2012, **84**, 487–515.
209. WHO, Medical devices, http://www.who.int/medical_devices/safety/en/.
210. L. Nyholm, *Analyst*, 2005, **130**, 599–605.
211. A. Bratov, J. Ramón-Azcón, N. Abramova, A. Merlos, J. Adrian, F. Sánchez-Baeza, M.-P. Marco and C. Domínguez, *Biosens. Bioelectron.*, 2008, **24**, 729–735.
212. S. J. Cameron, L. J. Sokoll, O. F. Laterza, S. Shah and G. B. Green, *Clin. Chim. Acta*, 2007, **376**, 168–173.
213. S. K. James, B. Lindahl, A. Siegbahn, M. Stridsberg, P. Venge, P. Armstrong, E. S. Barnathan, R. Califf, E. J. Topol, M. L. Simoons and L. Wallentin, *Circulation*, 2003, **108**, 275–281.
214. A. Tello-Montoliu, F. Marin, V. Roldan, L. Mainar, M. T. Lopez, F. Sogorb, V. Vicente and G. Y. H. Lip, *J. Intern. Med.*, 2007, **262**, 651–658.

215. J. Wu, D. M. Cropek, A. C. West and S. Banta, *Anal. Chem.*, 2010, **82**, 8235–8243.
216. L. N. Feng, Z. P. Bian, J. Peng, F. Jiang, G. H. Yang, Y. D. Zhu, D. Yang, L. P. Jiang and J. J. Zhu, *Anal. Chem.*, 2012, **84**, 7810–7815.
217. A. Periyakaruppan, R. P. Gandhiraman, M. Meyyappan and J. E. Koehne, *Anal. Chem.*, 2013, **85**, 3858–3863.
218. M. R. Akanda, H. A. Joung, V. Tamilavan, S. Park, S. Kim, M. H. Hyun, M. G. Kim and H. Yang, *Analyst*, 2014, **139**, 1420–1425.
219. E. A. de Vasconcelos, N. G. Peres, C. O. Pereira, V. L. da Silva, E. F. da Silva Jr and R. F. Dutra, *Biosens. Bioelectron.*, 2009, **25**, 870–876.
220. J. H. Chua, R. E. Chee, A. Agarwal, M. W. She and G. J. Zhang, *Anal. Chem.*, 2009, **81**, 6266–6271.
221. B. V. M. Silva, I. T. Cavalcanti, A. B. Mattos, P. Moura, M. D. P. T. Sotomayor and R. F. Dutra, *Biosens. Bioelectron.*, 2010, **26**, 1062–1067.
222. R. A. S. Fonseca, J. Ramos-Jesus, L. T. Kubota and R. F. Dutra, *Sensors*, 2011, **11**, 10785–10797.
223. B. E. F. deÁvila, V. Escamilla-Gómez, S. Campuzano, M. Pedrero and J. M. Pingarrón, *Electroanalysis*, 2013, **25**, 51–58.
224. K. C. Lin, V. Kunduru, M. Bothara, K. Rege, S. Prasad and B. L. Ramakrishna, *Biosens. Bioelectron.*, 2010, **25**, 2336–2342.
225. J. R. Windmiller, S. Chinnapareddy, P. Santhosh, J. Halámek, M. C. Chuang, V. Bocharova, T. F. Tseng, T. Y. Chou, E. Katz and J. Wang, *Biosens. Bioelectron.*, 2010, **26**, 886–889.
226. J. Barallat, R. Olivé-Monllau, J. Gonzalo-Ruiz, R. Ramírez-Satorras, F. X. Muñoz-Pascual, A. G. Ortega and E. Baldrich, *Anal. Chem.*, 2013, **85**, 9049–9056.
227. F. Zhou, M. Lu, W. Wang, Z. P. Bian, J. R. Zhang and J. J. Zhu, *Clin. Chem.*, 2010, **56**, 1701–1707.
228. B. McDonnell, S. Hearty, P. Leonard and R. O'Kennedy, *Clin. Biochem.*, 2009, **42**, 549–561.
229. L. Abad, F. Javier del Campo, F. X. Muñoz, L. J. Fernández, D. Calavia, G. Colom, J. P. Salvador, M. P. Marco, V. Escamilla-Gómez, B. Esteban-Fernández de Ávila, S. Campuzano, M. Pedrero, J. M. Pingarrón, N. Godino, R. Gorkin and J. Ducrée, *Electrophoresis*, 2012, **33**, 3187–3194.
230. O. Laczka, R. Maria Ferraz, N. Ferrer-Miralles, A. Villaverde, F. Xavier Munoz and F. Javier del Campo, *Anal. Chim. Acta*, 2009, **641**, 1–6.
231. C. Fernandez-Sanchez, C. J. McNeil, K. Rawson and O. Nilsson, *Anal. Chem.*, 2004, **76**, 5649–5656.
232. X. Xu, S. Zhang, H. Chen and J. Kong, *Talanta*, 2009, **80**, 8–18.
233. M.-I. Mohammed and M. P. Y. Desmulliez, *Lab Chip*, 2011, **11**, 569–595.
234. M. Varshney and Y. Li, *Biosens. Bioelectron.*, 2009, **24**, 2951–2960.
235. A. C. Barton, S. D. Collyer, F. Davis, G.-Z. Garifallou, G. Tsekenis, E. Tully, R. O'Kennedy, T. Gibson, P. A. Millner and S. P. J. Higson, *Biosens. Bioelectron.*, 2009, **24**, 1090–1095.

236. A. Kumar, S. Aravamudhan, M. Gordic, S. Bhansali and S. S. Mohapatra, *Biosens. Bioelectron.*, 2007, **22**, 2138–2144.
237. M. Yamaguchi, Y. Matsuda, S. Sasaki, M. Sasaki, Y. Kadoma, Y. Imai, D. Niwa and V. Shetty, *Biosens. Bioelectron.*, 2013, **41**, 186–191.
238. A. Vasudev, A. Kaushik, Y. Tomizawa, N. Norena and S. Bhansali, *Sens. Actuators, B*, 2013, **182**, 139–146.
239. E. Prats-Alfonso, X. Sisquella, N. Zine, G. Gabriel, A. Guimerà, F. J. del Campo, R. Villa, A. H. Eisenberg, M. Mrksich, A. Errachid, J. Aguiló and F. Albericio, *Small*, 2012, **8**, 2106–2115.
240. M. Labib, N. Khan, S. M. Ghobadloo, J. Cheng, J. P. Pezacki and M. V. Berezovski, *J. Am. Chem. Soc.*, 2013, **135**, 3027–3038.
241. M. S. C. Lu, Y.-C. Chen and P.-C. Huang, *Biosens. Bioelectron.*, 2010, **26**, 1093–1097.

CHAPTER 3

Principles and Strategies for Microchip Electrophoresis with Amperometric Detection

DULAN B. GUNASEKARA,[a,b,†] MANJULA B. WIJESINGHE,[a,b] RACHEL A. SAYLOR[a,b] AND SUSAN M. LUNTE*[a,b,c]

[a] Ralph N. Adams Institute for Bioanalytical Chemistry, University of Kansas, Lawrence, KS, USA; [b] Department of Chemistry, University of Kansas, Lawrence, KS, USA; [c] Department of Pharmaceutical Chemistry, University of Kansas, Lawrence, KS, USA
*Email: slunte@ku.edu

3.1 Introduction

Microchip electrophoresis (ME) is a powerful separation technique that has been extensively employed in many lab-on-a-chip devices since its introduction in the mid-1990s.[1,2] Electrophoretic separations in ME are based on the same principles and strategies that are employed for capillary-based separations. Separations in ME are performed in channels of micron or submicron dimensions on planar substrates. The small footprint and planar channels of these devices open up new applications in areas where traditional capillary-based electrophoresis (CE) is difficult to apply.

Most detection methods employed with capillary electrophoresis, such as laser-induced fluorescence, capacitively coupled contactless conductivity

[†]Present address: Department of Chemistry, University of North Carolina, Chapel Hill, NC, USA.

RSC Detection Science Series No. 6
Electrochemical Strategies in Detection Science
Edited by Damien W. M. Arrigan

Published by the Royal Society of Chemistry, www.rsc.org

detection, and amperometry, have been successfully adapted to ME. However, while absorbance detection is widely used in CE, it is difficult to implement in the chip format due to the small dimensions of the separation channel and difficulty in integrating the optics within the chip format. Laser-induced fluorescence (LIF) is the most commonly applied detection strategy in ME due to the relative ease of focusing the laser beam in the separation channel and detecting the resulting fluorescence using a microscope. However, LIF has two disadvantages in ME. The first is that fluorescence detection normally requires analyte derivatization with a fluorophore prior to analysis. Secondly, the associated optics and instrumentation for LIF detection are much larger than the chip used for analysis, negating some of the benefits of miniaturization that are characteristic of the microchip format, such as portability.

Electrochemical detection (EC) methods, based on either conductivity or amperometry, are becoming increasingly popular for ME. A major advantage of electrochemical methods is that the electrodes used for both detection and separation can be directly integrated into the microchip. Microelectrodes can be employed as working electrodes without a loss in sensitivity in comparison to their macroelectrode counterparts. Another advantage is that, unlike LIF detection, substrates do not need to be optically transparent. In addition, the associated electronics (potentiostats) can be easily miniaturized and battery powered, making portable analysis systems possible. Lastly, many analytes of interest, including many biologically relevant molecules, are electroactive and do not require derivatization prior to their detection. ME coupled to electrochemical detection (ME–EC) has been employed for a wide variety of applications, and there are several excellent reviews on the topic.[3–9]

In this chapter, we focus specifically on amperometric detection for microchip electrophoresis since that is the most common form of EC detection used in ME. The basic principles of electrophoretic separations in the microchip format will be presented first, followed by a discussion of theoretical and practical considerations of coupling microchip electrophoresis to amperometric detection. The use of different electrode materials and configurations to increase the selectivity of ME will be described as well as some representative applications of the technique. Future directions in instrument development and applications of ME–EC will then be discussed.

3.2 Principles of Microchip Electrophoresis Separations

The most commonly employed chip design in microchip electrophoresis experiments, the simple-t, is shown in Figure 3.1(A). For electrophoresis experiments, all the channels are filled with a conductive buffer. Sample is placed in the sample reservoir (S) and, following sample injection, separations are accomplished by applying a voltage across the channel.

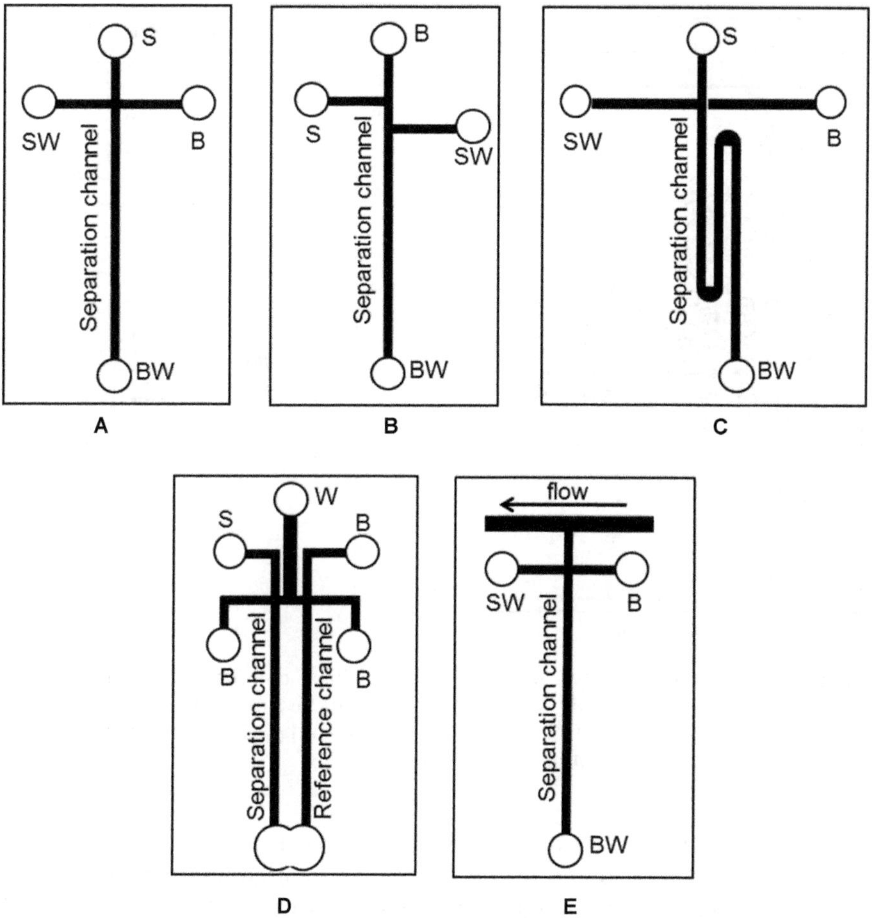

Figure 3.1 Schematics of microchip designs used for microchip electrophoresis: (A) simple-t design; (B) offset-t design; (C) serpentine design; (D) dual-channel microchip; (E) double-t design. S-sample, B-buffer, SW-sample waste, BW-buffer waste.

Glass and silica are the most common substrates for ME-based separations, and the silanol groups present on the channel surface play a role in the separation. When run buffers of a pH greater than 3 are employed, the surface of the channels becomes negatively charged due to the presence of ionized silanol groups. Upon application of an electric field (for simplicity, positive high voltage is considered here), this results in the production of electroosmotic flow (EOF), which moves all analytes regardless of charge toward the cathode. In the case of polymer substrates, the surface is often chemically modified to generate a negative surface charge to obtain EOF.

The production of EOF in a channel is due to the generation of a potential field at the surface of the channel known as the zeta potential (Figure 3.2A).

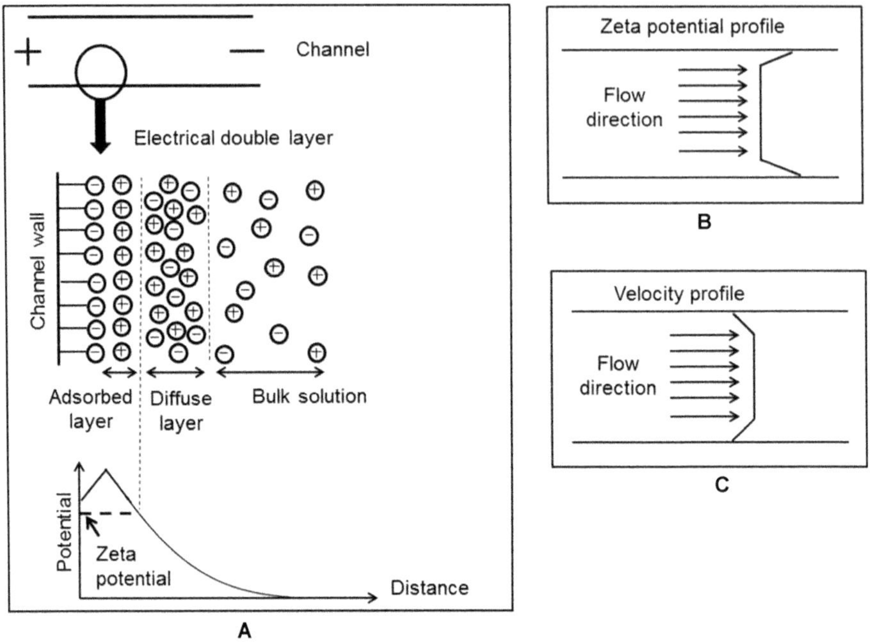

Figure 3.2 (A) The charged wall in a glass/silica channel, schematic of the electrical double layer, and potential distribution as a function of distance. (B) Zeta potential distribution and (C) EOF velocity distribution across the separation channel.

As a result of this zeta potential, cations such as H^+ and Na^+, in the buffer will electrostatically adhere to the surface of the channel in an attempt to neutralize the negative charge at the channel surface. However, these adsorbed ions cannot completely neutralize the charge, and a second diffuse layer, consisting of cations that are closely associated with the surface, is formed to balance the remaining surface charge.[10,11] This results in an electrical double layer as shown in Figure 3.2(A). Once a positive high voltage is applied, the positively charged hydrated ions in the diffuse layer are drawn toward the cathode (negative/ground electrode), creating electroosmotic flow. The EOF carries all analytes, regardless of charge, toward the cathode. In addition, since the flow is generated along the channel surface, it creates a plug flow profile (Figure 3.2C) that exhibits significantly less band-broadening than pressure-driven flows.[10,11]

In ME, the separation potential is applied across a channel by placing electrodes (normally Pt) into two buffer-filled reservoirs at either end of the channel. A high voltage is applied to one electrode while the other electrode is held at ground. This produces a linear voltage drop across the channel that is dependent on the resistance of the run buffer and the channel dimensions. Since current flows between the electrodes, redox reactions occur at both the anode and cathode.[8] In aqueous run buffers, these reactions are usually the reduction and oxidation of water at the cathode and anode, respectively:[12]

Cathode: $2H_2O + 2e^- \rightleftharpoons H_2 + 2OH^-$ (−0.828 *versus* NHE)
Anode: $O_2 + 4e^- + 4H^+ \rightleftharpoons 2H_2O$ (+1.229 *versus* NHE)

To achieve a separation in ME, it is necessary to produce differences in analyte migration in the electric field. In the simplest form of ME, free-zone electrophoresis, analytes are separated based on differences in the ratio of their charge-to-hydrodynamic radii. This separation is based on differences in migration caused by the electrophoretic mobility of an analyte, which is a measure of its mobility toward the electrode of opposite charge under the separation conditions. The electrophoretic mobility is dependent on a combination of two different forces on a charged molecule in an electric field. These are the electrostatic force (Eq) that is experienced by the molecule toward the electrode of opposite charge, which is balanced by the frictional force of the solution. The frictional force is dependent on the velocity of the ion in the field and its molecular volume (eqn (3.1)). The ion quickly reaches a steady-state velocity based on these two opposing forces:[11]

$$\nu_{ep} = Eq/6\pi\eta a \tag{3.1}$$

where E is the electric field gradient, q is the charge of the analyte, η is the viscosity of the run buffer, a is the hydrodynamic radius of the analyte, and ν_{ep} is the analyte velocity. In eqn (3.1), the factor $q/6\pi\eta a$ is defined as the electrophoretic mobility (μ_{ep}) and is a constant for a specific analyte in a given buffer system (η is constant). Therefore, μ_{ep} depends only on the ratio of charge-to-hydrodynamic radius, as mentioned above. Analytes must exhibit different electrophoretic mobilities under the separation conditions in order to be adequately resolved. The speed of analysis is partly dependent on the velocity of an ion in the channel (ν_{ep}), which is a function of the electric field strength (E) and the electrophoretic mobility (eqn (3.2)).[11] In ME, very fast separations have been achieved by using high field strengths (500–1000 V cm^{-1}):

$$\nu_{ep} = \mu_{ep} E_{ep} \tag{3.2}$$

Separations accomplished through the application of a positive voltage gradient across the negatively charged channel are referred to as "normal polarity" separations (Figure 3.3A). In normal polarity, the electrophoretic mobility of a cation is positive, and it is attracted to the cathode. The electroosmotic flow is also in the direction of the cathode. Therefore, positive ions will migrate out of the capillary first, with their migration time being a function of the combination of their electrophoretic mobility and the magnitude of the electroosmotic flow:[10,11]

$$\mu_{app} = \mu_{ep} + \mu_{eo} \tag{3.3}$$

where μ_{app} is the apparent (observed) electrophoretic mobility, μ_{ep} is the actual electrophoretic mobility, and μ_{eo} is the electroosmotic mobility.

Anions, on the other hand, are attracted to the anode and exhibit negative electrophoretic mobilities. In most cases, using positive polarity, the EOF is stronger than the μ_{ep} for large anions, so they still migrate toward the cathode. Exceptions are very small anions that have high negative electrophoretic mobilities or separations that are performed under conditions of low electroosmotic flow. In free-zone electrophoresis, neutral species migrate with the EOF and are not separated.

For small negative ions with high negative electrophoretic mobilities, "reverse polarity" is often used. To accomplish a reverse polarity separation, the electrophoresis channel is modified with a positive charge either by covalent modification or adsorption of a positively charged surfactant such as tetradecyltrimethylammonium bromide (TTAB). A comparison of normal and reverse polarity is shown in Figures 3.3A and B, respectively. In this separation mode, the order of migration is opposite that of normal polarity separations, and the buffer anions in the diffusive layer move toward the

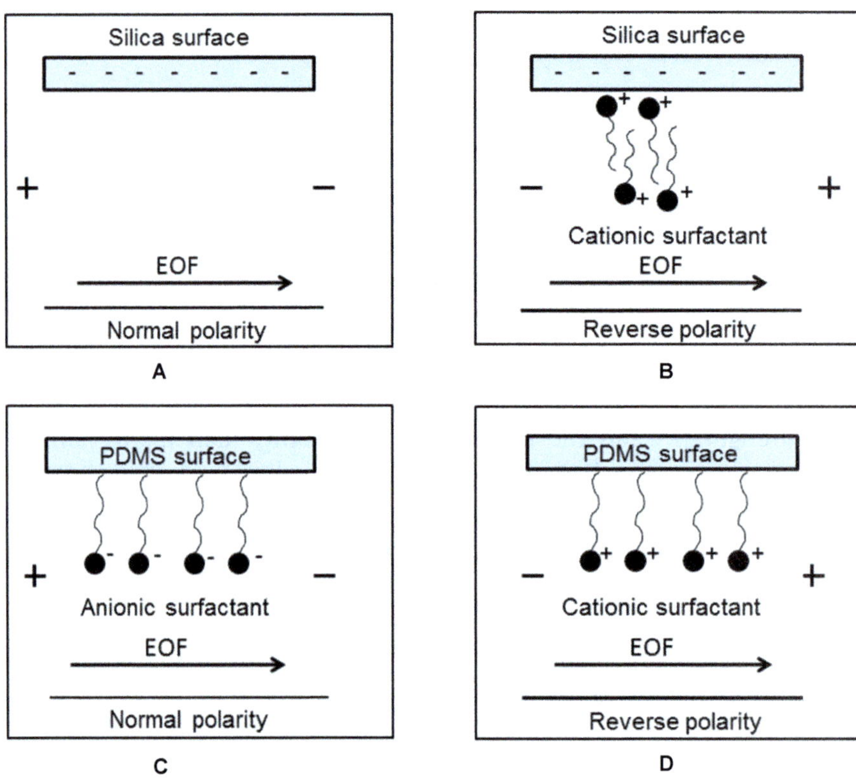

Figure 3.3 Charge at the surface of (A) unmodified glass or silica channel for normal polarity separation, (B) surfactant-modified glass or silica channel for reverse polarity separations, (C) PDMS or polymer channel modified with an anionic surfactant for normal polarity separations, and (D) PDMS or polymer channel modified with a cationic surfactant in reverse polarity separations.

anode (ground electrode) following the application of a negative high voltage gradient between the two channel ends.[10,11] Therefore, the EOF is in the opposite direction, and anions will migrate first, followed by neutrals and then cations.

Microchip electrophoresis separations can be performed using various materials including glass, polydimethylsiloxane (PDMS), poly(methyl methacrylate) (PMMA), and ceramics.[13] As indicated above, glass surfaces generate significant EOF at pH values greater than 3. However, many polymer chips, such as PDMS, do not natively possess a surface charge, so the substrate is often modified to generate an EOF. This can be accomplished by plasma oxidation, dynamic modification by run buffer components (Figure 3.3C and D), or the introduction of charged functional groups into the polymer itself.[14]

3.3 Microchip Designs

A significant advantage of the microchip format is the ability to integrate multiple functions, including sample preparation, preconcentration, and detection, into a single chip. The simple-t microchip configuration is the most common design used for ME. However, double-t, offset-t, serpentine channels, and dual-channel microchip designs have also been developed (Figure 3.1). In most devices, fluid is manipulated within the chip using electroosmotic flow.

In the simple-t, serpentine, and dual-channel designs, sample introduction into the separation channel is normally achieved using an electrokinetically gated injection as shown in Figure 3.4(A)–(C).[15–17] In this injection format, the high voltage applied at the buffer reservoir is floated for a short amount of time (Figure 3.4B), allowing sample to enter the separation channel. When the voltage is reestablished across the separation channel (Figure 3.4C), analytes are separated. While this injection strategy is simple to employ, as only applied voltages are manipulated, there is an electrokinetic bias on analytes entering the separation channel. For example, in positive polarity, smaller cations will be preferentially injected into the channel. The number of moles of each species that is injected in this mode is also dependent on the electrophoretic mobility of the compound due to the electrokinetic bias. This can sometimes be used to an advantage to selectively enhance the injection of species of a specific charge in the presence of interferences with an opposite charge or larger molecules of the same charge. Additionally, the volume of sample that is injected depends on the injection time, applied voltages, and buffer ionic strength; therefore, slight changes in conditions result in differing amounts of sample introduced into the channel. The volume of the sample injected can be calculated using the velocity of the EOF, injection time, and the area of the channel.

Another benefit of this injection strategy for ME is that application of the electric fields during the injection process generates a reproducible signature in the electropherogram, due to charging current on the electrode.

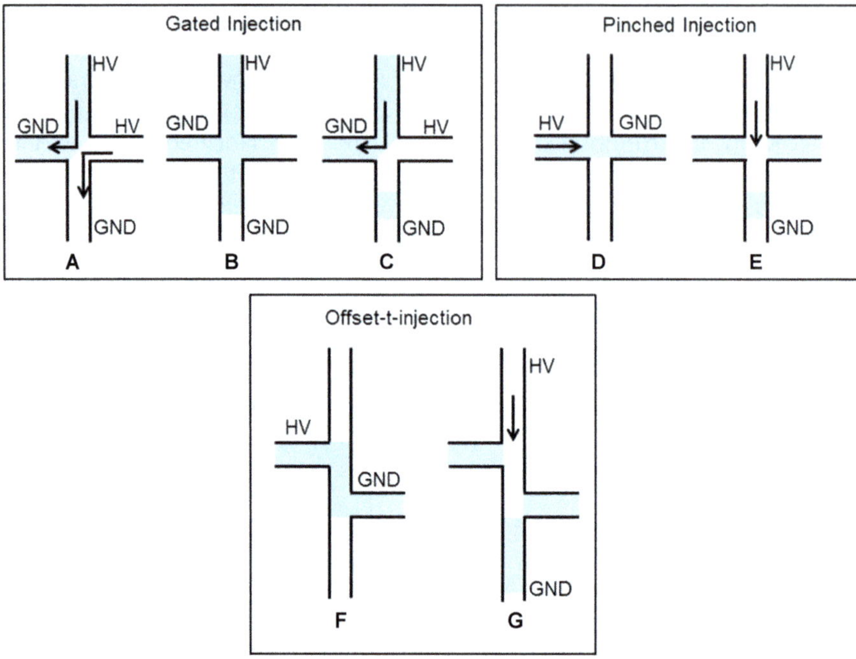

Figure 3.4 Sample injection for microchip electrophoresis. (A)–(C) Gated injection in a simple-t or serpentine channel device; (D) and (E) pinched injection in a simple-t or serpentine device; (F)–(G) defined volume sample injection using an offset-t device.

This can conveniently be used to mark time zero in the electropherogram. The field strength can be calculated using the junction voltage, or voltage at the channel intersection. Using Kirchhoff's laws and modeling the channels as resistors, the junction potential can be determined (Figure 3.5).[18]

The pinched injection mode can also be used in the simple-t and simple-t serpentine device designs (Figure 3.1A and C). In this case, sample is placed in one of the reservoirs in the channel perpendicular to the separation channel and a high voltage is applied across that channel, causing sample to fill the entire channel (Figure 3.4D). For separation, the run buffer is placed in the separation channel and the high voltage is then applied across the separation channel (Figure 3.4E). This permits a plug of sample corresponding to the size of the intersection to enter into the separation channel. One advantage of this injection approach is that one high voltage power supply is adequate to perform both injection and separation, while in gated injection mode two separate high voltage power supplies are required. Additionally, a defined volume is injected into the separation channel using this scheme.

Offset-t microchip designs also allow the injection of a finite and known volume into the separation channel (Figure 3.4F). The volume that is injected depends on the length of the offset of the perpendicular channel in relation to the separation channel. In this injection scheme, the offset

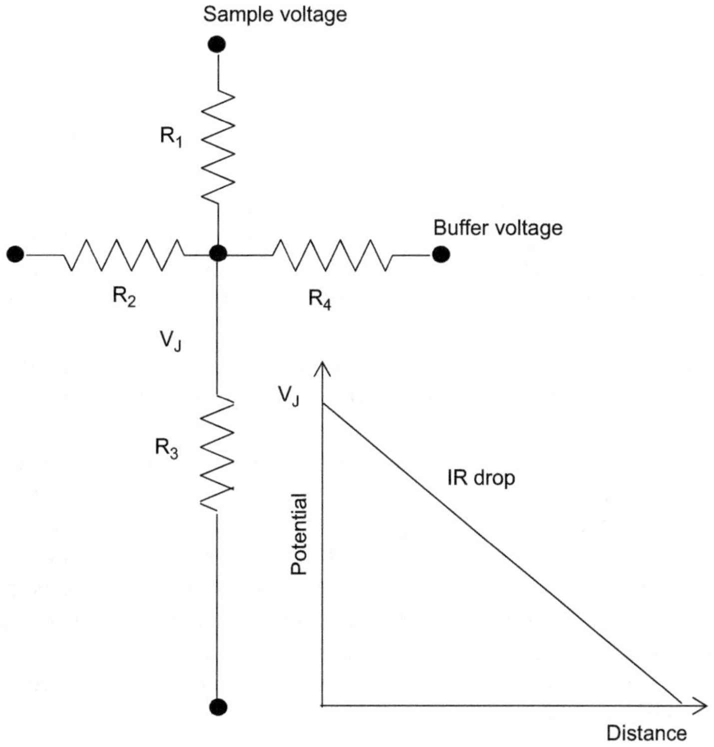

Figure 3.5 Equivalent electric circuitry that can be used for field strength calculations using the junction voltage (V_J).

channel is filled by applying a high voltage between two ends of the channel as shown in Figure 3.4(F). The volume trapped in the offset portion of the channel is then injected by applying a voltage across the separation channel (Figure 3.4G).

The serpentine design microchip (Figure 3.1C) arose out of a need for a longer separation channel to achieve better resolution of complex mixtures while still conforming to manufacturing constraints. Many devices are designed around the 4-inch silica wafers that are used in microelectronic device manufacturing; thus, it is not possible to fabricate long, straight channel configurations using conventional photolithographic methods. In designing serpentine microchips for gated injection, care must be taken to extend not only the separation channel itself but also the sidearms to enable good electrokinetic gating. A possible problem with serpentine channels is that using a separation channel that has the same width in the straight part as for the turns can cause band-broadening due to the "race track effect," leading to decreased resolution. This can be mitigated by using serpentine chips in which there is a decrease in channel width around the turns.[19]

The dual-channel design shown in Figure 3.1(D) has been used with in-channel amperometric detection. In this design, reference and working electrodes are placed in isopotential positions within the two separate channels for background subtraction.[20,21] This dual-channel design uses two electrokinetically gated intersections that allow the same amount of sample to be introduced into two distinct separation channels. This approach is discussed in more detail in Sections 3.6.3 and 3.9.2.2.

To integrate continuous flow streams with microchip electrophoresis separations, the double-t design was developed.[22,23] This design has been modified for continuous analysis of microdialysis samples by ME (Figure 3.1E).[24] A plug of sample is introduced into the separation channel by means of a flow-gated interface, which is established by the hydrodynamic pressure of microdialysis sampling and the electrokinetic flow from the application of a voltage. For sample introduction, as in the electrokinetic gating, the high voltage is floated and then reestablished. This design allows the integration of microdialysis sampling with microchip electrophoresis, permitting both *in vivo* and *in vitro* monitoring.[4]

Alternatively, valves have been employed to couple continuous-flow streams, such as microdialysis sampling, to microchip electrophoresis.[25] These PDMS-based valves are pneumatically driven and allow discrete sample injection into the separation channel. In contrast to the flow-gated design discussed above, the valving devices do not have an electrokinetic bias on sample introduction. Electrochemical detection has been integrated into these devices; in fact, the first application of microdialysis sampling coupled to microchip electrophoresis with electrochemical detection employed pneumatic valves.[26]

3.4 Electrochemical Detection

The most popular electrochemical detection method for ME is amperometric detection, where a constant potential is applied to the working electrode.[7,27] Amperometry is generally preferred over potential scanning methods such as cyclic voltammetry due to lower background currents, ease of operation, lower LODs, and higher sensitivity. With single electrode amperometric detection, selectivity is dependent on the redox characteristics of the analyte of interest as well as the applied potential at the working electrode. Unlike voltammetric detection methods, single electrode amperometric detection does not provide any voltammetric information about the analyte unless multiple injections of the same sample are analyzed at different potentials. However, selectivity can be enhanced through judicious choice of the applied working electrode potential, the use of selective/modified electrodes, and multiple electrode configurations. In addition, both voltammetric and redox information can be obtained within a single analysis if two or more electrodes are used for amperometric detection.

3.4.1 Amperometry

In amperometry, a constant potential that is dependent on the redox potential of the analyte of interest is applied to the working electrode. The optimal potential is normally just above the peak potential (E_p). The E_p can be estimated by running a cyclic voltammogram in bulk solution using the same electrode material and run buffer that will be used in the chip. A better, but more tedious, method is to construct a hydrodynamic voltammogram (HDV) for the analyte of interest. This is accomplished by measuring the peak height (current) obtained for a given analyte (following the electrophoretic separation) at several different working electrode potentials. A plot of peak current *versus* potential is then plotted for each analyte of interest. Using an applied potential that is above the E_p (on the current-limiting plateau) will improve the precision of the current response because a slight decrease in potential will not dramatically change the current output.

In ME–EC, the diffusion layer thickness at the working electrode is constant and the current response is limited by the mass transport of species to the working electrode, which is dependent on both the electrophoretic mobility of the analyte and the electroosmotic flow. Under these "stirred" conditions (analytes are constantly being moved over the electrode), voltammograms exhibit a sigmoidal shape.[28] The potential at which the analyte reaches a current-limiting plateau is then determined from the HDV as described above. While, as mentioned previously, it is possible to perform off-line cyclic voltammetry to determine the optimal potential for detection, this is not always ideal due to the effects of the separation field on the working electrode potential.

In ME-EC, typically only 20–30% of the total molecules in the sample plug are oxidized or reduced.[29] The maximum current that can be achieved under these hydrodynamic conditions can be calculated using a modified version of the Faraday equation:[28]

$$I_{\text{limit}} = nF(C_i - C_f)u \tag{3.4}$$

In this equation, I_{limit} is the current at the current-limiting plateau, n is the number of electrons involved in the redox process, F is the Faraday constant, C_i is the initial analyte concentration or the concentration before oxidation/reduction, C_f is the analyte concentration after the oxidation/reduction takes place, and u is the linear velocity of the analyte. To calculate I_{limit}, it is necessary to know the analyte concentration after the oxidation or reduction, C_f. The value of C_f can be calculated using the conversion efficiency of the electrode. To calculate the conversion efficiency, the current or charge of species observed due to oxidation/reduction of a species at the electrode is divided by the total current or charge that can be generated for the same amount of analyte, which can be calculated using the Faraday equation (eqn (3.5)).[28] The actual conversion efficiency depends on many factors, including electrode alignment, type of injection and reproducibility, and changes in flow. Therefore, care needs to be taken when calculating the conversion efficiency for different microchip configurations.[28]

If the conversion efficiency reaches 100%, the detection method is known as coulometry. In this situation, the moles of analyte (N) that were injected can be calculated directly from the peak area in coulombs (Q) using

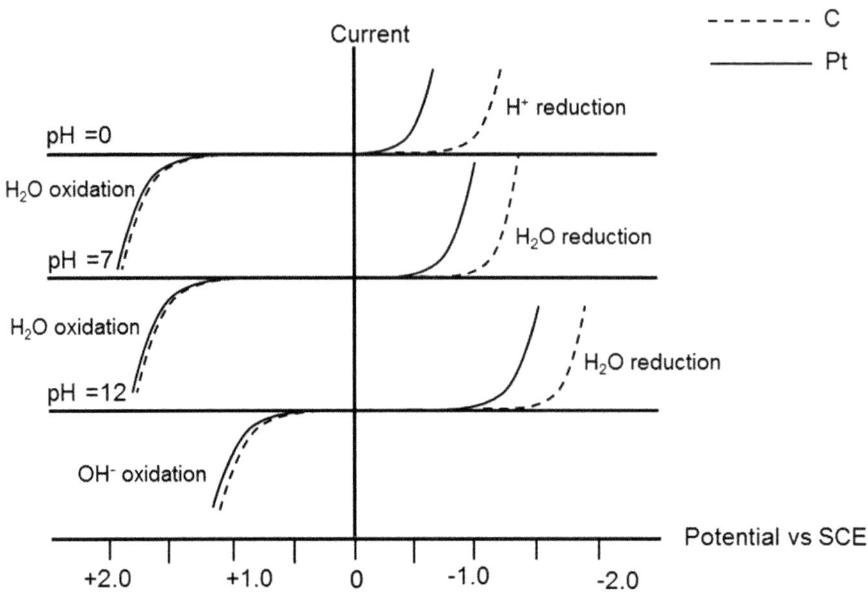

Figure 3.6 pH dependence on the potential window for carbon and platinum electrodes in aqueous media.
Data for this figure were obtained from ref. 12, 30–32.

the Faraday equation (eqn (3.5)), as long as the number of electrons involved in the redox process (n) is known:

$$Q = nFN \tag{3.5}$$

Amperometric detection for ME–EC is normally conducted in the oxidative mode, although the reductive mode has also been used. A challenge with reductive amperometric detection is an increase in background current due to the reduction of oxygen in the run buffer. To avoid this problem, the run buffer and sample must be thoroughly degassed prior to analysis. The potential window that can be used for amperometric detection is highly dependent on the electrode material and the buffer pH. Figure 3.6 shows the accessible potential windows for carbon and metal electrodes at various pH values.[12,30–32] The width of this potential window is usually defined by both the reduction of oxygen and the oxidation and reduction of water. Since aqueous buffers are most commonly employed for ME, the width of this potential window is also pH-dependent. The potential window can be expanded by using nonaqueous buffers for the separation.

3.5 Interaction of Separation Field with the Working Electrode

In the case of ME-LIF, the laser beam can be focused directly onto the separation channel to excite the analytes of interest as they pass by. An advantage of this approach is that detection can be performed virtually anywhere within the separation channel. However, with ME–EC, the electrodes cannot be placed deep inside the channel due to interactions of the separation field with the detection electrodes. Therefore, several different electrode configurations have been developed, as well as specialized potentiostats, to avoid and/or minimize interaction of the separation voltage with amperometric detection in ME. Before discussing these configurations and how they work, it is important to understand the theoretical basis of the interaction of an external (separation) field with a working electrode placed in that field. For this reason, the best available model to explain these interactions is presented prior to the more practical discussion of electrode configurations.

An electric field is required to accomplish separations in ME with typical field strengths between 100 and 1500 $V cm^{-1}$. In addition, the electrophoretic process normally generates microamperes of current, while only nano- or picoamperes of current are measured by the potentiostat. The electric field and resulting current can affect the working electrode when it is placed in the separation channel. It is this interaction that places fundamental constraints on the implementation of electrochemical detection for ME.[33,34] Fluctuations in the voltage and hence the separation current can lead to background noise at the detector and higher LOD. Additionally, when a high voltage is applied across a thin film metal electrode deposited on glass, the currents and heat generated by the separation voltage can damage the electrodes. Even more importantly, if the high currents generated by the electrophoretic separation ground through the working electrode, the potentiostat electronics can be destroyed.

In addition to the above considerations, the separation voltage will also induce a shift in the apparent potential of the working electrode if it is placed inside the separation channel. These half-wave potential shifts have been reported for both CE–EC and ME–EC.[33,34] A model for describing the effect of an applied electric field on the apparent potential of an electrode placed in that field has been described by both the Bohn and Wightman groups.[35,36] Figure 3.7(A) and (B) shows a potential *versus* distance diagram for normal and reverse polarity separations using the Bohn model.[35] As can be seen in this figure, the separation voltage decreases linearly across the separation channel as a function of the solution resistance. In these theoretical configurations, the separation ground and reference electrodes are assumed to be parallel to one another (in the detection reservoir) so that there is no effect of the separation field on the reference electrode. In addition, since the oxidative mode is considered, the positive potential represents lower energy and the negative potential represents higher energy.[35,36]

In normal polarity (Figure 3.7A), the separation voltage causes a decrease in the energy of the molecules near the electrode surface. Therefore, the electrode needs to drive to a more positive potential for electron transfer from molecules to the electrode (oxidation) to occur.[34] That is, the apparent half-wave potential for analytes is shifted in a positive direction when the

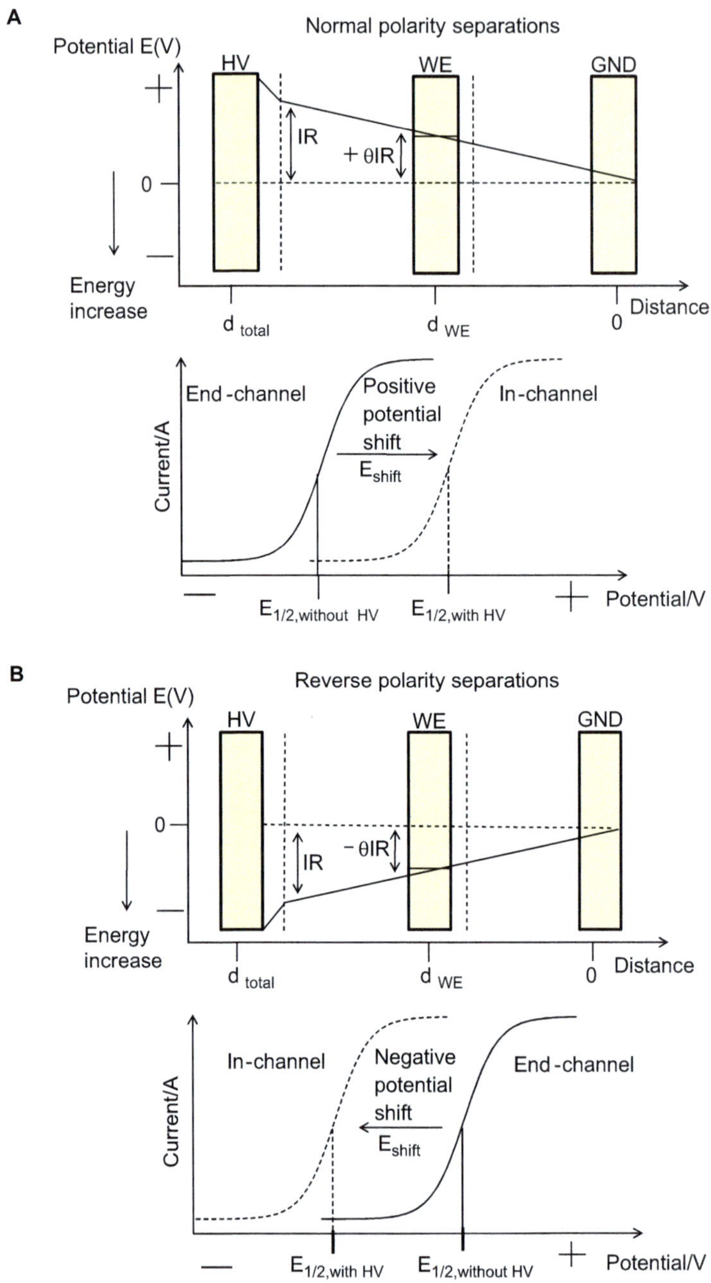

Figure 3.7 Potential *versus* distance diagram for normal and reverse polarity separations. The potential drop across the channel and representative potential shift in the HDV for (A) a normal polarity and (B) reverse polarity separation.
(Reproduced with permission from ref. 35, with kind permission from Springer Science and Business Media.)

electrode is placed inside the channel.[35,36] In the case of reverse polarity (Figure 3.7B), the separation voltage causes an increase in the energy of the molecules near the electrode surface. Therefore, less positive potentials need to be applied to the working electrode for oxidation to occur.[33]

This model is based on the premise that heterogeneous electron transfer occurs between the solution and the electrode; the solution and working electrode potential are distinct. Therefore, the effective potential of the electrode can be defined as the sum of both of these components:[35,36]

$$E_{\text{eff,WE}} = E_{\text{app,WE}} + E_{\text{shift}} \tag{3.6}$$

where $E_{\text{eff,WE}}$ is the effective working electrode potential, E_{shift} is the potential across the buffer above the electrode, and $E_{\text{app,WE}}$ is the potential applied to the working electrode by the potentiostat. The value of E_{shift} is dependent on the IR drop across the channel and the location of the working electrode within the channel. The total voltage drop across the channel is IR, where I is the separation current and R is the resistance in the microchannel.[35,36] E_{shift} is positive when positive polarity is used for separations because this is the extra potential that needs to be applied to the working electrode (since molecules in the solution have lower energy under these conditions). Therefore, the in-channel HDV is shifted in the positive direction compared to the end-channel HDV (Figure 3.7A). The potential of the solution above the electrode is then defined as:

$$E_{\text{shift}} = \theta IR \tag{3.7}$$

where $\theta = d_{\text{WE}}/d_{\text{total}}$ and d_{WE} is the distance between the separation ground electrode and the working electrode and d_{total} is the total distance between the high voltage electrode and ground. Using these parameters, the half-wave potential shift experienced by the working electrode can be defined as:

$$E_{\frac{1}{2},\text{ with HV}} = E_{\frac{1}{2},\text{ without HV}} + E_{\text{shift}} \tag{3.8}$$

The energy diagram for reverse polarity is shown in Figure 3.7(B); in this case, E_{shift} is negative because molecules in the solution have higher energy under these conditions. Therefore, the in-channel HDV is shifted in the negative direction compared to the end-channel HDV (Figure 3.7B).

This shift in half-wave potential has been observed for both normal and reverse polarity when the working electrode is placed inside or very near the end of the separation channel. For studies in which the electrode is placed in the channel, isolated or floating potentiostats have been used to protect the electronics.[33,34] More information about the in-channel configuration is provided in Section 3.6.3.

3.6 Electrode Configurations in Microchip Electrophoresis

Electrode alignment in ME is categorized into three different configurations depending on the placement of the electrode relative to the end of the separation channel. This categorization also shows the different strategies for preventing or minimizing the interaction of the separation field with the working electrode. The main three electrode configurations are end-, off-, and in-channel electrode alignments; these are shown in Figure 3.8.[37]

3.6.1 End-channel Detection

A common strategy to prevent the interaction of the separation field with the electrochemical detector is to place the electrode in the ground reservoir approximately 5–15 μm away from the end of the separation channel, where the electric field from the separation has dissipated.[9,37] This strategy is known as the end-channel electrode configuration (Figure 3.8A). In this configuration, analytes diffuse into the waste reservoir as they exit the channel prior to their detection, which can lead to band-broadening.[9,37] This band broadening leads to a loss of separation efficiency and a reduction in sensitivity. However, the noise due to the separation field is minimized, leading to lower limits of detection than for some of the other configurations. To more effectively dissipate the separation field and minimize the amount of analyte dispersion, Henry's group developed a "bubble cell." In this chip, the channel is widened at the end before going into the waste reservoir, and the working electrode is placed in this wider channel where the separation field is significantly decreased compared to the separation channel.[38] Major advantages of the end-channel approach are that it is compatible with commercially available (grounded) potentiostats and no

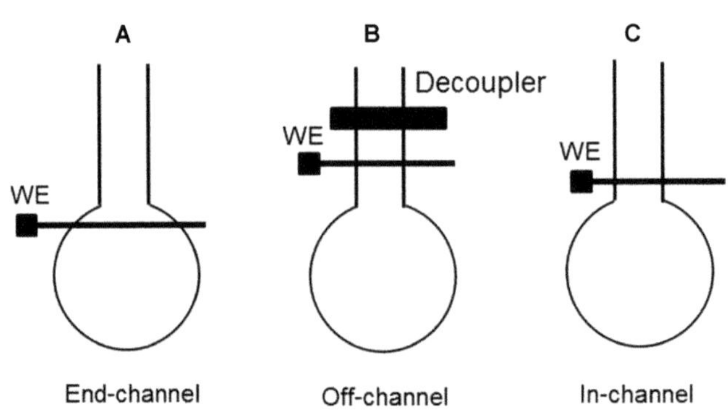

Figure 3.8 Electrode alignments in microchip electrophoresis. (A) End-channel detection, (B) off-channel detection using a decoupler, and (C) in-channel detection using an isolated potentiostat.

additional fabrication steps are needed to decouple the separation field from the working electrode.

3.6.2 Off-channel Detection

Off-channel detection involves placing the working electrode inside the channel, but the separation field is grounded prior to the working electrode using a metal decoupling electrode (Figure 3.8B).[9,39] Band electrodes fabricated from Pd or Pt are commonly employed for this purpose since they can adsorb the hydrogen gas produced at the cathode, minimizing bubble formation.[39,40] However, this approach has not been successfully employed in the reverse polarity mode because, under these conditions, O_2 is produced, which is not as actively adsorbed by the metal electrodes. Another challenge with the off-channel configuration is that the microfabricated metal band electrodes are not always stable in the presence of high separation fields. To circumvent this problem, metal wire electrodes embedded in an epoxy substrate have been used as the decoupler instead of microfabricated electrodes.[41,42]

One reported method for decoupling of the separation field from the working electrode that has been used in CE is to make a fracture on the capillary, which is then covered by an ion-permeable membrane such as cellulose acetate or Nafion and placed in the ground reservoir.[43–45] The separation current is then grounded through this fracture. Analyte is pushed from the fracture (decoupler) to the working electrode by the EOF that is produced in the separation part of the capillary. A similar approach has also been developed for ME using a cellulose acetate decoupler that is placed over a part of the separation channel that has been etched to provide an ion path. Using this configuration, very low noise (<1 pA) and an excellent LOD for dopamine (25 nM) were observed.[46]

The primary advantage of using the off-channel configuration is that the detector is isolated from the noise induced from fluctuations in the separation field. However, this approach also requires a strong EOF so that the analytes are pushed to the detection electrode with minimum band broadening. If low pH run buffers are employed with glass or fused silica substrates, band broadening can occur at the detection electrode because the flow profile changes from plug to parabolic flow after the decoupler.[37] The distance between the decoupler and the working electrode should also be kept to a minimum.

3.6.3 In-channel Detection

The third approach to performing amperometric detection in ME, in-channel detection, requires an isolated or "floating" potentiostat that is not electrically grounded. If an ungrounded potentiostat is available, it is possible to place the working electrode directly inside the channel (Figure 3.8C). However, it is important to note that most common commercially available potentiostats are grounded and cannot be employed in this configuration, since the high voltage will ground through the working electrode and destroy the electronic components of the potentiostat.

In-channel amperometric detection using an isolated potentiostat has been reported for both normal and reverse polarity separations.[33,34] The advantage of this configuration is that it minimizes band-broadening because there is no band dispersion in the channel and, therefore, high separation efficiencies and sensitivity can be obtained. However, because the electrode is placed in the separation field, there can be both potential shifts (as discussed in Section 3.5) and noise due to separation voltage fluctuations which can affect the LODs that can be achieved. To address the LOD concern, a dual channel/dual electrode approach has been reported for background subtraction (Figure 3.1D).[21,47] This microchip contains two separation channels; one is used as a separation channel and the second is used as a reference channel. The symmetry of dual-channel microchips allows symmetrical voltage distribution through the two channels and, therefore, a similar volume of sample and buffer can be injected into each channel. Working and reference electrode are placed inside the separation and reference channels, respectively. The noise due to fluctuations in the separation voltage will be the same at both electrodes and thus can be cancelled out.[21,47]

3.7 Instrumentation for Amperometric Detection in ME

Conventional three-electrode potentiostats have been the most commonly employed system for electrochemical detection in ME. In these conventional three-electrode systems, the current flows through the counter (auxiliary) electrode and working electrode, and there is negligible current flowing through the reference electrode. Typically in these potentiostats, the counter electrode potential is changed by the electronic circuitry such that the potential on the working electrode with respect to the reference electrode is at the required potential. The current produced by the oxidation or reduction of an analyte is measured between the counter and working electrode while the applied voltage is measured between the reference and working electrodes. Since there is negligible current flow through the reference electrode in this configuration, the potential of the working electrode does not shift due to an ohmic drop (*IR* drop). However, there is a distance-related ohmic drop that depends on the distance between the reference and working electrodes and the solution conductivity. Therefore, the working and reference electrodes need to be placed as close to each other as possible. In the case of two-electrode systems, the voltage is applied between the working electrode and reference/counter electrode. In this configuration, either large current measurements or high solution resistance can cause an ohmic drop, resulting in a potential change at the working electrode. Additionally, the large currents generated in this system can cause a change in the thermodynamic potential of the reference electrode. Therefore, two-electrode potentiostats are limited to low current measurements (low nA and pA), corresponding to low analyte concentrations.

For in-channel detection, an isolated or "floating" potentiostat must be used to avoid the separation voltage grounding through the potentiostat and destroying the electronics.[33,34] In contrast to most commercially available instruments, these potentiostats are completely electrically isolated and are usually battery-powered. Conventional commercially available grounded potentiostats cannot be employed for in-channel detection. Two prototype isolated potentiostats have been evaluated by the Lunte group for in-channel detection in ME. The early version employed a three-electrode system, while a newer prototype employs a two-electrode configuration.[34] The latter was developed in collaboration with Pinnacle Technology, Inc.[33,49]

When the two-electrode system is used with in-channel detection, one of the drawbacks is that it is difficult to determine the exact potential that is applied to the working electrode, due to the combined effects of the separation voltage and the lack of a counter electrode.[33,34] Therefore, an HDV must be obtained for each analyte under the separation conditions of interest to identify the optimum detection potential. The interaction of the separation voltage with the working electrode is also possible in the end- and off-channel configurations. Therefore, it is useful to generate HDVs for analytes in these electrode configurations as well to determine the optimum potential.

There are several commercially available potentiostats that have been used in ME–EC. The Gamry Femtostat has been used for in-channel detection by keeping both the reference and working electrode at isopotential with a three-electrode configuration.[21] The BASi Petit Ampere is battery powered, and could theoretically be used in the in-channel configuration. Additionally, modified BASi 4C-LC potentiostats have replaceable op-amps in case the separation current is grounded through the potentiostat.[48] If the separation voltage is grounded through the potentiostat, only the op-amp needs to be replaced. MicruX Technologies has developed hand-held battery powered instrumentation to perform both separation and detection for ME–EC. This is an ideal instrument to introduce ME–EC to undergraduate laboratories.[50,51] When using other commercially available grounded potentiostats, care needs to be taken to avoid grounding the separation voltage through the potentiostat. Damage to the instrumentation caused by a small oversight in electrode alignment can be very costly.

3.8 Signal, Noise, and Limits of Detection for ME–EC

3.8.1 Signal

One important benefit of amperometric detection for ME is that (under ideal conditions) the signal can be directly related to analyte concentration without the need for a calibration curve or standards. Signals generated by

ME–EC are current *versus* time profiles (*i–t* curves). Upon integration of peaks from *i–t* curves, charge (Q) is obtained. Using Faraday's equation (eqn (3.5)) and knowing how many electrons are transferred in the oxidation/reduction, as well as taking into account the conversion efficiency, the number of moles of analyte that was injected into the system can be calculated.[28,37] If the volume of the injection plug is known, concentration can then be easily determined.

The current response and potential needed for an analyte in ME–EC is highly dependent on the type of working electrode that is employed. The molecular composition of the working electrode can have significant effects on the electron transfer kinetics. Additionally, the peak height and separation efficiencies are highly dependent on the electrode alignment with respect to the channel (in-, end-, or off-channel) and band-broadening.

3.8.2 Noise

The high voltage power supply used for the electrophoretic separation is the main source of noise in ME–EC. This noise is due to fluctuations in the potential applied by the high voltage power supply across the channel as well as improper grounding of the separation voltage prior to detection.[21,48,52,53] The noise generated in an ME experiment is usually greater than 25 pA; however, the noise level of a typical amperometric detector is less than 1 pA.[21,48] The common components of noise observed during a typical ME–EC experiment are baseline drift, spikes, and oscillations.[52]

Low conductivity buffers, such as Good's buffers, are the best choice for ME–EC experiments. High conductivity buffers can contribute to non-Faradaic current that can cause additional noise. Fluctuations in the separation field can cause noise that increases as a function of voltage. This type of noise is most noticeable with the in-channel configuration. End- and off-channel configurations can also be affected by noise generated by the power supply if the working electrode is not fully decoupled from the separation field.

Any electrochemically active impurity in the sample or background electrolyte can cause generation of current at the working electrode and contribute to background noise. Therefore, the noise is also dependent on the working electrode potential. In general, low positive working electrode potentials result in lower noise levels compared to high positive working electrode potentials. At electrode potentials above +1.00 V *versus* Ag/AgCl, oxidation of water and production of oxygen at metal working electrodes can cause increases in background current. The background current generated by solvent oxidation is highly dependent on pH and the type of the electrode that is used for amperometric detection.[8] The potential window for metal and carbon electrodes was shown previously in Figure 3.6. In the reductive mode, dissolved oxygen and hydrogen evolution from the reduction of water can generate high background currents and noise. For this reason, removal of oxygen is important in these applications.

The electrode configuration used for ME–EC also affects the amount of noise observed in the electropherogram. The in-channel configuration suffers from higher noise compared to the other two configurations, due to the interaction of the working electrode with the separation field as discussed in Section 3.5. When an electrode is placed in the electric field, the voltage drop across the electrode ΔE can be calculated using the following equation:[53]

$$\Delta E = (E/L)w \tag{3.9}$$

where E is the separation voltage, L is the channel length, and w is the width of the electrode. Therefore, E/L is the separation field strength in V cm^{-1}. For 40 and 10 μm electrodes with a −300 V cm^{-1} separation field, the voltage drops across the solution above the electrodes are −1.2 and −0.45 V, respectively. This translates to an induced voltage of +1.2 V for the 40 μm electrode and +0.45 V for the 10 μm electrode. Under these conditions, oxidation of water can now occur at the 40 μm electrode and generate noise.[53]

To obtain better limits of detection for the in-channel configuration, a dual-channel microchip that contains separate reference and separation channels was reported by Hahn and coworkers (Figure 3.1D).[20,21] As discussed previously, this dual-channel configuration makes it possible to cancel the noise due to the separation field since the separation voltage fluctuations at both the working and reference electrodes are identical. The noise levels observed using this configuration were low (around 4 pA).[20,21,48]

In the end-channel configuration, the working electrode is placed in the ground reservoir. For the off-channel configuration, the separation field is grounded before the working electrode. Therefore, the amount of noise that is observed is much lower in these configurations. The sensitivity of end-channel and off-channel configurations is lower than that of in-channel. However, lower LODs can be achieved with the end-channel and off-channel configuration due to the relatively low noise compared to the in-channel configuration.[37] Hydrogen gas generated at Pt or Pd decouplers in off-channel configuration can eventually lead to bubbles that can generate noise at the working electrode or completely stop the separation process.

3.8.3 Signal-to-noise Ratio and LOD

To obtain the best LOD for an analyte of interest, both the signal and the noise need to be taken into consideration. Obviously, the ideal conditions are where there is a high signal and low noise so that S/N is at a maximum. The signal can be either current or charge corresponding to peak height or area, respectively. The measurement of signal is straightforward in ME–EC. However, noise level measurements can be ambiguous. Figure 3.9 shows a signal and two types of noise measurements, root-mean noise (N_{RMN}) and peak-to-peak noise (N_{P-P}), that can be used to calculate S/N and LOD. Spikes and drifts in the baseline can be removed by filtering and/or baseline subtraction methods before calculating the LOD.[54,55] For a realistic determination of the LOD for an analyte, the best approach is to measure a

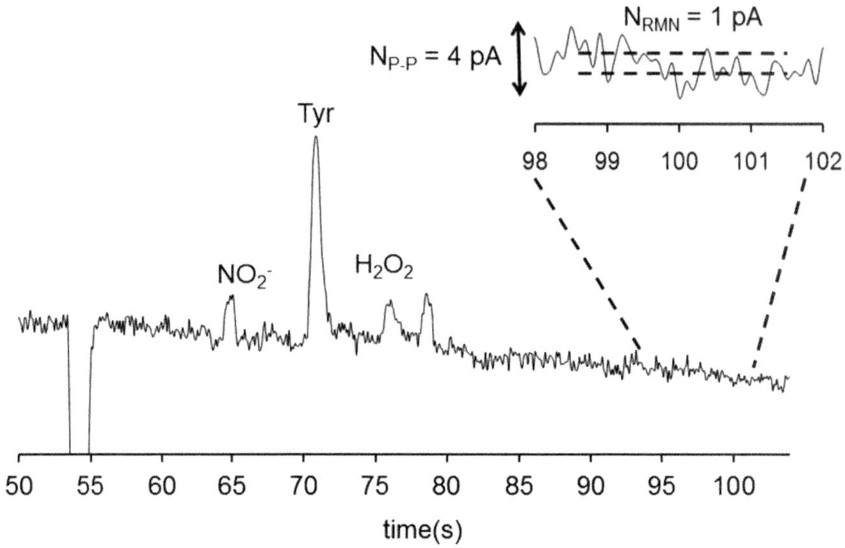

Figure 3.9 The signal and two types of noise measurement, root-mean noise (N_{RMN}) and peak-to-peak noise (N_{P-P}). This electropherogram was obtained for an equimolar (3.1 μM) mixture of nitrite, Tyr, H_2O_2, and an unknown, under reverse polarity conditions.
(Reproduced with permission from ref. 48.)

concentration that is close to the LOD and ensure that there is a S/N ratio greater than 3. This is important because as the signal decreases the noise becomes more apparent, and it is easy to underestimate the noise if a peak with an S/N ratio greater than 10 is used. At analyte concentrations close to the LOD, drifts in the baseline, spikes, and baseline oscillations can be significant. These sources of noise are often not noticeable when higher concentrations (signals) are used to estimate the LOD.

3.9 Types of Electrodes

Analytical performance of an amperometric detector that is integrated with microchip electrophoresis is highly dependent on the working electrode composition (*e.g.*, carbon- or metal-based), electrode size, the number of electrodes (*e.g.*, single electrode *versus* electrode array), configuration (*e.g.*, in-channel, end-channel, or off-channel), and whether a two- or three-electrode system is used.

3.9.1 Microelectrodes

For ME–EC measurements, many different types and sizes of microelectrodes have been employed. The current response obtained in any electrochemical system depends on the dimensions of the electrodes used for the measurements. Width is the most critical dimension for the band electrodes that are more commonly employed with ME–EC. When the width of a band electrode is smaller than 12.5 μm, the electrode is considered to be an ultramicroelectrode. Electrodes with widths of less than or equal to 100 nm are termed nanoelectrodes. As the size of the electrode decreases, mass transport to the electrode surface increases but the geometric area of the electrode decreases and therefore the signal decreases. The *IR* drop, along with the background currents, is also significantly decreased.[56] Compared to the signal, noise decreases at a higher rate as the electrode becomes smaller, resulting in improved S/N ratios when using microelectrodes.[57]

3.9.2 Multiple Electrodes

Multiple electrode detection systems using micro- or nanoelectrodes can be used in an array format for signal enhancement or for peak identification using dual electrodes. These systems can provide additional information about the identity of the analyte and/or better sensitivities.

3.9.2.1 Electrode Arrays

Microelectrodes as working electrodes for ME are of considerable importance because of their ultrafast mass transport due to small electrode dimensions, reduced *IR* drop, and improved S/N ratio. If a collection of microelectrodes is used in an array, it has a higher S/N than does a single electrode of total geometric area equal to that of the array. This is because capacitive currents (overall noise) are proportional to the active surface (sum of the area of individual electrodes) of the electrode array while the overall response is proportional to the geometric area (area of the electrode array). Arrays using carbon ink electrodes were employed in ME–EC by the Martin group to enhance the LOD obtained for catecholamines.[41,58]

3.9.2.2 Dual Electrodes

Two types of dual electrode configurations are commonly employed for amperometric detection: dual-parallel and dual-series. Dual-parallel electrodes are configured so that solution flows over both electrodes at the same time. In the series configuration, one electrode follows the other, so the solution "sees" the first electrode before the second.

Both dual-series and dual-parallel electrode configurations have been widely employed in LC and CE with electrochemical detection for voltammetric identification of analytes.[59–62] Dual-series electrodes have also been employed in ME in the generation–collection mode, where analytes are identified based on chemically reversible redox processes.[63–65] In this mode, the first electrode is commonly utilized for generation of the oxidation product of a reduced compound. The product is then reduced back to the original reduced species at the second electrode. The reduction usually occurs at a more selective detection potential than the oxidation since there are few reducible compounds in biological samples. Determination of catecholamines is a common application of this approach.[63–65] Catechol-containing compounds are oxidized to their corresponding ortho-quinones at the first electrode, and then reduced back to the catechol at the second electrode. Selectivity is obtained at the second electrode due to the low reduction potential that needs to be employed to detect the ortho-quinone product.

Collection efficiency, the percentage of species reduced at the second electrode compared to the amount of species oxidized at the first electrode, can also be used for identification of analytes. This is because the collection efficiency is different for different species and dependent on the reversibility of the electrochemical reaction. For example, phenol, catechol, and trihydroxyphenol all display different degrees of reversibility in this system and, therefore, different current ratios. The dual-series configuration has been used extensively for the identification of phenolic acids and catecholamines in ME.[63–65]

Even though the dual-parallel configuration has been successfully employed for voltammetric characterization of analytes through current ratios in CE,[66] this configuration has not been used in ME–EC. This is due to difficulties in the fabrication and design of two microelectrodes in a parallel configuration in the narrow electrophoresis channel. Recently, Gunasekara *et al.* have developed a dual-parallel electrode configuration using the dual-channel microchip design (Figure 3.1D). They have used this dual-parallel configuration to identify species in contaminated peroxynitrite samples and macrophage cell lysates.[67,68]

Voltammetry could also possibly be used for identification of compounds; however, potential scanning techniques such as cyclic voltammetry and square wave voltammetry are difficult to implement in ME. Some of the challenges include the high background noise that results from the voltage scanning due to capacitance (charging current), low temporal resolution of slow scanning techniques, high LOD, and reduced sensitivity.[69–71]

3.9.3 Electrode Materials

As stated earlier, the selection of electrode material is an important consideration in ME–EC. Most electrodes can be classified into two types: carbon-based electrodes and metal-based electrodes.

3.9.3.1 Carbon-based Electrodes

Carbon-based electrodes have many advantages over metal-based electrodes, and have been used extensively in microchip electrophoresis systems. Practically, carbon electrodes are relatively inexpensive and are simple to fabricate. Electrochemically, carbon electrodes have a large potential window, low background noise, low over-potential, and less fouling than their metal-based counterparts.[12] Additionally, many biologically relevant analytes are organic compounds and, therefore, generate a better response at carbon-based electrodes. Many different types of carbon have been integrated into microchip electrophoresis devices, including carbon fiber, carbon paste, carbon ink, pyrolyzed photoresist film carbon, carbon nanotubes, and boron-doped diamond electrodes. While carbon electrodes are an excellent choice for microchips made from PDMS and other polymers, plastics, or hybrid chips, they are not currently compatible with fabrication of an all-glass chip.

Carbon fiber electrodes are often employed for amperometric detection in microchip electrophoresis.[72] To integrate these fibers into a microchip device, a trench that is the width of the fiber is typically created in a PDMS substrate. The carbon fiber is then laid on the trench and pushed into it, taking extreme care not to break the fiber. To make connections to the fiber, a copper wire is epoxied onto the substrate, and the fiber and wire are connected using colloidal silver. While this procedure produces a fairly rugged electrode substrate that can, with care, last for months or years, the process for inserting the fiber into the trench can be difficult and time-consuming for a novice. Additionally, these electrodes are limited in size by what is commercially available, generally 7 and 33 μm in diameter.

Following their introduction by Ralph Adams in the 1950s,[73] carbon paste electrodes have been employed for many years for electrochemical applications. They were first integrated into ME devices in 2001 by the Lunte group.[74] These electrodes are more easily fabricated than carbon fibers; however, they also cannot be produced by classic lithographic techniques. To create a carbon paste electrode for ME, a trench is built in a substrate (PDMS, PMMA, glass) and a mixture of a carbon-based material (graphite, carbon nanotubes, *etc.*) and a binding agent (oil) is pressed into the trench.[75] One of the benefits of this procedure is that the carbon paste can also be integrated into more rigid substrates, such as plastics or glass, whereas carbon fibers can only be placed in the more malleable PDMS. For example, carbon paste electrodes have been integrated into PMMA substrates for the detection of dopamine and catechol following ME using a PDMS/PMMA hybrid simple-t device.[75]

Carbon ink electrodes have also been employed with ME–EC. These electrodes are fabricated by micromolding or screen printing and can be integrated on to various substrates, including PDMS,[37] glass,[76] and epoxy.[41] Wang's group employed an external screen-printed carbon ink electrode for end-channel detection with all-glass microchip.[77]

Pyrolyzed photoresist film carbon electrodes offer the sizeable advantage that they can be manufactured using classic photolithography techniques.[78] These electrodes are manufactured by depositing photoresist in the pattern of the desired electrode onto a glass plate using photolithography. The glass plates with photoresist are placed under inert conditions at temperatures up to 1100 °C to create the final electrode. These electrodes have been shown to have near-atomic flatness and good electrochemical properties similar to those of traditional glassy carbon electrodes; however, they do have a lower oxygen/carbon ratio, leading to slower electron transfer kinetics for some analytes such as dopamine.[79]

Surface electron transfer kinetics with carbon electrodes is highly dependent on the arrangement of carbon atoms in different planes.[80] Carbon nanotubes, one of the important allotropes of carbon, generate a large surface area, exhibit reduced surface fouling, and have significant electrocatalytic activity. Additionally, by treating the carbon surface either electrochemically or chemically, oxygen groups can be introduced as electrocatalytic sites. Signal enhancement is also possible with carbon nanotubes due to the large surface area available for the redox process. In general, the use of carbon nanotubes in ME–EC leads to low limits of detection by increasing the signal more than the noise.[81]

Boron-doped diamond electrodes are also important for electrochemical detection as they offer less capacitive currents, less surface fouling, and a broad potential window.[82,83] Diamonds are a crystalline allotrope of carbon with sp^3 hybridization that have higher electrical resistivity and higher thermal conductivity than most forms of carbon. Doping a diamond with boron provides it with p-type semiconductor properties.[84] These electrodes are prepared by depositing a thin film of boron-doped diamond on silicon wafers using a high-pressure plasma-assisted chemical vapor deposition system with both a carbon and a boron source. After the deposition, chemical etching is used to dissolve the Si substrate and obtain a boron doped-diamond thin film. A thin film of desired length can then be attached to substrates with an adhesive.[83] Boron-doped diamond electrodes have been used for ME–EC analysis of nitroaromatic explosives, organophosphate nerve agents, and phenols. The main advantages of this electrode material for ME–EC are its ruggedness and reproducibility.[85]

3.9.3.2 Metal-based Electrodes

Pt, Pd, Au, Cu, and Ag electrodes have all been employed for ME–EC. Metal electrodes have advantages over carbon electrodes in that they can be fabricated using classic lithography techniques, are amenable to mass production (commercially, MicruX Technologies and MicroLIQUID are two examples), and can be used in all-glass devices.[49,86,87] However, these electrodes can be fairly expensive to purchase or fabricate. Additionally, depending on the metal employed, a major disadvantage of these electrodes for ME–EC is that most metals require an adhesion layer (typically titanium, tantalum, or nichrome) to adhere to glass. This adhesion layer can generate an insulating oxide layer through grain boundary diffusion and limit the lifetime of the electrode when exposed to the separation field. In addition, undesired reactions such as adsorption of gases, formation of oxide layers, and adsorption of organics are drawbacks of metal electrodes.

To fabricate a metal electrode on a glass substrate, the desired electrode design is drawn using commercially available drawing software such as

AutoCAD and printed on a transparency sheet with sufficient resolution to block UV light through the printed areas during lithography. Glass substrates coated with chrome and positive photoresist are used as a substrate. The desired electrode geometry is then patterned on the substrate using the transparency mask and a UV flood source. Next, the photoresist is developed and the underlying chrome layer is etched away using chrome etchant. Using a buffered oxidant, the glass is etched to produce a trench. Lastly, a thin layer of an adhesion metal is deposited in the trench followed by the desired electrode material using a metal deposition system. The excess metal, photoresist, and chrome layer are removed to obtain a metal electrode supported on glass.[49,86]

As a substitute for metal deposition, a trench on a PDMS substrate similar to that used for integration of carbon fiber electrodes can also be used to integrate metal wires or the electrodes can simply be glued into the substrate.[88] In addition, metal wire electrodes embedded in an epoxy substrate have been used in ME.[41]

3.10 Applications

Microchip electrophoresis with electrochemical detection has been employed for various applications. Many features of ME–EC such as its small size, portability, availability, ability to obtain specific designs, low reagent consumption, and ability to perform analyses with limited sample volumes are difficult or impossible to achieve using traditional benchtop analytical instrumentation. Here, we will briefly discuss applications of this technique in biological, environmental, and food-based applications to give readers a sense of the wide range of applicability of this technique.

3.10.1 Biological Applications

ME–EC devices have been used for both *in vivo* and *in vitro* studies of biological systems, including on-animal sensing, analysis of reactive nitrogen species in cells, and release of catecholamines from PC-12 cells.[4,89] Analytes of interest in these studies can be broadly categorized as neurotransmitters,[37] reactive oxygen and nitrogen species,[33,86,90,91] cellular and non-cellular antioxidants,[86,92] amino acids,[93,94] and pharmaceuticals. Bioanalytical applications of ME–EC have been reviewed widely, and readers can find additional information in several review articles.[3–8] In this chapter, representative applications of ME–EC to microdialysis sampling, cellular analysis, and amino acids are presented. In addition, a brief account of carbohydrate detection is provided at the end of this section.

Microdialysis (MD) sampling has been coupled to ME–EC for both *in vitro* and *in vivo* analysis of biological systems.[4] The combination of these two techniques allow fast, near real-time monitoring of dynamic biological events such as neurotransmitter release and drug metabolism. In the first coupling of MD–ME–EC, Mecker and Martin employed a pneumatically driven valving system to inject discrete microdialysis samples into the separation channel. This device was used to monitor the stimulated release of dopamine from PC-12 cells *in vitro.*[26] MD–ME–EC also permits *in vivo* monitoring of drug metabolism. Due to the small size and portability of the ME–EC devices, on-animal sensing is possible. The Lunte group developed a double-t all-glass microchip with integrated platinum electrodes for coupling MD to ME–EC (Figure 3.10A).[49] While the first application of this device was to monitor an enzymatic reaction *in vitro* (Figure 3.10B),[49] it has since been employed for on-sheep analysis to monitor subcutaneous nitrite production following nitroglycerin perfusion.[95] This exciting application paves

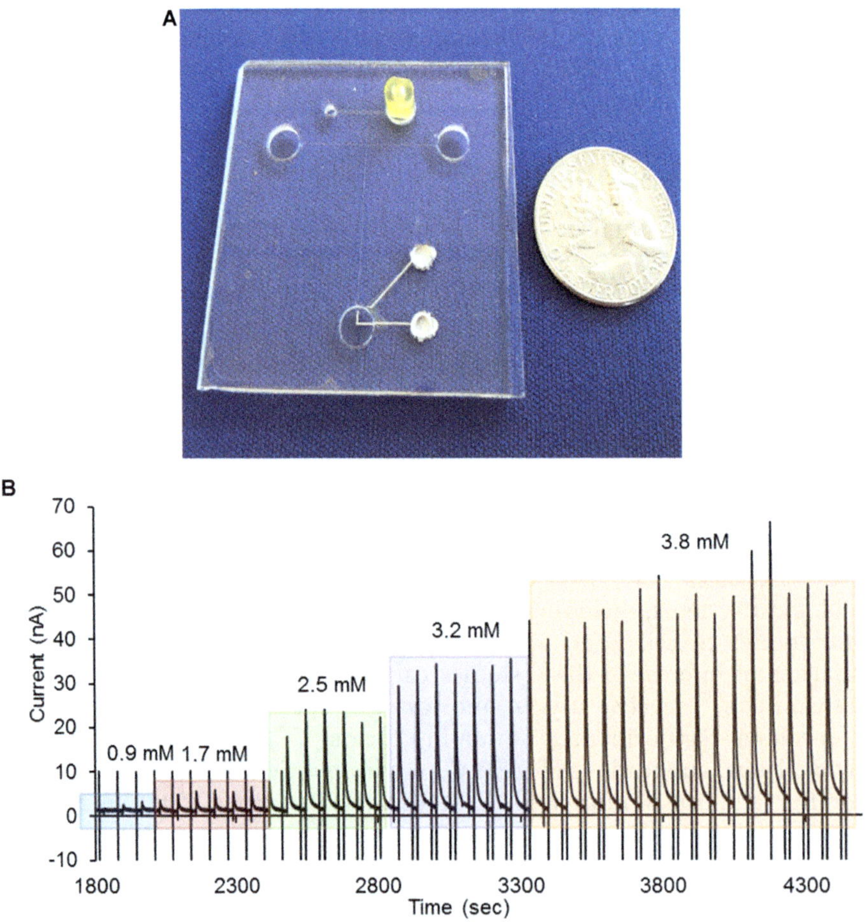

Figure 3.10 (A) Double-t all glass microchip with integrated platinum electrodes for coupling microdialysis sampling to ME–EC. (B) Continuous monitoring of H_2O_2 production following addition of glucose to a solution of glucose oxidase monitored by a linear microdialysis probe. (Reproduced with permission from ref. 49.)

the way for the development of on-animal sensors for neurotransmitters that can be used in freely roaming, freely behaving animals to correlate neurochemistry with behaviors.

Cellular analysis has also been performed using ME–EC, and was reviewed previously.[3] In addition to the MD–ME–EC-based cell analysis highlighted above, the Martin group developed a microchip device that was capable of immobilizing PC-12 cells on a collagen-coated micropallet, cell stimulation, sample injection and separation, and detection of dopamine and norepinephrine at a carbon ink microelectrode, all on-chip.[96] An integrated polystyrene and PDMS hybrid device was recently developed by the same group that has the ability to perform on-chip cell culturing and detection of cellular release with ME–EC.[97]

The quantification of pro-oxidants and antioxidants in cells reveals the cellular redox status and thus facilitates the understanding of diseases and disorders caused by the imbalance between pro-oxidants and antioxidants. The short channels and fast separations characteristic of ME allows the detection of reactive species such as peroxynitrite and nitric oxide before they degrade significantly.[33,90,91] ME also permits the separation of the metabolites of these reactive species and cell antioxidants. Recently, ME–EC was used for the analysis of NO_2^-, NO, and antioxidants in macrophage cells by employing a simple-t device with in-channel detection and a Pt working electrode. (Figure 3.11).[86]

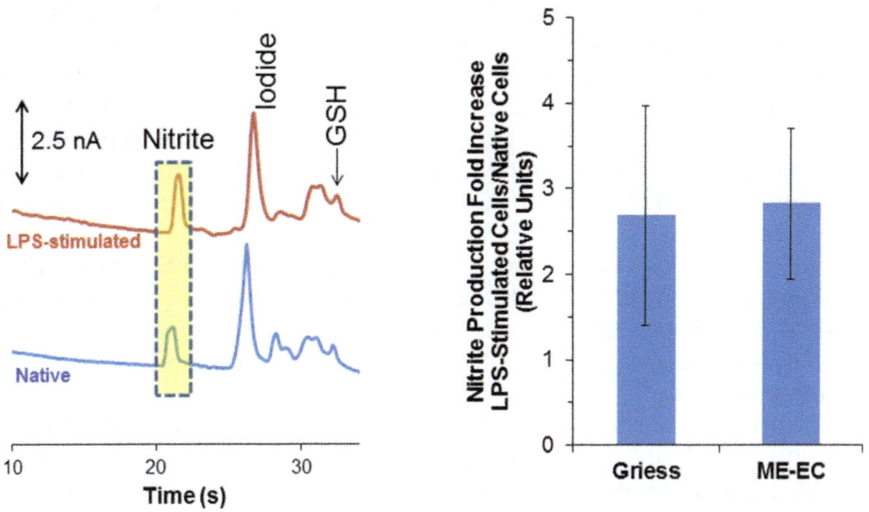

Figure 3.11 Separation and detection of intracellular nitrite and glutathione (GSH), in addition to an internal standard (iodide) in bulk cell lysates of native and LPS-stimulated macrophages. The bar graphs show a comparison of fold-increase production of nitrite levels following LPS stimulation measured using ME–EC and the Griess assay.
(Reproduced from ref. 86.)

Amino acid analysis, while generally performed by precolumn derivatization and ME-LIF, has also been accomplished using ME–EC. Using ME, Schwarz *et al.* successfully separated tryptophan, histidine, and glycine without derivatization and performed direct electrochemical detection with a copper-coated Pt electrode.[94] An alternative strategy for the detection of amino acids by ME–EC is to derivatize the compounds with an electroactive tag. The products of both *o*-phthaldialdehyde (OPA) and naphthalene-2,3-dicarboxaldehyde (NDA) are isoindoles that are electrochemically active. ME was used to separate NDA/CN-derivatized citrulline and glycine, which were then detected using the end-channel configuration with a carbon paste electrode.[74] Wang and coworkers employed on-chip derivatization using (OPA)/2-mercaptoethanol followed by separation and detection using ME at a gold screen-printed working electrode in the end-channel configuration to detect histidine, valine, isoleucine, leucine, glutamic acid, aspartic acid, arginine, and lysine.[98] Carbohydrates have also been detected directly using pulse amperometric detection and a gold electrode or, under high pH conditions, by amperometry.[99]

3.10.2 Environmental Applications

The portability of ME–EC allows on-site testing, which has proven beneficial for environmental applications. ME–EC-based devices have been employed for the determination of analytes such as explosives, chemical warfare agents, and pesticides, in addition to many other compounds.

The need for sensitive, fast, portable, and field-executable (fields such as airport security and custom and border regions) analytical devices capable of detecting explosives has increased dramatically during the last two decades. Toward this end, the Wang group developed an ME–EC system with an external glassy carbon electrode for the separation and detection of nitrite ester explosives. This class of chemicals is often used in homemade explosives, and the ME–EC device was capable of analyzing a mixture of four nitrite ester explosives in under 3 min (Figure 3.12).[100] Additionally, Hilmi and Luong developed an ME–EC device capable of separating and detecting several different nitroaromatic explosives (including TNT). These researchers employed ME–EC with an external detection cell with a gold working electrode.[101,102]

The detection of chemical warfare agents has also been accomplished with ME–EC. The ability to quickly detect warfare agents, such as nerve gases, would give soldiers and first responders information about the environment, allowing them to make more informed decisions regarding response. V-type nerve agents produce stable thiol degradation products that have been analyzed by ME–EC using screen-printed carbon electrodes.[103] ME–EC devices have also been developed to perform analyses in remote locations using lab-on-a-robot applications.[104]

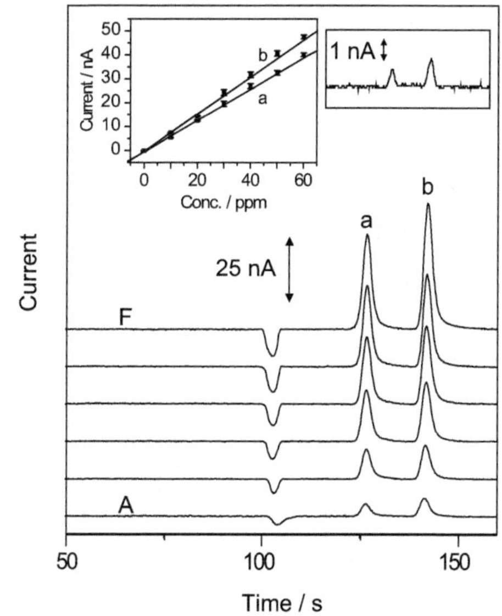

Figure 3.12 ME–EC analysis of nitrate explosives: (a) propylene glycol dinitrate and (b) glyceryl trinitrate.
(Reproduced from ref. 100.)

3.10.3 Food Applications

ME–EC provides an interesting possibility for quality control in the food industry as the devices are portable and could be used to monitor food throughout the journey from farm to the store shelf. Additionally, the small sample size and reagent volume requirements make this technique attractive. Food analysis by microchip electrophoresis has been reviewed previously.[6] ME–EC has been employed for the detection of heavy metals in vegetable juices,[105] nitrite in meats,[106] biogenic amines in beers,[107] and mycotoxin and zearalenone in infant foods,[108] among other applications.

Qualitative and quantitative analysis of antioxidants is a major application of ME–EC. Both natural and synthetic supplements containing antioxidants have been given significant attention recently. ME–EC has been successfully used for quantification of natural antioxidants, phenolic acids, and flavonoids (*e.g.*, catechin, rutin, and quercetin, and chlorogenic, ferulic, caffeic, protocatechuic, vanillic, and gallic acids) in foods. This technique could also be used to obtain a "class-selective electrochemical index" for the purpose of determining the major antioxidant classes (*e.g.*, flavonoids and phenolic acids) in a food product in a short time period.[109]

A comprehensive survey of applications of microchip electrophoresis coupled with amperometric detection is provided in Table 3.1. This table gives the type of sample/standard, electrode material, electrode alignment, and LOD.

Table 3.1 Applications of ME–EC.

Biological applications						
Compound	Sample/standards	Electrode	Electrode alignment	LOD	Notes/comments	Ref.
Catecholamine neurotransmitters (dopamine, norepinephrine, and catechol)	Standards	Pyrolyzed photoresist film carbon/carbon fiber/carbon ink/Pd	End-channel/ in-channel/ off-channel	73 nM Dopamine-end channel, pyrolyzed photoresist film carbon electrode	Different electrode materials and alignments	37
RNOS and antioxidants (nitrite, glutathione, ascorbic acid, and tyrosine)	Standards	Pt film	In-channel	2.6 μM Nitrite	Wireless isolated potentiostat	33 and 86
Peroxynitrite	Standards	Pd film	End-channel	2.4 μM Peroxynitrite	First use of ME–EC to detect peroxynitrite	91
Biochemicals/Pharmaceuticals (uric acid and interfering compounds, ascorbic acid, paracetamol, epinephrine, xanthine and theophylline)	Urine samples	Pt film	End-channel	4 μM Epinephrine 10 μM uric acid	MicruX portable electrophoresis instrument	110
Vitamins (pyridoxine, ascorbic acid, and folic acid)	Four different pharmaceuticals (Vitafardi, Aspol, Becozyme C Forte, and Hidrosil)	Modified glassy carbon and screen-printed electrodes (SCE) with multi-walled carbon nanotubes	End-channel	8 μm Pyridoxine 11 μm ascorbic acid 8 μm folic acid with modified SCE	Carbon nanotubes	111
Catechol estrogen-derived DNA adducts (8-Hydroxy-deoxyguanosine (8-OH-dG), 4-OHE2-1-N3Ade, 4-OHE2-1-N7Gua, and 4-OHE2-NacCys)	Standards	Au micro-electrode array (Pd–decoupler)	Off-channel	0.4 μM	PDMS-glass hybrid microchip	112
L-Tryptophan metabolites (5-hydroxy-L-tryptophan, 5-hydroxytryptamine, and 5-hydroxy-indole-3-acetic acid)	Standards	Pt wire	End-channel	0.84 μM 5-Hydroxy-L-tryptophan 0.67 μM 5-hydroxytryptamine 4.2 μM 5-hydroxy-indole-3-acetic acid	Filmy channel microchip	113

Cholesterol	Bovine serum	Au wire	End-channel	1 nM	Enzymatic assay	114
Underivatized carbohydrates, amino acids, and sulfur-containing antibiotics (fructose, sucrose, lactose, maltose, cysteine, histidine, ampicillin, glucose and penicillin)	Standards	Au wire	End-channel	10 µM Cysteine 350 µM histidine 5 µM ampicillin 5 µM penicillin	Pulsed-amperometric detection	99
Mixture of thiols (homocysteine, cysteine glutathione and *N*-acetylcysteine)	Standards	Palladium/mercury amalgam electrode	Off-channel (Pd decoupler)	9.2 µM	First report of the usage of Pd/Hg for thiols	92
Amino acids (arginine, proline, histidine, valine, and serine)	Standards	Cu wire	End channel	7 µM Arginine 6 µM proline 5 µM histidine 6 µM valine 5 µM serine	Detection of non-electroactive amino acids	115

Environmental applications						
Compound	Sample/standards	Electrode	Electrode alignment	LOD	Notes/comments	Ref.
Explosives (ethylene glycol dinitrate, pentaerythritol tetranitrate, propylene glycol, and dinitrate glyceryl trinitrate)	Standards	Glassy carbon	End-channel	0.3 ppm Glyceryl trinitrate 0.5 ppm propylene glycol dinitrate	On-site explosive analysis	100
Chemical warfare agents (2(dimethylamino)-ethanethiol (DMAET), 2-(di-ethylamino)ethanethiol (DEAET), and 2-mercaptoethanol (ME))	Untreated tap and river water samples	Carbon ink	End-channel	5 µM 2-(dimethylamino)-ethanethiol 8 µM 2-(di-ethylamino)-ethanethiol	Off-chip/on-chip derivatization of nerve agents with *o*-phthaldialdehyde	103
Pollutants (*p*-aminophenol, *o*-aminophenol, and *m*-aminophenol)	Standards	Au film	In-channel	14.6 nM *p*-Aminophenol 10.6 nM *o*-aminophenol 15.3 nM *m*-aminophenol	Dual-channel microchip for noise reduction	20
Anions (chloride, bromide, iodide and thiocyanide)	Standards	Silver film	End-channel	1.23 µM Chloride 1.56 µM bromide 0.56 µM iodide 2.03 µM thiocyanide	PDMS/glass hybrid microchip	116

Table 3.1 *(Continued)*

Food applications

Application	Sample	Electrode	Electrode alignment	LOD	Notes/comments	Ref.
Heavy metals (lead, cadmium, and copper)	Standards	Screen-printed carbon	End-channel	1.74 μM Pb^{2+} 0.73 μM Cd^{2+} 0.13 μM Cu^{2+}	First report of heavy metal analysis by ME–EC	105
Phenol compounds (tyrosol, hydroxytyrosol, and oleuropein glucoside)	Olive oil	Au wire	End-channel	15 μM Tyrosol 36 μM hydroxytyrosol 17 μM oleuropein glucoside	Novel method to quantify phenolics in olive oil	117
Biogenic amines (ethanolamine, tryptamine, and tryptophan)	Standards	Ruthenium oxide/hexacyanoruthenate polymeric film modified glassy carbon	End-channel	23 μM Ethanolamine 27 μM tryptamine 34 mM tryptophan	Electrocatalytic oxidation ability by a inorganic complex	107
Mycotoxin (zearalenone)	Powdered baby food, cereal, milkshakes	Glassy carbon electrode	End-channel	0.4 μg L^{-1} zearalenone	Electro-kinetic magnetic bead-based immune assay on a microchip	108
Nitrite	Ham and sausage samples	Copper (3-mercaptopropyl)-trimethoxysilane [Cu(II)-MPS] complex-modified carbon paste electrode	End-channel	0.35 ppm Nitrite	Modified electrode	106
Plant phenols (catechin, rutin, and quercetin, and chlorogenic, ferulic, caffeic, protocatechuic, vanillic, and gallic acids)	Apples, pears, red and white wines, and green tea	Glassy carbon	End-channel	≤6 μM Class selective approach <8 μM individual antioxidant detection	Class selective and individual analysis	109
Monosaccharides (glucose and fructose)	Honey samples	Copper nanowires	End-channel	4 μM glucose 23 μM fructose	Usage of nano-materials	118

3.11 Future Directions

Microchip electrophoresis with electrochemical detection is a powerful technique that is being employed for fast, near real-time, on-site analysis of many different compounds. The microchip format allows multiple processes to be integrated on-chip prior to separation and detection. Additionally, the small microchip format permits portable analysis. In the future, these benefits will continue to be exploited for applications such as cell cytometry, on-animal sensing, and portable diagnostics.

Chemical cytometry with LIF detection has been used for peptide-based single cell assays and single cell nitric oxide assays.[119–121] Electrochemical cytometry has also been reported by the Ewing group; however, in that report, a short capillary connected to a microfluidic device was used to transport isolated vesicles into an electrode for coulometric detection of the vesicles' content.[122] This is an important development for understanding neurotransmitter release, but the report has not used the full potential of electrochemical cytometry. In the future, electrochemical cytometry could be used for measuring multiple cellular species, their byproducts, drugs and their metabolites, and enzymatic activity.

The small and portable nature of microchip electrophoresis with electrochemical detection allows the device to be placed on-animal. Many behavioral studies employing animals involve tethering or restricting the animal to a small cage, seriously limiting the behaviors that can be studied. By placing the sampling, separation, and detection system on-animal, the animal can be freely roaming in its natural environment, enabling better studies of behaviors. On-animal sensing of analytes such as neurotransmitters will permit better correlation between animal behavior and neurotransmitter concentration. This miniaturization capability of ME–EC will lead to the development of novel devices based on ME–EC in the future for the detection of explosives at airport security or food quality control in supermarkets. This portable nature can also be exploited for biomedical applications, such as of point-of-care diagnostics. The ability to monitor analytes in near real-time will make it possible to make faster diagnosis of disease in both the doctor's office and at remote locations in developing countries.

Acknowledgements

The authors gratefully acknowledge NIH and NSF (NINDS R01 NS042929 and R21 NS061202 and NSF CHE-1411993) for their financial support for many of the studies that were discussed in this chapter. We would also like to acknowledge Nancy Harmony for editorial assistance in the preparation of this manuscript.

References

1. A. Manz, N. Graber and H. M. Widmer, *Sens. Actuators, B*, 1990, **1**, 244–248.

2. S. C. Jacobson, R. Hergenroder, L. B. Koutny and J. M. Ramsey, *Anal. Chem.*, 1994, **66**, 1114–1118.
3. A. S. Johnson, A. Selimovic and R. S. Martin, *Anal. Bioanal. Chem.*, 2013, **405**, 3013–3020.
4. R. A. Saylor and S. M. Lunte, *J. Chromatogr. A*, 2015, **1382**, 48–64.
5. G. Chen, Y. Lin and J. Wang, *Curr. Anal. Chem.*, 2006, **2**, 43–50.
6. A. Martin, D. Vilela and A. Escarpa, *Electrophoresis*, 2012, **33**, 2212–2227.
7. J. J. P. Mark, R. Scholz and F.-M. Matysik, *J. Chromatogr. A*, 2012, **1267**, 45–64.
8. P. Kuban and P. C. Hauser, *Electrophoresis*, 2009, **30**, 3305–3314.
9. N. A. Lacher, K. E. Garrison, R. S. Martin and S. M. Lunte, *Electrophoresis*, 2001, **22**, 2526–2536.
10. H. Whatley, *Clinical and Forensic Applications of Capillary Electrophoresis*, ed. J. R. Peterson and A. A. Mohammad, Humana Press Inc., Totowa, NJ, 2001, ch. 2, pp. 21–58.
11. J. C. Giddings, *Unified Separation Science*, John Wiley & Sons, Inc., New York, 1st edn, 1991.
12. A. J. Bard and L. R. Faulkner, *Electrochemical Methods: Fundamentals and Applications*, John Wiley & Sons, Inc., New York, 2nd edn, 2001.
13. P. N. Nge, C. I. Rogers and A. T. Woolley, *Chem. Rev.*, 2013, **113**, 2550–2583.
14. H. Makamba, J. H. Kim, K. Lim, N. Park and J. H. Hahn, *Electrophoresis*, 2003, **24**, 3607–3619.
15. B. E. Slentz, N. A. Penner and F. Regnier, *Anal. Chem.*, 2002, **74**, 4835–4840.
16. S. V. Ermakov, S. C. Jacobson and J. M. Ramsey, *Anal. Chem.*, 2000, **72**, 3512–3517.
17. J. M. Karlinsey, *Anal. Chim. Acta*, 2012, **725**, 1–13.
18. K. Seiler, Z. H. Fan, K. Fluri and D. J. Harrison, *Anal. Chem.*, 1994, **66**, 3485–3491.
19. S. K. Griffiths and R. H. Nilson, *Anal. Chem.*, 2001, **73**, 272–278.
20. C. Chen and J. Hahn, *Environ. Chem. Lett.*, 2011, **9**, 491–497.
21. C. Chen and J. H. Hahn, *Anal. Chem.*, 2007, **79**, 7182–7186.
22. S. Attiya, A. B. Jemere, T. Tang, G. Fitzpatrick, K. Seiler, N. Chiem and D. J. Harrison, *Electrophoresis*, 2001, **22**, 318–327.
23. Y.-H. Lin, G.-B. Lee, C.-W. Li, G.-R. Huang and S.-H. Chen, *J. Chromatogr. A*, 2001, **937**, 115–125.
24. B. H. Huynh, B. A. Fogarty, R. S. Martin and S. M. Lunte, *Anal. Chem.*, 2004, **76**, 6440–6447.
25. M. W. Li and R. S. Martin, *Electrophoresis*, 2007, **28**, 2478–2488.
26. L. C. Mecker and R. S. Martin, *Anal. Chem.*, 2008, **80**, 9257–9264.
27. W. R. Vandaveer, S. A. Pasas-Farmer, D. J. Fischer, C. N. Frankenfeld and S. M. Lunte, *Electrophoresis*, 2004, **25**, 3528–3549.
28. P. T. Kissinger and W. R. Heineman, *Laboratory Techniques in Electroanalytical Chemistry*, Marcel Dekker., Inc, New York, 2nd edn, 1996.
29. M. K. Hulvey, S. M. Lunte, D. J. Fischer and C. D. Kuhnline, *Encyclopedia of Analytical Chemistry*, Wiley, web edn, 2010.

30. S. Creager, *Handbook of Electrochemistry*, ed. C. G. Zoski, Elsevier, San Diego, 1st edn, 2007, ch. 3, pp. 57–72.
31. P. H. Rieger, *Electrochemistry*, Chapman & Hall, New York, 2nd edn, 1994.
32. R. D. Rocklin, *A Practical Guide to HPLC Detection*, ed. D. Parriott, Academic Press, Inc., San Diego, 1993, ch. 6, pp. 145–173.
33. D. B. Gunasekara, M. K. Hulvey and S. M. Lunte, *Electrophoresis*, 2011, **32**, 832–837.
34. R. S. Martin, K. L. Ratzlaff, B. H. Huynh and S. M. Lunte, *Anal. Chem.*, 2002, **74**, 1136–1143.
35. N. M. Contento and P. W. Bohn, *Microfluid. Nanofluid.*, 2014, 1–10.
36. S. P. Forry, J. R. Murray, M. L. Heien, L. E. Locascio and R. M. Wightman, *Anal. Chem.*, 2004, **76**, 4945–4950.
37. D. J. Fischer, M. K. Hulvey, A. R. Regel and S. M. Lunte, *Electrophoresis*, 2009, **30**, 3324–3333.
38. Q. Guan and C. S. Henry, *Electrophoresis*, 2009, **30**, 3339–3346.
39. N. A. Lacher, S. M. Lunte and R. S. Martin, *Anal. Chem.*, 2004, **76**, 2482–2491.
40. C.-C. Wu, R.-G. Wu, J.-G. Huang, Y.-C. Lin and H.-C. Chang, *Anal. Chem.*, 2003, **75**, 947–952.
41. A. Selimovic, A. S. Johnson, I. Z. Kiss and R. S. Martin, *Electrophoresis*, 2011, **32**, 822–831.
42. A. S. Johnson, A. Selimovic and R. S. Martin, *Electrophoresis*, 2011, **32**, 3121–3128.
43. D. M. Osbourn and C. E. Lunte, *Anal. Chem.*, 2001, **73**, 5961–5964.
44. S. Park, S. M. Lunte and C. E. Lunte, *Anal. Chem.*, 1995, **67**, 911–918.
45. T. J. O'Shea, R. D. Greenhagen, S. M. Lunte, C. E. Lunte, M. R. Smyth, D. M. Radzik and N. Watanabe, *J. Chromatogr. A*, 1992, **593**, 305–312.
46. D. M. Osbourn and C. E. Lunte, *Anal. Chem.*, 2003, **75**, 2710–2714.
47. C.-P. Chen, W. Teng and J.-H. Hahn, *Electrophoresis*, 2011, **32**, 838–843.
48. D. Meneses, D. B. Gunasekara, P. Pichetsurnthorn, J. A. Fracassi da Silva, F. C. de Abreu and S. M. Lunte, *Electrophoresis*, 2014, **36**, 441–448.
49. D. E. Scott, R. J. Grigsby and S. M. Lunte, *ChemPhysChem*, 2013, **14**, 2288–2294.
50. A. Fernández-la-Villa, D. Sánchez-Barragán, D. F. Pozo-Ayuso and M. Castaño-Álvarez, *Electrophoresis*, 2012, **33**, 2733–2742.
51. A. Fernández-la-Villa, D. F. Pozo-Ayuso and M. Castaño-Álvarez, *Electrophoresis*, 2010, **31**, 2641–2649.
52. R. Kuldvee, M. Kaljurand and H. Smit, *J. High Resolut. Chromatogr.*, 1998, **21**, 169–174.
53. W. Lu and R. M. Cassidy, *Anal. Chem.*, 1994, **66**, 200–204.
54. N. D. Laude, C. W. Atcherley and M. L. Heien, *Anal. Chem.*, 2012, **84**, 8422–8426.
55. R. Bharadwaj, J. G. Santiago and B. Mohammadi, *Electrophoresis*, 2002, **23**, 2729–2744.

56. P. S. Cahill, Q. D. Walker, J. M. Finnegan, G. E. Mickelson, E. R. Travis and R. M. Wightman, *Anal. Chem.*, 1996, **68**, 3180–3186.
57. R. M. Wightman, *Science*, 2006, **311**, 1570–1574.
58. L. C. Mecker, L. A. Filla and R. S. Martin, *Electroanalysis*, 2010, **22**, 2141–2146.
59. C. E. Lunte, T. H. Ridgway and W. R. Heineman, *Anal. Chem.*, 1987, **59**, 761–766.
60. C. E. Lunte, P. T. Kissinger and R. E. Shoup, *Anal. Chem.*, 1985, **57**, 1541–1546.
61. D. A. Roston, R. E. Shoup and P. T. Kissinger, *Anal. Chem.*, 1982, **54**, 1417A–1434A.
62. M. Zhong and S. M. Lunte, *Anal. Chem.*, 1999, **71**, 251–255.
63. D. J. Fischer, W. R. I. V. Vandaveer, R. J. Grigsby and S. M. Lunte, *Electroanalysis*, 2005, **17**, 1153–1159.
64. R. S. Martin, A. J. Gawron, S. M. Lunte and C. S. Henry, *Anal. Chem.*, 2000, **72**, 3196–3202.
65. L. C. Mecker and R. S. Martin, *Electrophoresis*, 2006, **27**, 5032–5042.
66. M. K. Dorris, E. W. Crick and C. E. Lunte, *Electrophoresis*, 2012, **33**, 2725–2732.
67. S. M. Lunte, D. B. Gunasekara, P. Pichetsurnthorn and D. Meneses, *Proceedings of μTAS*, The Chemical and Biological Microsystems Society, 2014, pp. 39–42.
68. D. B. Gunasekara, Ph.D. Thesis, University of Kansas, 2014.
69. S. S. Ferris, G. Lou and A. G. Ewing, *J. Microcolumn Sep.*, 1994, **6**, 263–268.
70. F. D. Swanek, G. Chen and A. G. Ewing, *Anal. Chem.*, 1996, **68**, 3912–3916.
71. H. Fang, T. L. Vickrey and B. J. Venton, *Anal. Chem.*, 2011, **83**, 2258–2264.
72. A. J. Gawron, R. S. Martin and S. M. Lunte, *Electrophoresis*, 2001, **22**, 242–248.
73. R. N. Adams, *Anal. Chem.*, 1958, **30**, 1576.
74. R. S. Martin, A. J. Gawron, B. A. Fogarty, F. B. Regan, E. Dempsey and S. M. Lunte, *Analyst*, 2001, **126**, 277–280.
75. A. Regel and S. Lunte, *Electrophoresis*, 2013, **34**, 2101–2106.
76. M. L. Kovarik, N. J. Torrence, D. M. Spence and R. S. Martin, *Analyst*, 2004, **129**, 400–405.
77. J. Wang, B. Tian and E. Sahlin, *Anal. Chem.*, 1999, **71**, 5436–5440.
78. S. Ranganathan, R. McCreery, S. M. Majji and M. Madou, *J. Electrochem. Soc.*, 2000, **147**, 277–282.
79. S. Ranganathan and R. L. McCreery, *Anal. Chem.*, 2001, **73**, 893–900.
80. M. Pumera and A. Escarpa, *Electrophoresis*, 2009, **30**, 3315–3323.
81. J. Wang, *Electroanalysis*, 2005, **17**, 1133–1140.
82. J. W. Strojek, M. C. Granger, G. M. Swain, T. Dallas and M. W. Holtz, *Anal. Chem.*, 1996, **68**, 2031–2037.
83. D. Shin, D. A. Tryk, A. Fujishima, A. Muck, Jr., G. Chen and J. Wang, *Electrophoresis*, 2004, **25**, 3017–3023.

84. G. M. Swain and R. Ramesham, *Anal. Chem.*, 1993, **65**, 345–351.
85. J. Wang, G. Chen, M. P. Chatrathi, A. Fujishima, D. A. Tryk and D. Shin, *Anal. Chem.*, 2003, **75**, 935–939.
86. D. B. Gunasekara, J. M. Siegel, G. Caruso, M. K. Hulvey and S. M. Lunte, *Analyst*, 2014, **139**, 3265–3273.
87. A. T. Woolley, K. Lao, A. N. Glazer and R. A. Mathies, *Anal. Chem.*, 1998, **70**, 684–688.
88. C. Henry, S. Lunte and J. Santiago, *Anal. Commun.*, 1999, **36**, 305–307.
89. S. A. Pasas, N. A. Lacher, M. I. Davies and S. M. Lunte, *Electrophoresis*, 2002, **23**, 759–766.
90. D. B. Gunasekara, M. K. Hulvey, S. M. Lunte and J. A. Fracassi da Silva, *Anal. Bioanal. Chem.*, 2012, **403**, 2377–2384.
91. M. K. Hulvey, C. N. Frankenfeld and S. M. Lunte, *Anal. Chem.*, 2010, **82**, 1608–1611.
92. C. Antwi, A. S. Johnson, A. Selimovic and R. S. Martin, *Anal. Methods*, 2011, **3**, 1072–1078.
93. M. A. Schwarz and P. C. Hauser, *Anal. Chem.*, 2003, **75**, 4691–4695.
94. M. A. Schwarz, B. Galliker, K. Fluri, T. Kappes and P. C. Hauser, *Analyst*, 2001, **126**, 147–151.
95. D. E. Scott, S. D. Willis, S. Gabbert, D. Johnson, E. Naylor, E. M. Janle, J. E. Krichevsky, C. E. Lunte and S. M. Lunte, *Analyst*, 2015, **140**(11), 3820–3829.
96. A. L. Bowen and R. S. Martin, *Electrophoresis*, 2010, **31**, 2534–2540.
97. A. S. Johnson, B. T. Mehl and R. S. Martin, *Anal. Methods*, 2015, **7**, 884–893.
98. J. Wang, M. P. Chatrathi and B. Tian, *Anal. Chem.*, 2000, **72**, 5774–5778.
99. C. D. García and C. S. Henry, *Anal. Chem.*, 2003, **75**, 4778–4783.
100. E. Piccin, N. Dossi, A. Cagan, E. Carrilho and J. Wang, *Analyst*, 2009, **134**, 528–532.
101. A. Hilmi and J. H. T. Luong, *Anal. Chem.*, 2000, **72**, 4677–4682.
102. A. Hilmi and J. H. T. Luong, *Environ. Sci. Technol.*, 2000, **34**, 3046–3050.
103. J. Wang, J. Zima, N. S. Lawrence, M. P. Chatrathi, A. Mulchandani and G. E. Collins, *Anal. Chem.*, 2004, **76**, 4721–4726.
104. J. L. Felhofer, L. Blanes and C. D. Garcia, *Electrophoresis*, 2010, **31**, 2469–2486.
105. O. Chailapakul, S. Korsrisakul, W. Siangproh and K. Grudpan, *Talanta*, 2008, **74**, 683–689.
106. M. J. A. Shiddiky, K.-S. Lee, J. Son, D.-S. Park and Y.-B. Shim, *J. Agric. Food Chem.*, 2009, **57**, 4051–4057.
107. N. Dossi, R. Toniolo, A. Pizzariello, S. Susmel and G. Bontempelli, *Electrophoresis*, 2011, **32**, 906–912.
108. M. Hervas, M. A. Lopez and A. Escarpa, *Analyst*, 2011, **136**, 2131–2138.
109. N. Kovachev, A. Canals and A. Escarpa, *Anal. Chem.*, 2010, **82**, 2925–2931.
110. A. Fernandez-la-Villa, V. Bertrand-Serrador, D. F. Pozo-Ayuso and M. Castano-Alvarez, *Anal. Methods*, 2013, **5**, 1494–1501.

111. A. G. Crevillén, M. Pumera, M. C. González and A. Escarpa, *Electrophoresis*, 2008, **29**, 2997–3004.
112. A. Bani-Yaseen, T. Kawaguchi, A. Price, C. Culbertson and R. Jankowiak, *Anal. Bioanal. Chem.*, 2011, **399**, 519–524.
113. W. Wang, F. F. u. Fu, X. Xu, J.-M. Lin and G. Chen, *Electrophoresis*, 2009, **30**, 3932–3938.
114. N. Ruecha, W. Siangproh and O. Chailapakul, *Talanta*, 2011, **84**, 1323–1328.
115. R. Liang, L. Wang, X. Meng, J. Wang and J. Qiu, *Microfluid. Nanofluid.*, 2011, **11**, 227–233.
116. J. Yan, X. Yang and E. Wang, *Electroanalysis*, 2005, **17**, 1222–1226.
117. M. d. P. Godoy-Caballero, M. I. Acedo-Valenzuela, T. Galeano-Diaz, A. Costa-Garcia and M. T. Fernandez-Abedul, *Analyst*, 2012, **137**, 5153–5160.
118. M. Garcia and A. Escarpa, *Electrophoresis*, 2014, **35**, 425–432.
119. M. L. Kovarik, A. J. Dickinson, P. Roy, R. A. Poonnen, J. P. Fine and N. L. Allbritton, *Integr. Biol.*, 2014, **6**, 164–174.
120. M. L. Kovarik, P. K. Shah, P. M. Armistead and N. L. Allbritton, *Anal. Chem.*, 2013, **85**, 4991–4997.
121. E. C. Metto, K. Evans, P. Barney, A. H. Culbertson, D. B. Gunasekara, G. Caruso, M. K. Hulvey, J. A. Fracassi da Silva, S. M. Lunte and C. T. Culbertson, *Anal. Chem.*, 2013, **85**, 10188–10195.
122. D. M. Omiatek, Y. Dong, M. L. Heien and A. G. Ewing, *ACS Chem. Neurosci.*, 2010, **1**, 234–245.

CHAPTER 4

Scanning Electrochemical Microscopy (SECM): Fundamentals and Applications in Life Sciences

ANGELIKA HOLZINGER, CHARLOTTE STEINBACH AND CHRISTINE KRANZ*

Institute of Analytical and Bioanalytical Chemistry, University of Ulm, 89069 Ulm, Germany
*Email: christine.kranz@uni-ulm.de

4.1 Introduction

The different techniques nowadays applied in life sciences may be considered as individual instruments in a symphony orchestra, each providing different valuable information. Fundamental questions are addressed regarding biomolecules, biomolecule-modified surfaces, live cells and complex biological functions such as cell signaling cascades, influences on cell proliferation, gene expression and cell death. Techniques such as optical microscopy, electrophoresis, chromatographic techniques bulk or on-chip electrochemical measurements and spectroscopic techniques are among the approaches providing bulk information usually averaging over a large number of biological entities. However, for most of the listed techniques either modification or complexing agents may be necessary and/or the obtained information cannot be correlated to structural changes. Fluorescence-based

RSC Detection Science Series No. 6
Electrochemical Strategies in Detection Science
Edited by Damien W. M. Arrigan

Published by the Royal Society of Chemistry, www.rsc.org

and high-resolution optical techniques provide spatially resolved information down to individual molecules (*e.g.*, single molecule fluorescence) but usually require labeling steps.[1] Scanning probe microscopy (SPM) techniques such as atomic force microscopy (AFM),[2] scanning electrochemical microscopy (SECM)[3] and scanning ion conductance microscopy (SICM)[4] yield valuable information when investigating biological samples in respect to topographical and structural analysis of, for example, cells, yet some of them lack chemical and molecular specificity. In particular electrochemical methods[5,6] play a dominant role in studying signaling processes as many transmitter molecules are either electroactive molecules (*e.g.*, catecholamines)[7] or can be selectively determined using biosensors.[8] Ideally, the detection of specific constituents and the response to stimulation and/or changes of the biological sample should be obtained in a temporally and spatially resolved manner. SECM, as introduced by Bard and co-workers,[9] is an attractive scanning probe technique for life sciences and related research areas, which was already demonstrated by early investigations on biological samples[10,11] and first enzyme activity-related investigations presented in 1992.[12] Since then, SECM evolved into an increasingly popular technique for studying biochemical and bio-related processes. Significant progress has been made over the years in instrumental developments, by introducing new imaging modes and establishing comprehensive theoretical models. While the early years of SECM were certainly shaped by the team of A. Bard and the research groups emerging from this nucleus, not much later research groups in Japan[13–15] and Europe[16–24] contributed to SECM research in the field of life sciences. In the early twenty-first century, SECM was improved in respect to resolution, introducing new imaging modalities and SECM research expanded to the investigation of DNA,[25–27] cells,[28,29] membranes[30,31] and neurons.[32] Returning to the metaphor of an orchestra, the musical development in allegro was not just limited to its leitmotif of SECM, but combinations with other scanning probe techniques such as AFM and SICM or optical techniques enriched the Symphony. Within this chapter an overview on SECM is provided along with the imaging modalities on biologically relevant applications in the life sciences and related research areas with selected examples. As this chapter cannot be comprehensive, the interested reader is directed further to the seminal book *Scanning Electrochemical Microscopy*.[33]

4.2 Instrumentation

In general, the SECM hardware for investigating biologically relevant samples consists of the same basic components as for other SECM applications; however, if needed the instrument can be located in an incubator environment, *e.g.*, for studying live cells over a longer period of time. Details on SECM instrumentation are described in detail elsewhere,[34] yet basic components are summarized herein. The hardware consists of the following components: a positioning system, the electrochemical cell with the investigated sample implemented in the cell base plate and a (bi)potentiostat for controlling the electrochemical experiment. Positioning of the micro- or nanosized electrode, frequently termed as the 'probe' (this term will be used throughout the chapter), above a sample is controlled by a stepper motor and/or piezo elements. Next to the sample, the SECM cell consists of probe, counter and reference electrodes; typically a bipotentiostat is used to control the probe and sample potential (if the experiment requires a biased sample). Positioning and scanning the micro- or nanoelectrode across the sample surface and correlating the recorded electrochemical signals with the position of the probe is achieved by an appropriate software package. For imaging biological samples, in particular cells, SECM can be combined with an inverted optical microscope (as shown in Figure 4.1A), which allows location of the biological entity,[35] or with fluorescence microscopy,[36] providing complementary information. For combined optical microscopy-SECM, Schuhmann and co-workers introduced the term "Bio-SECM".[37]

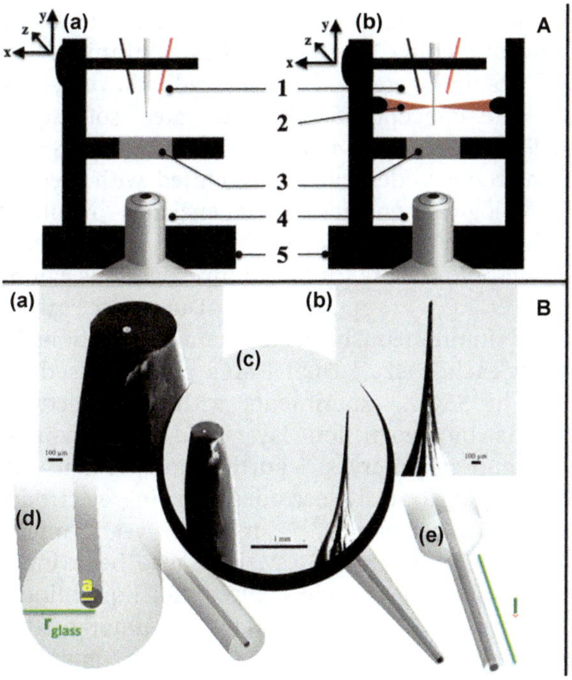

Figure 4.1 Scheme of Bio-SECM using constant height and constant distance approach. (A) (a) Schematic view of the Bio-SECM setup in constant height, and (b) in constant distance mode using optical shear force detection. 1: electrochemical cell with reference electrode, counter electrode and microelectrode as working electrode; 2: optical path of shear force detection; 3: sample; 4: objective of the inverted optical microscope; 5: SECM mount. (B) Different microelectrode geometries: (a) and (d) disc UME, indicating the RG of the electrode, (b) and (e) fiber-shaped electrode with characteristic length (c) comparison of the different electrode designs. (C) Imaging regimes: (a) constant height mode; (b) constant distance mode.

The unique behavior of microelectrodes, including reduced double-layer charging effects, well-defined steady-state currents and a reduced *iR* drop, allows their application as scanning probes.[38–40] The geometrical characteristics of the micro- or nanoelectrodes (electrode radius, shape, radius of insulation sheet and RG value: ratio of the radius of the insulating sheath and the electrode radius, see Figure 4.1B) are of particular importance in respect to the achievable lateral resolution. Gold, platinum and carbon are the predominantly used electrode materials for SECM probes. Usually, disc microelectrodes with electroactive radii ranging from 2.5 to 12.5 μm are readily fabricated by sealing the microwire or fiber in glass or embedding in resin using standard fabrication processes (Figure 4.1B (a) and (d)).[38] For dimensions smaller than 25 μm (diameter), the expression ultramicroelectrode (UME) is used within the SECM community. Carbon-fiber disc microelectrodes are frequently used for imaging biological processes as the electrode surface can be activated for species which require an adsorption step for electron transfer.[41] Sharpening methods such as using flame,[42] spark,[43] electrochemical etching,[44] argon ion beam thinning,[45] or gallium ion beam milling[46] result in electrodes with radii down to the nanometer range. Recently, the concept of microfabricated soft stylus microelectrodes and microelectrode arrays has been introduced for SECM imaging.[47,48] Additionally, microelectrodes can be modified with electrocatalytic layers or with biological entities, *e.g.*, enzymes resulting in microbiosensors.[49,50] If the enzymes are located inside a micropipette, non-electrode immobilized microbiosensors for SECM investigations can be obtained,[51] enabling a wide range of applications. Dual disc microelectrodes[52] are used to detect simultaneously electrochemical reactions by applying different potential at each UME.[53] Dual UMEs are also used for positioning microbiosensors in SECM experiments with one electrode serving as transducer for the biorecognition layer and the second—bare—electrode for recording approach curves.[54] Furthermore, various different electrode geometries such as conical electrodes and ring electrodes have also been applied in SECM experiments.[53–60] In recent years, micropipettes with orifices <300 nm adapted from SICM have also been introduced for SECM experiments as ion-selective electrodes,[61] for liquid/liquid interface studies of ion transfer or electron transfer[62] or—among other applications—in so-called scanning micropipette contact methods (SMCM),[63] and scanning electrochemical cell microscopy (SECCM).[64]

Microfibers may be also used in NSOM (near-field scanning optical microscopy) as fiber optics,[65] which are modified with an additional microelectrode, for example, as ring-shaped electrode located around the fiber, thereby combining SECM with near-field optical microscopy.[66–68] Matsue and co-workers reported on scanning chemiluminescence microscopy (SCLM)[69] and combined SECM/SCLM,[70] while Ueda *et al.*[71] showed a triple combination of scanning techniques merging SECM, AFM and NSOM functionality into a single probe. The combination techniques

SICM–SECM and AFM–SECM will be discussed in detail in Sections 4.5.1 and 4.5.2, respectively.

4.2.1 Positioning Modes in SECM

Traditionally, the SECM probe is positioned at a distance of several electrode radii above the sample surface by recording current–distance curves (approach curves) with a redox mediator added to solution exhibiting fast electron transfer characteristics. For imaging experiments, the probe is then scanned in a fixed *x,y*-plane across the sample surface, which is frequently called constant height mode (Figure 4.1C (a)). The working distance in SECM is strongly linked to the radius of the SECM probe as the signal generation is based on diffusion behavior towards the electrode influenced by the presence of the sample. For example, imaging DNA-modified sample surface using thiol-chemistry on a gold surface is readily achieved in constant height mode.[26] For samples with considerable changes in height and large aspect-ratio features like cells, the constant height mode may lead to falsified interpretation of the electrochemical results due to a convolution of topographical information and electrochemical activity. To avoid such convoluted signals, several current-independent positioning modes have been introduced over the years, which was recently reviewed.[72] For constant distance SECM imaging, feedback control mechanisms as derived from other scanning probe techniques such as NSOM, AFM and SICM have been implemented to maintain a constant probe-to-sample distance. The first approach was derived from shear force-based NSOM measurements,[73] which could be adapted to SECM measurements.[74,75] Fiber-shaped SECM probes (Figure 4.1.B (b) and (e))[74,76] are required for shear force detection as the electrode is agitated to horizontal vibration and the change of amplitude in close distance to the sample is used as the feedback signal. Optical-based shear force mode[51] and non-optical shear force mode using tuning fork[77,78] or piezoelectric plates[79] for detecting shear forces have been optimized over the years. For imaging biological samples, conventional and shear force mode SECM was combined with optical microscopy as shown in Figure 4.1.A (a) and (b).[35,37,80,81] Combined optical microscopy-SECM was demonstrated, *e.g.*, with optical shear force positioning, where a laser beam is focused at the tapered end of the microelectrode and a photodiode is used to record the vibration amplitude of the SECM tip (schematically shown in Figure 4.1A (b)).[37] In contrast to conventional SECM, the sample was moved in the *x,y*-plane and an additional piezo-driven positioner was used to move the tip in the *z*-direction. The optical shear force control suffers from stability issues, which may lead to loss of the feedback signal. Hence, most shear force-based SECM are nowadays based on a tuning fork or using two piezo plates for shear force detection. Besides shear force-based positioning, other approaches such as constant current[82] or constant impedance mode[83] and combinations of SECM with complementary scanning probe techniques (predominantly AFM and SICM) have been reported for studies of biologically relevant samples. Shear force-based constant distance mode SECM was used to visualize the oxygen consumption of single living cells in correlation with the distance of the tip to the cell surface.[84] Although hopping intermittent contact-SECM[85] or soft stylus electrodes[47] and electrode arrays[48] have not yet been demonstrated for imaging life cells, these approaches certainly have potential and are attractive for constant-distance SECM measurements.

Bio-SECM for investigating living cells, an organism or delicate samples such as enzymes may require controlled environmental conditions including temperature control,[86–88] fixed CO_2 content or oxygen-reduced atmospheres.[89] In addition, stimulation experiments such as changes in pH,[90] the addition of stimulating agents,[35,80] and mechanically induced stress,[91,92] require extensions of the SECM setup with micropipettes and/or cell/tissue stretching devices as shown in Figure 4.2A. Enzymatic reactions and cellular processes are typically temperature dependent. Hirano *et al.*[86] conducted cell measurements to study how antifreeze proteins can protect cells from hypothermic damage. They monitored the cell morphology over 24 h, while the cells were incubated at 4 °C. Schuhmann and co-workers also implemented a module for varying and controlling the temperature over a range of 0–100 °C during SECM measurements.[87] The temperature module consists of a Peltier element, a temperature controller and sensor (Figure 4.2B). By shielding the heating plate with a μ-metal foil, electronic noise impeding the measurements was reduced. However, the heating pulses still led to pulsation with the measured amperometric current and variation in the tip-to-sample distance, which required a synchronized data recording in respect to the temperature pulses. The enzymatic activity of glucose oxidase (GOx) was exemplarily shown in the sample generation/tip collection mode detecting hydrogen peroxide (H_2O_2) locally at enzyme spots in the temperature range 12–37 °C. The H_2O_2 concentration is independent of the glucose concentration due to a saturated substrate concentration in solution, and hence, can be directly correlated to the enzyme activity. Though a lower temperature results in increasing resolution, the enzyme activity is significantly increased with temperature (Figure 4.2B). As reported by Amemiya and co-workers, temperature control is also a prerequisite for accurate positioning of nanoelectrodes suppressing thermal drifts in SECM experiments.[88]

4.3 SECM Imaging Modes for Life Sciences

Several operation modes have been introduced in SECM, focusing on local surface modification[93] (direct mode) or mapping the (electro)activity of the investigated sample surface or determining homogenous and heterogeneous electron transfer rates. Feedback (FB)[94] and generation–collection (GC)[95] mode are the two predominant modes for investigating biologically relevant samples ranging from DNA-modified surfaces to mapping enzyme activity and studying life cells. Quite recently, for mapping local oxygen consumption, a new mode referred to as redox competition mode has been introduced.[96]

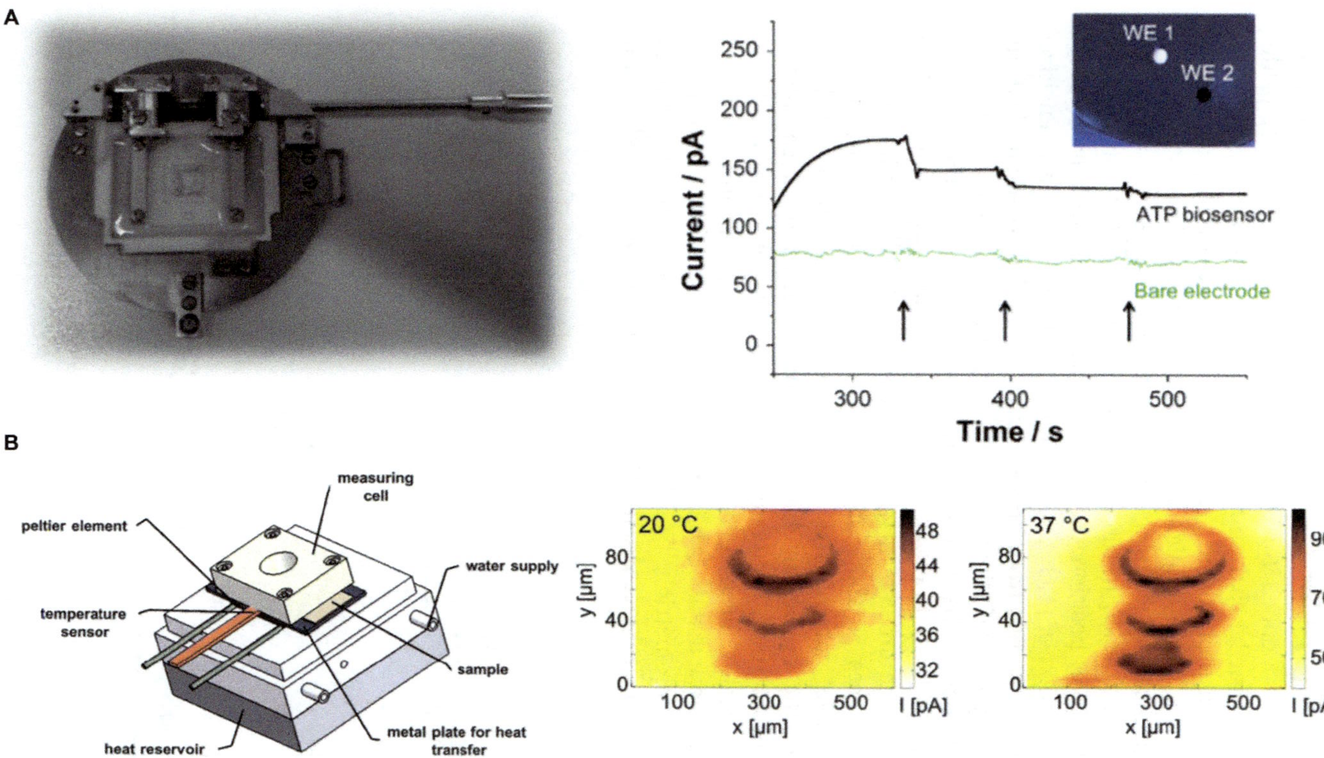

Figure 4.2 (A) Motorized stretching device that can be implemented in a SECM setup. Exemplary current *versus* time curve simultaneously recorded in dependence of different stretching levels at MC3T3-E1 osteoblastic cells with a dual biosensor assembly. (Reprinted from ref. 92 with permission from Elsevier.) The current recorded at the bare electrode is not changing significantly, thereby suggesting that no other redox-active molecules are co-released. (B) Scheme of the thermoelectrically controlled SECM cell; generator–collector mode SECM images recorded at different temperatures at GOx spots immobilized at silanized glass. (Reprinted from ref. 87 with permission from the Royal Society of Chemistry.)

4.3.1 Generation–Collection Mode

Static generation–collection experiments using a UME were among the first SECM-type experiments.[97,98] Substrate generation–tip collection (SG/TC) mode is frequently applied for imaging the enzyme activity, *e.g.*, oxido-reductases or enzyme-labeled immunosorbent assays (ELISA) (Figure 4.3A). In SG/TC mode, the probe detects redox- or electroactive species generated at the sample surface. For example, immobilized glucose oxidase catalyzes the oxidation of glucose to gluconolactone using oxygen as an electron acceptor, which is reduced to H_2O_2 in the enzymatic conversion. H_2O_2 can then be oxidized at the UME positioned at the enzyme spot. As initially no background concentration of H_2O_2 is present, the SG/TC mode is very sensitive; however, it usually suffers from limited lateral resolution, particularly if the enzyme spots are not sufficiently spaced.[99] In addition, quantitative treatment of SG/TC data is based on steady state conditions of the recorded signal and, hence, is only possible if the imaged structures (micro-sized domains) are of the order of magnitude of the UME used for imaging.[100] Another issue arises at short probe-to-sample distances ($d < 5r$), as the recorded faradaic current may reflect contributions from feedback currents when the biological entities are immobilized on a conductive substrate.[101] In addition, the collection efficiency is low compared to tip generation–substrate collection (TG/SC) mode, which shows an efficiency of almost 100% at close distances. Listing these limitations, it has to be noted that, besides amperometric[12,19] probes, UMEs modified with enzymatically active layers (microbiosensors)[49,54,102] electrocatalytic layers[103] and

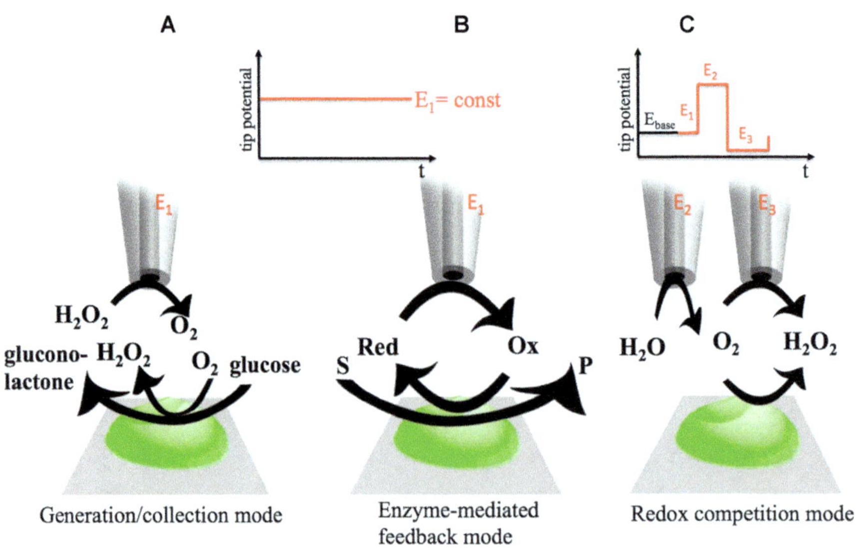

Figure 4.3 Schemes of SECM modes in life science applications. (A) Generation–collection mode; (B) enzyme-mediated feedback mode; (C) redox competition mode.

potentiometric[104–106] probes can be employed. In particular, potentiometric probes are suitable for SG/TC mode as the probe is a passive sensor. An approximation for the sensitivity in GC mode mapping the enzyme activity has been given by Horrocks and Wittstock:[101]

$$k_{\mathrm{cat}}\Gamma_{\mathrm{enz}} \geq \frac{Dc'}{r_{\mathrm{s}}} \tag{4.1}$$

where k_{cat} is the turnover number of the enzymatic conversion, Γ_{enz} is the surface concentration of the enzyme, D is the diffusion coefficient of the generated electroactive species, c' is the collection efficiency of the probe and r_{s} is the radius of the sample.

Although some limitations are listed here, the generation–collection mode is frequently used in SECM mapping of enzyme activity, membrane transport, release of signaling molecules at single cells or respiratory activity (oxygen consumption) of living cells.

4.3.2 Feedback Mode

Feedback mode is the most commonly applied operation mode in SECM. Positioning of the probe above the specimen is achieved in a feedback-type experiment recording the current in response to the *z*-distance to the sample surface.[94,107] Electron transfer studies, characterization of conductive and non-conductive samples, corrosion studies, membranes and coatings as well as electrocatalytic layers have been investigated in feedback mode. In feedback experiments, a redox mediator is added to solution in its oxidized or reduced form and a sufficient potential is applied to the probe to convert the electroactive species. Due to hemispherical diffusion towards the UME, the steady state current is given by:

$$i_{T\infty} = 4nFDcr \tag{4.2}$$

where $i_{T\infty}$ is steady state current, n is the number of transferred electrons, F is the Faraday constant, D and c are the diffusion coefficient and concentration of the redox mediator, respectively, and r is the radius of the disc microelectrode. If the UME is moved towards the sample surface, the nature of the sample and the distance will influence the concentration profile at the UME, leading either to a reduced current (negative feedback effect) or increased current due to regeneration of the redox mediator at the sample surface (positive feedback effect).[94]

For mapping the morphology of living cells, the negative feedback effect has been employed using a hydrophilic redox mediator that does not penetrate the cell membrane and does not react with cell metabolites. Changes in the distance between the investigated biological entity (*e.g.*, cell with a certain height) and the probe, which is scanned in constant height across the sample surface, will lead to a negative feedback effect reflecting the cell topography. Although this has been experimentally demonstrated, notably, the achievable resolution is still not comparable with other scanning probe microscopy techniques (*e.g.*, AFM) nowadays used in life sciences, therefore limiting the utility to mere studies on cell topology. A significant advantage of SECM compared to other SPM techniques, however, is that SECM allows the detection of metabolic species, and in addition the probe does not contact the sample. This is particularly interesting for delicate proteins such as enzymes, the activity of which may suffer significantly as a result of an immobilization process. Early studies of mapping enzyme activity of oxidoreductases in feedback mode have been presented by the Bard group.[12,108] The enzyme-mediated feedback mode (Figure 4.3B) takes the kinetics of the mediator regeneration based on the Butler–Volmer equation and the kinetics of the enzymatic reactions described by the Michaelis–Menten kinetics into account. Due to the limited sensitivity in feedback mode, only highly active enzymes or surfaces with high enzyme coverage can be imaged in FB mode. Increasing the concentration of the mediator is therefore limited by the catalytic reaction rate according to eqn (4.3),[12] as well as by the size of the UME:

$$k_{cat}\Gamma_{enz} \geq 10^{-3}\frac{Dc}{r_T} \tag{4.3}$$

where k_{cat} is the turnover number of the enzymatic conversion, Γ_{enz} is the surface concentration of the enzyme, D and c are the diffusion coefficient and concentration of the redox mediator, respectively, and r is the radius of the probe. The kinetics are still dependent on enzymatic reaction rate, yet related to the detection limit of the redox-active species, which is detected at the UME. Although the resolution is improved compared to GC mode, the FB mode can only be used for imaging oxidoreductases.

Besides mapping enzyme activity, the feedback mode has been demonstrated in numerous applications including imaging relevant cellular function of living cells.

4.3.3 Redox Competition Mode

Schuhmann and co-workers introduced the so-called redox competition (RC) mode, which is based on the competition reaction of probe and sample for the same species.[96] Predominantly, this mode is used for screening catalytic activity, *e.g.*, for heterogeneous oxygen reduction, where high background signals due to large sample areas present an experimental challenge. Imaging the respiratory activity of single cells also reflects a fairly complex problem, in particular if the cells are imaged in constant height mode. Changes in the oxygen concentration within the gap between probe and cell are influenced by several factors such as depletion and change of the sample-to-probe distance.[109] RC mode overcomes such problems as oxygen is generated at the UME on applying a potential pulse initiating water electrolysis followed by a pulse sequence for the redox competition. Hence both sample and probe consume the generated species and the time dependent current decay curve recorded at the probe provides information on the local oxygen consumption (Figure 4.3C). In this mode, the current response is no longer dependent on the sample size (in comparison to SG–TC mode) but is dominated by the probe radius.

4.4 Applications

As covered in several recent reviews,[18,21,110,111] SECM is an attractive analytical technique in the life sciences providing spatial and temporal resolution of biological relevant processes without requiring labeling. In addition, SECM is a non-invasive technique, which does not induce unwanted processes due to, for example, force interaction. Furthermore, experiments can be conducted in buffered solution (*i.e.,* at physiologically relevant conditions) in controlled environments. From early SECM experiments investigating modified carbon paste electrodes in 1989,[10] SECM has evidently matured into a routine tool applied to studying biological systems.

4.4.1 Mapping Enzyme Activity

Glucose oxidase, EC 1.1.3.4 is one of the most commonly investigated enzymes in SECM experiments. The first SECM experiments of mapping GOx activity were published by Bard and co-workers using the enzyme-mediated feedback mode.[12] The FB mode can only be applied for oxidoreductases and in case of high turnover rates (see Section 4.3.1). Over the years, a multitude of enzymes, multi-enzyme structures such as alkaline phosphatase, urease, galactosidase, glucose dehydrogenase, diaphorase, horseradish peroxidase and lactate oxidase, have been studied in FB and GC mode (Table 4.1 below). In addition to mapping enzyme activity immobilized at a solid substrate, enzymes can also be immobilized at the surface of an UME, resulting in microbiosensors, which then may be used as probes in SECM experiments for species requiring high selectivity, or for species that cannot be directly detected at a bare/unmodified electrode.

Enzymes are widely used in enzyme-based sensing platforms such as electrochemical, optical or mass sensitive biosensors, but also in other analytically relevant formats such as enzyme-linked immunosorbent assays. For sensing approaches, the immobilization process is a crucial step as it may lead to deactivation of the enzymatic component and/or may lead to an inhomogeneous distribution of the enzyme within the sensor structure. Enzymes have been immobilized at various substrates using physisorption, embedding in organic polymers, hydrogels and inorganic matrices (entrapment), microencapsulation, crosslinking and *via* covalent attachment. The direct adsorption of lactate oxidase onto glassy carbon electrodes or highly ordered pyrolytic graphite (HOPG) using AFM for characterizing the morphology and SECM for mapping enzyme activity has been studied by Parra *et al.*[112] For example, GOx was immobilized at active Langmuir–Blodgett films crosslinked with glutaraldehyde to glass slides,[113] and at

Table 4.1 Selected examples of enzyme-related SECM experiments.[a]

Enzyme	Substrate	Imaging mode	Positioning mode/experimental details (if not otherwise noted positioning was in CH mode)
Glucose oxidase (GOx) EC 1.1.3.4	Gluc	GC	Patterning and imaging of enzyme spots,[19] enzyme-filled micropipettes as biosensors,[51] polymer-entrapped biosensor,[131] imaging co-immobilized structures,[119] mapping activity at different temperatures[87]
		FB	GOx immobilized at insulating substrates,[12] GOx immobilized at functionalized polypyrrole[16]
		FB and GC	Patterning and imaging of enzyme spots[132]
Horseradish peroxidase (HRP) (EC 1.11.1.7)	H_2O_2	FB	HRP immobilized at pores of polycarbonate membrane,[133] microstructured enzyme spots,[134–136] HRP as label for antibody[137,127] and DNA sensing[138]
		GC	Immobilization *via* SAM,[118] multi-enzyme devices,[139] HRP as label for antibody[14,125,140] and DNA sensing,[141] imaging microbiosensor[49]
Microperoxidase (MP-11)	H_2O_2	FB	Microstructured enzyme spots, electron-transfer[17]
Urease (EC 3.5.1.5)	Urea/NH_4^+	GC	Potentiometric antimony UME,[104] ion-selective UME[105]
Laccase (EC 1.3.10.2)	O_2	RC	Mediatorless bioelectrocatalytic dioxygen reduction[142]
		FB	Immobilized at sol–gel processed silicate films[89]
		GC	"Wired enzymes" *via* redox polymers[143,144]
PGG-dependent glucose dehydrogenase (EC 1.1.99.17)	Gluc/malt	FB	Microstructured enzyme spots[145]
	PAPG	GC/FB	Co-immobilization with galactosidase,[121] immobilized at magnetic microbeads[146]
Alcohol dehydrogenase (EC 1.1.1.1)	Ethanol	FB	Immobilized *via* a redox hydrogel[147]
Diaphorase (EC 1.6.99.1)	NADH	FB	Microstructured enzyme spots,[13] CD mode: immobilized at Au[148] and Pt array microelectrodes[78]
Galactosidase (EC 3.2.1.93)	PAPG	GC	Immobilized at magnetic beads,[99,149] label for DNA sensing[150]
Alkaline phosphatase (EC 3.1.3.1)	*para*-Aminophenol	GC	Label for antibody[123,124] and DNA sensing[151]
		FB	Label for DNA sensing[152]
Lactate oxidase (EC 1.3.10.2)	Lactate	FB	Immobilization at glassy carbon electrodes[153]

[a]CH: constant height mode; CD: constant distance mode; GC: generation/collection mode, FB: feedback mode, RC: redox competition mode. Gluc: glucose; Malt: maltose; H_2O_2: hydrogen peroxide, NH_4^+: ammonium ion, PAPG: *p*-aminophenyl-β-D-galactopyranoside, NADH: nicotinamide adenine dinucleotide.

Au electrodes using conductive polymer films.[16] The high chemical affinity between sulfur and gold is exploited for covalent attachment of sulfur-containing biological entities. Self-assembled monolayers (SAM) are frequently applied for immobilizing enzymes due to the high stability and ordered orientation. SAM layers were formed from alkoxysilanes,[13] alkanethiols[19] and other organo-sulfur compounds.[114–116] In addition, thin films of silicate have been used for immobilization. Laccase EC 1.10.3.2 and bilirubin oxidase (BOD) EC 1.3.3.5, which catalyze the reduction of oxygen, have been immobilized in a sol–gel process.[89] The activity of the enzymes was studied in FB mode using $K_4[Fe(CN)_6]$ as redox mediator. Individual spots of immobilized laccase and bilirubin oxidase activity (spot size approx. 50 μm in diameter; distance between spots approx. 130 μm) could be investigated in as single experiment (Figure 4.4).

In particular, GOx is frequently used as a model enzyme immobilized *via* self-assembly. For example, microstructured GOx spots were generated by SECM-based electrochemical desorption of a long-chain alkanethiolate layer, which was then modified with cystaminium dihydrochloride (ω-functionalized thiol) for selective binding of GOx.[117] SECM in generator–collector mode was used to map the enzyme activity of the generated microspots. GOx

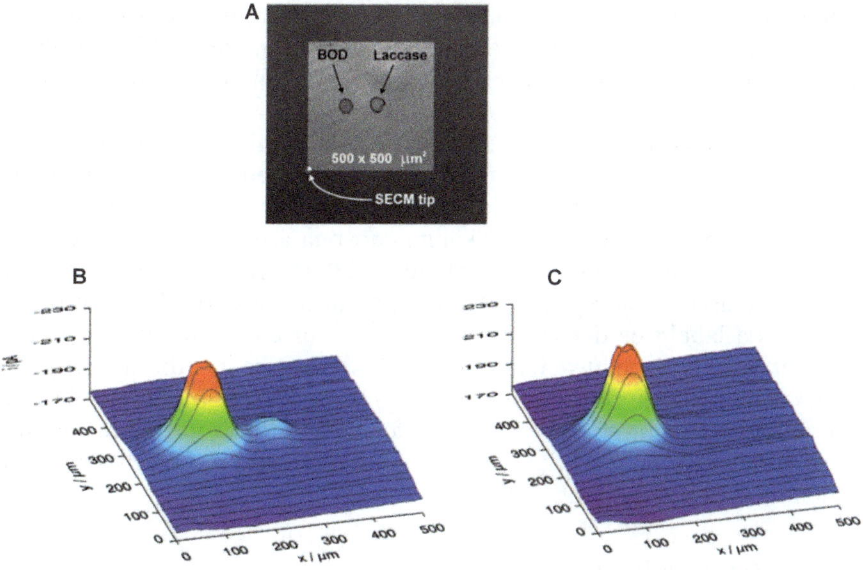

Figure 4.4 Confocal laser scanning microscope (CLSM) and SECM images of BOD and laccase microspots immobilized *via* a sol–gel process. (A) Reflection mode CLSM image with the UME visible in the bottom left corner of the image. (B) SECM feedback images of both enzyme spots in aerated 0.1 mM $K_3[Fe(CN)_6]$/0.1 M phosphate buffer solution (pH 4.8). (C) SECM feedback images of the BOD spot in aerated 0.1 mM $K_3[Fe(CN)_6]$/0.1 M phosphate buffer solution (pH 7.0). Images were recorded with a gold UME (25 μm in diameter), which was biased at 0.0 V *versus* Ag/AgCl/KCl (3 M).
(Reprinted from ref. 89 with permission from the Royal Society of Chemistry.)

and horseradish peroxidase (HRP) EC 1.11.1.7 were immobilized *via* aminopropyltriethoxysilane (APTES) on titanium dioxide layers.[118] Microstructured surfaces containing different enzymes (GOx and HRP) have been generated in a combination of microcontact printing and local electrochemical desorption.[119] Multi-enzyme structures have also been demonstrated *via* spotting small enzyme droplets using a flow-through microdispenser,[120] whereby different immobilization schemes were used such as covalent attachment using functionalized self-assembled thiol monolayers or enzyme multilayers by crosslinking or entrapping the enzymes in a polymer film. A successful combination of FB and GC mode was demonstrated by Zhao and Wittstock,[121] improving the detection of galactosidase (Gal) EC 3.2.1.93 activity using glucose dehydrogenase (GDH) EC 1.1.99.17 for an amplification effect. It is difficult to map the activity of Gal, which is frequently used for labeling,[122] with sufficient lateral resolution as the enzyme-mediated FB mode cannot be used. In GC mode, the activity of Gal is detected *via* the conversion of *p*-aminophenyl-β-D-galactopyranoside (PAPG) into *p*-aminophenol (PAP). PAP is oxidized to *p*-quinone imine (PQI) at an applied potential of 400 mV *versus* Ag/AgCl at the UME. As PQI is a cofactor of the enzyme GDH, co-immobilized GDH converts PQI into PAP in the presence of D-glucose. The measured current at the UME reflects the oxidation of PAP from the Gal-catalyzed reaction (GC mode) and the contribution from the GDH-catalyzed reaction (FB mode), which results in signal amplification due to the FB contribution, which only occurs in a small region defined by the location of the GDH.

ELISAs play an important role in biomedical applications. Similar to the development of biochips containing immobilized enzyme spots, mapping the enzyme activity of enzyme-labeled proteins has also been successfully demonstrated with SECM.[14,123] Alkaline phosphatase,[124] which catalyzes the hydrolysis of *p*-aminophenyl-phosphate (PAPP) to PAP and HRP,[14,125–128] is frequently used as label in SECM-ELISA studies. Recently, SECM was proposed as a label-free detection method[129,130] for clinically relevant targets with different molecular weights such as PSA (prostate specific antigen), NTx (marker for collagen degradation) and ciprofloxacin (antibiotics). Individual dotted arrays of biotinylated antibodies for each of the targets were obtained *via* biotinylated poly(ethylene imine) and neutravidin, which covalently binds the biotinylated antibodies.

4.4.2 DNA Hybridization

The development of DNA microarrays and their high-throughput read-out play an important role in disease diagnostic, sequence mapping and gene expression. The standard read-out of DNA microchips is based on fluorescence-labeling or radio-marker labeling. However, within recent years electrochemical methods have been proposed as a viable alternative due to their reduced costs, sensitivity and possibility of label-free detection. DNA hybridization has also been studied by SECM using nanoparticle labels,[154,155] enzyme labels[141,150,151] (GC mode), redox-labels[156] and electroactive intercalators.[157,158] In addition, SECM has the advantage of label-free detection based on, for example, direct oxidation of guanine base,[159] using repulsive interaction between a negatively charged redox species and DNA leading to a negative feedback effect.[160–162] Different immobilization schemes for the capture DNA at solid substrates, including thiol-chemistry on gold, amine-tethering on glass, and immobilization using conductive polymers, have been applied. Szunerits and co-workers[138] combined fabrication (direct mode SECM) and detection steps by copolymerization of pyrrole/pyrrole-oligonucleotide (ODN) strands. Hybridization was monitored using a molecular assembly approach including the biotinylated complementary strand, which then reacts with streptavidin and biotinylated HRP forming a sandwich assay. In the presence of 4-chloro-1-naphthol, insoluble 4-chloro-1-naphthon is formed driven by the enzymatic catalyzed conversion of H_2O_2, which consecutively leads to a change of the conductivity at the DNA spots due to the precipitation 4-chloro-1-naphthon, and hence to a negative feedback effect. Another SECM investigation of a sandwich-type DNA microarray with improved sensitivity and sequence specific hybridization was shown by Fan *et al.*[141] The capture probe was immobilized at aldehyde covered glass slides *via* microprinting and hybridization with target DNA and biotinylated indicator probes resulted in a "sandwich" DNA structure. SiO_2 nanoparticles modified with streptavidin and HRP served as the detection element. The enzyme-catalyzed reduction of H_2O_2 in the presence of hydroquinone (H_2Q) leads to the formation of benzoquinone (BQ), which then can be re-oxidized at the SECM probe leading to an enhanced current.

Kraatz and Bard[161] investigated the electron transfer rate constant for ferricyanide, a negatively charged redox species, across a self-assembled monolayer of thiol-modified DNA duplexes immobilized at a gold electrode in the absence and presence of different bivalent ions (Zn^{2+}, Ca^{2+}, Mg^{2+}). An enhancement was not only observed with Zn^{2+}, which in fact is known to form metallized DNA, but also for earth alkali ions. Kraatz and co-workers also published the detection of single nucleotide mismatches in DNA arrays[162] based on the increased repulsion of the negatively charged redox mediator in the absence of base-pair mismatches. SECM imaging and approach curves were recorded at the double stranded (ds)-DNA spots, thereby reflecting the complementary sequence containing single-nucleotide mismatches or a combination of multiple mismatches (Figure 4.5A). The apparent electron transfer rate constants were determined based on modeling of the experimental data using a finite element method. The lowest k values were observed for the fully matched ds-DNA spots.

The same group also demonstrated the discrimination between mitochondrial (mt)-DNA of different animals[163] and even the discrimination between animals of the same subfamily using SECM and electrochemical impedance spectroscopy.[164] Two 5′-disulfide-modified 51-*mer* mt-DNA fragments of the Cox1 gene of *Bos taurus* (Holstein cow) and *Bison bison* (North American buffalo) were used as the capture strands. After pre-hybridization

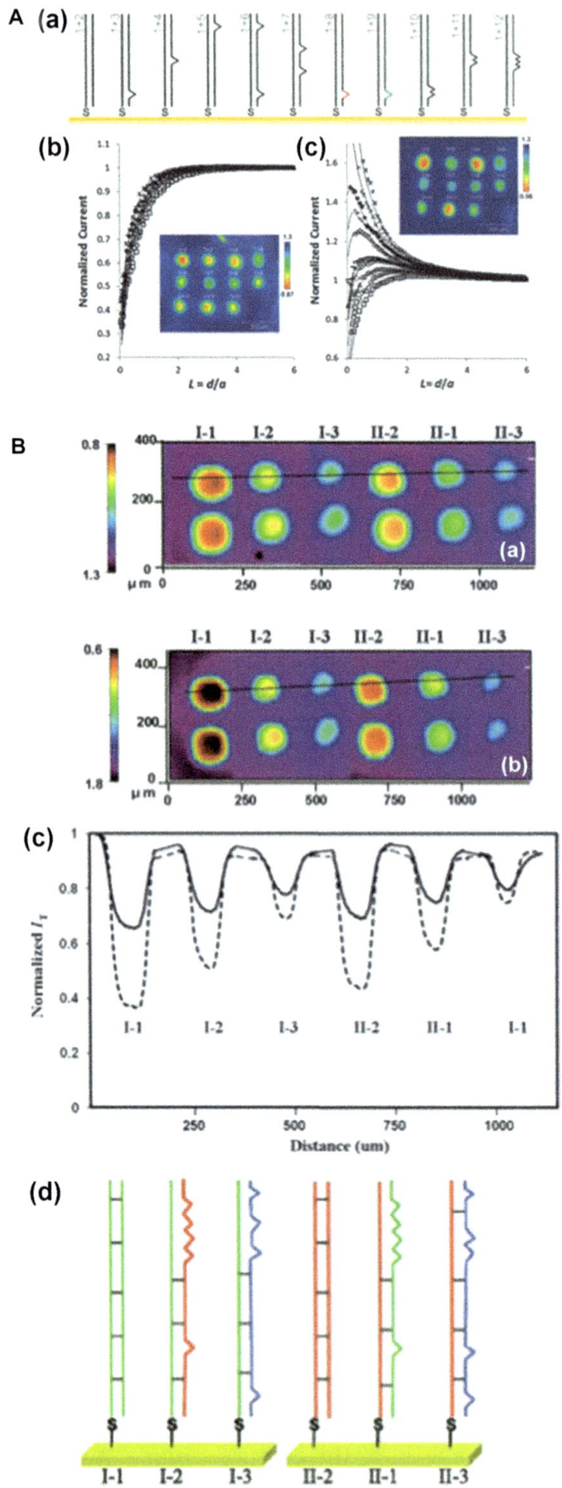
A
(a)
(b)
Normalized Current
L = d/a
(c)
Normalized Current
L = d/a
B
I-1
I-2
I-3
II-2
II-1
II-3
(a)
μm
(b)
(c)
Normalized I_T
Distance (um)
(d)

with the complementary strands of the *Bos taurus, Bison bison* and *Bison bonasus* (European buffalo), the different ds-DNA were spotted onto a clean gold surface. Figure 4.5B shows the corresponding SECM images in the presence (a) and absence of Zn^{2+} ions (b), line scans (c) and sequences (d). Clearly, the fully matched duplexes revealed the lowest current response. The presence of Zn^{2+} ion amplifies the obtained current signal. It was also possible to differentiate between positions of mismatches of base pairs in ds-DNA with equal[165] and unequal[166] strand sizes.

4.4.3 SECM Investigations of Cells

Since the introduction of SECM, a multitude of different studies involving live cells have been demonstrated with this scanning probe technique. As already pointed out in the introduction of this chapter, SECM has several unique advantages for studying live cells and cellular processes. Besides its non-invasive character and the fact that cells can be studied in buffered solution in controlled environments, localized micro-electrochemistry allows study not only of cell morphology, but also of the release of signaling molecules, respiratory activity, cell metabolism, membrane transport and monitoring responses towards drug administration. The investigated cells range from bacteria to plant, animal and human cells. Significant effort has been targeted towards the investigation of cell viability by monitoring the respiratory activity or distinguishing normal cells from tumor cells. In recent years, several excellent reviews and book chapters have been published studying cells with SECM.[18,111,167–169] Table 4.2 gives an overview of selected SECM investigations of different mammalian cell types.

Figure 4.5 SECM imaging of DNA hybridization. (A) (a) Scheme of mismatches and their relative positions. (b) Normalized approach curves observed above individual ds-DNA spots in the absence of Zn^{2+} and (c) normalized approach curves observed above individual ds-DNA spots in the presence of Zn^{2+}. Solid lines represent simulated approach curves. Respective SECM images recorded over ds-DNA microarrays are shown as insets in (b) and (c). (Reprinted from ref. 163 with permission from the Royal Society of Chemistry.) (B) SECM images recorded at homo (I-1), and (II-2) and hetero DNA duplexes (I-2, I-3), (II-1), and (II-3) before (a) and after (b) the addition of Zn^{2+}. The color scale bar shows the intensity of the current. (c) Current profile extracted from the ds-DNA spots before (solid) and after (dashed) addition of Zn^{2+} (indicated by the black line in (a) and (b)). (d) Schematic of the homo (matched) and hetero (mismatched) DNA duplexes immobilized on gold surfaces *via* Au–S linkage. Homo (I-1) *Bos taurus* probe and complementary strand, hetero (I-2) *Bos taurus* probe with *Bison bison* complementary, hetero (I-3) *Bos taurus* probe with *Bison bonasus* complementary, homo (II-2) *Bison bison* probe and complementary, hetero (II-1) *Bison bison* probe with *Bos taurus* complementary, and hetero (II-3) *Bison bison* probe with *Bison bonasus* complementary strands.
(Reprinted from ref. 164 with permission from the Royal Society of Chemistry.)

Table 4.2 Examples of live cell studies using SECM.[a]

Cell type	Positioning mode	Imaging mode	Experimental details
Epithelial-like human colon cells (SW-480), single cell	CH	GC	Mapping respiratory activity[173]
Breast epithelial (MCF-10A) normal and metastatic breast cells (MDA-MB-231), carcinoma (MCF-7) multiple cells, single cells	CH	FB	Transmembrane charge transfer;[29,177,178] intracellular measurements with nanoelectrodes[179]
Neurons (PC12) adrenal medulla single/multiple cells	CH, CD	GC, FB, CC, AC	Neurotransmitter secretion,[187] morphology mapping,[32,83] respiratory activity,[172] FSCV, respiratory activity[176,188]
Erythroleukemia cell line (K562/ADM), multiple cells	CH	GC	Drug resistance, fluorescence microscopy[170]
Cervix carcinoma (HeLa), single/multiple cells	CH	GC, FB, CC	Respiratory activity,[174,189] permeability of Ag NP[190]
Human umbilical vein endothelial cells (T-HUVEC), single cells	CH, CD	GC, RC	NO release,[191,192] respiratory activity[109]
Bladder (T24), renal carcinoma (A498), single cell	CH	GC	Release of ROS[193]
Epidermoid carcinoma (A431), single cell	CH	GC, FB	Imaging of membrane proteins on cell surface[194]
Gastric carcinoma (BGC-823), single cell	CH	GC	*In situ* imaging of membrane glycan motifs[195]
Hepatoblastoma (HepG2), single cell	CH	GC	Topography in absence of NfeAFP,[86] induced apoptosis[196]
Human embryonic kidney cells (HEK293), multiple cells	CD	RC	Visualization of oxygen consumption[84]
Mouse leukemia macrophage cell macroRAW 264.7, multiple cells	CH	GC	Cell adhesion promotion on surfaces[197]
Adrenal gland cell, single animal cell	CD	GC	Improved imaging of cells with carbon fiber electrodes[37]
Newborn rat osteoclasts	CH	FB	Calcium disposal during osteoclast activity[198]
Animal embryonic stem cells, single cell, multiple cells	CH	GC	Evaluation of differentiation status[199]

Table 4.2 (*Continued*)

Cell type	Positioning mode	Imaging mode	Experimental details
Animal fibroblasts (CRL-2752, Cos-7)	CD, CH	GC, AC	MP-supported ITIES for evaluation of silver ion toxicity in living cells,[200] topography of cells[80]
Madin–Darby canine kidney strain II, multiple cells	CH	GC	Imaging passive transport across cell monolayers with sub-cellular resolution[201]
Rat kidney epithelial cell (NRK-52E), multiple cells, single cell	CH	FB, LSCM	Imaging passive transport across cell monolayers,[201] imaging and chemical attack of single living cell[90]
Single bovine embryos, multiple cells	CD	GC	Respiratory activity[175,202]

[a]AC: alternating current, CC: constant current, CD: constant distance, CH: constant height, FB: feedback, GC: generation/collection, MP: micropipette, LSCM: laser scanning confocal microscopy, ROS: reactive oxygen species.

Live cells have been investigated in generation–collection mode, feedback mode and redox competition mode in constant height and in constant distance regime, respectively. Note that most cells exhibit topologies in the micrometer regime with cell heights that, in addition, may change over the time course of the experiment. Hence, SECM in constant distance mode is preferred for imaging experiments, although both constant height and constant distance approaches can be found in the literature. Whereas most constant distance SECM studies of living cells are performed in shear force mode, impedance-based positioning of the SECM probe has also been demonstrated.[80,83] Matsue and co-workers embedded cells in microwells, which allows not only high-throughput but may also minimize effects due to cell topology in constant height mode operation.[170] Cell measurements have been performed in static experiments (Figure 4.6A, B) or in imaging experiments (Figure 4.6C, D). Thereby the release of messenger molecules above the cell is investigated, or cellular processes within the cell are monitored using nanoelectrodes, or the topography, or cell reactivity of individual cells or cell assemblies, is mapped, respectively. Frequently, SECM cell measurements are combined with optical and fluorescence microscopy as described in Section 4.2 to locate the micro- or nanoelectrode at individual cells or to obtain simultaneously optical information using fluorescence techniques.

4.4.3.1 Imaging Respiratory Activity and Redox Activity

Respiratory activity and photosynthetic activity have been extensively studied by SECM. First reports investigating *Ligustrum sinensis*, Elodea[11] and guard

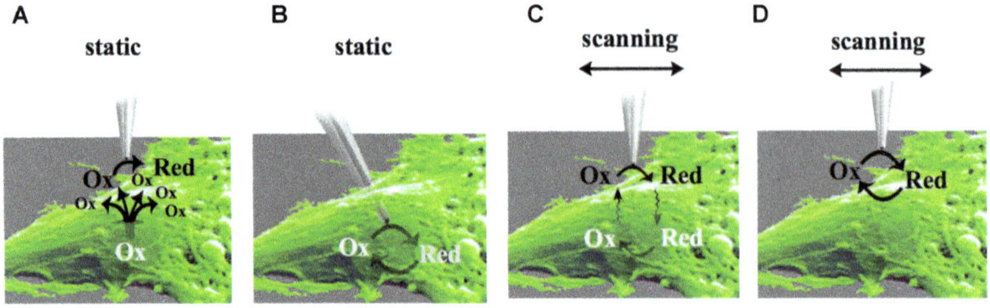

Figure 4.6 Schemes of SECM experiments for investigating live cells. (A) Static mode with the SECM probe positioned at the cell surface recording molecules released from the cell. (B) Static mode with a pointed SECM nanoprobe positioned within the cell recording redox activity. (C) Imaging of cellular activity with redox mediators, which can cross the cell membrane. (D) Imaging of cell topography in feedback mode with redox mediators that cannot cross the cell membrane (negative FB).

cells of *Tradescantia flumentis*[171] were demonstrated by Bard and co-workers. Oxygen consumption is an important parameter for cell viability that is capable of providing information on the health state of individual cells or the effectiveness in drug screening experiments. For example, Matsue and co-workers published a series of papers investigating the respiratory activity of different cell types such as PC12 cells,[172] SW-480 cells,[173] HeLa cells,[174] human erythroleukemia cell line (K562) and the same drug resistant cell line (ADM/K562)[170] and single bovine embryos.[175] Oxygen consumption is usually measured by monitoring the oxygen reduction current at the SECM probe, which is scanned in close proximity across the cells. The current value decreases at the UME due to the oxygen uptake by the cells and serves as the measure for respiratory activity. Recently, Schuhmann and co-workers addressed issues associated with mapping the respiratory activity in constant height mode when the oxygen reduction is measured at a constant SECM probe potential.[84] High background signals associated with only small current variations impede an accurate measurement. In addition, the generated concentration gradients due to the oxygen reduction at the UME may induce transmembrane diffusion of oxygen across the cell membrane leading to an altered local oxygen concentration. To overcome such problems, the authors conducted shear force-based constant distant measurements at human embryonic kidney cells (HEK293) using the redox competition mode to avoid the depletion of oxygen in the small gap between the cell surface and the SECM probe.[84,109] Figure 4.7 shows a constant height mode SECM image of adherently grown Retzius cells imaged in redox competition mode.[109]

The oxygen consumption of PC12 cells as model system has been investigated in alternating current (AC)-SECM in combination with fast scan cyclic voltammetry.[176] In general, the redox activity of cells can be used as a measure to distinguish between nontransformed and metastatic human breast cells with SECM as demonstrated in a collaborative effort by Mirkin and Rotenberg.[29,177–179] Measurements were performed in feedback mode

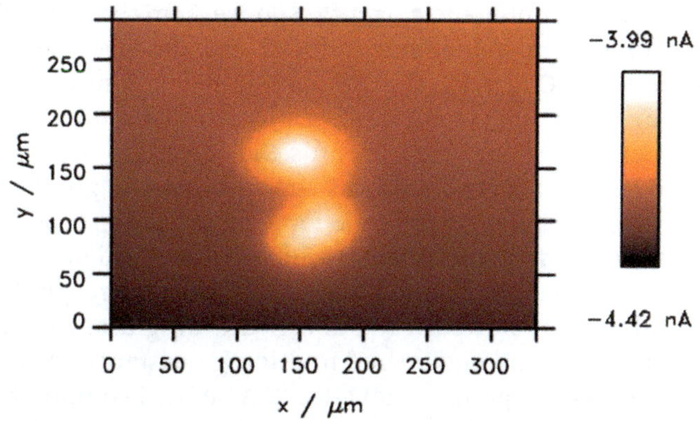

Figure 4.7 Redox competition SECM image recorded in constant height mode at growing Retzius cells (adherent cells) from medicinal leech. (Reprinted from ref. 109 with permission from the Royal Society of Chemistry.)

and approach curves were recorded to investigate mechanistic pathways and to determine apparent electron transfer rates for different redox mediators. For example, the reduction of menadione at the SECM probe generated menadiol, which is a lipophilic redox species able to cross the cell membrane. Inside the cell, the mediator can be re-oxidized by intercellular redox processes. Hence, in comparison to a hydrophilic redox mediator like ferrocene carboxylate, which results in a negative feedback due to hindered diffusion, menadione results in an increased current when scanning the tip across the cell surface. Interestingly, a smaller positive feedback effect and associated smaller apparent electron transfer rate (obtained from evaluation of approach curves) was observed when comparing a metastatic breast cancer cell with a nontransformed cell, which was attributed to high endogenous levels of protein kinase Cα and allows identification of tumorous cells. However, the measurements were performed in oxygen-reduced solution, which may affect on the cell behavior; in addition, the cytotoxic effect of menadione was not investigated in depth during these studies. The same team demonstrated measurements within individual cells using sharp fiber-shaped nanoelectrodes (radii of tens of nanometers and RG values of 3) with hydrophobic and hydrophilic redox species.[179] Mass/charge-transfer rates for hydrophobic redox mediators across the cell membrane could be determined. In addition, nanoelectrodes were used to image the cell morphology of human breast epithelial cells (MCF-10A) in constant current mode, attempting to achieve high-resolution images similar to AFM measurements. Generally, sensing intracellular redox enzyme activity or cell surface enzyme activity provides important information on the metabolic vitality of cells, which was the focus of several studies.[35,180–184] For example, Matsue and co-workers used a double mediator system to determine the intracellular NAD(P)H : quinone oxidoreductase (NQO) activity of single HeLa cells using menadione and ferrocyanide as redox mediators.[183] The double-mediator system was used to amplify the current response from the intracellular NQO activity and to reduce menadione-induced cell damage. Besides the investigation of the NQO activity, the optimum in menadione concentration in respect to minimizing its cytotoxic effect was also determined.

Significant efforts are targeted towards the screening of drugs and chemotherapeuticals in cancer therapy. As shown by several groups, SECM—as a non-invasive technique—is attractive for such screening experiments (*e.g.*, studying anticancer drug resistance), which has been demonstrated on a single and multiple cell level mainly in monitoring the respiratory activity. Mauzeroll and co-workers recently published a drug assessment study investigating HeLa cells and engineered multidrug resistance related protein 1 (MRP1) overexpressing variants (HeLa-R).[185] A defined co-immobilization of both cell lines was secured by using a structured substrate to ensure the same environmental conditions for all cells, and to allow scanning of the same cell spots (Figure 4.8A and B). The redox mediator ferrocenemethanol

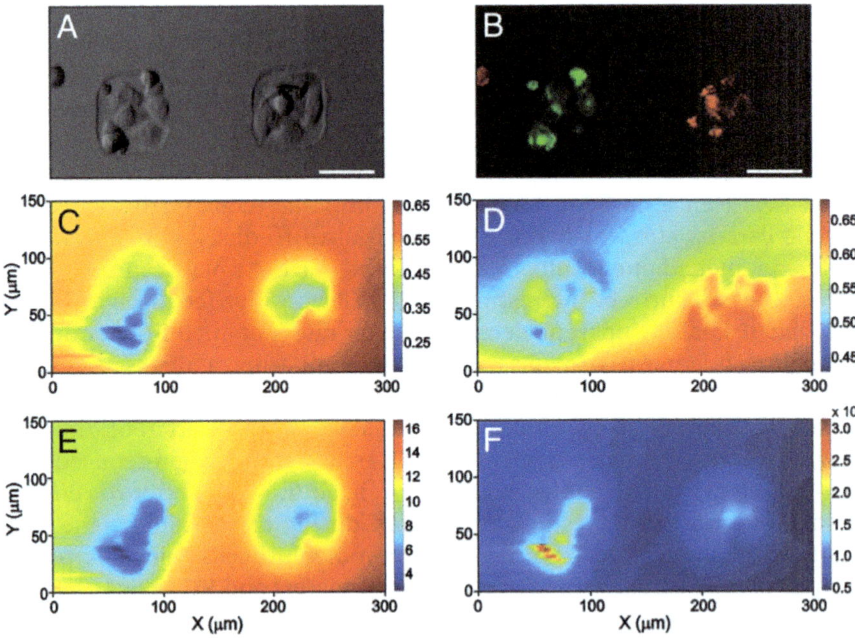

Figure 4.8 SECM images displaying the decoupled feedback response for a HeLa and HeLa-R cell co-culture substrate. (A) Optical image of a co-culture pattern containing HeLa-R cells (left) and HeLa cells (right). (B) Fluorescence image of the same sample shown in (A), with HeLa-R cells stained green and HeLa cells stained red (C and D, respectively). Normalized SECM currents recorded at the sample shown in (A) with the UME positioned at a distance of 12 μm above the substrate in 1 mM $[Ru(NH_3)_6]$ (C) and 1 mm $FcCH_2OH$ (D). (E) Extracted normalized tip-to-substrate distance profile. (F) Profile of the extracted apparent heterogeneous rate constant (cm s^{-1}). (Scale bar: 50 μm)
(Reprinted from ref. 185 with permission from the National Academy of Sciences of the United States of America.)

($FcCH_2OH$) was chosen to quantify MRP1 activity of these two cell lines. Multidrug resistance involves the overexpression of transmembrane proteins P-glycoprotein and (MRP1), which act as molecular "pumps" actively removing therapeutic agents from cancer cells. The current measured at the SECM probe when scanning the cell spots is related to the local concentration of $FcCH_2OH$, which is influenced by the topography of the cell (diffusion limitation), the permeability of the cell membrane to $FcCH_2OH$, and the glutathione-dependent regeneration of $FcCH_2OH$. $FcCH_2OH$ inside the cells changes the intracellular glutathione disulfide levels, producing an excess of glutathione (GSH), which is expelled from the cell by MRP1 and further increases the local concentration of $FcCH_2OH$ by reducing at the tip generated $[FcCH_2OH]^+$. The influence of the topography on the overall current response was evaluated by imaging the same cell spots with a redox mediator that cannot cross the cell membrane.

Earlier work focused on constant height imaging of PC12 cells using negative feedback mode before and after exposing them to nerve growth factor, hypotonic or hypertonic buffered conditions. The cells differentiate into a neuron phenotype by growing narrow neurites, which could be visualized using carbon fiber electrodes. In addition, changes in height of the cells due to variations in the tonicity could be observed.[32] The changes in cell morphology of rat kidney epithelial cells (NRK-52E) was monitored in constant height mode with sub-micrometer Pt electrodes.[186] The selective attack of single cells with local generated hydroxide ions resulted in a change of the morphology by necrosis due to the pH increase whereas the reference cells remained unchanged.

4.4.3.2 *Release of Signaling Molecules*

Microelectrochemistry has a long history in life sciences for *in vivo* measurements of oxygen and later on for *in vivo* and *in vitro* studies of signaling molecules. Cellular functions are regulated by messenger molecules such as hormones, neurotransmitters, growth factors, *etc.*, which typically alter the intracellular level of secondary messenger molecules inducing further signaling cascades. Messenger molecules are mainly released *via* exocytosis, thereby delivering zepto- to attomoles of these species at a millisecond time scale. Molecules are frequently stored within vesicles, which then fuse with the cell membrane, subsequently releasing their content into the extracellular environment. Upon stimulation, the time scale of such exocytotic events is dependent on the cell type and ranges from milliseconds (*e.g.*, neurotransmitters) to minutes (*e.g.*, release of surfactant molecules from lung epithelial cells). Most of the messenger molecules are either electroactive or detectable using appropriately modified electrodes. Amperometric and voltammetric measurements are typically obtained with carbon-fiber micro or nanoelectrodes, (less at Pt micro or nanoelectrodes), or chemically modified microelectrodes (electrocatalytic or biosensing layers), which are positioned in the vicinity of the cell surface. Hence, measurements are typically carried out with the microelectrode stationary above the cell surface. In contrast to such stationary experiments, SECM enables the mapping of the cell topography, as well as simultaneously recording release events. However, notably, for fast exocytotic release, SECM experiments are also carried out in a stationary fashion but with the advantage that the distance between probe and cell can be accurately determined. In terms of mapping the release of signaling molecules, SECM has been used for studying, for example, catecholamines such as dopamine, nitric oxide (NO), reactive oxygen and nitrogen species (ROS and RNS, respectively) and adenosine triphosphate (ATP). As discussed in the previous section of this chapter, SECM is a highly attractive analytical technique to study cellular processes such as the export of cytotoxic constituents from cells.[180] In another study, shear force mode was optimized to map the cell morphology of PC12 and Retzius cells using carbon fiber UMEs;[187] however, notably, in comparison to AFM the achievable resolution was not sufficient to record details of the cell morphology. After recording the morphology, the carbon fiber microelectrode was positioned above the cell center and the secretion of neurotransmitter measured amperometrically in a stationary experiment. Through an elevated K^+ concentration, the depolarization of the cell membrane and an influx of Ca^{2+} ions was induced, which consecutively led to the discrete secretion of neurotransmitters (in the case of PC12 cells a mixture of adrenaline, dopamine, noradrenaline). Differentiated and undifferentiated PC12 model neurons were imaged with improved lateral resolution in SECM constant current or constant impedance[83] mode. Again, after recording the topography the carbon ring electrode (1 μm in diameter) was positioned above the cell and 105 mM K^+ was added to depolarize the cell membrane, evoking vesicular release. Changing the ionic strength impedes the impedance-based imaging and therefore constant distance imaging could not be maintained during stimulation experiments. Hence, the release of neurotransmitter was recorded during recording line scans in constant height mode.

The importance of nitric oxide (NO) as an endothelial-derived relaxation factor (EDRF) is documented by the Nobel Prize in Physiology or Medicine awarded to R. G. Furchgott in 1998. Altered NO levels are involved in hypertension, hyperglycaemia, arteriosclerosis, Parkinson's disease, Alzheimer's disease, arthritis, reperfusion injury and cancer.[203] NO can be directly oxidized at platinum electrodes using a Clark's type electrode.[204] Nowadays electrocatalytically modified microelectrodes are among the most attractive approaches for measuring NO in biological systems, which was first demonstrated by Malinski and Taha by coating a carbon fiber electrode with nickel porphyrin for *in situ* amperometric detection of NO.[205] A comprehensive review on electrochemical NO detection was recently published by Bedioui and Griveau.[206]

Schuhmann and collaborators published a series of papers mapping the NO release at human umbilical vein endothelial cells (HUVEC) with SECM.[191,192,207–209] NO measurements at HUVEC cells have been initially performed with a UME modified with pyrrole-functionalized Mn(II) trimethoxyphenylcarboxyphenyl porphyrin (Mn(II)triOMePCPPyP).[207] The modified microelectrode was positioned *via* a micromanipulator above the cells and the release was monitored in a stationary experiment. More accurate positioning was obtained with a dual microelectrode approach where one electrode was used for positioning (recording the oxygen reduction current) while the second electrode was modified with Ni(4-*N*-tetramethyl)pyridyl porphyrin immobilized in negatively charged electrodeposition paint. After positioning the electrode assembly HUVEC cells were stimulated with bradykinin administered with a micropipette and NO release was detected at different distances.[208] The combination of such a dual microelectrode assembly with an automatic robotic system allowed on-line drug screening within 96-well microtiter plates.[210] As the sensitivity of the reported modified NO sensors is limited, most experiments have been performed with microelectrodes (approx. 50 μm in diameter). Significantly smaller electrodes allowed NO detection at the single cell level using shear force-based SECM. Disc-shaped carbon-fiber electrodes (7–9 μm in diameter) were modified with platinum and consecutively with a layer of Ni tetrasulfonate phthalocyanine tetrasodium salt (Ni-TSPc). The selectivity was improved by coating the electrodes with a Nafion film, thereby repelling negatively charged ions such as nitrite, which is among the main interfering species for NO measurements in biological systems. These electrodes have been successfully applied in measuring the NO release at transformed HUVEC cells.[209] A further miniaturization of a NO sensor was demonstrated by Shim and Lee.[211] They fabricated cone-shaped nanopore Pt electrodes with a pore opening radius smaller than 1 μm and pore depth of approximately 30 μm. The etched Pt pores were further electrochemically modified with Pt and poly(5-amino-1-naphthol) to achieve the required sensitivity and selectivity for NO detection. The performance of the NO sensor as SECM probe was evaluated in imaging the local NO concentration of an NO-emitting microstructured film (diazeniumdiolate-doped silicone rubber) on a glass substrate.

Besides NO, ATP is a ubiquitous signaling molecule inducing autocrine and paracrine responses in almost all mammalian cells.[212] As the direct electrochemical conversion of ATP requires high potentials and leads to electrode fouling, ATP biosensors have been introduced for sensitive and selective ATP measurements. Miniaturized ATP sensors have been demonstrated using bienzymatic assays either employing glycerol kinase and glycerol-3-phosphate oxidase[213] or a competitive assay based on GOx and hexokinase,[50] which was derived from an approach introduced by Pfeiffer and Scheller using a Clark-type electrode modified with an enzyme-containing membrane.[214] ATP transport through track-etched polycarbonate membranes was imaged with an ATP microbiosensor based on the competitive assay.[54] Positioning of the ATP microbiosensor was based on a dual electrode assembly with one bare microelectrode for recording approach curves and the second electrode modified with the biosensor layer. The mechanically induced release of ATP was measured at osteoblastic cells (MC3T3-E1) in a stationary SECM experiment (see Figure 4.2A).[92] Localized

ATP measurements at carotid bodies isolated from young rats could also be demonstrated during normoxic hypercapnia and hypoxia.[215]

4.4.4 Biofilms and Bacteria

SECM studies also focus on bacteria or biofilms; however, less literature is published in this domain. Gram positive and negative bacteria such as *Staphylococcus aureus,*[216] *Rhodobacter sphaeroides,*[217] *Salmonella typhimurium*[218] and *Escherichia coli*[219–222] have been studied. For example, SECM studies of *E. coli* focused on respiratory activity of collagen-embedded bacteria. The antimicrobial effect of silver ions was studied mapping the respiratory activity of *E. coli* with a Clark microelectrode.[219] In a later study Bard and co-workers investigated the mechanism of silver ion uptake in collagen-embedded *E. coli* and also fibroblast cells.[221] Silver is one of the most widespread used nanomaterials in consumer products due to its antibacterial properties. However, concerns have been raised regarding the resistance of bacterial strains against silver ions. The uptake of silver ions was enhanced when administering 4-aminopyridine, a voltage-sensitive potassium ion channel blocker. As a result, the cell viability was reduced, which was determined by the respiratory activity of the investigated cells. A micropipette-based Ag^+ ion-selective electrode positioned at the vicinity of the bacterial cells allowed mapping of the Ag^+-uptake. Bard and co-workers also investigated the hydrogen peroxide concentration above bacterial biofilms found in dental plaque. Co-cultured biofilms of *Streptococcus gordonii* (Sg), *Aggregatibacter actinomycetemcomitans* (Aa) and mutant Aa (deficient in catalyze KatA) were investigated in generator–collector mode.[223] Sg is a Gram-positive bacterium, which converts sugars into lactic acid, producing hydrogen peroxide in the presence of oxygen. Aa is a Gram-negative bacterium, which is associated with localized aggressive periodontitis. Given the fairly large spots with millimeter dimensions, only line scans were recorded across the biofilms (Figure 4.9B). The authors could demonstrate that Aa cocultured with Sg detoxifies the Sg-produced hydrogen peroxide levels due to KatA, a protein of the catalyse enzyme family. Line scans clearly revealed a decrease in H_2O_2 levels above the Aa spot and in comparison higher levels of hydrogen peroxide above Aa mutants (current decrease of approx. 0.66 above Aa and only 0.99 decrease in current above mutant Aa in comparison to the current for the Sg region, see Figure 4.9E). Interestingly, the recorded H_2O_2 levels in the vicinity of the biofilms determined by SECM were more than an order of magnitude higher (mM range) compared to the H_2O_2 concentrations measured in the supernatant solution using a fluorometric assay.

Recent SECM studies have focused on the investigation of soil bacteria such as *Rahnella aquatilis* RA1,[224] and aquatic bacteria such as *Marinobacter aquaeolei*[225] and *Vibrio fischeri, a* bioluminescent marine bacterium.[226] Microbial metal respiration is central to a wide variety of environmentally significant processes, including the biogeochemical cycling of metals, trace elements and phosphate, degradation of natural and contaminant organic

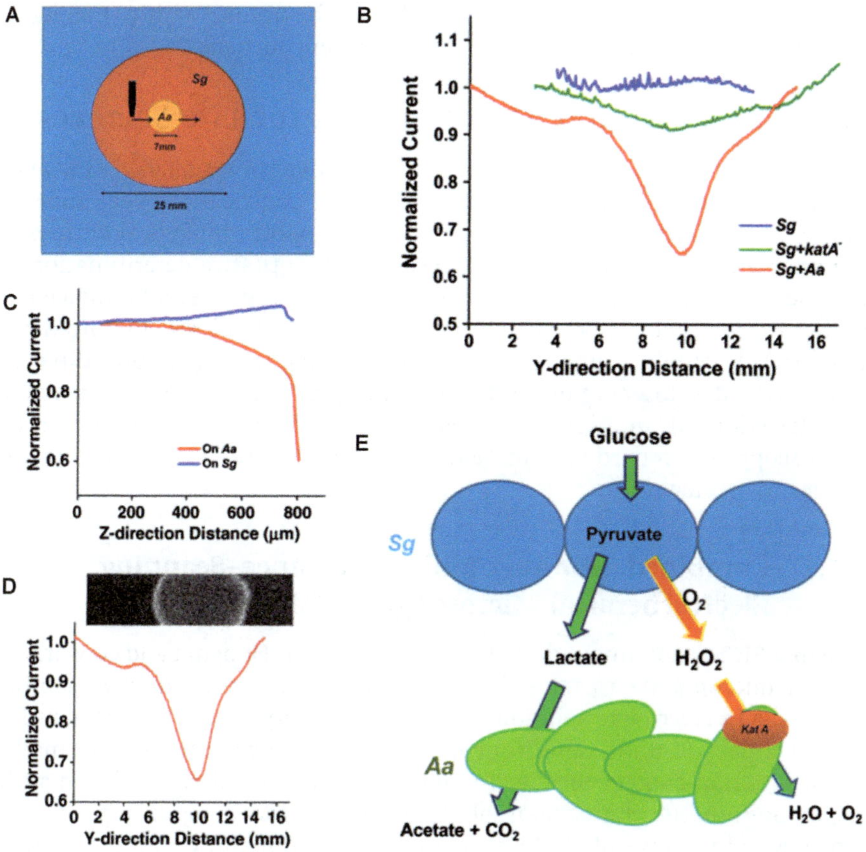

Figure 4.9 (A) Schematic view of a biofilm containing mixed species indicating the SECM probe scan direction. (B) Normalized current changes of a SECM *y*-scan over Sg alone, mutant Aa in an Sg film, and wild-type Aa in an Sg film, respectively, at 37 °C with the probe biased at 0.8 V *versus* Ag/AgCl. Green is pure Sg, blue is mutant Aa with Sg, red is wild Aa with Sg. (C) Approach curves for pure Sg and wild-type Aa region in an Aa + Sg biofilm (B, red curve). (D) Normalized current changes recorded in a line *y*-scan across Aa + Sg biofilm. Light production (in white) is indicative of promoter response to hydrogen peroxide. (E) Model for the role of hydrogen peroxide in a Sg and Aa co-cultured biofilm.
(Reprinted from ref. 223 with permission from the National Academy of Sciences of the United States of America.)

matter and weathering of clays and minerals. Etienne and co-workers investigated the local pH changes at the mineral-*Rahnella aquatilis* RA1/air interface determining acidification kinetics.[224] Denuault and co-workers studied the influence of *M. aquaeolei* on the electro-activity of FeS_2 and $CuFeS_2$ obtained from the Trans-Atlantic Geotraverse (TAG) hydrothermal mound.[225] Tip–substrate voltammetry was used to study the redox processes of the minerals. The catalase activity was investigated of the bioluminescent marine bacterium *Vibrio fischeri*, which can be found within the light-emitting organs of sepiolid squids and monocentrid fishes.

4.5 Combined SECM Techniques for Life Sciences

Scientific challenges emerge from the need to determine qualitatively and/or quantitatively determine species in complex matrices. Hence, analysis of the interactions of molecules at, with and/or within biological entities is a fundamental challenge in life sciences. The quantitative determination of key molecules and also tracking and imaging molecular events, molecular pathways, and molecular signals at a single cell level are important to understand cellular functions. Ongoing efforts are therefore directed towards the development of combined techniques and strategies that enable the detection of multiple parameters. In addition, in scanning probe microscopy and in particular in AFM and SECM this trend is clearly evident, as recently reviewed.[72,227]

4.5.1 Combined Scanning Ion Conductance–Scanning Electrochemical Microscopy (SICM–SECM)

The first SICM experiment dates back to 1989, when Hansma and co-workers used a micropipette to map the topography of a non-conductive sample immersed in electrolyte solution.[4] This SPM technique then lay dormant for almost a decade until significant instrumental improvements were made. SICM is a non-contact SPM technique, with the spatial resolution mainly depending on the dimensions of the orifice of the imaging pipette, *i.e.*, typically in the range of 30–300 nm (see Figure 4.10A and B). An ion current based on a small potential difference applied between a reference electrode inside an electrolyte-filled micro- or nanopipette and a reference electrode immersed in the bath solution is measured in dependence on the pipette position. A strong dependence of the recorded ion current in dependence on the tip–sample separation is observed, which is governed by the overall resistance dominated by the pipette resistance and the access resistance between the pipette tip and the sample surface. The ion current serves as input signal to an electronic feedback circuit, maintaining the tip at a constant distance to the sample surface. Several imaging modes have been developed (*e.g.*, hopping mode, Figure 4.10C), which have been recently reviewed by Baker and co-workers.[228] Nowadays, SICM has evolved into an attractive scanning probe technique for the life sciences with topographical resolutions at soft samples comparable to AFM and the distinct advantage of a non-invasive imaging procedure. Such micropipettes can be easily modified with a conductive layer, which is then insulated and opened so that a ring electrode around the pipette opening is exposed as shown by the Bard group for micro-sized probes[229] and Comstock *et al.*[230] and Takahashi *et al.*[231] with reduced electrode dimension around 300 nm. Modification with a carbon ring electrode around the orifice of the pipette has been shown by Thakar

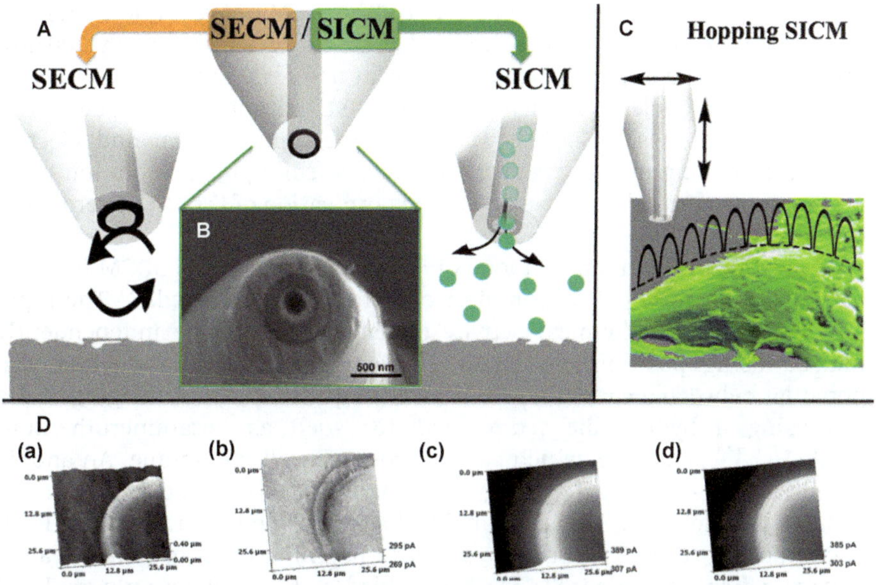

Figure 4.10 Combined SICM–SECM: (A) Scheme of combined SICM–SECM principle; (B) SEM image of a combined SICM–SECM probe with ring electrode. (C) Principle of SICM hopping mode. (D) SICM–SECM images of immobilized GOx: (a) and (b) topography and electrochemical images before and (c) and (d) after adding 20 mM glucose in 0.50 mM $FcCH_2OH + 0.1$ M KCl. The SECM nanoring and SICM nanopipette electrodes were held at 500 and 200 mV *versus* Ag/AgCl, respectively, for (a)–(c), while the potential of the SICM nanopipette electrode was changed to −200 mV *versus* Ag/AgCl for recording the image shown in (d). Scan range: 32 μm × 32 μm.
(Reprinted from ref. 231 with permission from the American Chemical Society.)

et al.[232] SICM–SECM probes with nanoscopic disc electrodes have been produced from pulled theta-capillaries with one barrel afterwards filled with carbonaceous material by pyrrolytic decomposition.[233,234] In a collaborative effort, Matsue and co-workers and Korchev and co-workers[231] demonstrated high-resolution imaging of immobilized HRP and GOx spots in hopping mode SICM and GC and FB SECM mode, thereby providing nanometer resolution in respect to topography and enzyme activity. Figure 4.10D shows high-resolution images of the GOD spot before and after the addition of glucose. The SICM–SECM images were obtained in a phosphate buffered saline (PBS) solution containing 0.50 mM ferrocene-methanol ($FcCH_2OH$). The dual probe was biased at 200 mV (SICM Ag/AgCl electrodes) and 500 mV *versus* Ag/AgCl (SECM ring electrode), sufficiently to oxidize $FcCH_2OH$ at the ring electrodes to $[FcCH_2OH]^+$ that serves in the enzymatic catalyzed reaction as electron acceptor. Figure 4.10D (a) and (b) show the SICM and SECM images before adding glucose to the buffered solution. In the absence of glucose, the SECM image reveals small changes in the current at the edge, probably due to a negative feedback effect. The SICM images show a ring-like topography of the GOx spot with an edge height of approximately 200 nm. After adding glucose, an increase in current associated with the GOD-catalyzed reaction is clearly evident (Figure 4.10.D (c) and (d)). Double-barrel carbon nanoprobes (DBCNPs) were used for topographical and electrochemical imaging of living sensory neurons revealing high-resolution topography and a current response due to the permeation of $FcCH_2OH$ across the cell membrane.[234]

Derived from the hopping mode of SICM, high-resolution SECM imaging with single nanoscopic carbon electrodes has been achieved.[235] The topography and electrochemical activity of live cells could be independently imaged using position modulation in combination with switching the potential between a cathodic potential for negative feedback imaging of the cell using a hydrophilic redox mediator such as hexamineruthenium ($[Ru(NH_3)_6]^{3+}$) chloride, which cannot cross the cell membrane. An anodic potential was used for detecting electroactive species associated with the cellular function (*e.g.*, signaling molecule) in GC mode. The negative feedback imaging reflects a constant current imaging approach. This so-called voltage-switching mode (VSM-SECM) enabled high-resolution topographical imaging of various cells (*e.g.*, boar sperm cell, differentiated PC12 cells, epidermoid carcinoma (A431) cells, cardiac myocyte, and hair cells) with nanoscopic carbon electrodes (radius in the range 6.5–100 nm).[235] By switching the potential the epidermal growth factor labeled alkaline phosphatase at A431 cells could be imaged.

4.5.2 Combined Atomic Force–Scanning Electrochemical Microscopy (AFM–SECM)

In 1986 Binnig, Quate and Gerber published the first paper on AFM,[2] which became a routine tool also in modern life sciences. Commercially available AFM instruments in combination with an incubator offer suitable solutions for imaging live biological systems at controlled temperature and CO_2 conditions in buffered/physiological solution. Besides mapping morphological changes, information on local concentration changes of certain molecules at biological samples is of ubiquitous interest. Hence, the unique advantage of combining AFM with SECM is the direct correlation of topological information with chemical surface activity, both at an excellent lateral resolution. From the SECM point of view, such combined probes allow accurate positioning and constant distance operation of sub-micro- and nanoelectrodes. The operation principle of AFM is based on the force interaction of a sharp tip at the end of a flexible cantilever (with force constants down to $0.01\ N\ m^{-1}$ for soft sample imaging) and the sample. Based on the distance and nature of the sample, these attractive or repulsive forces deflect the cantilever, which is usually detected *via* optical read out using a laser beam, reflected from the end of the cantilever and detected at a four-segmented photodiode. The recorded deflection signal is used in a feedback circuit for regulating the distance between the tip and the sample with nanoscale accuracy.

The combination of AFM and SECM usually relies on sophisticated bifunctional probes, yet with a variety of different geometries. Such probes have been fabricated using different approaches including individually made cantilever-shaped electrodes with conical or spherical electrodes located at the apex of the tip (Figure 4.11A),[236,237] or recessed disc electrodes,[238] batch-fabricated AFM probes with conical electrodes also constituting the AFM tip[239–242] and with recessed ring[243] or disc[244] electrodes. Furthermore, modification of commercially available AFM probes is possible with a conductive AFM tip,[245] or with integrated recessed electrodes and a non-conductive AFM tip (Figure 4.11A (a))[246–248] as well as carbon-nanotube modified probes.[249,250] Recently, a combined AFM–SECM probe with a conductive polymer tip for potential electrical stimulation and force spectroscopy under potential control has been demonstrated.[251] Although a significant number of papers have been published concerning fabrication of such probes and imaging model samples, only a few applications using such probes for actually imaging bio-related samples have been demonstrated. Imaging of soft sample requires that the AFM is operated in dynamic mode minimizing the force interaction. The first combined AFM–SECM measurements in dynamic mode were demonstrated by the team of Kranz and co-workers.[252] In addition, they could achieve simultaneously obtained topographical and electrochemical information on immobilized enzyme spots using AFM–SECM probes with recessed frame electrodes.[253] Measurements were obtained either in dynamic mode AFM and GC mode SECM for imaging GOx immobilized in soft polymeric spots (as shown in Figure 4.11B)[254] or contact mode AFM and GC mode SECM for HPR immobilized *via* thiol chemistry.[253] An advantage of the probe design with electrode located below the non-conductive AFM tip is that the electrode can be modified with enzyme layers or electro-catalytic films. The integration of a miniaturized glucose biosensor into an AFM probe allowed imaging glucose transport through a polycarbonate membrane with 200 nm diameter pores.[30] The topography of the membrane pores obtained in dynamic mode operation correlated well with the simultaneously recorded enhanced current above pores due to glucose transport, which was oxidized *via* the enzyme-catalyzed reaction resulting in the formation of H_2O_2. The concentrations determined above the pores fitted well with theoretical calculations. Label-free detection of HCV-1 DNA through hybridization of short (18-*mer*) DNA sequences was also demonstrated at an AFM tip–integrated electrode based on the electrostatic modulation of the ion-exchange kinetics of a conducting polymer.[255] Although these microfabricated probes can be easily modified with a sensing scheme, to date the achievable resolution cannot compete with approaches demonstrated by Demaille and co-workers. They developed hand-made cantilever-shaped probes with spherical and conical electrodes, which were then modified with redox moieties (*e.g.,* ferrocene) *via* poly(ethylene glycol) (PEG) chains, in what was termed TARM (tip-attached redox mediator)/AFM/SECM.[256,257] The redox-labeled polymer chains can also be attached to the sample surfaces, termed molecule touching (Mt)-AFM/SECM,[258] (*e.g.*, gold nanoparticles[259]). Both approaches supersede the addition of a free diffusing redox mediator, thereby facilitating high-resolution electrochemical imaging. (Mt)-AFM/SECM was successfully applied to image Fc-PEG-labeled mouse/antimouse immunoglubin G (IgG) immobilized at gold and HOPG.[260] Structured proteins spots were fabricated using bead lithography, which also enabled modification of the protein-free conductive substrate with long PEG chains blocking the electron transfer. Measurements were carried out in dynamic mode AFM and FB mode SECM, respectively. The redox-labeled flexible chains enable electron shuttling between the conductive AFM–SECM tip and the sample resulting in a positive feedback current at the redox-labeled protein spots as shown in Figure 4.11C.

The obtained resolution was in the range of 100 nm.[260] Notably, a labeling step similar to fluorescence microscopy is required for the shown high-resolution imaging (Figure 4.11B). Recently, wire-shaped conical gold electrodes were used as AFM probes to detect the activity of redox active enzymes

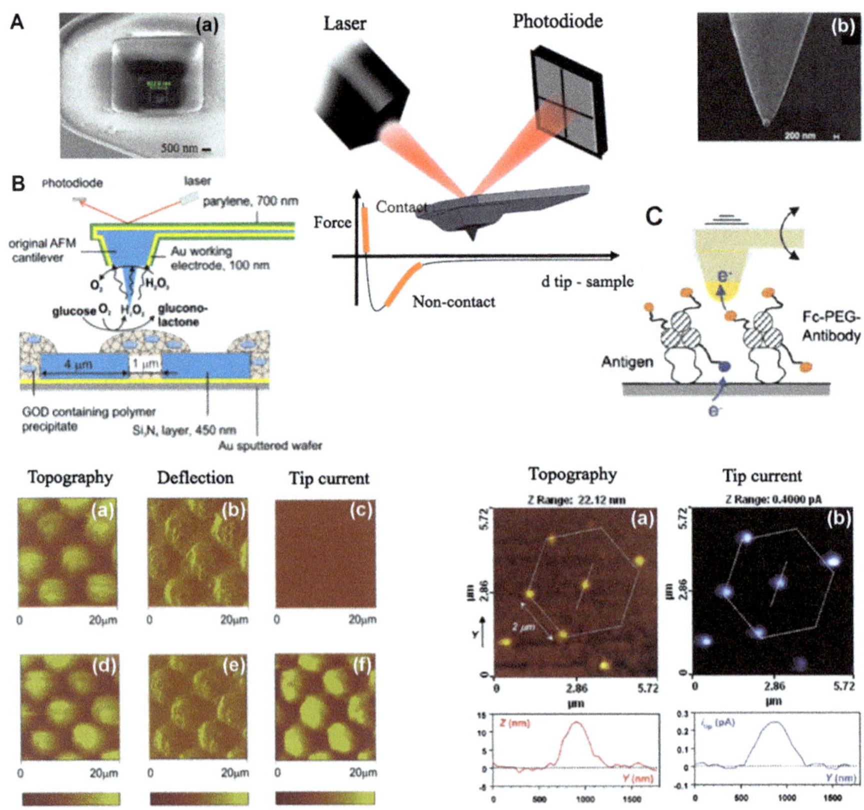

expressed at the surface of genetically modified *E. coli*.[261] Thus engineered bacteria expressed alcohol dehydrogenase II (ADHII) at the cell surface. Quinone as electron transfer mediator was covalently attached site specifically to the displayed ADHII. The conductive probe was used to locate a single bacterium at the sample surface and attach it to the tip *via* thiol chemistry. In this configuration, the electron transfer of different mutants between the NAD^+ binding site in alcohol dehydrogenase II and quinone could be studied depending on the distance.

4.5.3 Additional Combined SECM Approaches

For applications of SECM in the life sciences, further combined approaches are of interest such as integrating fluorescence techniques, near-field optical microscopy, surface plasmon resonance, vibrational spectroscopy and mass spectrometry, to potentially provide complementary localized molecular information. The first approaches to combining SECM with scanning optical microscopy date back to 2002, when Bard and co-workers modified an optical fiber with a ring electrode.[66] The first combination of SECM with fluorescence microscopy was demonstrated by Heinze and co-workers for

Figure 4.11 Combined AFM–SECM: (A) Scheme of force interaction curve and deflection-based positioning including SEM images of two different probe designs: AFM tip with integrated recessed frame electrode (a) and hand-made cantilever-shaped probe with conical electrode at the apex of the tip (b). (Reprinted from ref. 237 with permission from the American Chemical Society.) (B) Scheme representing simultaneous AFM–SECM imaging of soft polymer spots containing GOx in GC mode SECM. Topography of the enzyme-containing polymer spots and current image of GOx activity imaged in GC mode SECM and AFM tapping mode with free drive amplitude of 5 V and drive frequency of 32.4 kHz. Height (a) and (d), corresponding amplitude (b) and (e), and recorded current (c) and (f) images recorded in air-saturated phosphate buffer (0.1 mol L^{-1}, pH 7.4) in the absence (a)–(c) and in the presence (d)–(f) of glucose (50 mmol L^{-1}). The tip was held at a potential of 0.75 V (*versus* Ag/AgCl Quasi-REF). (Reprinted from ref. 254 with permission from John Wiley & Sons.) (C) Scheme of Mt/AFM–SECM detection of proteins immobilized onto a conducting surface. The target protein is recognized *via* a specific antibody (IgG) labeled by flexible PEG, which bears redox ferrocene (Fc) heads. Mt/AFM–SECM tapping mode images of a gold surface with an ordered array of adsorbed mouse IgG (antigen) dots fabricated *via* bead lithography (2 μm beads): (a) topography and (b) simultaneously acquired current image. The almost hexagonal symmetry of the dot array is underlined. Cross sections of the topography and current images along the short white line are shown, respectively. Images were recorded in 0.1 M phosphate buffer (pH 8). The tip and substrate potentials were biased at $E_{tip} = 0.30$ V *versus* SCE and $E_{sub} = -0.05$ V *versus* SCE, respectively. The probe is oscillated at its fundamental flexural frequency at 2.82 kHz; ~25% damping; imaging rate, 0.3 Hz. (Reprinted from ref. 260 with permission from the American Chemical Society.)

mapping the pH-gradient at micro- and nanoelectrodes.[36] Recently, several studies used fluorescence microscopy in combination with SECM for investigating cells. Fluorescence microscopy is among the most commonly used techniques for investigating dynamic processes in biological systems; however, this requires labeling with specific fluorophores, which have appropriate excitation and emission characteristics for the targeted species and structures. In addition, photobleaching or other photophysical effects adversely affect the biological sample. Combined SECM–fluorescence microscopy was proposed for mapping the release of reactive oxygen species (ROS) from prostate cancer cells (PC3).[262] Correlation studies between fluorescence microscopy visualizing intracellular ROS content and SECM line scans mapping the release of ROS were obtained. Recently, high-resolution SECM in combination with fluorescence microscopy has been demonstrated, imaging hippocampus neurons with FM-43-1 stained synaptic vesicles.[235] Fluorescence images revealed vesicles on the varicosity, which corresponded to the SECM topographical image recorded in constant current mode.

The combination of three scanning probes techniques—AFM–SECM–NSOM—has been applied for imaging PC12 cells as presented by Suzuki and co-workers.[71] By AFM tapping mode of a PC12 cell and in a second scan in constant height mode simultaneous SECM and NSOM measurements were obtained from differentiated Fluo-4-labeled neurites revealing the Ca^{2+} distribution. The SECM image, however, only corresponds to the topography based on the hindered diffusion of redox species towards the combined probe.

4.6 Outlook

A distinct trend of SECM in the life sciences is clearly targeted towards high-resolution chemical imaging with accurate control of the sample-to-probe distance. In recent years, significant progress has been achieved in sophisticated probe development, but also in positioning techniques enabling reliable distance control for nanoscopic probes. Last but not least, combined scanning probe techniques contribute significantly to this trend. Elucidating complex cellular functions and advancing fundamental knowledge on the governing biological processes remains essential for advancing new strategies in early recognition, mitigation or treatment of diseases. Hence, high spatial and temporal resolution of molecular events, molecular pathways and molecular signals are required. Ongoing efforts in advancing scanning probe techniques in general such as the combination with spectroscopy provides access to molecular information with high resolution, *e.g.*, tip-enhanced Raman spectroscopy (TERS) or AFM-combined with mass spectrometry. In SECM, combinations with spectroscopic techniques are still in their infancy, but are expected to evolve in the near future to provide molecular specific information in relevant biological processes.

Acknowledgements

The authors would like to acknowledge the Baden-Württemberg Stiftung within the project "Methoden für die Lebenswissenschaften" and the Boehringer Ingelheim-Ulm University Biocenter (BIU) for funding of related research at IABC.

References

1. X. Michalet, A. N. Kapanidis, T. Laurence, F. Pinaud, S. Doose, M. Pflughoefft and S. Weiss, *Annu. Rev. Biophys. Biomol. Struct.*, 2003, **32**, 161–182.
2. G. Binnig, C. F. Quate and C. Gerber, *Phys. Rev. Lett.*, 1986, **56**, 930–933.
3. A. J. Bard, F.-R. F. Fan, D. T. Pierce, P. R. Unwin, O. Wipf, David and F. Zhou, *Science*, 1991, **254**, 68–74.
4. P. K. Hansma, B. Drake, O. Marti, S. A. Gould and C. B. Prater, *Science*, 1989, **243**, 641–643.
5. C. A. Marsden, M. H. Joseph, Z. I. Kruk, N. T. Maidment, R. D. O'Neill, J. O. Schenk and J. A. Stamford, *Neuroscience*, 1988, **25**, 389–400.
6. K. L. Adams, M. Puchades and A. G. Ewing, *Annu. Rev. Anal. Chem.*, 2008, **1**, 329–355.
7. R. M. Wightman, L. J. May and A. C. Michael, *Anal. Chem.*, 1988, **60**, 769A–779A.
8. N. Dale, S. Hatz, F. Tian and E. Llaudet, *Trends Biotechnol.*, 2005, **23**, 420–428.
9. A. J. Bard, F.-R. F. Fan, J. Kwak and O. Lev, *Anal. Chem.*, 1989, **61**, 132–138.
10. J. Wang, L.-H. Wu and R. Li, *J. Electroanal. Chem.*, 1989, **272**, 285–292.
11. C. Lee, J. Kwak and A. J. Bard, *Proc. Natl. Acad. Sci. U. S. A.*, 1990, **87**, 1740–1743.
12. D. T. Pierce, P. R. Unwin and A. J. Bard, *Anal. Chem.*, 1992, **64**, 1795–1804.
13. H. Shiku, T. Takeda, H. Yamada, T. Matsue and I. Uchida, *Anal. Chem.*, 1995, **67**, 312–317.
14. H. Shiku, T. Matsue and I. Uchida, *Anal. Chem.*, 1996, **68**, 1276–1278.
15. T. Matsue, *Anal. Sci.*, 2013, **29**, 171–179.
16. C. Kranz, G. Wittstock, H. Wohlschläger and W. Schuhmann, *Electrochim. Acta*, 1997, **42**, 3105–3111.
17. C. Kranz, T. Lötzbeyer, H.-L. Schmidt and W. Schuhmann, *Biosens. Bioelectron.*, 1997, **12**, 257–266.
18. A. Schulte, M. Nebel and W. Schuhmann, *Annu. Rev. Anal. Chem.*, 2010, **3**, 299–318.
19. G. Wittstock and W. Schuhmann, *Anal. Chem.*, 1997, **69**, 5059–5066.
20. G. Wittstock, T. Wilhelm, S. Bahrs and P. Steinrücke, *Electroanalysis*, 2001, **13**, 669–675.
21. G. Wittstock, M. Burchardt, S. E. Pust, Y. Shen and C. Zhao, *Angew. Chem., Int. Ed.*, 2007, **46**, 1584–1617.

22. P. R. Unwin, J. V. Macpherson, M. A. Beeston, N. J. Evans, D. Littlewood and N. P. Hughes, *Adv. Dent. Res.*, 1997, **11**, 548–559.
23. I. Turyan, T. Matsue and D. Mandler, *Anal. Chem.*, 2000, **72**, 3431–3435.
24. S. Nugues and G. Denuault, *J. Electroanal. Chem.*, 1996, **408**, 125–140.
25. K. Yamashita, M. Takagi, S. Takenaka, K. Uchida and H. Kondo, *Analyst*, 2001, **126**, 1210–1211.
26. F. Turcu, A. Schulte, G. Hartwich and W. Schuhmann, *Angew. Chem., Int. Ed.*, 2004, **43**, 3482–3485.
27. E. Fortin, Y. Defontaine, P. Mailley, T. Livache and S. Szunerits, *Electroanalysis*, 2005, **17**, 495–503.
28. T. Yasukawa, T. Kaya and T. Matsue, *Electroanalysis*, 2000, **12**, 653–659.
29. B. Liu, S. A. Rotenberg and M. V. Mirkin, *Proc. Natl. Acad. Sci. U. S. A.*, 2000, **97**, 9855–9860.
30. A. Kueng, C. Kranz, A. Lugstein, E. Bertagnolli and B. Mizaikoff, *Angew. Chem., Int. Ed.*, 2005, **44**, 3419–3422.
31. J. Zhang and P. R. Unwin, *J. Am. Chem. Soc.*, 2002, **124**, 2379–2383.
32. J. M. Liebetrau, H. M. Miller, J. E. Baur, S. A. Takacs, V. Anupunpisit, P. A. Garris and D. O. Wipf, *Anal. Chem.*, 2003, **75**, 563–571.
33. *Scanning Electrochemical Microscopy*, ed. A. J. Bard and M. V. Mirkin, CRC Press, Boca Raton, 2nd edn, 2012.
34. F.-R. F. Fan, A. J. Bard, and P. He, in *Scanning Electrochemical Microscopy*, ed. A. J. Bard and M. V. Mirkin, CRC Press, Boca Raton, 2nd edn, 2012, pp. 15–24.
35. J. Mauzeroll and A. J. Bard, *Proc. Natl. Acad. Sci. U. S. A.*, 2004, **101**, 7862–7867.
36. F.-M. Boldt, J. Heinze, M. Diez, J. Petersen and M. Boersch, *Anal. Chem.*, 2004, **76**, 3473–3481.
37. L. P. Bauermann, W. Schuhmann and A. Schulte, *Phys. Chem. Chem. Phys.*, 2004, **6**, 4003–4008.
38. R. M. Wightman, S. Pons, D. R. Rolison, and P. P. Schmidt, *Ultramicroelectrodes*, Datatech Systems, Morganton, NC, 1987.
39. C. G. Zoski, *Electroanalysis*, 2002, **14**, 1041–1051.
40. J. Heinze, *Angew. Chem., Int. Ed. Engl.*, 1993, **32**, 1268–1288.
41. J. G. Roberts, B. P. Moody, G. S. McCarty and L. A. Sombers, *Langmuir*, 2010, **26**, 9116–9122.
42. T. G. Strein and A. G. Ewing, *Anal. Chem.*, 1992, **64**, 1368–1373.
43. J. Millar and C. W. A. Pelling, *J. Neurosci. Methods*, 2001, **110**, 1–8.
44. S. Chen and A. Kucernak, *Electrochem. Commun.*, 2002, **4**, 80–85.
45. X. Zhang, W. Zhang, X. Zhou and B. Ogorevc, *Anal. Chem.*, 1996, **68**, 3338–3343.
46. R. Thakar, A. E. Weber, C. A. Morris and L. A. Baker, *Analyst*, 2013, **138**, 5973–5982.
47. F. Cortes-Salazar, M. Träuble, F. Li, J.-M. Busnel, A.-L. Gassner, M. Hojeij, G. Wittstock and H. H. Girault, *Anal. Chem.*, 2009, **81**, 6889–6896.

48. F. Cortés-Salazar, D. Momotenko, A. Lesch, G. Wittstock and H. H. Girault, *Anal. Chem.*, 2010, **82**, 10037–10044.
49. B. R. Horrocks, D. Schmidtke, A. Heller and A. J. Bard, *Anal. Chem.*, 1993, **65**, 3605–3614.
50. A. Kueng, C. Kranz and B. Mizaikoff, *Biosens. Bioelectron.*, 2004, **19**, 1301–1307.
51. A. Hengstenberg, C. Kranz and W. Schuhmann, *Chem. - Eur. J.*, 2000, **6**, 1547–1554.
52. F.-M. Matysik, *Electrochim. Acta*, 1997, **42**, 3113–3116.
53. T. Yasukawa, T. Kaya and T. Matsue, *Anal. Chem.*, 1999, **71**, 4637–4641.
54. A. Kueng, C. Kranz and B. Mizaikoff, *Biosens. Bioelectron.*, 2005, **21**, 346–353.
55. W. Zhan and A. J. Bard, *Anal. Chem.*, 2006, **78**, 726–733.
56. Y. Lee, S. Amemiya and A. J. Bard, *Anal. Chem.*, 2001, **73**, 2261–2267.
57. P. Liljeroth, C. Johans, C. J. Slevin, B. M. Quinn and K. Kontturi, *Anal. Chem.*, 2002, **74**, 1972–1978.
58. P. Liljeroth, C. Johans, C. J. Slevin, B. M. Quinn and K. Kontturi, *Electrochem. Commun.*, 2002, **4**, 67–71.
59. M. V. Mirkin, F.-R. F. Fan and J. Bard, *J. Electroanal. Chem.*, 1992, **328**, 47–62.
60. C. G. Zoski and M. V. Mirkin, *Anal. Chem.*, 2002, **74**, 1986–1992.
61. S. Amemiya and A. J. Bard, *Anal. Chem.*, 2000, **72**, 4940–4948.
62. Y. Shao and M. V. Mirkin, *J. Phys. Chem. B*, 1998, **102**, 9915–9921.
63. C. G. Williams, M. A. Edwards, A. L. Colley, J. V. Macpherson and P. R. Unwin, *Anal. Chem.*, 2009, **81**, 2486–2495.
64. N. Ebejer, M. Schnippering, A. W. Colburn, M. A. Edwards and P. R. Unwin, *Anal. Chem.*, 2010, **82**, 9141–9145.
65. E. Betzig, J. K. Trautman, T. D. Harris, J. S. Weiner and R. L. Kostelak, *Science*, 1991, **251**, 1468–1470.
66. Y. Lee, Z. Ding and A. J. Bard, *Anal. Chem.*, 2002, **74**, 3634–3643.
67. K. Maruyama, H. Ohkawa, S. Ogawa, A. Ueda, O. Niwa and K. Suzuki, *Anal. Chem.*, 2006, **78**, 1904–1912.
68. Y. Takahashi, Y. Hirano, T. Yasukawa, H. Shiku, H. Yamada and T. Matsue, *Langmuir*, 2006, **22**, 10299–10306.
69. S. Kasai, H. Zhou and T. Matsue, *Chem. Lett.*, 2000, 200–201.
70. S. Kasai, Y. Hirano, N. Motochi, H. Shiku, M. Nishizawa and T. Matsue, *Anal. Chim. Acta*, 2002, **458**, 263–270.
71. A. Ueda, O. Niwa, K. Maruyama, Y. Shindo, K. Oka and K. Suzuki, *Angew. Chem., Int. Ed.*, 2007, **46**, 8238–8241.
72. C. Kranz, *Analyst*, 2014, **139**, 336–352.
73. E. Betzig, P. L. Finn and J. S. Weiner, *Appl. Phys. Lett.*, 1992, **60**, 2484–2486.
74. M. Ludwig, C. Kranz, W. Schuhmann and H. E. Gaub, *Rev. Sci. Instrum.*, 1995, **66**, 2857–2860.
75. P. I. James, L. F. Garfias-Mesias, P. J. Moyer and W. H. Smyrl, *J. Electrochem. Soc.*, 1998, **145**, L64–L66.

76. B. Ballesteros Katemann and W. Schuhmann, *Electroanalysis*, 2002, **14**, 22–28.
77. M. Buchler, S. C. Kelley and W. H. Smyrl, *Electrochem. Solid-State Lett.*, 2000, **3**, 35–38.
78. H. Yamada, H. Fukumoto, T. Yokoyama and T. Koike, *Anal. Chem.*, 2005, 77, 1785–1790.
79. B. Ballesteros Katemann, A. Schulte and W. Schuhmann, *Chem. – Eur. J.*, 2003, **9**, 2025–2033.
80. P. M. Diakowski and Z. Ding, *Phys. Chem. Chem. Phys.*, 2007, **9**, 5966–5974.
81. Y. Takahashi, H. Shiku, T. Murata, T. Yasukawa and T. Matsue, *Anal. Chem.*, 2009, **81**, 9674–9681.
82. D. O. Wipf, A. J. Bard and D. E. Tallman, *Anal. Chem.*, 1993, **65**, 1373–1377.
83. R. T. Kurulugama, D. O. Wipf, S. A. Takacs, S. Pongmayteegul, P. A. Garris and J. E. Baur, *Anal. Chem.*, 2005, **77**, 1111–1117.
84. M. Nebel, S. Grützke, N. Diab, A. Schulte and W. Schuhmann, *Angew. Chem., Int. Ed.*, 2013, **52**, 6335–6338.
85. R. A. Lazenby, K. McKelvey and P. R. Unwin, *Anal. Chem.*, 2013, **85**, 2937–2944.
86. Y. Hirano, Y. Nishimiya, K. Kowata, F. Mizutani, S. Tsuda and Y. Komatsu, *Anal. Chem.*, 2008, **80**, 9349–9354.
87. D. Schäfer, A. Puschhof and W. Schuhmann, *Phys. Chem. Chem. Phys.*, 2013, **15**, 5215–5223.
88. J. Kim, M. Shen, N. Nioradze and S. Amemiya, *Anal. Chem.*, 2012, **84**, 3489–3492.
89. W. Nogala, K. Szot, M. Burchardt, F. Roelfs, J. Rogalski, M. Opallo and G. Wittstock, *Analyst*, 2010, **135**, 2051–2058.
90. S. Bergner, J. Wegener and F.-M. Matysik, *Anal. Chem.*, 2011, **83**, 169–174.
91. E. Hecht, P. Knittel, E. Felder, P. Dietl, B. Mizaikoff and C. Kranz, *Analyst*, 2012, **137**, 5208–5214.
92. E. Hecht, A. Liedert, A. Ignatius, B. Mizaikoff and C. Kranz, *Biosens. Bioelectron.*, 2013, **44**, 27–33.
93. C. Kranz, M. Ludwig, H. E. Gaub and W. Schuhmann, *Adv. Mater.*, 1995, 7, 38–40.
94. J. Kwak and A. J. Bard, *Anal. Chem.*, 1989, **61**, 1221–1227.
95. C. Lee, J. Kwak and F. C. Anson, *Anal. Chem.*, 1991, **63**, 1501–1504.
96. K. Eckhard, X. Chen, F. Turcu and W. Schuhmann, *Phys. Chem. Chem. Phys.*, 2006, **8**, 5359–5365.
97. R. C. Engstrom, M. Weber, D. J. Wunder, R. Burgess and S. Winquist, *Anal. Chem.*, 1986, **58**, 844–848.
98. R. C. Engstrom, T. Meaney, R. Tople and R. M. Wightman, *Anal. Chem.*, 1987, **59**, 2005–2010.
99. C. Zhao, J. K. Sinha, C. A. Wijayawardhana and G. Wittstock, *J. Electroanal. Chem.*, 2004, **561**, 83–91.
100. C. M. Sanchez-Sanchez, J. Rodriguez-Lopez and A. J. Bard, *Anal. Chem.*, 2008, **80**, 3254–3260.

101. B. R. Horrocks and G. Wittstock, in *Scanning Electrochemical Microscopy*, ed. A. J. Bard and M. V. Mirkin, CRC Press, Boca Raton, 2nd edn, 2012, pp. 317–378.
102. M. Ciobanu, D. E. Taylor Jr., J. P. Wilburn and D. E. Cliffel, *Anal. Chem.*, 2008, **80**, 2717–2727.
103. O. G. Voronin, A. Hartmann, C. Steinbach, A. A. Karyakin, A. R. Khokhlov and C. Kranz, *Electrochem. Commun.*, 2012, **23**, 102–105.
104. B. R. Horrocks, M. V. Mirkin, D. T. Pierce and A. J. Bard, *Anal. Chem.*, 1993, **65**, 1213–1224.
105. C. Wei, A. J. Bard, G. Nagy and K. Toth, *Anal. Chem.*, 1995, **67**, 1346–1356.
106. D. O. Wipf, F. Ge, T. W. Spaine and J. E. Baur, *Anal. Chem.*, 2000, **72**, 4921–4927.
107. J. L. Amphlett and G. Denuault, *J. Phys. Chem. B*, 1998, **102**, 9946–9951.
108. D. T. Pierce and A. J. Bard, *Anal. Chem.*, 1993, **65**, 3598–3604.
109. M. Nebel, S. Grützke, N. Diab, A. Schulte and W. Schuhmann, *Faraday Discuss.*, 2013, **164**, 19–32.
110. M. V. Mirkin, W. Nogala, J. Velmurugan and Y. Wang, *Phys. Chem. Chem. Phys.*, 2011, **13**, 21196–21212.
111. S. Bergner, P. Vatsyayan and F.-M. Matysik, *Anal. Chim. Acta*, 2013, **775**, 1–13.
112. A. Parra, E. Casero, L. Vázquez, J. Jin, F. Pariente and E. Lorenzo, *Langmuir*, 2006, **22**, 5443–5450.
113. S. Sun, P.-H. Ho-Si and D. J. Harrison, *Langmuir*, 1991, **7**, 727–737.
114. S. Ferretti, S. Paynter, D. A. Russell, K. E. Sapsford and D. J. Richardson, *TrAC, Trends Anal. Chem.*, 2000, **19**, 530–540.
115. G. Wittstock, *Fresenius' J. Anal. Chem.*, 2001, **370**, 303–315.
116. N. K. Chaki and K. Vijayamohanan, *Biosens. Bioelectron.*, 2002, **17**, 1–12.
117. G. Wittstock and W. Schuhmann, *Anal. Chem.*, 1997, **69**, 5059–5066.
118. E. M. Oliveira, S. Beyer and J. Heinze, *Bioelectrochemistry*, 2007, **71**, 186–191.
119. T. Wilhelm and G. Wittstock, *Angew. Chem., Int. Ed.*, 2003, **42**, 2248–2250.
120. S. Gáspár, M. Mosbach, L. Wallman, T. Laurell, E. Csöregi and W. Schuhmann, *Anal. Chem.*, 2001, **73**, 4254–4261.
121. C. Zhao and G. Wittstock, *Angew. Chem., Int. Ed.*, 2004, **43**, 4170–4172.
122. E. Burestedt, C. Nistor, U. Schagerlöf and J. Emnéus, *Anal. Chem.*, 2000, **72**, 4171–4177.
123. G. Wittstock, K. Yu, H. B. Halsall, T. H. Ridgway and W. R. Heineman, *Anal. Chem.*, 1995, **67**, 3578–3582.
124. C. A. Wijayawardhana, G. Wittstock, H. B. Halsall and W. R. Heineman, *Electroanalysis*, 2000, **12**, 640–644.
125. S. Kasai, A. Yokota, H. Zhou, M. Nishizawa, K. Niwa, T. Onouchi and T. Matsue, *Anal. Chem.*, 2000, **72**, 5761–5765.
126. T. Yasukawa, Y. Hirano, N. Motochi, H. Shiku and T. Matsue, *Biosens. Bioelectron.*, 2007, **22**, 3099–3104.

127. W. S. Roberts, F. Davis, J. L. Holmes, S. D. Collyer, L. D. Larcombe, S. L. Morgan and S. P. J. Higson, *Biosens. Bioelectron.*, 2013, **41**, 282–288.
128. W. Song, Z. Yan and K. Hu, *Biosens. Bioelectron.*, 2012, **38**, 425–429.
129. J. L. Holmes, F. Davis, S. D. Collyer and S. P. J. Higson, *Anal. Chim. Acta*, 2011, **689**, 206–211.
130. J. L. Holmes, F. Davis, S. D. Collyer and S. P. J. Higson, *Anal. Chim. Acta*, 2012, **741**, 1–8.
131. C. Kurzawa, A. Hengstenberg and W. Schuhmann, *Anal. Chem.*, 2002, **74**, 355–361.
132. T. Wilhelm, G. Wittstock and R. Szargan, *Fresenius' J. Anal. Chem.*, 1999, **365**, 163–167.
133. J. Zhou, C. Campbell, A. Heller and A. J. Bard, *Anal. Chem.*, 2002, **74**, 4007–4010.
134. W. S. Roberts, F. Davis, S. D. Collyer and S. P. J. Higson, *Analyst*, 2011, **136**, 5287–5293.
135. X. Li, Q. Geng, Y. Wang, Z. Si, W. Jiang, X. Zhang and W. Jin, *Electroanalysis*, 2007, **19**, 1734–1740.
136. X. Li, Q. Geng, Y. Wang, Z. Si, W. Jiang, X. Zhang and W. Jin, *Electrochim. Acta*, 2007, **53**, 2016–2024.
137. G. Sciutto, S. Prati, R. Mazzeo, M. Zangheri, A. Roda, L. Bardini, G. Valenti, S. Rapino and M. Marcaccio, *Anal. Chim. Acta*, 2014, **831**, 31–37.
138. E. Fortin, P. Mailley, L. Lacroix and S. Szunerits, *Analyst*, 2006, **131**, 186–193.
139. M. Burchardt and G. Wittstock, *Langmuir*, 2013, **29**, 15090–15099.
140. F. Conzuelo, L. Stratmann, S. Grützke, J. M. Pingarrón and W. Schuhmann, *Electroanalysis*, 2014, **26**, 481–487.
141. H. Fan, X. Wang, F. Jiao, F. Zhang, Q. Wang, P. He and Y. Fang, *Anal. Chem.*, 2013, **85**, 6511–6517.
142. W. Nogala, K. Szot, M. Burchardt, M. Joensson-Niedziolka, J. Rogalski, G. Wittstock and M. Opallo, *Bioelectrochemistry*, 2010, **79**, 101–107.
143. J. L. Fernández, N. Mano, A. Heller and A. J. Bard, *Angew. Chem., Int. Ed.*, 2004, **43**, 6355–6357.
144. P. Scodeller, R. Carballo, R. Szamocki, L. Levin, F. Forchiassin and E. J. Calvo, *J. Am. Chem. Soc.*, 2010, **132**, 11132–11140.
145. E. M. Hussien, T. Erichsen, W. Schuhmann and M. Maciejewska, *Anal. Bioanal. Chem.*, 2008, **391**, 1773–1782.
146. C. Zhao and G. Wittstock, *Anal. Chem.*, 2004, **76**, 3145–3154.
147. M. Niculescu, S. Gáspár, A. Schulte, E. Csöregi and W. Schuhmann, *Biosens. Bioelectron.*, 2004, **19**, 1175–1184.
148. D. Oyamatsu, Y. Hirano, N. Kanaya, Y. Mase, M. Nishizawa and T. Matsue, *Bioelectrochemistry*, 2003, **60**, 115–121.
149. C. Nunes Kirchner, M. Träuble and G. Wittstock, *Anal. Chem.*, 2010, **82**, 2626–2635.

150. C. Nunes Kirchner, S. Szunerits and G. Wittstock, *Electroanalysis*, 2007, **19**, 1258–1267.
151. S. Neugebauer, A. Zimdars, P. Liepold, M. Gebala, W. Schuhmann and G. Hartwich, *ChemBioChem*, 2009, **10**, 1193–1199.
152. I. Palchetti, S. Laschi, G. Marrazza and M. Mascini, *Anal. Chem.*, 2007, **79**, 7206–7213.
153. A. Parra, E. Casero, L. Vázquez, J. Jin, F. Pariente and E. Lorenzo, *Langmuir*, 2006, **22**, 5443–5450.
154. J. Wang, F. Song and F. Zhou, *Langmuir*, 2002, **18**, 6653–6658.
155. N. F. Ferreyra, S. Bollo and G. A. Rivas, *J. Electroanal. Chem.*, 2010, **638**, 262–268.
156. H. Song, P. M. Diakowski, R. H. E. Hudson and H.-B. Kraatz, *J. Inorg. Organomet. Polym. Mater.*, 2012, **22**, 178–182.
157. A. J. Wain and F. Zhou, *Langmuir*, 2008, **24**, 5155–5160.
158. A. A. Gorodetsky, W. J. Hammond, M. G. Hill, K. Slowinski and J. K. Barton, *Langmuir*, 2008, **24**, 14282–14288.
159. J. Wang and F. Zhou, *J. Electroanal. Chem.*, 2002, **537**, 95–102.
160. F. Turcu, A. Schulte, G. Hartwich and W. Schuhmann, *Angew. Chem., Int. Ed.*, 2004, **43**, 3482–3485.
161. B. Liu, A. J. Bard, C.-Z. Li and H.-B. Kraatz, *J. Phys. Chem. B*, 2005, **109**, 5193–5198.
162. P. M. Diakowski and H.-B. Kraatz, *Chem. Commun.*, 2009, 1189–1191.
163. P. M. Diakowski and H.-B. Kraatz, *Chem. Commun.*, 2011, **47**, 1431–1433.
164. M. H. Shamsi and H.-B. Kraatz, *Analyst*, 2011, **136**, 4724–4731.
165. M. N. Alam, M. H. Shamsi and H.-B. Kraatz, *Analyst*, 2012, **137**, 4220–4225.
166. M. H. Shamsi and H.-B. Kraatz, *Analyst*, 2013, **138**, 3538–3543.
167. I. Beaulieu, S. Kuss, J. Mauzeroll and M. Geissler, *Anal. Chem.*, 2011, **83**, 1485–1492.
168. S. Amemiya, J. Guo, H. Xiong and D. A. Gross, *Anal. Bioanal. Chem.*, 2006, **386**, 458–471.
169. J. Mauzeroll and S. B. Schougaard, in *Scanning Electrochemical Microscopy*, ed. A. J. Bard and M. V. Mirkin, CRC Press, Boca Raton, 2nd edn, 2012, pp. 379–415.
170. Y.-S. Torisawa, T. Kaya, Y. Takii, D. Oyamatsu, M. Nishizawa and T. Matsue, *Anal. Chem.*, 2003, **75**, 2154–2158.
171. M. Tsionsky, Z. G. Cardon, A. J. Bard and R. B. Jackson, *Plant Physiol.*, 1997, **113**, 895–901.
172. Y. Takii, K. Takoh, M. Nishizawa and T. Matsue, *Electrochim. Acta*, 2003, **48**, 3381–3385.
173. T. Yakusawa, Y. Kondo, I. Uchida and T. Matsue, *Chem. Lett.*, 1998, 767–768.
174. T. Kaya, Y.-S. Torisawa, D. Oyamatsu, M. Nishizawa and T. Matsue, *Biosens. Bioelectron.*, 2003, **18**, 1379–1383.
175. H. Shiku, T. Shiraishi, H. Ohya, T. Matsue, H. Abe, H. Hoshi and M. Kobayashi, *Anal. Chem.*, 2001, **73**, 3751–3758.

176. J. A. Koch, M. B. Baur, E. L. Woodall and J. E. Baur, *Anal. Chem.*, 2012, **84**, 9537–9543.
177. B. Liu, S. A. Rotenberg and M. V. Mirkin, *Anal. Chem.*, 2002, **74**, 6340–6348.
178. W. Feng, S. A. Rotenberg and M. V. Mirkin, *Anal. Chem.*, 2003, **75**, 4148–4154.
179. P. Sun, F. O. Laforge, T. P. Abeyweera, S. A. Rotenberg, J. Carpino and M. V. Mirkin, *Proc. Natl. Acad. Sci. U. S. A.*, 2008, **105**, 443–448.
180. J. Mauzeroll, A. J. Bard, O. Owhadian and T. J. Monks, *Proc. Natl. Acad. Sci. U. S. A.*, 2004, **101**, 17582–17587.
181. Y.-S. Torisawa, N. Ohara, K. Nagamine, S. Kasai, T. Yasukawa, H. Shiku and T. Matsue, *Anal. Chem.*, 2006, **78**, 7625–7631.
182. K. Nagamine, Y. Takahashi, K. Ino, H. Shiku and T. Matsue, *Electroanalysis*, 2011, **23**, 1168–1174.
183. Y. Matsumae, Y. Takahashi, K. Ino, H. Shiku and T. Matsue, *Anal. Chim. Acta*, 2014, **842**, 20–26.
184. N. Gao, X. Wang, L. Li, X. Zhang and W. Jin, *Analyst*, 2007, **132**, 1139–1146.
185. S. Kuss, D. Polcari, M. Geissler, D. Brassard and J. Mauzeroll, *Proc. Natl. Acad. Sci. U. S. A.*, 2013, **110**, 9249–9254.
186. S. Bergner, J. Wegener and F.-M. Matysik, *Anal. Chem.*, 2011, **83**, 169–174.
187. A. Hengstenberg, A. Blöchl, I. D. Dietzel and W. Schuhmann, *Angew. Chem., Int. Ed.*, 2001, **40**, 905–908.
188. D. S. Schrock and J. E. Baur, *Anal. Chem.*, 2007, **79**, 7053–7061.
189. M. Nishizawa, K. Takoh and T. Matsue, *Langmuir*, 2002, **18**, 3645–3649.
190. Z. Chen, S. Xie, L. Shen, Y. Du, S. He, Q. Li, Z. Liang, X. Meng, B. Li, X. Xu, H. Ma, Y. Huang and Y. Shao, *Analyst*, 2008, **133**, 1221–1228.
191. A. Pailleret, J. Oni, S. Reiter, S. Isik, M. Etienne, F. Bedioui and W. Schuhmann, *Electrochem. Commun.*, 2003, **5**, 847–852.
192. S. Borgmann, I. Radtke, T. Erichsen, A. Blöchl, R. Heumann and W. Schuhmann, *ChemBioChem*, 2006, **7**, 662–668.
193. X. Zhao, S. Lam, J. Jass and Z. Ding, *Electrochem. Commun.*, 2010, **12**, 773–776.
194. Y. Takahashi, T. Miyamoto, H. Shiku, R. Asano, T. Yasukawa, I. Kumagai and T. Matsue, *Anal. Chem.*, 2009, **81**, 2785–2790.
195. Y. Xue, L. Ding, J. Lei, F. Yan and H. Ju, *Anal. Chem.*, 2010, **82**, 7112–7118.
196. Y. Hirano, K. Kowata, M. Kodama and Y. Komatsu, *Bioelectrochemistry*, 2013, **92**, 1–5.
197. N. Ktari, P. Poncet, H. Sénéchal, L. Malaquin, F. Kanoufi and C. Combellas, *Langmuir*, 2010, **26**, 17348–17356.
198. C. E. Berger, H. Rathod, J. I. Gillespie, B. R. Horrocks and H. K. Datta, *J. Bone Miner. Res.*, 2001, **16**, 2092–2102.
199. Y. Matsumae, T. Arai, Y. Takahashi, K. Ino, H. Shiku and T. Matsue, *Chem. Commun.*, 2013, **49**, 6498–6500.

200. D. Zhan, X. Li, W. Zhan, F.-R. F. Fan and A. J. Bard, *Anal. Chem.*, 2007, **79**, 5225–5231.
201. S. Bergner, J. Wegener and F.-M. Matysik, *Anal. Methods*, 2012, **4**, 623–629.
202. H. Shiku, T. Shiraishi, S. Aoyagi, Y. Utsumi, M. Matsudaira, H. Abe, H. Hoshi, S. Kasai, H. Ohya and T. Matsue, *Anal. Chim. Acta*, 2004, **522**, 51–58.
203. D. G. Hirst and T. Robson, *Methods Mol. Biol.*, 2011, **704**, 1–13.
204. K. Shibuki, *Neurosci. Res.*, 1990, **9**, 69–76.
205. T. Malinski and Z. Taha, *Nature*, 1992, **358**, 676–678.
206. F. Bedioui and S. Griveau, *Electroanalysis*, 2013, **25**, 587–600.
207. J. Oni, N. Diab, I. Radtke and W. Schuhmann, *Electrochim. Acta*, 2003, **48**, 3349–3354.
208. S. Isik, M. Etienne, J. Oni, A. Blöchl, S. Reiter and W. Schuhmann, *Anal. Chem.*, 2004, **76**, 6389–6394.
209. S. Isik and W. Schuhmann, *Angew. Chem., Int. Ed.*, 2006, **45**, 7451–7454.
210. S. Borgmann, I. Radtke, T. Erichsen, A. Blöchl, R. Heumann and W. Schuhmann, *ChemBioChem*, 2006, **7**, 662–668.
211. J. H. Shim and Y. Lee, *Anal. Chem.*, 2009, **81**, 8571–8576.
212. R. Corriden and P. A. Insel, *Sci. Signal.*, 2010, **3**, 1–12.
213. E. Llaudet, S. Hatz, M. Droniou and N. Dale, *Anal. Chem.*, 2005, **77**, 3267–3273.
214. F. Scheller and D. Pfeiffer, *Anal. Chim. Acta*, 1980, **117**, 383–386.
215. J.-F. Masson, C. Kranz, B. Mizaikoff and E. B. Gauda, *Anal. Chem.*, 2008, **80**, 3991–3998.
216. K. Nagamine, T. Kaya, T. Yasukawa, H. Shiku and T. Matsue, *Sens. Actuators, B*, 2005, **108**, 676–682.
217. C. Cai, B. Liu, M. V. Mirkin, H. A. Frank and J. F. Rusling, *Anal. Chem.*, 2002, **74**, 114–119.
218. N. Matsui, T. Kaya, K. Nagamine, T. Yasukawa, H. Shiku and T. Matsue, *Biosens. Bioelectron.*, 2006, **21**, 1202–1209.
219. K. B. Holt and A. J. Bard, *Biochemistry*, 2005, **44**, 13214–13223.
220. T. Kaya, D. Numai, K. Nagamine, S. Aoyagi, H. Shiku and T. Matsue, *Analyst*, 2004, **129**, 529–534.
221. D. Zhan, F.-R. F. Fan and A. J. Bard, *Proc. Natl. Acad. Sci. U. S. A.*, 2008, **105**, 12118–12122.
222. K. Nagamine, S. Onodera, A. Kurihara, T. Yasukawa, H. Shiku, R. Asano, I. Kumagai and T. Matsue, *Biotechnol. Bioeng.*, 2006, **96**, 1008–1013.
223. X. Liu, M. M. Ramsey, X. Chen, D. Koley, M. Whiteley and A. J. Bard, *Proc. Natl. Acad. Sci. U. S. A.*, 2011, **108**, 2668–2673.
224. D. Oulkadi, S. Banon, C. Mustin and M. Etienne, *Electrochem. Commun.*, 2014, **44**, 1–3.
225. M. Müller, R. A. Mills, R. B. Pearce, J. A. Milton, P. J. Statham, J. R. Lloyd, A. Mujahid and G. Denuault, *Geomicrobiol. J.*, 2014, **31**, 373–382.

226. E. Abucayon, N. Ke, R. Cornut, A. Patelunas, D. Miller, M. K. Nishiguchi and C. G. Zoski, *Anal. Chem.*, 2014, **86**, 498–505.
227. A. Eifert and C. Kranz, *Anal. Chem.*, 2014, **86**, 5190–5200.
228. C.-C. Chen, Y. Zhou and L. A. Baker, *Annu. Rev. Anal. Chem.*, 2012, **5**, 207–228.
229. D. A. Walsh, L. Ferna, J. Mauzeroll and A. J. Bard, *Anal. Chem.*, 2005, 77, 5182–5188.
230. D. J. Comstock, J. W. Elam, M. J. Pellin and M. C. Hersam, *Anal. Chem.*, 2010, **82**, 1270–1276.
231. Y. Takahashi, A. I. Shevchuk, P. Novak, Y. Murakami, H. Shiku, Y. E. Korchev and T. Matsue, *J. Am. Chem. Soc.*, 2010, **132**, 10118–10126.
232. R. Thakar, A. E. Weber, C. A. Morris and L. A. Baker, *Analyst*, 2013, **138**, 5973–5982.
233. K. Ino, K. Ono, T. Arai, Y. Takahashi, H. Shiku and T. Matsue, *Anal. Chem.*, 2013, **85**, 3832–3835.
234. Y. Takahashi, A. I. Shevchuk, P. Novak, Y. Zhang, N. Ebejer, J. V. MacPherson, P. R. Unwin, A. J. Pollard, D. Roy, C. A. Clifford, H. Shiku, T. Matsue, D. Klenerman and Y. E. Korchev, *Angew. Chem., Int. Ed.*, 2011, **50**, 9638–9642.
235. Y. Takahashi, A. I. Shevchuk, P. Novak, B. Babakinejad, J. Macpherson, P. R. Unwin, H. Shiku, J. Gorelik, D. Klenerman, Y. E. Korchev and T. Matsue, *Proc. Natl. Acad. Sci. U. S. A.*, 2012, **109**, 11540–11545.
236. J. V. Macpherson and P. R. Unwin, *Anal. Chem.*, 2000, **72**, 276–285.
237. J. Abbou, C. Demaille, M. Druet and J. Moiroux, *Anal. Chem.*, 2002, **74**, 6355–6363.
238. A. Davoodi, J. Pan, C. Leygraf and S. Norgren, *Electrochem. Solid-State Lett.*, 2005, **8**, B21–B24.
239. P. S. Dobson, J. M. R. Weaver, M. N. Holder, P. R. Unwin and J. V. Macpherson, *Anal. Chem.*, 2005, **77**, 424–434.
240. R. J. Fasching, Y. Tao and F. B. Prinz, *Sens. Actuators, B*, 2005, **108**, 964–972.
241. P. L. T. M. Frederix, M. R. Gullo, T. Akiyama, A. Tonin, N. F. De Rooij, U. Staufer and A. Engel, *Nanotechnology*, 2005, **16**, 997–1005.
242. A. J. Wain, A. J. Pollard and C. Richter, *Anal. Chem.*, 2014, **86**, 5143–5149.
243. H. Shin, P. J. Hesketh, B. Mizaikoff and C. Kranz, *Anal. Chem.*, 2007, **79**, 4769–4777.
244. M. Salomo, S. E. Pust, G. Wittstock and E. Oesterschulze, *Microelectr. Eng.*, 2010, **87**, 1537–1539.
245. Y. Hirata, S. Yabuki and F. Mizutani, *Bioelectrochemistry*, 2004, **63**, 217–224.
246. C. Kranz, G. Friedbacher, B. Mizaikoff, A. Lugstein, J. Smoliner and E. Bertagnolli, *Anal. Chem.*, 2001, **73**, 2491–2500.
247. A. Eifert, W. Smirnov, S. Frittmann, C. Nebel, B. Mizaikoff and C. Kranz, *Electrochem. Commun.*, 2012, **25**, 30–34.
248. E. Lee, M. Kim, J. Seong, H. Shin and G. Lim, *Phys. Status Solidi*, 2013, 7, 406–409.

249. D. P. Burt, N. R. Wilson, J. M. R. Weaver, P. S. Dobson and J. V. Macpherson, *Nano Lett.*, 2005, **5**, 639–643.
250. A. V. Patil, A. F. Beker, F. G. M. Wiertz, H. A. Heering, G. Coslovich, R. Vlijm and T. H. Oosterkamp, *Nanoscale*, 2010, **2**, 734–738.
251. P. Knittel, M. J. Higgins and C. Kranz, *Nanoscale*, 2014, **6**, 2255–2260.
252. A. Kueng, C. Kranz, B. Mizaikoff, A. Lugstein and E. Bertagnolli, *Appl. Phys. Lett.*, 2003, **82**, 1592–1594.
253. C. Kranz, A. Kueng, A. Lugstein, E. Bertagnolli and B. Mizaikoff, *Ultramicroscopy*, 2004, **100**, 127–134.
254. A. Kueng, C. Kranz, A. Lugstein, E. Bertagnolli and B. Mizaikoff, *Angew. Chem., Int. Ed.*, 2003, **42**, 3238–3240.
255. C. dos Santos Riccardi, C. Kranz, J. Kowalik, H. Yamanaka, B. Mizaikoff and M. Josowicz, *Anal. Chem.*, 2008, **80**, 237–245.
256. A. Anne, E. Cambril, A. Chovin, C. Demaille and C. Goyer, *ACS Nano*, 2009, **3**, 2927–2940.
257. A. Anne, C. Demaille and C. Goyer, *ACS Nano*, 2009, **3**, 819–827.
258. A. Anne, E. Cambril, A. Chovin and C. Demaille, *Anal. Chem.*, 2010, **82**, 6353–6362.
259. K. Huang, A. Anne, M. A. Bahri and C. Demaille, *ACS Nano*, 2013, **7**, 4151–4163.
260. A. Anne, A. Chovin, C. Demaille and M. Lafouresse, *Anal. Chem.*, 2011, **83**, 7924–7932.
261. L. Alfonta, B. Meckes, L. Amir, O. Schlesinger, S. Ramachandran and R. Lal, *Anal. Chem.*, 2014, **86**, 7674–7680.
262. S. E. Salamifar and R. Y. Lai, *Anal. Chem.*, 2013, **85**, 9417–9421.

CHAPTER 5

Electrochemical Detection of Nanoparticles

HER SHUANG TOH AND RICHARD G. COMPTON*

Department of Chemistry, Physical and Theoretical Chemistry Laboratory, Oxford University, South Parks Road, Oxford OX1 3QZ, UK
*Email: Richard.Compton@chem.ox.ac.uk

5.1 Nanoparticles and their Properties

Nanoparticles are a special class of material of any shape with at least one dimension between 1 and 100 nm based on the guidelines of IUPAC.[1] They are important because their properties may deviate significantly from that of the bulk material. One of the most studied aspects is their optical characteristics.[2] Taking the nanoplasmonic properties of silver nanoparticles as an example, spherical silver nanoparticles are known to have a yellow colour and a surface plasmon absorption around 400 nm.[3] As the size of the silver decreases to ultra-small (usually clusters below 2 nm in diameter), fluorescent properties start to develop.[4] Bulk silver shows different optical properties[5] and it does not fluoresce. For such metallic nanoparticles, fluorescence develops at a size smaller than $\sim$2 nm because the continuum of energy levels found in bulk metal that allows the free flow of electrons across the material breaks down and the energy levels become more distinct at this size.[6] For other types of nanoparticles such as quantum dots, fluorescence may arise at sizes greater than 2 nm due to the electronic structure present in the semiconductor materials.[7]

Apart from the prominent field of nanoplasmonics, the physical properties of the particles may also change at the nanoscale. First, the percentage

RSC Detection Science Series No. 6
Electrochemical Strategies in Detection Science
Edited by Damien W. M. Arrigan

Published by the Royal Society of Chemistry, www.rsc.org

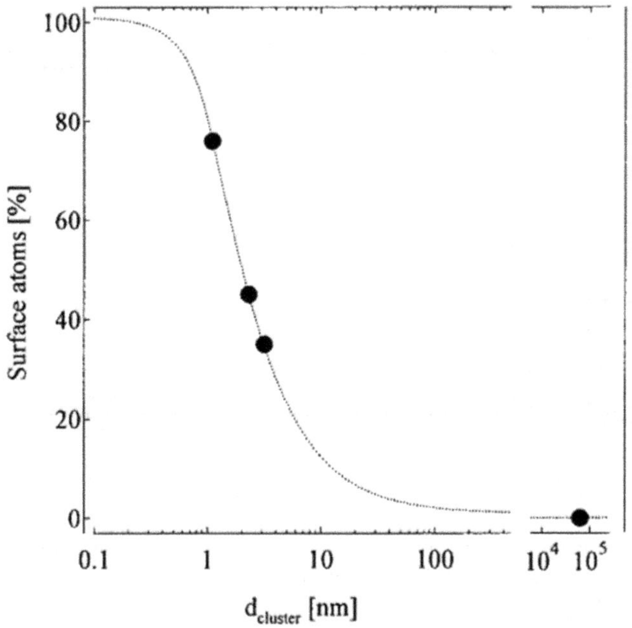

Figure 5.1 Schematic diagram showing how the percentage of surface atoms varies with the particle diameter for palladium.
(Reproduced from ref. 8 with kind permission from Springer Science and Business Media.)

of the surface atoms increases.[8] Figure 5.1 shows that a 1 nm diameter palladium nanoparticle has about 80% of the atoms on the surface.[8] This increases the surface area to volume ratio leading to their possible application as effective catalysts.[9–11] Second, the decrease in nanoparticle size increases the degree of surface curvature, resulting in a higher energy.[12] This leads to phenomena such as lowering of the melting temperature[13] and alteration of the metal oxidation potential.[14–16] In Figure 5.2, it is seen that a difference of no less than 500 °C in the melting temperature is found for cadmium sulfide nanoparticles of radii 25 and 50 Å.[13]

The properties of nanoparticles are also heavily influenced by the presence of capping agents. They reduce agglomeration of the nanoparticles by using different mechanisms (*i.e.* electrostatic forces, steric repulsions).[17–19] However, the capping agents are used for much more than stabilisation. The shape of silver nanoparticles synthesised can be controlled through the use of organic capping agents such as poly(vinyl pyrrolidone) and sodium citrate.[20] In turn, the surface plasmon peak occurs at different wavelengths depending on the particle shape.[21] With the variation of triethanolamine, oleic acid or thioglycerol as capping agents, the luminescence properties of zinc oxide nanoparticles differ.[22] Likewise, the extinction coefficient of gold nanoparticles changes as the citrate capping agent is replaced by 1-decanethiol or oleylamine.[23]

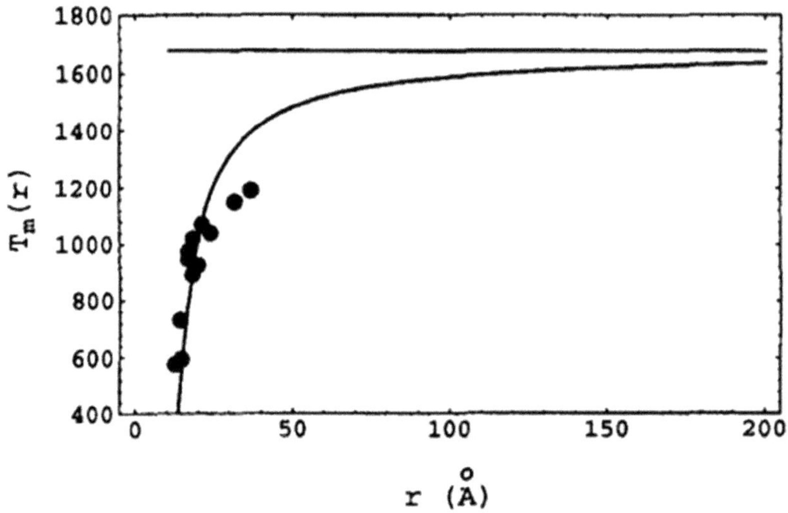

Figure 5.2 Melting temperature of CdS nanoparticles as a function of size. Reproduced from ref. 13 with permission from Cambridge University Press.

The novel properties of nanoparticles, such as a large surface area, can bring advantages that greatly benefit the society through the form of improved or new industrial and consumer products. For example, elemental silver is known to react with thiol groups and inhibit enzymes with thiol groups, hence granting silver anti-bacterial activity.[24] Silver nanoparticles are known to generate reactive oxygen species,[25,26] which put stress on cells.[27,28] Thus, they are now incorporated into products such as carpets,[29] socks[30] and fabric softener[31] as an anti-bacterial agent.[25,32–35] The capability of zinc oxide nanoparticles to remain invisible while blocking a wide spectrum of harmful ultraviolet rays makes them popular in sunscreens.[36] The tiny size of the nanoparticles allows them to remain transparent compared to their bulk counter parts that appear white.[37] Janus particles, which have at least two surfaces of different chemistry and/or polarity,[38] can be used for water repellent fabric manufacturing.[39] The hydrophilic side attaches to the fabric, leaving the exposed hydrophobic portion to repel water.[39] These products are just a few examples of the thousands of items that employ 'nanotechnology'. This industry was estimated to be worth US$20.7 billion in 2012 and predicted to expand to US$48.9 billion in 2017 according to BCC Research.[40] By October 2013, 1628 nanotechnology-enabled consumer products had been registered with the 'Consumer Products Inventory' on 'The Project on Emerging Nanotechnologies'.[31] The number of consumer products has increased rapidly over recent years. The most popular nanomaterials, according to their major elements, are silver, followed by titanium (including titanium dioxide and other forms), carbon (including fullerene, carbon nanotubes, graphene, graphite and other forms), silicon (found as silica and other forms), zinc (including zinc oxide and other forms) and gold.[31]

Given the flourishing growth of nanotechnology based consumer products, there are also *possible* negative implications. It is suggested that for pregnant mice injected with 100 μg of titanium dioxide nanoparticles, the nanoparticles are passed to their offspring.[41] Oral ingestion is also a possible route for nanoparticles to pass from the parent mice to the offspring.[42] A study in 2012 reported that the exposure of an average American to titanium dioxide nanoparticles is less than 2 mg per kg of body mass per day.[43] A study on carbon-based nanomaterials showed that they might activate inflammatory genes in pulmonary tissues in mice if their lung tissues are directly exposed to 125 μg of nanoparticles.[44] Given the possible health implications, nanomaterial usage likely needs regulation and laws to construct a safety net.

Current regulations are at an early stage but many countries and associations, including the European Union (EU), China, India, Taiwan and Thailand, are making significant efforts in nanoparticle research so as to make better legislation.[45] For the EU, there are no current regulations *specific* to nanotechnology.[46] Instead, they actively promote research and discussions. The programme REACH (Registration, Evaluation, Authorisation and Restriction of Chemicals) has stringent and broad regulations on chemicals and drugs to cover the risk of nanotechnology in the EU.[47] In 2012, France passed a decree for mandatory reporting of products containing nanomaterials.[48] China contributes significantly to the global joint effort to understand nanoparticles so as to make better guidelines.[49] Internationally, many properties (like concentrations, size distributions, impurities present, *in vivo* and *in vitro* toxicology effect) have been recommended as should be reported and diverse techniques like microscopy, elemental analysis, centrifugation and spectroscopy are recommended for nanoparticle analysis.[50]

5.2 Non-electrochemical Techniques for Nanoparticle Detection

To characterise nanoparticles, the International Life Science Institute Research Foundation/Risk Science Institute suggests a wide variety of laboratory methods for analysis.[50] These include microscopy techniques such as transmission electron microscopy (TEM) and scanning electron microscopy (SEM). Laser techniques, such as nanoparticle tracking analysis (NTA) and dynamic light scattering (DLS) are also recommended. Elemental analysis can be performed through inductively coupled plasma mass spectroscopy (ICP-MS). A wide range of surface analysis techniques such as X-ray diffraction (XRD) and X-ray photon spectroscopy (XPS) also play main roles in nanoparticle characterisation. Ultra-violet visible (UV-vis) spectroscopy, capable of measuring surface plasmon resonance, is also among the suggested techniques. As reported in Table 5.1, different characteristics

Table 5.1 Range of analytical techniques used for nanoparticle characterisation. (Reproduced from ref. 50 under the terms of Creative Commons Attribution 4.0 International. permits use, distribution and reproduction in any medium, provided the original work is properly cited and is not used for commercial purposes.)[a]

Analytical technique							
		Transmission Electron Microscopy (TEM)	Scanning Electron Micro-scopy (SEM)	X-Ray Diffrac-tion (XRD)	X-ray Photon Spectro-scopy (XPS)	Auger Spectro-scopy (AES)	Secondary Ion Mass Spectro-scopy (SIMS)
Physico-chemical characteristic	Size distribution (primary particles)	▲	●	●			
	Shape	▲	●				
	Surface area	●	◇				
	Composition	●	●	●			▲
	Surface chemistry	●	✓		●	●	✓
	Surface contamination	✓			●	●	
	Surface charge – suspension/ solution						
	Surface charge – power (use bio fluid surrogate)						
	Crystal structure	●	◇	▲			
	Particle physicochemical structure	▲	●				✓
	Agglomeration state	▲	●	✓			
	Porosity	◇					
	Heterogeneity	▲	●				

[a]Other applicable techniques are available that have not been listed. ▲ Highly applicable. ● Capable of providing information in some cases. ✓ Capable of providing information in some cases, with validation from more accurate/applicable techniques. ◇ Capable of providing qualitative or semi-quantitative information.

of the nanoparticles such as size, concentration, elemental composition, surface properties and shape can be measured by various methods.[50] In general, however, a combination of techniques is required to fully characterise the nanoparticles. In the paragraphs below, the various characterisation methods mentioned earlier are surveyed for their relative merits.

Microscopy techniques such as SEM and TEM provide powerful magnification. SEM uses an electron beam to scan and generate images and is capable of resolving detail down to 3 nm.[51] The ability of SEM to 'see' the nanoparticles provides information on their size, shape and surface area. However, SEM is incapable of differentiating between any capping agent and nanoparticle surface features.[6] Moreover, to be examined by SEM, the samples have to be conductive. Thus, a thin layer of sputtered metal coating is often used for non-conducting samples.[52] In addition, the SEM sample studied has to be non-volatile and capable of tolerating vacuum

Scanning probe microscopy	Dynamic Light Scattering (DLS)	Zeta potential	Size exclusion chromato-graphy	Analytical ultracen-trifugation	Differen-tial Mobility Analysis (DMA)	Isothermal adsorption (*e.g.* BET)	Spectro-scopic technique (UV vis. IR, Raman, NMR)	Elemental analysis (*e.g.* ICP-MS/AA *etc.*)
●	●		●	●	●			
●	✓				✓			
◇	◇			◇	✓	●		
							●	▲
✓		◇					✓	
✓							✓	
		▲						
		▲						
●	✓		✓	✓	●			
					●	●		
◇							✓	

conditions.[53] SEM also has a possible destructive nature as the electron beams may damage the sample.[54]

The other powerful magnification technique, TEM, uses an electron beam to tunnel through the sample for imaging.[55] TEM is able to detect particles as small as 0.1 nm.[51,55] With aberration correction, TEM is capable of reaching a higher resolution where single atoms can be seen.[56] Figure 5.3 shows the high resolution TEM image of a silver nanoparticle deposited onto an amorphous carbon film.[57] The lattice structure is atomically resolved. Each individual bright spot represents a single atomic column.[57] Similar to SEM, TEM is able to visualise and characterise nanoparticles. Viable samples for TEM are subjected to strict criteria; they must be electron transparent and able to withstand the vacuum environment.[53,55,58] In TEM, the energy from the electron beam may damage the samples to different extents (especially organic samples).[54] Various sample preparations have been

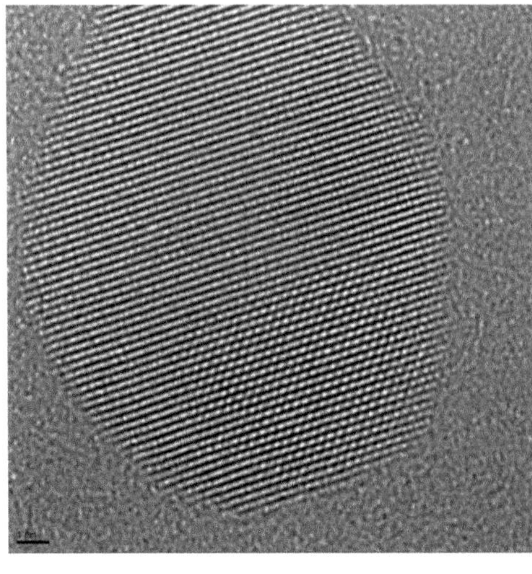

Figure 5.3 High resolution TEM image of a silver nanoparticle. Scale bar of 1 nm. (Adapted from ref. 57 with permission.)

improvised to reduce the sample damage (*e.g.* TEM sample freezing with liquid nitrogen, electrochemical pre-treatment of metal nanoparticles, *etc.*).[59,60]

The usage of lasers gives rise to the two different techniques of NTA and DLS. NTA uses a laser to illuminate particles and a microscope to track the individual particles.[61] By tracking the Brownian motion of the particles, the diffusion coefficient (D) is obtained. The hydrodynamic radius of the particle (r) is then be calculated with the use of the Stokes–Einstein equation:[62]

$$D = \frac{k_B T}{6\pi\eta r} \tag{5.1}$$

where k_B is the Boltzmann constant of 1.38×10^{-23} J K^{-1} per molecule, T is the temperature in kelvin and η is the viscosity of the liquid sample. The use of the Stokes–Einstein equation assumes that the measured particle is a perfect sphere. In reality, nanoparticle samples often contain particles close to a spherical shape rather than being perfect spheres. Figure 5.4 shows a typical sample of spherical silver nanoparticles imaged by TEM, most of which are not perfect spheres.[63] However, NTA is not without advantages. This technique is easy to use and only a small volume of sample is required (around 0.3 mL) each time.[64] The sample does not have to undergo any special pre-treatment apart from dilution to the appropriate concentration. Nonetheless, NTA has a narrow optimal concentration range of 10^7–10^9 particles per mL in order to track sufficient particles to measure a proper size distribution.[64] Therefore, an estimation of the sample concentration is usually required before NTA is performed.

DLS measures the hydrodynamic radius through the light scattered by the particles.[65] A shorter wavelength is favourable for analysis as it is scatters

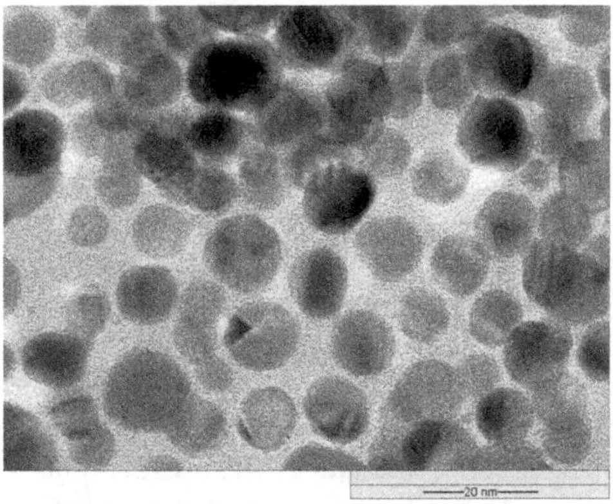

Figure 5.4 TEM brightfield image of a silver nanoparticle sample. Scale bar of 20 nm.
(Reprinted from ref. 63 with permission from Elsevier.)

more efficiently, thus lowering the limit of particle size measurable.[66] According to the DLS manufacturer Malvern, the technique is capable of detecting a large range of sizes from 0.3 nm to 10 μm.[67] Pre-knowledge of the reflective index and absorbed wavelength of the sample are required for size analysis.[68] Similar to NTA, DLS is easy to use and sample preparation includes simple filtration and dilution. As both NTA and DLS measure the particle hydrodynamic radii, a careful choice of capping agent of the nanoparticles is required as certain capping agents, such as poly(ethylene glycol) (PEG), can cause a significant difference between the hydrodynamic radius and the actual radius.[69]

The elemental analysis of nanoparticles is often performed through ICP-MS. An argon plasma is utilised to dissociate the analytes into fragments.[70] The fragmented sample is separated according to the mass-to-charge ratio and then detected through an electron pulse with an electron multiplier. ICP-MS is capable of detecting certain metallic elements down to 1 part per trillion.[70] Moreover, isotopic analysis and screening for multiple elements in one single measurement is possible.[70] Due to its high sensitivity, contamination is a possible major concern. Argon, which is abundant in the plasma, generates fragments such as $^{40}Ar^+$, $^{56}ArO^+$ and $^{80}Ar_2{}^+$, which can interfere with ^{40}Ca, ^{56}Fe and ^{80}Se detection, respectively; ^{56}Fe and ^{80}Se are found commonly as nanoparticles in the forms of iron oxide[67] and cadmium selenide,[71] respectively.

Surface analysis is commonly performed with XRD and XPS. XRD is based on the principle of Bragg's law; using a source of X-rays, crystalline samples are scanned at various angles and the surface properties are derived from the way the electromagnetic waves are diffracted.[72,73] It determines the structural composition of the sample and is also phase sensitive towards the different morphology of lattice planes.[72,73] However, XRD is limited to crystalline samples.[6] Another method that uses X-rays for surface analysis is XPS (X-ray photoelectron spectroscopy).[74] It reveals information on the elemental composition on the sample surface. It is very sensitive towards changes on the surface which, for example, allows the study of metal nanoparticle oxidation in air.[74] XPS is also capable of identifying the capping agent on nanoparticles.[75] Samples are processed after the XPS chamber is carefully brought to ultra-high vacuum.[74] XPS can operate as a non-destructive or destructive technique (through mechanical sectioning).[74] The sample holder usually has to be cleaned through heating and ion bombardment.[74] Although both XRD and XPS use X-rays for surface analysis, the key difference is that XRD deals primarily with physical properties (*e.g.* phase and orientation of lattice planes) while XPS essentially determines chemical properties (*e.g.* the chemical state of the elements).

Lastly, UV-vis spectroscopy measures the surface plasmon signals caused by the oscillating electron field of the nanoparticles.[76] Certain small changes on the nanoparticle surface can be detected through UV-vis spectroscopy. For example, a layer of oxide on silver and gold nanoparticles causes a redshift in the surface plasmon signal.[77] Changes in shape of the nanoparticles[17] or the presence of different capping agents[19] on the nanoparticles can modify the wavelength of the surface plasmon signal. Despite the capabilities of UV-vis spectroscopy, it is only able to give a rough gauge of the nanoparticle size.[3] In addition, the extinction coefficient changes with size, making it difficult to determine the concentration of nanoparticles since a typical sample contains a distribution of sizes.[3] Often, it is used to confirm the presence of nanoparticles instead of quantifying them.[78]

Although there are many approaches to detect and study nanoparticles, there is still *no* single ideal technique to characterise the nature of dynamic nanoparticles.[79] Usually, a combination of methods is required to obtain the full picture. Nonetheless, various methods may measure a different size average.[79] SEM and TEM, which visualise nanoparticles, determine the number average while NTA obtains the number average based on the diffusion coefficient. A sample bias towards a larger particle mean size is probable because larger nanoparticles are easier to detect.[79] The three techniques of SEM, TEM and NTA record the responses from individual particles. On the other hand, DLS measures the ensemble feedback from the nanoparticles. Hence, the average size measured by DLS is a weighted average based on the translational diffusional coefficient. Signals generated by smaller nanoparticles can be difficult to detect due to the masking by the larger ones.[79] Therefore, the different techniques of nanoparticle sizing may result in various measured average sizes, due to the different ways used to calculate the average, the polydispersity of the sample[79] and the capping agents present on the particles.[69] Moreover, with techniques like TEM and SEM that operate under vacuum, the removal of solvent from the nanoparticles can cause the sample to change; for example, agglomeration occurs during drying after silver nanoparticles are drop cast on the sample holder.[14] However, this may be overcome through freeze drying the sample or by solvent sublimation.[79] Aggregation and dissolution are the two main processes that can cause a difference between the measured size and the actual size.

One major field that has not been fully utilised for nanoparticle characterisation is electrochemistry. Given its fast, robust methodology and the recent development of nanoparticle–electrode impact experiments, electrochemical methods have considerable scope and promise for nanoparticle characterisation. These are discussed in Sections 5.3 and 5.4.

5.3 Stripping Voltammetry for Nanoparticle Detection

Electrochemical methods have the advantages of being fast, sensitive and robust. With the invention of disposable screen printed electrodes[80–83] and the use of small portable potentiostats,[84] it is possible to perform electrochemical experiments at almost any location. Blood glucose sensors,[85,86] gas sensors,[87–89] metal ions sensors[90,91] and chilli heat (capsaicin) sensors[92] are both successful point-of-use examples where electrochemistry is utilised. Given the advantageous properties of electrochemical methods, it is potentially beneficial to use them as nanoparticle detection techniques. Nanoparticles are often found in environmental settings such as river water or seawater but especially in the discharge of effluents.[93] Portable electrochemical systems would be suitable for quick onsite sampling of nanoparticles. Currently, most studies are at the fundamental levels where the nanoparticles are characterised and detected in laboratory settings. If the nanoparticles are redox active species, the simplest and most direct method of nanoparticle detection can be performed through stripping voltammetry.

Stripping voltammetry involves oxidising or reducing the nanoparticles immobilised on the electrode surface through electrochemical techniques such as linear sweep voltammetry (LSV), cyclic voltammetry (CV) and pulse voltammetry analysis (differential pulse voltammetry (DPV) and square wave voltammetry (SWV)) to generate an electrochemical signal.

CV changes the potential of the working electrode linearly in a forward and then a backward direction and records the current as the potential changes.[94] Small potential steps are made and the current, i, is measured as the potential is applied. The current is measured at one point along the step.[95] The current measurements at the steps, i, are plotted against the potential to give a voltammogram that looks like Figure 5.5.[14] Figure 5.5 shows the CV of the oxidative stripping of silver nanoparticles from a glassy carbon electrode.[14] The amount of nanoparticles present can be quantified through eqn (5.2):

$$N = \frac{Q}{nF} \tag{5.2}$$

The charge passed under the voltammetric signal, Q, will correspond directly to the moles of surface immobilised nanoparticles, N, through Faraday's first law where n is the number of moles of electrons involved in one mole of reaction and F is the Faraday constant of 96 485 C mol^{-1}. LSV

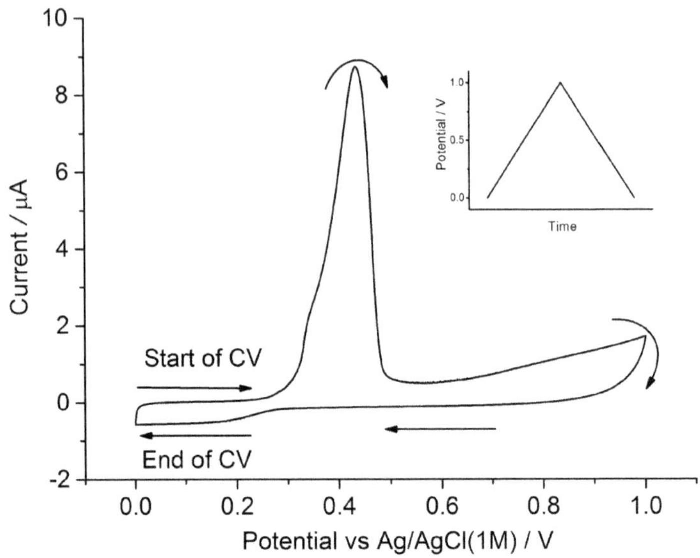

Figure 5.5 Oxidative stripping of silver nanoparticles from a glassy carbon electrode. Insert: the change in potential with regards to time. (Adapted from ref. 14 with permission.)

works in the same way as CV but the potential is only swept in one direction (half the cycle shown in Figure 5.5).

Pulse voltammetric methods such as DPV and SWV are receiving more attention as they are more sensitive.[96] The additional sensitivity is achieved by using a pulsed wave form in addition to a scanning potential.[97] The current differences between the pulsed steps are plotted against potential to give a voltammogram. Commonly, this subtraction allows the background current to be better offset, resulting in a greater resolution of the voltammetric peaks. As a result, the sensitivity towards the redox species is improved. Moreover, as the potential changes back and forth, the redox species is oxidised and reduced repeatedly if the redox reaction is reversible.[97] This gives a larger amount of charge measured under the voltammetric signal and eqn (5.2) cannot determine the amount of nanoparticles immobilised on the electrode surface. Instead, a calibration curve is often used for SWV and DPV.

Electrochemical methods are sensitive but nanoparticles often need to be analysed at very low concentrations. In many experiments, the total Faradaic charge passed or the voltammetric signal peak height are used to determine the amount of nanoparticles present. However, to achieve a quantifiable signal, the nanoparticle concentration required is much higher than those naturally found. Therefore, a pre-concentration step is part of the stripping voltammetry analysis. A wide assortment of tactics like drop casting and adsorption can be used. Figure 5.6 shows the three main types of pre-concentration strategies (adsorption, drop casting and affinity enhancement) before stripping voltammetry is performed.

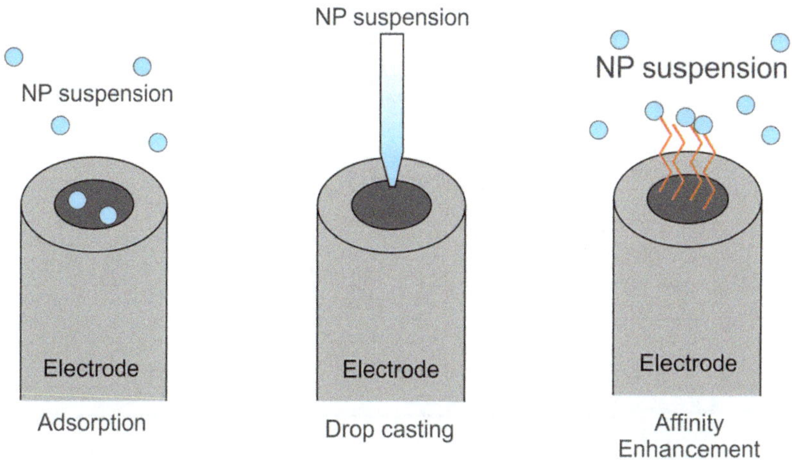

Figure 5.6 Main strategies for pre-concentration of nanoparticles. Drawings are not to scale.

First, the pre-concentration method of adsorption involves immersing the electrode into a nanoparticle suspension and waiting for adsorption to occur. Commonly, a potential is held at the electrode during adsorption to minimise the amount of nanoparticles desorbing.[98] Adsorption can also be performed at open-circuit potential.[99] One main advantage is that the adsorption period can be extended when a sample of very low nanoparticle concentration is analysed. After the adsorption step, stripping voltammetry analysis is performed to determine the amount of particles present through Faraday's first law (eqn (5.2)). One good demonstration of using adsorption as a strategy for pre-concentration is a study by Neumann *et al.*[98] In their work, a carbon microelectrode is held at −0.25 V *versus* a saturated calomel electrode (SCE) for a period of time to pre-concentrate the silver nanoparticles onto the electrode surface before stripping voltammetry is performed. Figure 5.7 shows the stripping peaks obtained after various periods of adsorption.[98] As the adsorption time lengthens, the amount of silver nanoparticles adsorbed increases and the voltammetric signal grows larger.

The correct choice of medium is essential to obtain quantitative data from stripping experiments. For example, Pumera *et al.* have oxidised adsorbed gold nanoparticles to the gold chloride complex $AuCl_4^-$ using strong hydrochloric acid.[99] Comparing the oxidation of gold nanoparticles to $AuCl_4^-$ and the reduction of $AuCl_4^-$ to metallic gold, the latter process allowed the concentration of the gold nanoparticles to be determined at a higher sensitivity through a better, well-defined signal. A similar concept has also been applied to gold nanoparticles and bromide ions.[100] Judging from the success of changing nanoparticles into other forms, cathodic stripping voltammetry (CSV) may be applicable to detection of metal oxide nanoparticles.[101–104] Typical systems would be the oxides of manganese[101,102] and lead.[103,104]

Another way to promote adsorption is to use migratory forces.[105] In one method, gold nanoparticles are synthesized *in situ* in the presence of very

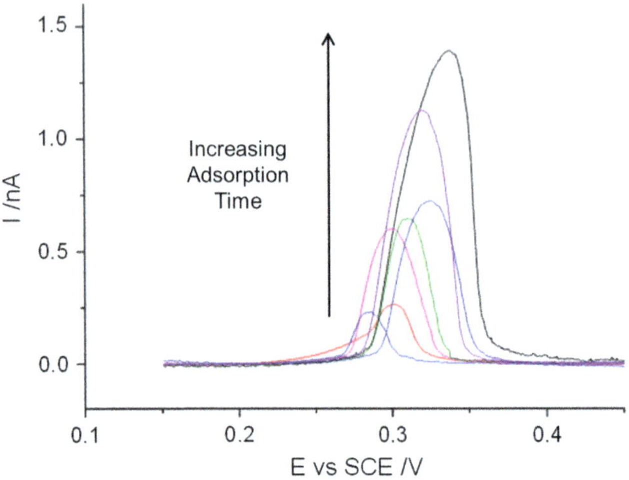

Figure 5.7 Stripping voltammograms obtained by pre-treating a micro carbon electrode in a silver nanoparticle suspension at −0.25 V *versus* SCE for various periods of time.
(Reproduced from ref. 98 with permission from John Wiley & Sons.)

little electrolyte.[105] A positive potential is held on the working electrode, attracting the negatively charged citrate capped gold nanoparticles. With the extra migratory forces, the efficiency of adsorption is improved.

In contrast, instead of adsorbing many particles on the electrode, Lakbub *et al.* have used a nanosized electrode for single nanoparticle adsorption.[106] As the single gold nanoparticle adsorbs onto the platinum electrode, the cyclic voltammetry performed records a signal corresponding to the reduction of gold oxide. Figure 5.8 shows that peak A, which corresponds to the reduction of gold oxide, can be used to infer the apparent nanoparticle size through the charge passed under the signal.[106]

The second pre-concentration strategy is drop casting. When the nanoparticles are isolated, they can be drop cast and immobilise onto the working electrode. This concept is illustrated in a report on poly(*N*-vinylcarbazole) (PVK) nanoparticles.[107] PVK undergoes a two-electron oxidation with the mechanism shown in Figure 5.9.[108] The amount of the redox active PVK can be determined through eqn (5.2). The synthesized PVK nanoparticles are drop cast and dried on a glassy carbon electrode. Oxidation signals are observed in a cyclic voltammogram and the charge passed under the voltammetric peak scales directly with the concentration of monomers present. It is reported that the moles of electrons transferred per mole of reactant, *n*, is close to 2.[107] With this, a linear calibration is obtained between the peak charge and the number of monomers present, allowing the determination of unknown monomer concentrations on the electrode surface.

Instead of using the charge under peak area, the voltammetric peak height can also be used to determine the amount of nanoparticles present on the

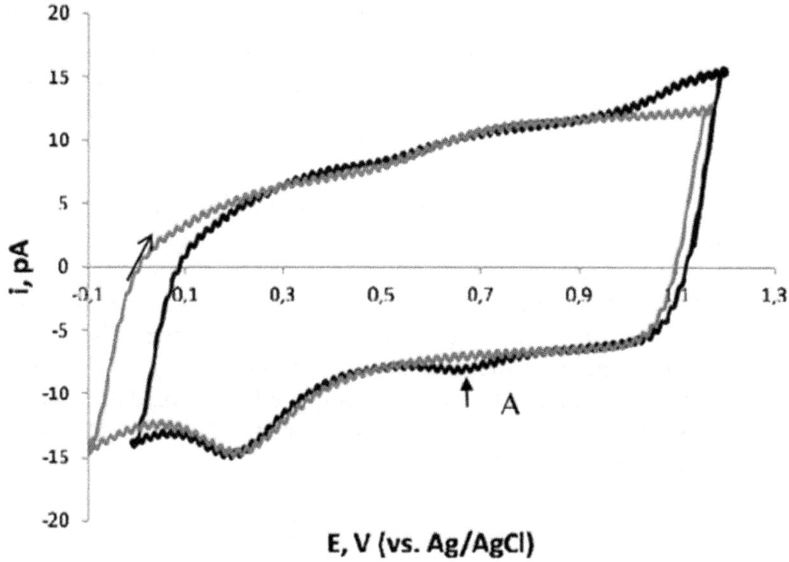

Figure 5.8 Stripping of gold oxide (peak A) generated from a single gold nanoparticle from a platinum electrode.
(Reproduced from ref. 106 with permission from John Wiley & Sons.)

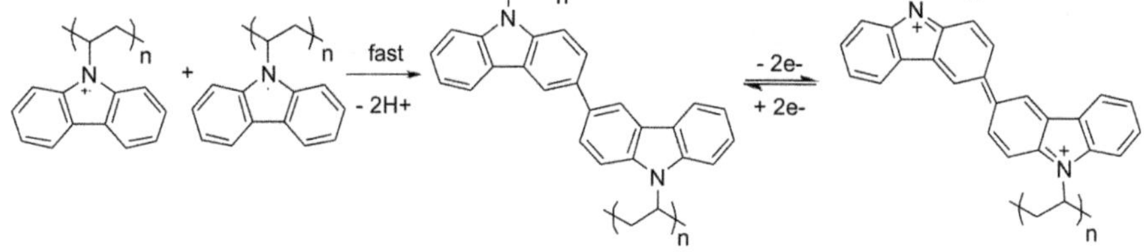

Figure 5.9 Poly(*N*-vinylcarbazole) (PVK) oxidation mechanism.
(Reproduced from ref. 108 with kind permission from Springer Science and Business Media.)

electrode surface. In work by Teo *et al.*, copper oxide nanoparticles were drop casted onto the working electrode and a cyclic voltammogram performed with the nanoparticle modified electrode.[109] The voltammetric peak height was linearly correlated to the amount of nanoparticles drop cast on the electrode. This method has been applied to nickel,[110] nickel oxide,[110] molybdenum[111] and silver[16] nanoparticle detection.

Apart from determining the concentration of nanoparticles, it is possible to size the nanoparticles through the measurement of the stripping peak potential. It has been demonstrated that size affects the redox peak potential[14–16] and it might be possible to size the nanoparticles through the shift in potential.[14] Taking silver as an example, Pumera and his co-workers reported that a less positive oxidation peak potential is observed as the size of the silver nanoparticles decreases.[16] This can be explained through two factors.

The first factor that causes the recorded oxidation peak potential to differ with various size is surface energy.[12] As the particles decrease in size, more atoms are on the surface of the particle and the curvature of the particle increases. Thus, the atoms are less stabilised, leading to an increase in surface energy. In turn, the oxidation peak potential lowers because it is easier to oxidise smaller nanoparticles with higher energy.[12] The relationship between the nanoparticle radius, *r*, and the formal oxidation potential of the nanoparticle, $E^{\theta}_{\mathrm{f(NP)}}$ can be expressed through Equation (5.3) (significant for particles of diameter less than 30 nm):[12,14,112]

$$E^{\theta}_{\mathrm{f(NP)}} = E^{\theta}_{\mathrm{f}} - \frac{1}{r} \cdot \frac{2\gamma M}{n\rho F} \tag{5.3}$$

where E^{θ}_{f} is the formal oxidation potential of bulk silver of +0.80 V *versus* SHE,[113] γ is the surface energy, M is the molecular weight of silver, n is the number of electrons involved per mole of reaction and ρ is the density of silver. As far as the surface energy of the nanoparticles is considered, the ease of silver oxidation increases with decreasing radius.

Second, the surface coverage of the particles may affect the oxidation potential. It depends on the diffusion regime applicable for the system.[14,114] A schematic diagram of the four different cases is given in Figure 5.10.[14]

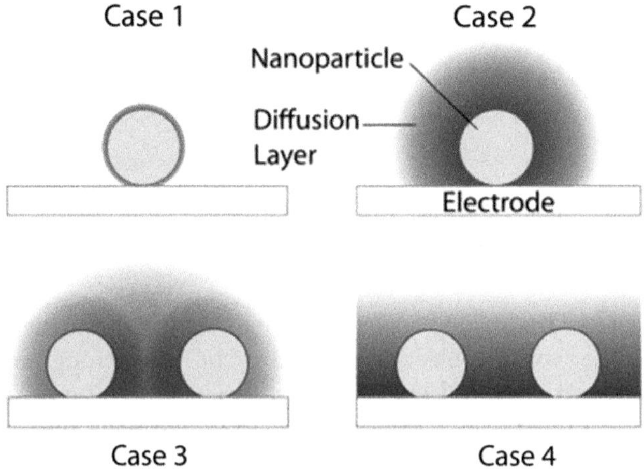

Figure 5.10 Schematic diagram showing the four cases of diffusional regimes at individual nanoparticle or nanoparticle assemblies.
(Reproduced from ref. 14 with permission.)

In Case 1, there is a thin, linear diffusion layer at each particle and this occurs within a short time scale of a few nanoseconds (very fast scan rates). Hence, it is not applicable to most electrochemical methods, which take seconds or minutes to complete. In Case 2, a radial diffusion builds up at the particle. Cases 1 and 2 are applicable when the nanoparticles are widely separated and there is no overlapping of diffusion layers between the neighbouring particles. In Case 3, the diffusional layers of neighbouring particles overlap partially to various extents. In Case 4, there is a complete overlap of the diffusional layers of the nanoparticles resulting in a linear diffusion at the electrode surface. Taking the reversible oxidation of silver as a case study, we can see how the oxidation peak potential is affected by various factors at different cases of diffusion regimes.[14,15] As the reversible oxidation of silver occurs, the concentration of silver(I) ions at the electrode surface influences the oxidation potential, E_{Ag/Ag^+}, through the Nernst equation:

$$E_{Ag/Ag^+} = E^{\theta}_{f(Ag/Ag^+)} + \frac{RT}{F}\ln\frac{[Ag^+]_0}{[\,]^{\theta}} \tag{5.4}$$

where $E^{\theta}_{f(Ag/Ag^+)}$ is the formal oxidation potential of silver under standard conditions (+0.8 V *versus* NHE) and $[\,]^{\theta}$ is the standard state of 1 mol dm^{-3}. As seen in Eqn (5.4), an increase in silver ion concentration in solution will lead to a higher oxidation potential of the Ag/Ag^+ redox couple. Thus, the speed of the silver(I) ions diffusing away from the nanoparticle surface will influence the oxidation potential. In Case 2, the oxidation potential is dependent on the particle size but not surface coverage as the diffusion is dependent on individual particles. Given that, in Case 2, the mass transport coefficient (k_{mt}) is related to the radius of the nanoparticle, r, through eqn (5.5):[14]

$$k_{mt} = \frac{D\ln 2}{r} \tag{5.5}$$

the mass transport of silver ions away from the electrode surface is inversely related to the radius of the nanoparticle; a larger nanoparticle would have a slower diffusion of silver(I) ions away from the surface, resulting in a higher oxidation peak potential. In Case 3, the oxidation potential is determined both by the surface coverage and nanoparticle size. The proportion of the influence of the two factors depends on the extent of the overlapping layers. At high surface coverage of silver nanoparticles (Case 4), the oxidation potential decreases with decreasing silver coverage. The particle size does not affect the oxidation potential as it does not influence the diffusion of silver(I) ions away from the surface. Therefore, during the evaluation of a change in oxidation peak potential, the diffusional regimes have to be taken into account.[14,15,98] In summary, the simple 'drop cast and strip' method allows one to identify the nanoparticles through the potential of the redox

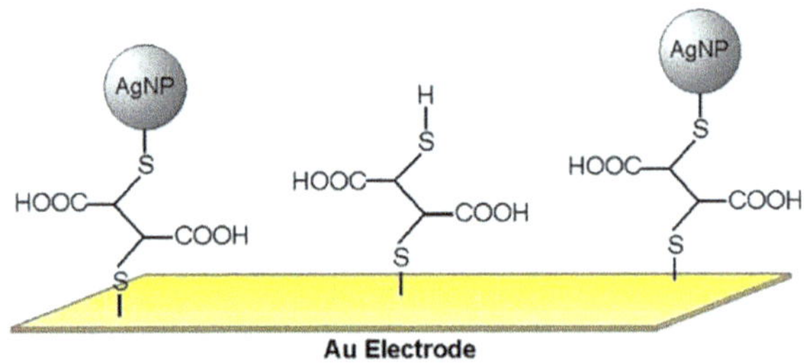

Figure 5.11 Schematic diagram of DMSA (*meso*-2,3-dimercaptosuccinic acid) anchoring silver nanoparticles near a gold electrode through the interactions with thiol groups.
(Reproduced from ref. 116 with permission from John Wiley & Sons.)

peaks and it holds the *possibility* of determining the size of nanoparticle when the conditions are very strictly monitored.

The third pre-concentration approach is electrode modification to enhance the affinity of the electrode surface for the nanoparticles. Making use of the strong interaction between silver and thiol, cysteine modified electrodes are used to improve the adsorption of silver nanoparticles.[115] A similar concept has also been applied to gold electrodes with the modification agent *meso*-2,3-dimercaptosuccinic acid (DMSA). The thiol group at both ends of DMSA helps to anchor the silver nanoparticle onto the gold electrode (Figure 5.11).[116] The efficiency of silver nanoparticles adsorbing onto the DMSA modified electrode is improved compared to an unmodified gold electrode. This shortens the time required for nanoparticles pre-concentrating onto the electrode surface. Presently, silver nanoparticles have been successfully captured on glassy carbon electrodes,[115] carbon screen printed electrodes,[117] gold electrodes[116] and home-made gold CD electrodes[118] under laboratory conditions[115,116] and in seawater.[117,118] Future developments include leaving one-shot disposable electrodes in the environment for extended periods of time to capture nanoparticles before they are collected for analysis.[118]

Apart from pure electrochemical methods, other techniques like microscopy and electrophoresis have been combined with electrochemistry for nanoparticle detection. In work by Batchelor-McAuley *et al.*, silver nanoparticles have been tracked with a microscope through the scattered light from a green laser.[119] Through the light scattered, holograms were constructed to determine the size and the pathway taken by the nanoparticles as they adsorbed on an electrode. Stripping voltammetry confirmed the amount of silver the particles contained (eqn (5.2)). A similar concept has been attempted by Munteanu *et al.* to track radical nanoparticles.[120] In a quite different work by Chua and Pumera, a lab on chip platform has been used to separate nanoparticles of different sizes and sized using electrophoresis and the measurement of retention time.[121]

Stripping voltammetry detects numerous nanoparticles simultaneously. In Section 5.4, experiments to detect single nanoparticles through impacts on the working electrode surface are surveyed.

5.4 Nanoparticle–Electrode Impacts for Single Nanoparticle Detection

The technique of studying the signal generated by a single nanoparticle impacting an electrode has gained much attention recently and it is one of the most promising methods for single nanoparticle detection.[122–125]

In the early 2000s, collisions between micron-sized particles and electrodes were observed as 'spikes' on voltammograms.[126,127] The difference in the current between the 'spikes' and the background current was related to the size of the particles.[128–131] These impacts were non-Faradaic. In a typical experiment, heptane droplets were mixed into an aqueous solution of perchloric acid and studied with a gold working electrode.[129] Figure 5.12 shows a schematic diagram of the heptane droplets hitting the electrode and causing a disturbance in the double layer.[129] As the heptane droplets hit the electrode surface, they momentarily block the electrode surface and disrupt the interface. After the impact, the double layer reforms. Figure 5.13 shows the non-Faradaic 'spikes' observed at different electrode potentials as heptane droplets impact a gold electrode in 0.1 M perchloric acid.[129] Depending on the potential held, it generates oxidative 'spikes' (potential held positive of the potential of zero charge (pzc)) or reductive 'spikes' (potential held negative of pzc) that are exclusively of a non-Faradaic nature. For this system, the pzc is measured to be at +0.01 ± 0.01 V *versus* Pd/H_2.

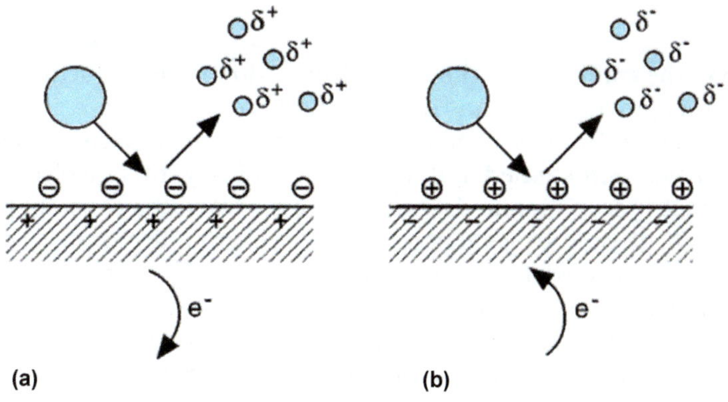

Figure 5.12 Schematic diagram showing the heptane droplets impacting the electrode and disrupting the double layer. (a) Potential applied > potential of zero charge; (b) potential applied < potential of zero charge. (Reprinted with permission from ref. 129. Copyright (2002) American Chemical Society.)

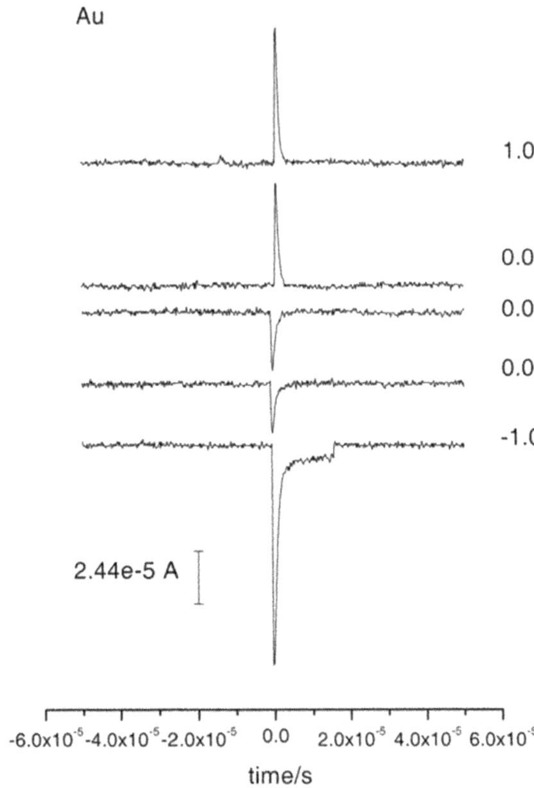

Figure 5.13 Spikes observed as heptane droplets of an average 3 μm diameter impacts the electrode at different potentials.
(Reprinted with permission from ref. 129. Copyright (2002) American Chemical Society.)

Therefore, the transition of the spikes from oxidative to reductive occurs at pzc and this provides a means of measuring pzc.

More recently, *Faradaic* charges resulting from the impacts of nano-sized particles have been studied and measured. Typically, this involves a chronoamperometric scan that holds the working electrode at a constant potential for a period of time to generate a stable background current. The nanoparticles are detected when they impact the electrode, giving a current transient. The change in current appears as a 'spike' or a 'step' in the current–time transient (chronoamperometry). It is usually performed on an electrode that is of a micrometre size to ensure a low background noise.

For impact experiments, the signal may arise directly from the redox reaction of the nanoparticles or the signal can result from an indirect redox reaction that happens on the nanoparticle surface. For indirect redox reactions, Wang *et al.* have sub-classified them into two categories of 'hit and run' and 'hit and stand'.[123] All of the three scenarios are illustrated in a

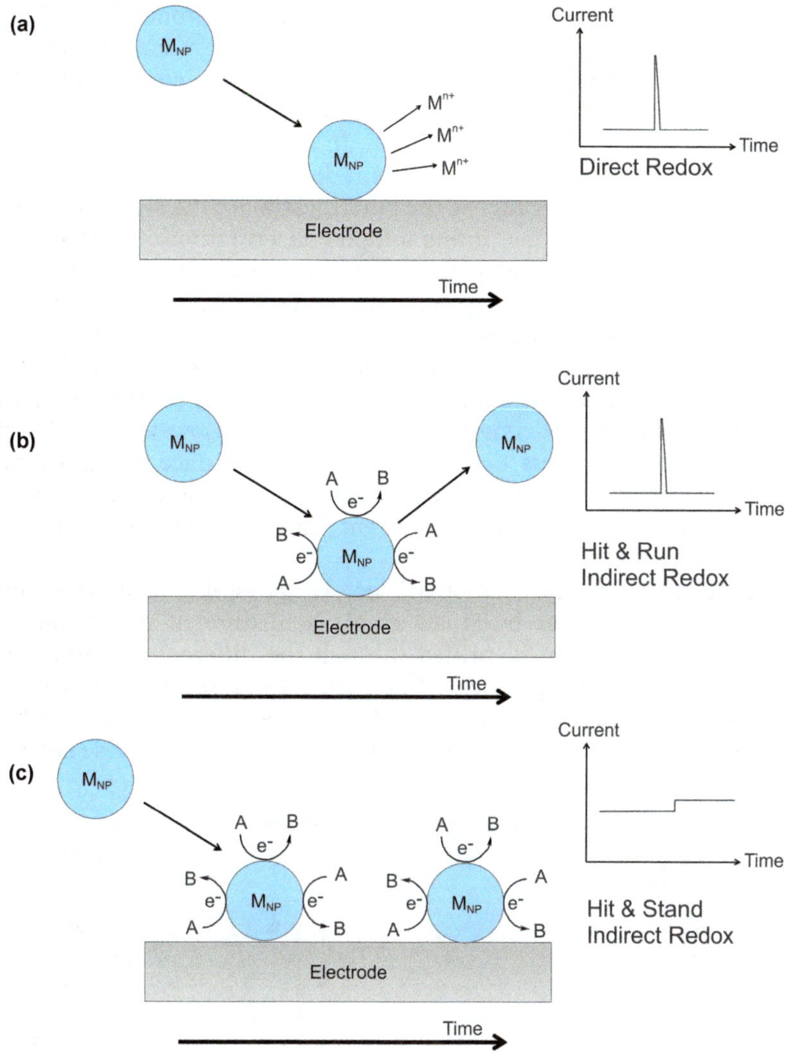

Figure 5.14 The three main scenarios of impact experiments. The voltammogram generated by each mechanism is next to the schematic diagram. (a) Direct redox; (b) indirect hit and run; (c) indirect hit and stand.

schematic diagram in Figure 5.14. The first scenario (Figure 5.14a) involves a direct redox reaction. As the nanoparticle M impacts the electrode, the potential held on the electrode oxidises it into M^{n+}, generating a sharp voltammetry signal ('spike'). This is also applicable to nanoparticles that are reducible to generate reductive 'spikes'. After the direct redox reaction, the soluble products diffuse away.

The second scenario (Figure 5.14b) involves a 'hit and run' and an indirect redox reaction. In this case, the nanoparticle M itself does not undergo any redox reaction. Instead, it catalyses a redox reaction to convert A into B in the solution phase. As the nanoparticle impacts the electrode ('hit'), the electrons involved in the transformation of A into B generate a current. As the nanoparticle moves away from the electrode ('run'), the conversion of A into B stops and no further change in faradaic current is recorded. This results in a 'spike' recorded in the voltammogram. In this scenario, the nanoparticles are not consumed, making this a non-destructive detection technique. An advantage of this technique is the amplification of the signal. Comparing the catalysis of A into B (assuming a high concentration of A) to the direct oxidation or reduction of the nanoparticles, more electrons are passed in the former case. Thus, the signal generated is often larger in magnitude.

The third working scenario of impact experiments (Figure 5.14c) involves a 'hit and stand' and an indirect redox reaction. Similar to the second scenario, this is also a non-destructive technique. The nanoparticle M again promotes the conversion of A into B. However, in the third scenario, once the nanoparticle hits the electrode surface ('hit'), it remains on the electrode surface for a considerable time ('stand') and continues to convert A into B. A change in current ('step') is observed instead of a 'spike'.

Identification of the nanoparticles is possible with the impact experiments. In a paper by Stuart *et al.*, a mixture of nickel and silver nanoparticles have been identified through the different oxidation potentials at which the 'spikes' occur.[132] Figure 5.15 shows the onset of silver oxidation 'spikes' at +0.45 V *versus* Ag/AgCl reference electrode whereas nickel oxidation 'spikes' only appeared beyond +1.55 V.[132] These apparently

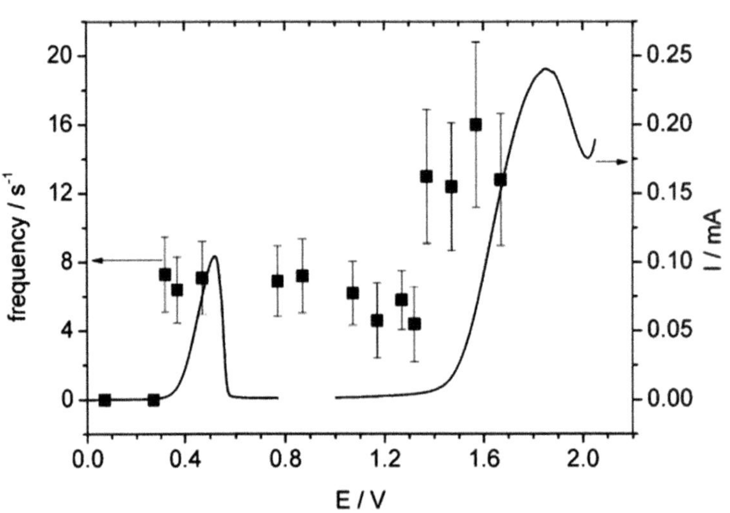

Figure 5.15 Overlap plot of stripping voltammograms of a silver nanoparticle modified GC electrode and a nickel nanoparticle modified GC electrode (right-hand axis) and the impact frequency (left-hand axis) (see text for details).
(Reproduced from ref. 132 with permission.)

matched the oxidation potentials of the nanoparticles in the stripping voltammogram.

Nanoparticles can be sized in two ways. First, in the case of direct redox reaction (first scenario), by relating the charge of a 'spike' through Faraday's first law to the number of atoms present in a nanoparticle the size of the nanoparticle can thus be studied. Second, for the indirect redox methods (second and third scenarios), the radius of the nanoparticle (assumed spherical) can be determined through the current for the conversion of A into B.

In the first scenario of direct redox reaction, many redox active species have been identified.[132–147] For example, various papers on single silver nanoparticle detection have been published.[132–142] Silver nanoparticles are added into the electrolyte and the impact of silver nanoparticles on a carbon microelectrode is recorded.[132–142] The signal results from the direct oxidation of the silver nanoparticles:

$$Ag_{(NP)} \rightarrow nAg^{+} + ne^{-} \tag{5.6}$$

where n is the number of silver atoms in a silver nanoparticle. Figure 5.16 depicts typical chronoamperometry obtained by the impact experiments.[133]

Each 'spike' relates to a single nanoparticle impacting and oxidising on the electrode and the size of the nanoparticles can be calculated through Equation (5.7), assuming the nanoparticles are spherical in shape:[132]

$$Q = \frac{4n\pi F\rho r_{NP}^{3}}{3A_{r}} \tag{5.7}$$

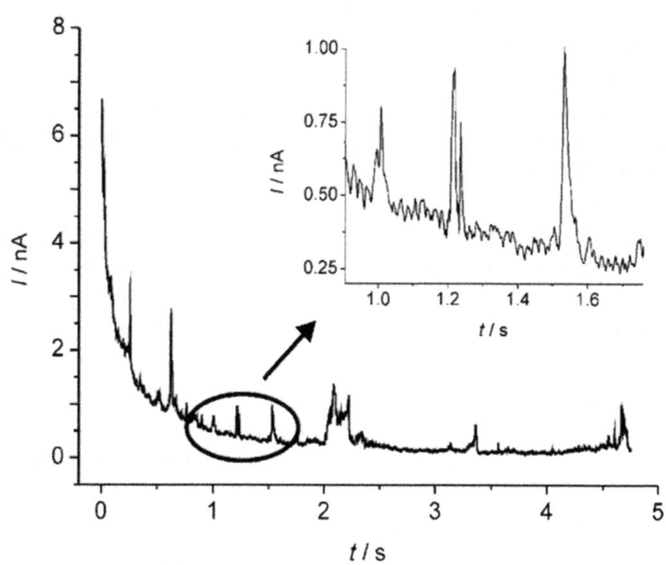

Figure 5.16 'Spikes' generated by the oxidation of silver nanoparticles after impacting a carbon electrode.
(Reproduced from ref. 133 with permission from John Wiley & Sons.)

where Q is the charge passed under the 'spike', n is the number of moles of electrons involved per mole of silver (*i.e.* 1), r_{NP} is the radius of the silver nanoparticle and A_r is the relative atomic mass of silver. It has been shown that the size of the silver nanoparticles derived from the 'spikes' matches with the sizing measured from SEM images.[135,148] In addition to determining the size of the silver nanoparticles, if the size of the nanoparticles is plotted against time, the aggregation kinetics of silver nanoparticles in electrolyte can be monitored.[135] In a work by Ellison *et al.*, the proportion of monomer, dimer, trimer, tetramer, pentamer and hexamer of silver nanoparticles have been determined with the impact experiments.[136] Lees *et al.* showed that aggregation is slowed down in the presence of a high amount of citrate, the capping agent of silver nanoparticles.[137] As an ultimate aim to detect nanoparticles in the environment matrix, Stuart *et al.* showed, through impact experiments, that the detection of laboratory synthesized silver nanoparticles in seawater is possible.[138] They have further shown that the commercially available silver nanoparticles can also be characterised in the environmental medium of seawater.[139] The direct 'hit and run' method has also been applied to other nanoparticles including iron oxide (Fe_3O_4),[143] nickel,[134,144,145] gold,[145–147] cadmium[149] and very recently organic nanoparticles such as indigo dye nanoparticles[148] and oil blue dye nanoparticles.[150]

As only a single nanoparticle undergoes the redox reaction each time upon impacting the microelectrode, the amount of Faradaic charge generated is relatively small. For example, the oxidation of a 20 nm radius silver nanoparticle will generate about 0.3 pC. Charges down to 1.22 fC (where 1 fC $= 10^{-15}$ C) have been detected for a nanoparticle with radius of 3.2 nm.[137] The 'spikes' typically last from 5 to 20 ms.

In a variant, a type of impact experiment known as 'tagged redox coulometry' has been developed. It follows the first scenario where there is a direct redox reaction. However, in 'tagged redox coulometry', the nanoparticles themselves are not oxidised. Instead, redox molecules 'tagged' on the nanoparticles undergo the reaction.[140,141,151] The principle was demonstrated by Zhou *et al.*, who attached 1,4-nitrothiophenol (NTP), which can undergo a four-electron reduction, onto silver nanoparticles.[140,141] As the nanoparticle hits the electrode surface, the NTP on the nanoparticle surface is reduced, generating a 'spike'. Thus, the 'tagged redox coulometry' method allows the detection of nanoparticles, which are not necessarily redox active, without destroying them.

The second scenario of indirect 'hit and run' involves three components: a redox reaction of A into B where A is abundant in the electrolyte, an electrode inert to the redox reaction and a nanoparticle that catalyses the process of A into B. In this case, the electrochemical signal is detected as the electrocatalytic process ($A \rightarrow B$ in Figure 5.14b) occurs when the catalytic nanoparticle hits the electrode surface.[144,152–155] The latter is inert to the A into B reaction at the potential used. The amount of reactant used in the electrocatalytic process in the electrolyte brings about a signal amplification compared to 'tagged redox coulometry'. One good example that demonstrates the second scenario is the catalysed reduction of hydrogen peroxide by silver nanoparticles on an inert carbon microfiber electrode.[152] As the silver nanoparticles hit the electrode surface, hydrogen peroxide is reduced on the nanoparticle surface, giving a 'spike' on the scan. The magnitude of the charge of the 'spikes' scales linearly with the hydrogen peroxide concentrations, showing that signal amplification is possible by increasing the reactant concentration. As the silver nanoparticle concentration increases, the frequency of impacts correlates linearly with it. This method has also been applied to numerous other redox systems such as water oxidation catalysed by iridium oxide nanoparticles,[153] proton reduction catalysed by nickel/nickel oxide nanoparticles,[144] hydrogen peroxide reduction by cerium(IV) oxide nanoparticles[156] and proton reduction promoted by platinum nanoparticles.[154] The possibility that proton reduction was catalysed by silver or gold nanoparticles of different sizes was studied by Kahk *et al.*, who showed that authentic nano-catalysis (kinetic acceleration due to small particle size) is observed for the studied silver nanoparticles.[155] Nano-catalysis was not observed on the gold nanoparticles.

The third scenario of indirect 'hit and stand' also depends on the nanoparticles used to catalyse a redox reaction. Platinum nanoparticles are widely studied under this scenario on electrodes of various materials (carbon[157] and gold[157–159]). In work by Xiao *et al.*, proton reduction is chosen as the redox reaction to be catalysed by platinum nanoparticles on a carbon electrode.[157] The platinum nanoparticles remain on the carbon electrode for a substantial amount of time after hitting the carbon electrode ('hit and stand'). Proton reduction occurs at a lower potential on platinum compared to carbon. Thus, as more and more platinum nanoparticles are adsorbed on the carbon surface, an increased amount of proton reduction occurs on the electrode surface, resulting in a staircase-like voltammogram. Figure 5.17 shows a zoomed version of a typical chronoamperometry scan.[157] The circled portion shows the current response when a single platinum nanoparticle hits the electrode surface. An increase in current is observed as the nanoparticle adds to the amount of electroactive surface where proton reduction can occur. This diffusional controlled process of converting A into B, assuming a perfect spherical particle on a perfectly flat electrode, relates the radius of the nanoparticle, r_{NP}, to the amplitude of the current 'step', I, through eqn (5.8):[157,160–162]

$$I = 4\pi(\ln 2)nFDCr_{NP} \tag{5.8}$$

where D is the diffusion coefficient of the reactant and C is the reactant concentration.

Another well-studied system to characterise platinum nanoparticles is through the catalysis of hydrazine oxidation to nitrogen and protons.[157,159] It is suggested that in hydrazine oxidation, at low electrolyte concentrations (*i.e.* 5 mM), the impact frequency increases due to migration effects of the negatively charged citrate capped platinum nanoparticles.[159] If a mercury drop electrode is used instead of a gold or carbon electrode for hydrazine oxidation, as the platinum nanoparticles hit the electrode the mercury

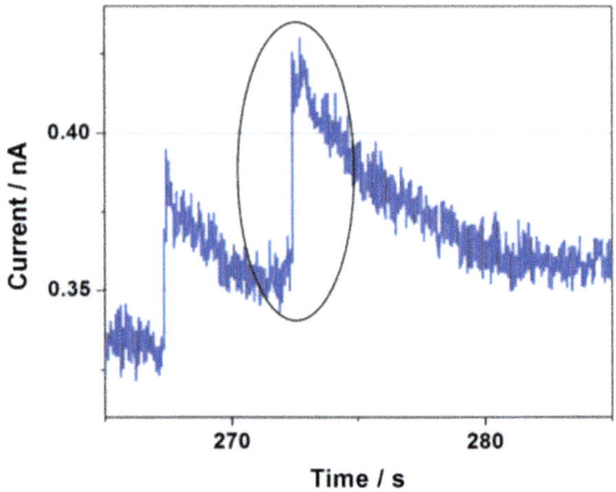

Figure 5.17 Zoomed in view of an impact experiment with the indirect 'hit and stand' mechanism. Platinum nanoparticles impact on the carbon electrode, resulting in an increase in current corresponding to the proton reduction occurring on the platinum nanoparticle surface. (Reprinted with permission from ref. 157. Copyright (2008) American Chemical Society.)

poisons the catalytic property of platinum, resulting in current 'spikes' observed instead of 'steps'.[163] However, the presence of hydrazine in the platinum nanoparticle suspension may promote aggregation, leading to measurement of bigger clusters.[164] Titanium dioxide nanoparticles have also been sized through similar experiments.[165] They photocatalyse the oxidation of methanol and, in the presence of light, 'steps' were observed in the chronoamperometric scan.

How do the three scenario compare to one another? All of them allow single nanoparticle detection and they are able to identify and size the nanoparticles.[132–139,141,143,146–148,150,152,155,157,163,164] All three experiments are also capable of studying nanoparticle aggregation.[136,164] The seeming advantages of indirect redox reactions (scenarios 2 and 3) over direct redox reaction (scenario 1) are that they are non-destructive and signal amplification can be attained. Thereby, it is easier to detect smaller nanoparticles that generate small signals using indirect redox methods. However, direct redox reactions are less prone to false positives and system fouling. Further, across the impact experiments, there is also an assumption in eqn (5.7) and (5.8) that all the nanoparticles are perfect spheres.

The impact experiments give rise to other thoughts. First, for direct processes, the possibility of under-sizing the nanoparticles due to partial reaction has been established as statistically unlikely in a paper by Dickinson *et al.*[166] By studying the interplay of the Brownian motion and the timescale on which impacts occur, it is concluded that under direct oxidation or reduction the nanoparticles are either 0 or 100% oxidised. Thus, partial reaction is statistically unlikely and hence the particles measured are not under-sized. Second, for indirect processes, the presence of large capping agents on the working electrode surface such as 12-mercaptodocecanic acid and 16-mercaptohexadecanoic acid can stop the platinum nanoparticles from promoting hydrazine oxidation.[167] Thus, the size of the capping agent used is important in controlling charge transfer from the nanoparticle to the electrode. The capping agent used should not be too bulky. Third, the factor limiting the minimum measurable size is the signal-to-noise ratio. A lower limit of detection is achievable if the ratio is improved. One of the approaches to this issue is to reduce the size of the electrode. In most of the literature work, microelectrodes are used to minimise noise. Kleijn *et al.* have developed a new prototype of nanoelectrode where they limit the size of the electroactive surface through controlling the amount of electrolyte in contact with the electrode.[168] Figure 5.18 shows a schematic diagram where the electrolyte is suspended above the electrode surface in a pipette tip.[168] Despite the electrode area being the entire rectangle at the base, only the portion in contact will act as the working electrode (highlighted in the box), limiting the electrode size and hence reducing the noise of the system. Lastly, the microelectrodes used in most experiments may experience shielding effects from the sheath surrounding the electrode.[169]

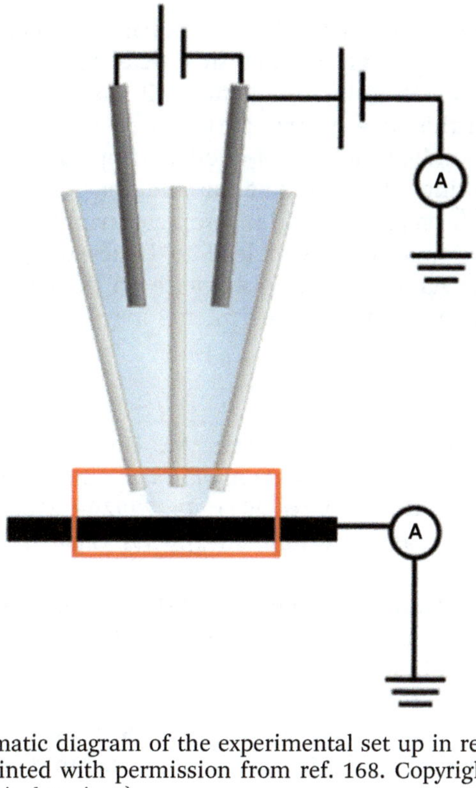

Figure 5.18 Schematic diagram of the experimental set up in ref. 163. (Reprinted with permission from ref. 168. Copyright (2012) American Chemical Society.)

The nanoparticles approaching the electrode may adsorb onto the glass before they reach the electrode surface. A micro-cylinder electrode, which consists of a wire protruding from the surface, reduces the shielding effect of the glass sheath, resulting in an increased impact frequency and allows detection of a lower nanoparticle concentration.[170]

Impact experiments have also been combined with other electrochemical techniques which allow a 'twin prong' strategy for nanoparticle detection. When combined with the technique of stripping voltammetry (Section 5.3), the sticking coefficient of nanoparticles can be calculated.[145,171] The electrode is suspended in a nanoparticle suspension for a period to allow adsorption. Then, stripping voltammetry is performed to obtain the number of nanoparticles adsorbed on the surface. After that, impact experiments are used to determine the number of nanoparticles hitting the electrode surface. The ratio of the number of nanoparticles sticking to the number of nanoparticles impacting the electrode surface is taken to be the sticking coefficient. This has been performed on silver,[171] nickel[145] and gold[145] nanoparticles, showing that about 10–20% of the nanoparticles stick to the carbon electrode surface.[145,171]

5.5 Conclusion

In conclusion, the novel method of 'nanoparticle–electrode impact' has scope to revolutionise single nanoparticle detection and characterisation. Though the work is still in its initial stages, it has already provided significant insights and this area of nanoparticle detection is predicted to grow rapidly. The well-studied method of anodic stripping voltammetry complements it by providing both fundamental and analytical information. However, recent work[172] has shown that incomplete stripping of nanoparticles in anodic stripping voltammetry can lead to issues of quantification when aggregation of the nanoparticles takes place on the electrode surface. For this reason, single nanoparticle experiments may be preferred.

References

1. M. Vert, Y. Doi, K.-H. Hellwich, M. Hess, P. Hodge, P. Kubisa, M. Rinaudo and F. Schué, *Pure Appl. Chem.*, 2012, **84**, 377–410.
2. M. I. Stockman, *Opt. Express*, 2011, **19**, 22029–22106.
3. D. D. Evanoff, Jr. and G. Chumanov, *J. Phys. Chem. B*, 2004, **108**, 13957–13962.
4. I. Diez and R. H. A. Ras, *Nanoscale*, 2011, **3**, 1963–1970.
5. A. Otto, *Z. Phys. A: Hadrons Nucl.*, 1968, **216**, 398–410.
6. C. N. R. Rao, G. U. Kulkarni, P. J. Thomas and P. P. Edwards, *Chem. Soc. Rev.*, 2000, **29**, 27–35.
7. J. Li and J.-J. Zhu, *Analyst*, 2013, **138**, 2506–2515.
8. C. Nützenadel, A. Züttel, D. Chartouni, G. Schmid and L. Schlapbach, *Eur. Phys. J. D*, 2000, **8**, 245–250.

9. K. Esumi, R. Isono and T. Yoshimura, *Langmuir*, 2004, **20**, 237–243.
10. N. Pradhan, A. Pal and T. Pal, *Colloids Surf., A*, 2002, **196**, 247–257.
11. L. Yin and J. Liebscher, *Chem. Rev.*, 2007, **107**, 133–173.
12. W. J. Plieth, *J. Phys. Chem.*, 1982, **86**, 3166–3170.
13. F. G. Shi, *J. Mater. Res.*, 2011, **9**, 1307–1314.
14. H. S. Toh, C. Batchelor-McAuley, K. Tschulik, M. Uhlemann, A. Crossley and R. G. Compton, *Nanoscale*, 2013, **5**, 4884–4893.
15. S. E. Ward Jones, F. W. Campbell, R. Baron, L. Xiao and R. G. Compton, *J. Phys. Chem. C*, 2008, **112**, 17820–17827.
16. M. Giovanni and M. Pumera, *Electroanalysis*, 2012, **24**, 615–617.
17. X. Li, J. J. Lenhart and H. W. Walker, *Langmuir*, 2010, **26**, 16690–16698.
18. A. M. El Badawy, K. G. Scheckel, M. Suidan and T. Tolaymat, *Sci. Total Environ.*, 2012, **429**, 325–331.
19. K. A. Huynh and K. L. Chen, *Environ. Sci. Technol.*, 2011, **45**, 5564–5571.
20. J. Zeng, Y. Zheng, M. Rycenga, J. Tao, Z.-Y. Li, Q. Zhang, Y. Zhu and Y. Xia, *J. Am. Chem. Soc.*, 2010, **132**, 8552–8553.
21. T. R. Jensen, M. D. Malinsky, C. L. Haynes and R. P. Van Duyne, *J. Phys. Chem. B*, 2000, **104**, 10549–10556.
22. A. K. Singh, V. Viswanath and V. C. Janu, *J. Lumin.*, 2009, **129**, 874–878.
23. X. Liu, M. Atwater, J. Wang and Q. Huo, *Colloids Surf. B*, 2007, **58**, 3–7.
24. *Progress in Medicinal Chemistry*, ed. G. P. Ellis and D. K. Luscombe, Elsevier, Amsterdam, 1992.
25. C. Batchelor-McAuley, K. Tschulik, C. C. M. Neumann, E. Laborda and R. G. Compton, *Int. J. Electrochem. Sci.*, 2014, **9**, 1132–1138.
26. C. Carlson, S. M. Hussain, A. M. Schrand, L. K. Braydich-Stolle, K. L. Hess, R. L. Jones and J. J. Schlager, *J. Phys. Chem. B*, 2008, **112**, 13608–13619.
27. H.-L. Su, C.-C. Chou, D.-J. Hung, S.-H. Lin, I.-C. Pao, J.-H. Lin, F.-L. Huang, R.-X. Dong and J.-J. Lin, *Biomaterials*, 2009, **30**, 5979–5987.
28. A. Lapresta-Fernández, A. Fernández and J. Blasco, *TrAC, Trends Anal. Chem.*, 2012, **32**, 40–59.
29. Silvercare, http://www.silvercare.co.za/ (Accessed 4th Feb 2014).
30. Magical socks nanosilver with silver nanoparticles, http://www.nanosilver.eu/Tema/Why-Nanosilver/Magical-Socks-Nanosilver-with-Silver-Nanoparticles (Accessed 4th Feb 2014).
31. Project on Emerging Nanotechnologies Analysis, http://www.nanotechproject.org/cpi/about/analysis/ (Accessed 4 Feb 2014).
32. C. Musante and J. C. White, *Environ. Toxicol.*, 2012, **27**, 510–517.
33. R. K. Kunkalekar, M. M. Naik, S. K. Dubey and A. V. Salker, *J. Chem. Technol. Biotechnol.*, 2013, **88**, 873–877.
34. Y. Park, H. J. Noh, L. Han, H.-S. Kim, Y.-J. Kim, J. S. Choi, C.-K. Kim, Y. S. Kim and S. Cho, *J. Nanosci. Nanotechnol.*, 2012, **12**, 7087–7095.
35. B. Vaseeharan, P. Ramasamy and J. C. Chen, *Lett. Appl. Microbiol.*, 2010, **50**, 352–356.

36. ZinClear™ IM Zinc Oxide Dispersions, http://www.dow.com/personalcare/technologies/uvtechnologies/zinclear.htm (Accessed 5th Feb 2014).
37. F. J. Moloney, S. Collins and G. M. Murphy, *Am. J. Clin. Dermatol.*, 2002, **3**, 185–191.
38. A. Walther and A. H. E. Müller, *Soft Matter*, 2008, **4**, 663–668.
39. A. Synytska, R. Khanum, L. Ionov, C. Cherif and C. Bellmann, *ACS Appl. Mater. Interfaces*, 2011, **3**, 1216–1220.
40. Nanotechnology: A Realistic Market Assessment, http://www.bccresearch.com/market-research/nanotechnology/nanotechnology-market-applications-products-nan031e.html (Accessed 4th February 2014).
41. K. Takeda, K.-I. Suzuki, A. Ishihara, M. Kubo-Irie, R. Fujimoto, M. Tabata, S. Oshio, Y. Nihei, T. Ihara and M. Sugamata, *J. Health Sci.*, 2009, **55**, 95–102.
42. Y. Lee, J. Choi, P. Kim, K. Choi, S. Kim, W. Shon and K. Park, *Toxicol. Res.*, 2012, **28**, 139–141.
43. A. Weir, P. Westerhoff, L. Fabricius, K. Hristovski and N. von Goetz, *Environ. Sci. Technol.*, 2012, **46**, 2242–2250.
44. K. Donaldson, D. Brown, A. Clouter, R. Duffin, W. MacNee, L. Renwick, L. Tran and V. Stone, *J. Aerosol Med.*, 2002, **15**, 213–220.
45. E. Mantovani, A. Porcari M. D. Morrison and R. E. Geertsma, *Developments in Nanotechnologies Regulation and Standards 2011 – Report of The Observatory Nano.*, 2011, http://www.steptoe.com/assets/htmldocuments/DevelopmentsInNanotechnologiesRegulationandStandards_20111.pdf (Accessed 4th February 2014).
46. MHRA – How we regulate nanotechnology, http://www.mhra.gov.uk/Howweregulate/Nanotechnology/ (Accessed 4th February 2014).
47. European Commission – Nanotechnology, http://ec.europa.eu/nanotechnology/policies_en.html#strategy (Accessed 4th February 2014).
48. Legifrance – Décret n° 2012-232 du 17 février 2012 relatif à la déclaration annuelle des substances à l'état nanoparticulaire pris en application de l'article L. 523-4 du code de l'environnement, http://www.legifrance.gouv.fr/affichTexte.do?cidTexte = JORFTEXT000025377246&dateTexte = &categorieLien = id (Accessed 4th February 2014).
49. D. S. L. Jarvis and N. Richmond, Regulation and Governance of Nanotechnology in China: Regulatory Challenges and Effectiveness, *Eur. J. Law Technol.*, 2011, **2**(3), 1–11.
50. G. Oberdörster, A. Maynard, K. Donaldson, V. Castranova, J. Fitzpatrick, K. Ausman, J. Carter, B. Karn, W. Kreyling, D. Lai, S. Olin, N. Monteiro-Riviere, D. Warheit and H. Yang, *Part. Fibre Toxicol.*, 2005, **2**, 8.
51. S. L. Flegler, J. W. Heckman Jr. and K. L. Klomparens, *Scanning and Transmission Electron Microscopy: An Introduction*, Oxford University Press, New York, 1993.
52. A. M. Collins, *Nanotechnology Cookbook*, Elsevier, Oxford, UK, 2012.

53. *Handbook of Charged Particle Optics*, ed. J. Orloff, CRC Press, Boca Raton, Florida, 2nd edn, 2008.
54. R. F. Egerton, P. Li and M. Malac, *Micron*, 2004, **35**, 399–409.
55. D. B. Williams and C. B. Carter, *Transmission Electron Microscope*, 2nd edn, Springer, New York, 1996.
56. P. E. Batson, N. Dellby and O. L. Krivanek, *Nature*, 2002, **418**, 617–620.
57. Image kindly provided by C. Damm, D. Pohl and B. Rellinghaus, Dept. Metastable and Nanostructured Materials, IFW Dresden, Germany.
58. L. Reimer and H. Kohl, *Transmission Electron Microscopy*, 5th edn, Springer, New York, 2008.
59. K. J. J. Mayrhofer, S. J. Ashton, J. C. Meier, G. K. H. Wiberg, M. Hanzlik and M. Arenz, *J. Power Sources*, 2008, **185**, 734–739.
60. R. F. Egerton, *Ultramicroscopy*, 1980, **5**, 521–523.
61. Nanosight – Nanoparticle Tracking Analysis, http://www.nanosight.com/technology/nanoparticle-tracking-analysis-nta (accessed 23 March 2014).
62. J. T. Edward, *J. Chem. Educ.*, 1970, **47**, 261.
63. H. S. Toh, C. Batchelor-McAuley, K. Tschulik, C. Damm and R. G. Compton, *Sens. Actuators, B*, 2014, **193**, 315–319.
64. Nanosight – FAQs, http://www.nanosight.com/resources/faqs (Accessed 23 March 2014).
65. *Dynamic Light Scattering: Applications of Photon Correlation Spectroscopy*, ed. R. Pecora, Springer, New York, 1985.
66. P. van der Meeren, H. Bogaert, J. Vanderdeelen and L. Baert, *Part. Part. Syst. Charact.*, 1992, **9**, 138–143.
67. Malvern Instruments Ltd – Zetasizer Nano Range, http://malvern.bsky.net/labeng/products/zetasizer/zetasizer_nano.htm (Accessed 28 March 2014).
68. *Principles, Methods and Application of Particle Size Analysis*, ed. J. P. M. Syvutski, Cambridge University Press, Cambridge, 1997.
69. J. Manson, D. Kumar, B. J. Meenan and D. Dixon, *Gold Bull.*, 2011, **44**, 99–105.
70. Perkin Elmer – The 30-Minute Guide to ICP-MS, http://www.perkinelmer.co.uk/CMSResources/Images/44-74849tch_icpmsthirtyminuteguide.pdf (Accessed 26 March 2014).
71. C. G. Herbert and R. A. W. Johnstone, *Mass Spectrometry Basics*, CRC Press LLC, Boca Raton, Florida, 2003.
72. Y. Waseda, E. Matsubara and K. Shinoda, *X-ray Diffraction Crystallography: Introduction, Examples and Solved Problem*, Springer-Verlag, Berlin, 2011.
73. B. E. Warren, *X-ray Diffraction*, Courier Dover Publication, New York, 2012.
74. *Practical Surface Analysis: Auger and X-ray Photoelectron Spectroscopy*, ed. D. Briggs and M. P. Seah, John Wiley and Sons Ltd, Chichester, 2nd edn, 1990.

75. Y. Tan, Y. Li and D. Zhu, *Langmuir*, 2002, **18**, 3392–3395.
76. K. L. Kelly, E. Coronado, L. L. Zhao and G. C. Schatz, *J. Phys. Chem. B*, 2003, **107**, 668–677.
77. A. Kuzma, M. Weis, S. Flickyngerova, J. Jakabovic, A. Satka, E. Dobrocka, J. Chlpik, J. Cirak, M. Donoval, P. Telek, F. Uherek and D. Donoval, *J. Appl. Phys.*, 2012, **112**, 103531.
78. E. Tomaszewska, K. Soliwoda, K. Kadziola, B. Tkacz-Szczesna, G. Celichowski, M. Cichomski, W. Szmaja and J. Grobelny, *J. Nanomater.*, 2013, **2013**, 313081.
79. R. F. Domingos, M. A. Baalousha, Y. Ju-Nam, M. M. Reid, N. Tufenkji, J. R. Lead, G. G. Leppard and K. J. Wilkinson, *Environ. Sci. Technol.*, 2009, **43**, 7277–7284.
80. A. Morrin, A. J. Killard and M. R. Smyth, *Anal. Lett.*, 2003, **36**, 2021–2039.
81. J. Wang, M. Pedrero, H. Sakslund, O. Hammerich and J. Pingarron, *Analyst*, 1996, **121**, 345.
82. V. K. Rao, M. K. Sharma, P. Pandey and K. Sekhar, *World J. Microbiol. Biotechnol.*, 2006, **22**, 1135–1143.
83. N. Thiyagarajan, J.-L. Chang, K. Senthilkumar and J.-M. Zen, *Electrochem. Commun.*, 2014, **38**, 86–90.
84. A. A. Rowe, A. J. Bonham, R. J. White, M. P. Zimmer, R. J. Yadgar, T. M. Hobza, J. W. Honea, I. Ben-Yaacov and K. W. Plaxco, *PLoS One*, 2011, **6**, e23783.
85. K. E. Toghill and R. G. Compton, *Int. J. Electrochem. Sci.*, 2010, **5**, 1246–1301.
86. J. Wang, *Chem. Rev.*, 2008, **108**, 814–825.
87. X.-J. Huang, L. Aldous, A. M. O'Mahony, F. J. del Campo and R. G. Compton, *Anal. Chem.*, 2010, **82**, 5238–5245.
88. M. C. Buzzeo, C. Hardacre and R. G. Compton, *Anal. Chem.*, 2004, **76**, 4583–4588.
89. S. Neethirajan, D. S. Jayas and S. Sadistap, *Food Bioprocess Technol.*, 2008, **2**, 115–121.
90. J. Wang, *Electroanalysis*, 2005, **17**, 1341–1346.
91. D. Q. Hung, O. Nekrassova and R. G. Compton, *Talanta*, 2004, **64**, 269–277.
92. R. T. Kachoosangi, G. G. Wildgoose and R. G. Compton, *Analyst*, 2008, **133**, 888–895.
93. S. A. Blaser, M. Scheringer, M. Macleod and K. Hungerbühler, *Sci. Total Environ.*, 2008, **390**, 396–409.
94. R. G. Compton and C. E. Banks, *Understanding Voltammetry*, 2nd edn, World Scientific, Singapore, 2011.
95. A. S. Barnes, I. J. Streeter and R. G. Compton, *J. Electroanal. Chem.*, 2008, **623**, 129–133.
96. E. Laborda, J. González and Á. Molina, *Electrochem. Commun.*, 2014, **43**, 25–30.
97. J. Wang, *Analytical Electrochemistry*, Wiley-VCH Verlag GmbH, Weinheim, 2nd edn, 2001.

98. C. C. M. Neumann, C. Batchelor-McAuley, K. Tschulik, H. S. Toh, P. Shumbula, J. Pillay, R. Tshikhudo and R. G. Compton, *ChemElectroChem*, 2014, **1**, 87–89.
99. M. Pumera, M. Aldavert, C. Mills, A. Merkoçi and S. Alegret, *Electrochim. Acta*, 2005, **50**, 3702–3707.
100. A. Gillespie, D. Jao, A. Andriola, T. Duda, C. F. Yang and L. Yu, *Anal. Lett.*, 2012, **45**, 1310–1320.
101. C. E. Banks, J. Kruusma, R. R. Moore, P. Tomčík, J. Peters, J. Davis, S. Komorsky-Lovrić and R. G. Compton, *Talanta*, 2005, **65**, 423–429.
102. A. J. Saterlay, J. S. Foord and R. G. Compton, *Analyst*, 1999, **124**, 1791–1796.
103. A. J. Saterlay, D. F. Tibbetts and R. G. Compton, *Anal. Sci.*, 2000, **16**, 1055–1060.
104. A. J. Saterlay, C. Agra-Gutiérrez, M. P. Taylor, F. Marken and R. G. Compton, *Electroanalysis*, 1999, **11**, 1083–1088.
105. Y. Wang, E. Laborda, B. J. Plowman, K. Tschulik, K. R. Ward, R. G. Palgrave, C. Damm and R. G. Compton, *Phys. Chem. Chem. Phys.*, 2014, **16**, 3200–3208.
106. J. Lakbub, A. Pouliwe, A. Kamasah, C. Yang and P. Sun, *Electroanalysis*, 2011, **23**, 2270–2274.
107. X.-F. Zhou, W. Cheng, C. Batchelor-McAuley, K. Tschulik and R. G. Compton, *Electroanalysis*, 2014, **26**, 248–253.
108. R. G. Compton, F. J. Davis and S. C. Grant, *J. Appl. Electrochem.*, 1986, **16**, 239–249.
109. W. Z. Teo, A. Ambrosi and M. Pumera, *Electrochem. Commun.*, 2013, **28**, 51–53.
110. M. Giovanni, A. Ambrosi and M. Pumera, *Chem. - Asian J.*, 2012, 7, 702–706.
111. M. Giovanni and M. Pumera, *Electrochem. Commun.*, 2011, **13**, 203–204.
112. O. S. Ivanova and F. P. Zamborini, *J. Am. Chem. Soc.*, 2010, **132**, 70–72.
113. *CRC Handbook of Chemistry and Physics 2012–2013*, ed. W. M. Haynes, CRC Press, Boca Raton, Florida, 93rd edn, 2012.
114. T. J. Davies and R. G. Compton, *J. Electroanal. Chem.*, 2005, **585**, 63–82.
115. K. Tschulik, R. G. Palgrave, C. Batchelor-McAuley and R. G. Compton, *Nanotechnology*, 2013, **24**, 295502.
116. E. J. E. Stuart, K. Tschulik, J. Ellison and R. G. Compton, *Electroanalysis*, 2014, **26**, 285–291.
117. W. Cheng, E. J. E. Stuart, K. Tschulik, J. T. Cullen and R. G. Compton, *Nanotechnology*, 2013, **24**, 505501.
118. E. J. E. Stuart, K. Tschulik, D. Lowinsohn, J. T. Cullen and R. G. Compton, *Sens. Actuators, B*, 2014, **195**, 223–229.
119. C. Batchelor-McAuley, A. Martinez-Marrades, K. Tschulik, A. N. Patel, C. Combellas, F. Kanoufi, G. Tessier and R. G. Compton, *Chem. Phys. Lett.*, 2014, **597**, 20–25.

120. S. Munteanu, J. P. Roger, Y. Fedala, F. Amiot, C. Combellas, G. Tessier and F. Kanoufi, *Faraday Discuss.*, 2013, **164**, 241–258.
121. C. K. Chua and M. Pumera, *Electrophoresis*, 2013, **34**, 2007–2010.
122. N. V. Rees, *Electrochem. Commun.*, 2014, **43**, 83–86.
123. W. Wang and N. Tao, *Anal. Chem.*, 2014, **86**, 2–14.
124. N. V. Rees, Y.-G. Zhou and R. G. Compton, *RSC Adv.*, 2012, **2**, 379.
125. W. Cheng and R. G. Compton, *TrAC, Trends Anal. Chem.*, 2014, **58**, 79–89.
126. F. Scholz, D. Hellberg, F. Harnisch, A. Hummel and U. Hasse, *Electrochem. Commun.*, 2004, **6**, 929–933.
127. D. Hellberg, F. Scholz, F. Schauer and W. Weitschies, *Electrochem. Commun.*, 2002, **4**, 305–309.
128. C. E. Banks, N. V. Rees and R. G. Compton, *J. Electroanal. Chem.*, 2002, **535**, 41–47.
129. C. E. Banks, N. V. Rees and R. G. Compton, *J. Phys. Chem. B*, 2002, **106**, 5810–5813.
130. N. V. Rees, C. E. Banks and R. G. Compton, *J. Phys. Chem. B*, 2004, **108**, 18391–18394.
131. A. D. Clegg, N. V. Rees, C. E. Banks and R. G. Compton, *ChemPhysChem*, 2006, 7, 807–811.
132. E. J. E. Stuart, Y.-G. Zhou, N. V. Rees and R. G. Compton, *RSC Adv.*, 2012, **2**, 6879–6884.
133. Y.-G. Zhou, N. V. Rees and R. G. Compton, *Angew. Chem., Int. Ed.*, 2011, **50**, 4219–4221.
134. Y.-G. Zhou, B. Haddou, N. V. Rees and R. G. Compton, *Phys. Chem. Chem. Phys.*, 2012, **14**, 14354–14357.
135. N. V. Rees, Y.-G. Zhou and R. G. Compton, *ChemPhysChem*, 2011, **12**, 1645–1647.
136. J. Ellison, K. Tschulik, E. J. Stuart, K. Jurkschat, D. Omanović, M. Uhlemann, A. Crossley and R. G. Compton, *ChemistryOpen*, 2013, **2**, 69–75.
137. C. Batchelor-McAuley, J. Ellison, K. Tschulik, P. L. Hurst, R. Boldt and R. G. Compton, *Analyst*, 2015, **140**, 5048–5054.
138. E. J. E. Stuart, N. V. Rees, J. T. Cullen and R. G. Compton, *Nanoscale*, 2013, **5**, 174–177.
139. E. J. E. Stuart, K. Tschulik, D. Omanović, J. T. Cullen, K. Jurkschat, A. Crossley and R. G. Compton, *Nanotechnology*, 2013, **24**, 444002.
140. Y.-G. Zhou, N. V. Rees and R. G. Compton, *Chem. Commun.*, 2012, **48**, 2510–2512.
141. N. V. Rees, Y.-G. Zhou and R. G. Compton, *Chem. Phys. Lett.*, 2012, **525–526**, 69–71.
142. Y.-G. Zhou, N. V. Rees and R. G. Compton, *ChemPhysChem*, 2011, **12**, 2085–2087.
143. K. Tschulik, B. Haddou, D. Omanović, N. V. Rees and R. G. Compton, *Nano Res.*, 2013, **6**, 836–841.

144. Y.-G. Zhou, N. V. Rees and R. G. Compton, *Phys. Chem. Chem. Phys.*, 2013, **15**, 761–763.
145. Y.-G. Zhou, E. J. E. Stuart, J. Pillay, S. Vilakazi, R. Tshikhudo, N. V. Rees and R. G. Compton, *Chem. Phys. Lett.*, 2012, **551**, 68–71.
146. Y.-G. Zhou, N. V. Rees, J. Pillay, R. Tshikhudo, S. Vilakazi and R. G. Compton, *Chem. Commun.*, 2012, **48**, 224–226.
147. D. Qiu, S. Wang, Y. Zheng and Z. Deng, *Nanotechnology*, 2013, **24**, 505707.
148. W. Cheng, X.-F. Zhou and R. G. Compton, *Angew. Chem., Int. Ed.*, 2013, **52**, 12980–12982.
149. Y.-G. Zhou, N. V. Rees and R. G. Compton, *Chem. Phys. Lett.*, 2011, **511**, 183–186.
150. X.-F. Zhou, W. Cheng and R. G. Compton, *Nanoscale*, 2014, **6**, 6873–6878.
151. G. P. Santos, A. F. A. A. Melo and F. N. Crespilho, *Phys. Chem. Chem. Phys.*, 2014, **16**, 8012–8018.
152. E. J. E. Stuart, N. V. Rees and R. G. Compton, *Chem. Phys. Lett.*, 2012, **531**, 94–97.
153. S. J. Kwon, F.-R. F. Fan and A. J. Bard, *J. Am. Chem. Soc.*, 2010, **132**, 13165–13167.
154. X. Xiao and A. J. Bard, *J. Am. Chem. Soc.*, 2007, **129**, 9610–9612.
155. J. M. Kahk, N. V. Rees, J. Pillay, R. Tshikhudo, S. Vilakazi and R. G. Compton, *Nano Today*, 2012, 7, 174–179.
156. N. P. Sardesai, D. Andreescu and S. Andreescu, *J. Am. Chem. Soc.*, 2013, **135**, 16770–16773.
157. X. Xiao, F.-R. F. Fan, J. Zhou and A. J. Bard, *J. Am. Chem. Soc.*, 2008, **130**, 16669–16677.
158. H. Zhou, J. H. Park, F.-R. F. Fan and A. J. Bard, *J. Am. Chem. Soc.*, 2012, **134**, 13212–13215.
159. J. H. Park, A. Boika, H. S. Park, H. C. Lee and A. J. Bard, *J. Phys. Chem. C*, 2013, **117**, 6651–6657.
160. I. J. Streeter and R. G. Compton, *J. Phys. Chem. C*, 2007, **111**, 18049–18054.
161. K. R. Ward, N. S. Lawrence, R. Seth Hartshorne and R. G. Compton, *J. Electroanal. Chem.*, 2012, **683**, 37–42.
162. P. A. Bobbert, M. M. Wind and J. Vlieger, *Physica A*, 1987, **141**, 58–72.
163. R. Dasari, D. A. Robinson and K. J. Stevenson, *J. Am. Chem. Soc.*, 2013, **135**, 570–573.
164. S. E. F. Kleijn, B. Serrano-Bou, A. I. Yanson and M. T. M. Koper, *Langmuir*, 2013, **29**, 2054–2064.
165. A. Fernando, S. Parajuli and M. A. Alpuche-Aviles, *J. Am. Chem. Soc.*, 2013, **135**, 10894–10897.
166. E. J. F. Dickinson, N. V. Rees and R. G. Compton, *Chem. Phys. Lett.*, 2012, **528**, 44–48.
167. X. Xiao, S. Pan, J. S. Jang, F.-R. F. Fan and A. J. Bard, *J. Phys. Chem. C*, 2009, **113**, 14978–14982.

168. S. E. F. Kleijn, S. C. S. Lai, T. S. Miller, A. I. Yanson, M. T. M. Koper and P. R. Unwin, *J. Am. Chem. Soc.*, 2012, **134**, 18558–18561.
169. S. Eloul and R. G. Compton, *ChemElectroChem*, 2014, **1**, 917–924.
170. J. Ellison, C. Batchelor-McAuley, K. Tschulik and R. G. Compton, *Sens. Actuators, B*, 2014, **200**, 47–52.
171. Y.-G. Zhou, N. V. Rees and R. G. Compton, *Chem. Phys. Lett.*, 2011, **514**, 291–293.
172. S. J. Cloake, H. S. Toh, P. T. Lee, C. Salter, C. Johnston and R. G. Compton, *ChemistryOpen*, 2015, **4**, 22–26.

CHAPTER 6

Nanoelectrodes in Electrochemical Analysis

AMÉLIE WAHL AND ALAN O'RIORDAN*

Nanotechnology Group, Tyndall National Institute – University College Cork, Dyke Parade, Cork, Ireland
*Email: alan.oriordan@tyndall.ie

6.1 Introduction

David Hubel is widely acknowledged as the first person (in the late 1950s) to have fabricated metal microelectrodes that consisted of an insulated tungsten wire with an exposed electrolytically sharpened tip ($\sim$0.4–10 μm).[1,2] In the intervening years, microfabrication techniques, borrowed from the microelectronics industry, have enabled routine and facile fabrication of electrodes with a critical dimension on the order of less than ten microns down to around a few hundreds of nanometres. These electrodes are generally classified as ultra-microelectrodes with a critical dimension of <25 μm. The fabrication and the investigation of electrochemical properties of these ultra-microelectrodes are well documented in the literature.[3–6] More recently, the trend to miniaturisation to realise further enhanced electrochemical performance has driven development and fabrication of electrodes with a critical dimension in the range of 1 to 100 nm—defined as nanoelectrodes. Early investigations into the use of nanoelectrodes in electrochemistry were based on advances in micro and ultra-micro fabrication techniques. For example, in the late 1980s Morris *et al.* first fabricated platinum and gold nanoband electrodes by depositing platinum or gold onto smooth mica followed by encapsulating all but one edge of the film in

RSC Detection Science Series No. 6
Electrochemical Strategies in Detection Science
Edited by Damien W. M. Arrigan

Published by the Royal Society of Chemistry, www.rsc.org

epoxy.[7] Since then, with ever-increasing sophistication of methods to produce materials at the nanoscale, nanoelectrodes have opened up new research domains and application opportunities in many areas of electroanalysis, in the environment,[8] health[9,10] and security[11] sectors.

6.1.1 Benefits of Nanoelectrodes

In recent years, the potential use of nanoelectrodes in a wide range of electrochemical applications has received increased interest worldwide due to the many accrued benefits from working at the nanoscale.[12] For instance, when compared to larger electrodes, nanoelectrodes are often found to exhibit enhanced, electrical,[13–15] physical,[16–18] chemical,[19–21] biological,[22–24] optical,[25–27] magnetic,[28–30] mechanical,[31–33] catalytic[34–36] or thermal[37–39] properties, which may in turn improve their electrochemical performance.[40–42] Electrochemical analysis is of particular interest as it enables the rapid detection of numerous analytes with high sensitivity and at low cost. As such, nanoelectrodes have the potential to push back traditional boundaries that are challenging (or even impossible) with classic micro and ultra-microelectrodes, thereby leading to new application opportunities in electrochemical analysis.

It is well known that the electrochemical response of an electrode is strongly dependent on size as the critical dimension influences mass transport behaviour taking place to the electrode surface, such that increasingly enhanced electrochemical behaviour is expected with decreasing electrode dimension.[6,12,43–49] When the critical dimension of an electrode becomes smaller than the diffusion layer thickness, radial (convergent, three-dimensional) diffusion becomes dominant, resulting in increased rates of mass transport to an electrode surface. Thus both ultra-microelectrodes and nanoelectrodes benefit from this diffusion regime where efficient mass transport allows the faradaic current to attain a time-independent steady-state value when the applied potential is swept linearly, at low-to-medium scan rates (under entirely quiescent conditions). As the scan rate increases, there is a corresponding decrease in thickness of the diffusion layer, such that at high-to-very-high scan rates the measured current at ultra-microelectrodes can become time-dependent. By contrast, given their very small critical dimension, nanoelectrodes have been shown to display time independent electrochemical responses with scan rates in excess of 50 V s^{-1}, as will be discussed further in Subsection 6.3.1.2.

Concerning pulse voltammetric techniques at single nanoelectrodes, when the potential is scanned in a series of steps, the obtained peak current is also maximised. In other words, larger faradaic current densities and much faster responses can be achieved at single nanoelectrodes. The magnitude of the steady-state or peak current is likely to be proportional to the analyte concentration, which is useful for electrochemical analysis. At nanoelectrode arrays, electrochemical behaviour is more complex as the mass transport and electrode geometry requires a 'design for application approach'.[50] Nevertheless, whether single or arrays of nanoelectrodes are employed, background charging currents are significantly reduced due to the very small electrode surface area.[51,52] Signal-to-noise ratios and sensitivity of nanoelectrodes are thus likely to be significantly greater than at ultra-microelectrodes.

Another interesting characteristic is the time taken to charge the double layer capacitance at nanoelectrodes, which is minimised because of their very small surface area and they consequently have much smaller cell time constants (*RC*, where *R* is the total cell resistance and *C* is the double layer capacitance). This is important as the potential at the electrode–solution interface only attains the applied potential value when the charging process is complete on timescales longer than about 5–10 *RC*.[12,47,52,53] This means that nanoelectrodes are able to respond more rapidly to changes in the applied potential than their ultra-micro counterparts, thereby extending the range of sample types and experimental timescales that are useful in electrochemistry. In addition, negligible ohmic drops of potential (*IR*, where *I* is the total current and *R* is the total cell resistance) are observed at nanoelectrodes, thereby eliminating the ohmic effects known to diminish the applied potential, *e.g.*, it eliminates distorted current responses and shifted peak potential in cyclic voltammetry.[53,54] This enables electrochemical analysis in more resistive solutions (*e.g.*, organic solutions) or in reduced or no supporting electrolyte solutions. Nanoelectrodes may also permit two electrode measurements, as the low faradaic currents are not sufficient to polarise (and thus change potential of) a reference electrode (which also behaves as a counter electrode in a two-electrode setup) and therefore a large electrode that acts as a current sink is not required. Finally, nanoscale dimensions enable analysis in much lower sample volumes and may also open the door to understanding more complex irreversible electrochemical reaction and electron transfer kinetics.[55]

However, clearly, the successful exploitation of all the above benefits depends not just on the sensitivity and reliability of the instrumentation employed for electrochemical analysis, but more importantly on nanoelectrode design and fabrication techniques employed/available. Examples of nanoelectrode design, nanofabrication and characterisation techniques are discussed in Section 6.2.

6.1.2 Challenges

Even though nanoelectrodes may provide numerous advantages in electrochemistry, a potential challenge associated with these electrodes is the low measurable signal (pA–nA) achieved, which may not be suitable for some applications, *e.g.*, remote sensing.[12,56] Additionally, measurements are often limited to the study of only one target analyte at a time since nanoelectrodes are commonly employed individually, *i.e.*, single nanoelectrode format.[12] These challenges can be addressed by the use of nanoelectrode arrays, whereby a large number of nanoelectrodes may be contained on a given footprint of an electrochemical sensor device (the smaller the electrode surface, the more electrodes could be integrated). Individual nanoelectrodes within an array can/may be designed to operate in parallel, thus amplifying the electrochemical signal while retaining the beneficial characteristics of single nanoelectrodes. Each nanoelectrode can/may also be (bio)chemically modified, thereby allowing the simultaneous study of multiple analytes, *i.e.*, multiplexed analysis. However, a key challenge to widespread take-up of nanoelectrodes is the requirement for a full understanding of the analyte diffusion profiles existing at nanoelectrode arrays, which will be further discussed in Subsection 6.3.1.3.

6.2 Nanoelectrodes Design, Fabrication and Characterisation

6.2.1 Nanoelectrode Design

6.2.1.1 Geometry

The electrochemical behaviour observed at nanoelectrodes (*e.g.*, magnitude of limiting currents achieved, steady-state behaviour) is strongly dependent on geometry, *i.e.*, both shape and dimension.[12,47,57–59] For example, the sensitivity of a gold nanowire electrode is such that small differences in nanowire dimensions, arising from small variances during nanofabrication, result in a detectable increase or decrease in the magnitude of steady-state currents measured.[46] Consequently, robust nanofabrication methods are required to ensure reproducible current responses. In addition, structural characterisation is required to determine the real electrochemical active surface area to normalise to current density to allow cross comparison between electrodes—these approaches are discussed in Section 6.2.3.

At the nanoscale, the most common geometries reported in the literature are nanobands, nanodisks, nanospheres and nanohemispheres,[53,58,60–63] while more complex geometries including nanopores, nanotubes, nanorings and nanowires are less commonly reported.[64,65] There are well-established analytical equations to determine the expected limiting faradaic current at the four most common nanoelectrode geometries for both sweep and pulse voltammetry techniques (Table 6.1).[12,44,47,66–69] However, at present there are no agreed analytical equations to determine the magnitude of current at the other more complex nanostructures. For this reason, the estimated limiting current at single more complex nanoelectrodes geometries may only be estimated by modifying or adapting equations applying to nanobands, nanodisks, nanohemispheres, and nanospheres with varying level of success. For arrays of nanoelectrodes, the limiting current depends on the diffusion profiles existing at each nanoelectrode and whether these profiles overlap or not within an array.[43,50,70] At diffusionally independent nanoelectrode arrays (using an applied sweep potential waveform) current responses are typically sigmoidal, reaching (quasi) steady-state values

Table 6.1 Common nanoelectrode geometries and their associated limiting current equations applying to sweep voltammetry techniques, where n is the number of electrons involved in the redox reaction, F is Faraday's constant (C mol^{-1}), D is the diffusion coefficient (cm^2 s^{-1}), C* is bulk concentration (mol cm^3), t is time, l (cm) is the electrode length, w (cm) is the electrode width and r (cm) is the electrode radius.

Design		Dimension	Limiting current
Nanoband	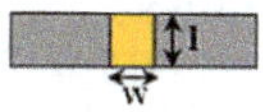	l: length w: width	Quasi-steady state $i_l = \dfrac{nFlC^*2\pi D}{\ln\left(\dfrac{64Dt}{w^2}\right)}$
Nanodisk		r: radius	Steady-state $i_l = 4nFDC^*r$
Nanohemisphere			Steady-state $i_l = 2\pi nFDC^*r$
Nanosphere			Steady-state $i_l = 4\pi nFDC^*r$

proportional to the number of nanoelectrodes. For nanoelectrodes with overlapping diffusion profiles, the array typically behaves as a larger electrode (equal in size to the total electrode footprint) and exhibit either sigmoidal current responses (but with magnitudes lower than multiples of the number of nanoelectrodes) or peak shaped diffusion limited responses—this is discussed further in Section 6.3. Concerning pulse voltammetry, the analytical equations at nanoelectrode arrays are not well established for any electrode geometry and full 3D modelling remains to be undertaken. For example, it was recently shown that when a square wave potential waveform was applied to arrays of nanowire electrodes, lower current values were recorded at nanoelectrodes with larger inter-electrode separations while higher current magnitudes were obtained at more closely spaced nanoelectrodes.[50] This is in direct contrast to the behaviour observed when using the same electrode with sweep voltammetry techniques. The diffusion profiles occurring at more complex electrodes geometries (for both sweep and pulse methods) are even more challenging to elucidate and are the focus of on-going work by several different research groups.[12,50,57,58,70–72]

6.2.1.2 Material

The electrochemical performance of nanoelectrodes is also strongly material dependent.[12,57] Over the last two decades, the most popular nanoelectrode materials reported have been platinum, gold and graphitic carbon while other materials used include mercury, nickel, iridium, silver and transition metal oxide materials.[73–77] The first three materials mentioned above demonstrate excellent electrical properties, relatively easy electron transfer and high catalytic activities for many chemical reactions, they are chemically inert, are biocompatible, may be used over a relatively large potential window and can be easily modified with a wide range of biomolecules and chemical ligands.[42,78–81] Consequently they are the materials of choice for electrochemical platforms being employed in chemical sensors, biosensing, catalysis and fuel cells. For instance, platinum, gold and graphitic carbon nanostructures have been used for the electrochemical detection of hydrogen peroxide,[56,82,83] dopamine,[13] glucose[10,84,85] and explosives[11] amongst others.

Advances in nanofabrication techniques and supramolecular chemistry have enabled development of hybrid nanomaterials integrated into devices that exhibit greater electrochemical performance.[86–90] This new class of nanomaterials exhibit several advantages: they retain the beneficial properties of two or more components but also provide the ability to tune these properties, provide larger electrochemically active surface areas while retaining nanoscale benefits, and improved biocompatibility. Examples include platinum nanoelectrodes coated with gold nanoparticles,[91] a glassy carbon electrode modified with gold–platinum alloy nanoparticles[92] and gold nanoparticles modified indium tin oxide electrode.[93] Platinum, gold and carbon nanoelectrodes have been modified with a range of other conductive or (semi)conductive materials such as silver,[94–96] copper,[97–99] silica[100,101] and nickel.[102]

New exciting hybrid nanostructures are appearing in the literature regularly with graphene-modified nanostructures being the newest and most pervasive.[103] This emergent material has received considerable interest due to its many purported enhanced properties including high surface area, low-cost, very high conductivity, high flexibility and high electrical mobility.[104–108] However, to date there are only a few studies on its potential use in electrochemical-based applications.[109–111] Therefore, research on the electrochemical behaviour and applications of graphene-modified nanoelectrodes is growing significantly.

6.2.2 Nanoelectrode Fabrication

A large number of bottom-up and top-down nanofabrication techniques have been reported in the literature. Many recent excellent reviews focus on nanoelectrodes integrated on silicon chip substrates and the reader is directed to these for more specific information.[12,57,112,113] These fabrication approaches are of particular importance for nanoelectrochemistry since nanoelectrodes of (almost) any geometry (including 3D structures) and (almost) any material may be fabricated. Chips can be designed so as to include several single nanoelectrodes and/or nanoelectrode arrays and be electrically addressed with microelectronic circuitry. Micron scale reference and counter electrodes may also be co-fabricated and electrically contacted on the same chip substrate using standard microelectronic fabrication techniques to yield fully integrated devices.[46,114] Such devices may be compatible with very low sample volumes, are robust and hence have little or no maintenance requirements, and enable detection limits that are comparable to or better than the laboratory-based equivalent. Furthermore, fully integrated devices may enable rapid (millisecond), real-time and multiplexed electrochemical analysis. Thus, on-chip nanoelectrodes are finding increased use for remote sensing applications required in environmental, security and health sectors.[57,115,116] State-of-the-art fabrication techniques such as hybrid electron-beam and optical lithographies are among the most flexible techniques for producing electrically contacted nanoelectrodes on chip substrates with extremely high resolution (sub-50 nm) and reproducibility. This approach enables routine fabrication and integration of huge variety of nanoelectrode configurations and geometries with highly reproducible dimensions while also being scalable to mass manufacture.

6.2.3 Surface Area Characterisation

The real surface area of an electrode corresponds to the area where electrochemical reactions take place, *i.e.*, the electrochemical active area. Determination of the real surface area is important as it is directly related to the magnitude of the faradaic currents produced by electrochemical reactions. Normalisation of the faradaic currents with respect to the electrochemical active area (current density) enables the direct comparison between the electrochemical performances of different electrodes. It requires the knowledge of both the geometric surface area and the roughness of an electrode. To this end, several techniques for the determination of electrodes electrochemical active area have been reported in the literature.[117–119] Traditionally these techniques have relied on the initial formation of a monolayer of hydride, oxide or deposited metal followed by its removal (by oxidation, reduction or metal stripping, respectively). The measured current associated with the formation/removal processes were then related back to the electrochemical active surface area using simple equations to yield a calculated electrochemical area. However, these techniques assumed monolayer formation only and 1:1 stoichiometric conditions, which are extremely difficult to achieve practically, at the nanoscale.[120,121] Consequently, reported electrochemical active area and surface roughness are often overestimated due to formation of multilayers.

Advanced microscopic techniques have been developed to overcome these limitations. Scanning or transmission electron microscopies may be used to characterise nanostructure dimensions with high resolution and hence enable the accurate determination of (planar) geometric area. Atomic force microscopy may be used to determine and calculate nanostructure roughness. Thus, a combination of these techniques enables the accurate determination of nanoelectrode electrochemical active area. A recent example of this approach was demonstrated whereby scanning electron and atomic force microscopies were combined to yield an accurate determination of surface roughness and consequently the electrochemical active area of nanowire electrodes[122] (Figure 6.1). An alternative approach is to employ scanning electrochemical microscopy for accurate electrochemical characterisation at nanoscale. However, the use of advanced microscopy may be cost-prohibitive, as it is complex and requires highly trained personnel.

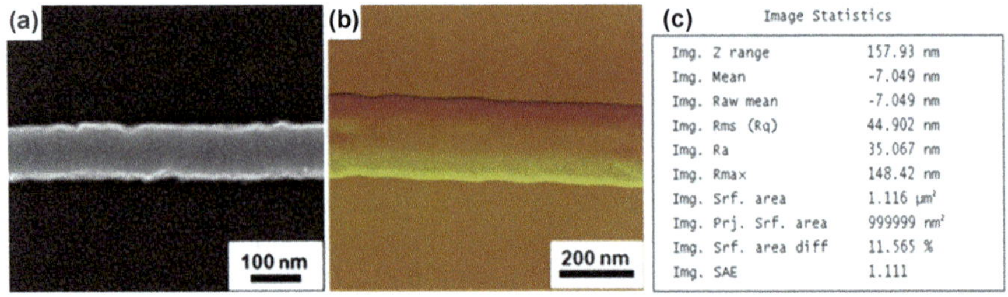

Figure 6.1 (a) SEM image of an individual gold nanowire electrode, average width 102 ± 3 nm. An ultra-thin gold layer was sputtered prior to analysis to reduce charging effects. (b) AFM micrograph (feedback signal) of a single gold nanowire electrode after the selective opening of the passivation overlayer. (c) Typical surface area data obtained from an AFM unprocessed height micrograph of a single gold nanowire electrode. (Reprinted and adapted with permission from ref. 122, copyright 2014 Elsevier Ltd.)

6.3 Electrochemical Analysis at the Nanoscale

The electrochemical response measured at an electrode is a combination of charging (capacitive) and faradaic current. Faradaic reactions involve either reduction or oxidation of redox species at an electrode–solution interface, which are typically represented by the following reaction:

$$O + ne^- \rightleftharpoons R \tag{6.1}$$

where O is the oxidised form of a redox species in solution, R is its reduced form and n is the number of electrons (e^-) transferred. The rate of reaction is controlled by several factors, including (i) transport of material from the bulk solution to the electrode–solution interface (mass transport, Subsection 6.3.1), (ii) adsorption/desorption (if relevant) and (iii) rate of electron transfer occurring in the interfacial region (electron transfer kinetics, Subsection 6.3.2). The slowest of these processes dominates the overall electrochemical reaction rate, and hence controls the resulting faradaic currents.[69]

6.3.1 Mass Transport to Nanoelectrodes

The mass transport behaviour of electrochemical systems occurs through three processes: convection, migration and diffusion. Convection is the movement of molecules arising from thermal gradients or external forces such as mechanical stirring, vibrations, pumping or fluid injection. Migration is the movement of charged species under the influence of an electric field. Diffusion is the random movement of redox molecules from regions of higher concentration (bulk solution) to regions of lower concentration (electrode–solution interface), *i.e.*, due to concentration gradients.[47,69] The diffusional flux of species, $J(x,t)$, is defined as the number of moles passing through a certain position with respect to the electrode surface, x, per unit time, t, as described by Fick's first law:

$$J(x,t) = -D\frac{\partial C(x,t)}{\partial t} \tag{6.2}$$

where D is the diffusion coefficient ($cm^2\ s^{-1}$) and $\partial C(x,t)/\partial x$ is the concentration gradient.[47,69]

Diffusion is frequently the only form of mass transport involved in electrochemical experiments, *i.e.*, mass transport effects arising from migration and convection may be negligible, *e.g.*, for measurements undertaken in the presence of excess supporting electrolyte, in static un-agitated environments, at constant temperature and over short periods of time. Thus, the faradaic current, $i(t)$, is often proportional to the flux of species at the electrode surface, $J(0,t)$:

$$i(t) = -nFAJ(0,t) \tag{6.3}$$

where n is the number of electrons involved in the redox reaction, F is Faraday's constant ($C\ mol^{-1}$) and A (cm^2) is the electrode electrochemical active area.[47,69]

6.3.1.1 Determination of the Diffusion Coefficient

Clearly, the study of mass transport of electrochemical systems requires that the diffusion coefficient, D, of the redox species of interest to be accurately determined. However, values of D are not readily found in the literature as they depend on several factors such as solution temperature, viscosity, supporting electrolyte concentration, *etc.* There are several electrochemical methods available, however, that can be used to determine unknown diffusion coefficients and are summarised in Table 6.2.[47] All of these techniques require knowledge of accurate electrode electrochemical active area (or its critical dimension), the number of electrons involved in electron transfer and the bulk concentration of redox species and electrolyte. The electrode electrochemical active area may be determined using microscopic techniques as discussed previously. Lists of n are available in the literature for a wide range of redox species or otherwise may be determined using, for example, coulometric or thin layer methods.[47] Regarding the accuracy of the bulk concentration, care must be taken to ensure that solutions are prepared with chemicals of high purity and that adequate volumetric techniques are employed. The determination of the diffusion coefficient also strongly relies on careful measurements and calibrated instruments and electrodes. Finally, it is recommended to employ several techniques to enhance the level of confidence in the experimental value of D.

Table 6.2 Summary of electrochemical methods for the determination of diffusion coefficients. (Reprinted and adapted from ref. 47, with permission from Elsevier.)

Method	Procedure	Equation	Comments
Methods with $D \propto i$			
Chronoamperometry at disk UME	Measure steady-state current	$D = \frac{i_{ss}}{4nFC^* r_0}$	RG $\gtrsim$ 100, $4Dt/r_0^2 > 500$ for $< 2\%$ planar diffusion contribution to i_{ss}
Chronoamperometry at hemispherical UME	Measure steady-state current	$D = \frac{i_{ss}}{2\pi nFC^* r_0}$	RG $\gtrsim$ 100, $Dt/r_0^2 > 800$ for $< 2\%$ planar diffusion contribution to i_{ss}
Linear scan voltammetry at disk UME	Measure steady-state current	$D = \frac{i_{ss}}{4nFC^* r_0}$	RG $\gtrsim$ 100, same time criterion as chronoamperometry, but use $t\|E_f - E_{1/2}\|/\nu$
Linear scan voltammetry at hemispherical UME	Measure steady-state current	$D = \frac{i_{ss}}{2\pi nFC^* r_0}$	RG $\gtrsim$ 100, same time criterion as chronoamperometry, but use $t\|E_f - E_{1/2}\|/\nu$
Methods with $D \propto i^{3/2}$			
Rotating disk electrode	Measure limiting current, i_l as a function of rotation rate ω and plot i_l *vs.* $\omega^{1/2}$	$D = \left(\frac{\text{slope}}{0.62nFAC^* \nu^{-1/6}}\right)^{3/2}$	Use with $10 < \omega < 10\,000$
Methods with $D \propto i^2$			
Chronoamperometry at planar electrode	Measure current as a function of a time, plot $i(t)$ *vs.* $t^{-1/2}$	$D = \frac{(\text{slope})^2 \pi}{(nFAC^*)^2}$	Avoid charging current measurements at short times ($t \lesssim 10\, R_s C_d$), avoid long times (convection, edge effects)
Normalized chronoamperometry at disk UME	Measure current as a function of time and normalize by steady-state current. Plot normalized current *vs.* $t^{-1/2}$	$D = \frac{4r_0^2}{(\text{slope})^2 \pi^3}$	Requires only r_0; avoid charging current measurements at short times ($t \lesssim 10\, R_s C_d$)
Chronopotentiometry at planar electrode	Measure transition time at several current densities, find average value of τ	$D = \frac{4 i_0^2 \tau}{\pi (nFC^*)^2}$	Avoid conditions with long transition times (convective effects); less accurate when there are large relative background contributions to the overall current (*i.e.*, low C^*, large C_d, short τ)
Linear scan voltammetry at a planar electrode	Measure peak current as a function of scan rate, plot i_p *vs.* $V^{1/2}$	$D = \frac{(\text{slope})^2}{n^3 (2.69 \times 10^5\, AC^*)^2}$	Nernstian systems only; avoid scan rates with appreciable nonlinear diffusion; should be considered an approximate method

6.3.1.2 Diffusion at Single Nanoelectrodes

Since convection and migration effects are often considered negligible as measurements are typically undertaken in excess supporting electrolyte, in static un-agitated environments, at constant temperature and over short periods of time; diffusion alone typically controls the analyte concentration profile created by electron transfer processes occurring at the nanoelectrode–solution interface, *i.e.*, the diffusion layer. The diffusion layer thickness, δ, varies considerably with the electrode critical dimension, r, such that two general cases are considered, *i.e.*, planar and radial diffusion (Figure 6.2).[6]

Planar diffusion arises when the analyte diffuses predominantly in one direction from the bulk solution to the electrode, usually in a direction normal to the electrode surface.[47,69] For sweep voltammetry techniques, planar diffusion typically results in the analyte being consumed faster than it is being replenished, especially when the applied potential reaches the formal potential of the redox couple of interest. This leads to analyte depletion at the electrode–solution interface and hence causes currents to decay with time, *i.e.*, diffusion-limited, peak-shaped and time-dependent electrochemical responses.[43] In other words, the diffusion layer thickness is small compared to the electrode critical dimension ($\delta < r$) such that planar diffusion profiles dominate.[6] This situation is typical at ultra-microelectrodes at low to medium scan rates but can also arise at nanoelectrodes during ultrafast analysis. In pulse voltammetric techniques, the potential waveform can be viewed as a staircase, whereby the potential is pulsed symmetrically (square wave and differential pulse voltammetry) or non-symmetrically (normal pulse and staircase voltammetry). For symmetric pulse methods cumulative positive and negative symmetric pulse cycles cause redox species to alternatively diffuse to and from an electrode surface such that the initial condition for each pulse cycle is the complex diffusion layer that has evolved from prior pulses.[66,123] Consequently, as the potential reaches the formal potential of a redox couple, the concentration of either oxidative or reductive species (depending on the potential pulse direction) tends to zero, resulting in broad voltammetric peaks as opposed to sigmoidal, steady-state responses. For normal pulse and staircase voltammetric techniques, although the initial supply of analyte is relatively poor due to planar diffusion, quasi sigmoidal responses may be achieved as non-symmetrical positive (or negative) pulses lead to the amplification of either oxidation or reduction.[123,124]

Concerning radial diffusion to an electrode, the diffusion layer is large compared to the electrode critical dimension ($\delta > r$).[47,69] As a result, the analyte diffuses equally in all directions to the electrode surface (3D convergence). In sweep voltammetry, it allows fresh analyte to be constantly supplied at the surface and thus sigmoidal electrochemical responses reaching steady-state and time-independent values proportional to the bulk concentration may be readily achieved.[6,48] For square wave and differential

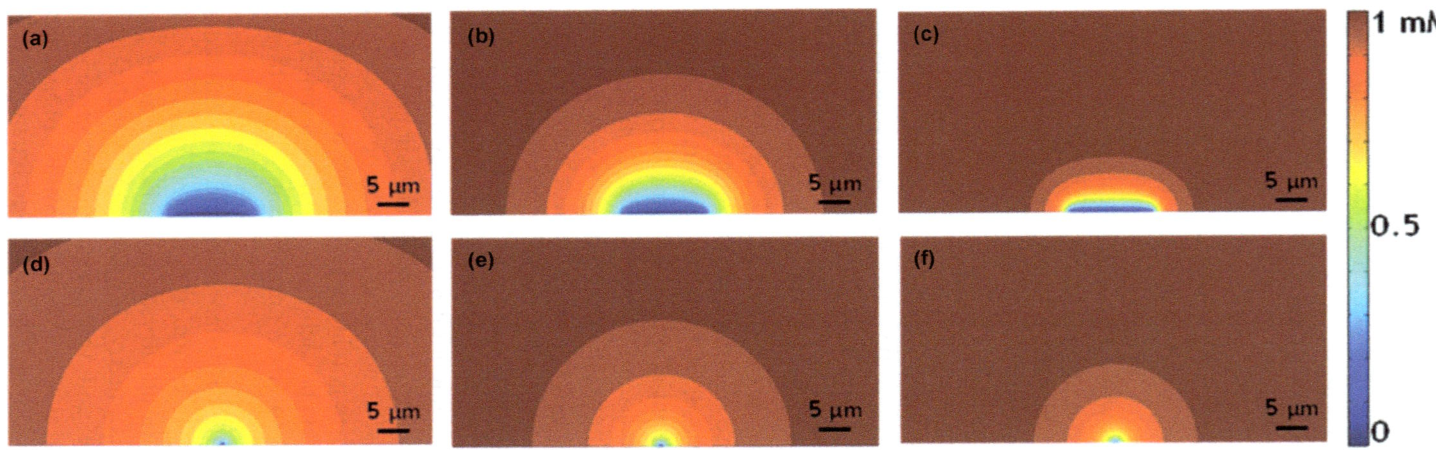

Figure 6.2 Simulated FcCOOH concentration profiles at a plane perpendicular to a single ultra-microdisc electrode, at scan rates of (a) 50, (b) 500 and (c) 5000 mV s^{-1}. $C_R = 1$ mM, $D_R = 5.4 \times 10^{-6}$ cm^2 s^{-1}, $E^\circ = 0.317$ V, $T = 298.15$ K; (d), (e) and (f) respectively are the equivalent FcCOOH concentrations profiles at a single nanowire electrode with $E^\circ = 0.155$ V. (Reprinted and adapted with permission from ref. 6, copyright 2014 Journal of The Electrochemical Society.)

pulse voltammetric methods, the initial supply of analyte is abundant such that the depletion of either oxidised or reduced form of the analyte (depending on the pulse direction) may only start when the applied potential reaches the formal potential of a redox couple—thus sharp peaks are obtained reaching a maximised current, *i.e.*, likely to be proportional to bulk concentration.[66,123] A key advantage of the presence of these sharp peaks is the ability to separate these peaks, thus enabling qualitative identification of molecular species with respect to their formal potential.[11] For normal pulse and staircase voltammetric techniques, limiting currents may reach steady-state and time-independent values as the initial supply of analyte is increased.[123,124] Analyte radial diffusion is thus more efficient than analyte planar diffusion and is likely to occur at electrodes with decreasing critical dimension, in particular it is characteristic for nanoelectrodes.

As mentioned above, both ultra-microelectrodes and nanoelectrodes benefit from a diffusion regime where efficient mass transport allows the faradaic current to attain a time-independent steady-state value when the applied potential is swept linearly, at low to medium scan rates (under entirely quiescent conditions). As the scan rate increases, there is a corresponding decrease in the thickness of the diffusion layer, so at high-to-very-high scan rates the measured current at ultra-microelectrodes can become time-dependent. By contrast, given the very small critical dimension of nanoelectrodes, these have been shown to remain time-independent with scan rates in excess of 50 V s^{-1}. This is shown in the simulation concentration profiles (and confirmed experimentally) of 1 mM FcCOOH in 10 mM PBS at the ultra-microdisc electrode (12.5 μm diameter) surface under cyclic voltammetric conditions at 50, 500 and 5000 mV s^{-1}, as presented in Figure 6.2(a)–(c), respectively. At 50 mV s^{-1}, the diffusion layer thickness appears uniform and larger than the electrode diameter, suggesting analyte mass transport behaviour is dominated by radial diffusion. At 500 mV s^{-1}, the inner diffusion layers become planar, suggesting a change in the diffusional mass transport behaviour from radial to planar, also known as a transition zone. At 5000 mV s^{-1}, the extent of the diffusion layer is of similar dimension than the electrode diameter, clearly showing that planar diffusion dominates at the electrode surface, corresponding to the analyte being consumed faster than it is replenished. This indicates that ultra-microdisc electrodes would exhibit steady-state currents at long analysis times (low scan rates) but diffusion limited peak currents at short analysis times (high scan rates). By comparison, Figure 6.2(d)–(f) show 2D concentration profiles of 1 mM FcCOOH in 10 mM PBS occurring at a single nanowire electrode under the same experimental conditions. For all potential scan rates, it is clear that the diffusion layer thickness is uniform and much larger than the nanowire electrode width. This shows that analyte mass transport behaviour is dominated by radial diffusion for all analysis times. Therefore, radial diffusion at nanoelectrodes may be observed even at high measurement rates, thereby considerably lowering electrochemical analysis time scales.

6.3.1.3 Diffusion at Nanoelectrode Arrays

Nanoelectrode arrays may be fabricated as ensembles, which typically consist of dense arrays of closely packed nanostructures decorating macro or micro electrodes to achieve high measureable currents (nA–μA). However, these higher currents arise from an increase in electroactive area rather than an improvement in analyte mass transport, as identified by diffusion-limited peak currents.

Conversely, nanoelectrode arrays may also be fabricated as ordered arrays consisting of well-separated nanoelectrodes using top-down fabrication techniques borrowed from the microelectronics industry. Unlike close packed ensembles, well-separated nanoelectrodes amplify electrochemical signals while also maintaining the benefits observed at single nanoelectrodes, especially the enhanced mass transport behaviour.[64,70,72] However, to benefit from enhanced mass transport behaviour, an in-depth understanding of the diffusion profiles existing at individual nanoelectrode elements is required. In this regard, Compton *et al.* have identified four types of diffusional mass transport behaviour to (nano)electrode arrays[48,125] (Figure 6.3). These regimes are described by the relationship between the thickness of the

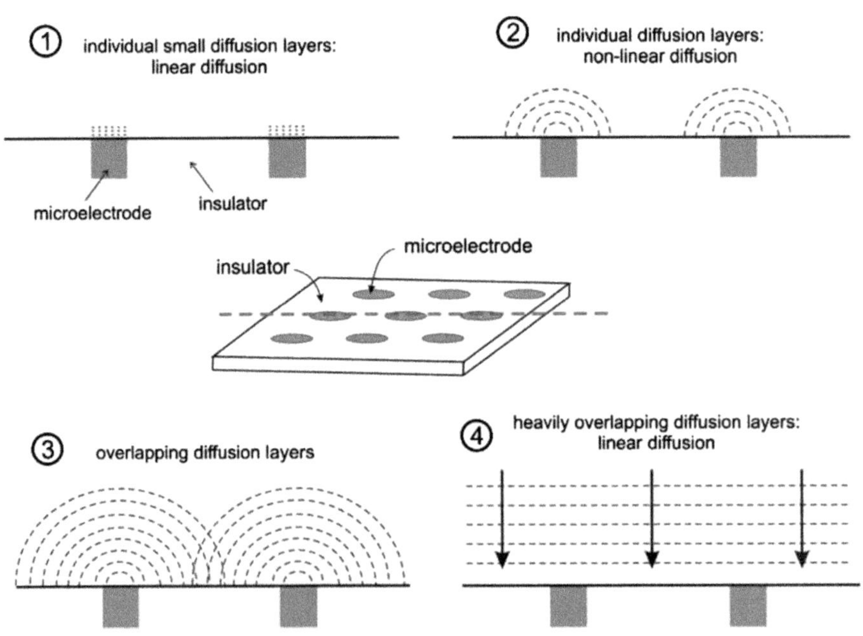

Figure 6.3 Schematic representation of the four different diffusion regimes that may occur at (nano)electrode arrays. (1) $\delta \ll r$, planar diffusion to each electrode, (2) $\delta = r < d$, radial diffusion to each electrode, (3) $r < \delta \leq d$, transition zone and (4) $\delta \gg d$, planar diffusion to the entire array. (Reprinted and adapted with permission from ref. 48 and 125, copyright 2005 Journal of Electroanalytical Chemistry.)

individual diffusion layer, δ, the electrodes' critical dimension, r, and the inter-electrode distance, d, with respect to the electrochemical analysis time: (i) at short times with well-separated electrodes, individual diffusion layers are small compared to the electrode critical dimensions ($\delta \ll r$) resulting in individual planar diffusion layers; (ii) at long times with well-separated electrodes, the diffusion layer thickness becomes larger than the electrode critical dimension and no overlap between adjacent layers occurs ($\delta = r < d$), thus individual radial diffusion layers are obtained; (iii) at long times for smaller inter-electrode separations, overlapping of individual diffusion layers is likely to happen ($r < \delta \leq d$), such that the closer the electrodes the greater the degree of overlap until reaching (iv) complete overlap resulting in planar diffusion to the whole array ($\delta \gg d$). Advantages associated with each different regime depend on the voltammetric techniques employed. Thus, a design for application approach is required for nanoelectrode arrays.

6.3.1.4 *Simulation of Diffusional Mass Transport*

Realistically, mass transport is a three-dimensional phenomenon, which may be time consuming and complicated to solve. However, it is sometimes possible to approximate it as a two-dimensional simulation problem.[126] This simplification is known as the diffusion domain approach, whereby the three-dimensional shape of an electrode may be divided into multiple, independent and highly symmetric two-dimensional slices. A key caveat, however, in making geometric simplifications is that symmetry is maintained over the entire timescale of the experiment. For instance, consider a nanoband electrode where the length maybe considered infinitely long compared to the width of the nanoband. In this case, only hemispherical diffusion to the electrode width may be considered (as a 2D slice across the entire nanoband), allowing the diffusional mass transport to the nanoelectrode to be treated as a 2D problem. Initially, this diffusion domain approach was defined, and applied to disks and bands by Amatore *et al.*[49] Recently, this approach has been simplified for a selection of geometries (both voltammetry and chronoamperometry techniques) and is fast becoming the preferred method for analysis of mass transport groups at nanoelectrodes and their arrays.[43,48,50,125,127]

Simulations are typically undertaken using commercial software to extract expected currents as well as providing a pictorial representation of concentration profiles. At single nanoelectrodes, simulations in conjunction with experimental results are principally used to provide important insight into analyte concentration profiles at the nanoelectrode–solution interface under the application of an external potential. However, for nanoelectrode arrays, simulations at arrays are essential to provide an in-depth understanding of diffusion regimes occurring at different electrode geometries. These are required to ensure, for example, diffusional independence, such that design of the electrodes may be optimised prior to fabrication.

6.3.2 Electron Transfer Kinetics at Nanoelectrodes

Nanoelectrodes may provide fundamental insight into electron transfer kinetics because this process becomes slower than the mass transfer rate. Effectively, since radial (three-dimensional) analyte diffusion dominates at nanoelectrodes, steady-state responses are achieved using sweep voltammetry techniques even at high measurement rates and hence electrochemical reactions are less likely to be limited by the rate of mass transport of fresh analyte to the electrode surface. Thus, nanoelectrodes permit kinetic measurements by steady-state experiments rather than by transient techniques, thereby enabling electron transfer kinetic studies of high speed electrochemical reactions. The rate at which electron transfer takes place across the electrode–solution interface is described by the heterogeneous electron-transfer rate constant. For example, the heterogeneous electron transfer-rate constant for the oxidation of ferrocene monocarboxylic acid in phosphate buffer saline solution at single gold nanowires was found to be about two orders of magnitude greater than previously determined at a 2 mm diameter platinum disk electrode.[56,128]

6.4 Recent Advances and Future Development

6.4.1 Design for Application Approach for Nanoelectrode Arrays

When designing nanoelectrode arrays, the electrochemical method, *e.g.*, scanning or pulse voltammetry that will be employed in the final application is probably the first factor to consider as this directly effects the necessary inter-electrode distance. Of note, for the same solution conditions and potential window, higher currents are typically achieved with pulse voltammetry techniques because of their greater speed, signal-to-noise ratio and sensitivity compared to sweep voltammetry techniques.

In sweep voltammetry, redox species are gradually consumed at the electrode–solution interface as the potential is swept linearly. At single nanoelectrodes exhibiting radial diffusion behaviour, the analyte is replenished as fast as it is consumed, thereby achieving steady-state, time-independent and sigmoidal electrochemical responses likely to be proportional to the bulk concentration. At nanoelectrode arrays, maximum efficiency may thus be obtained when each nanoelectrode exhibits radial diffusion behaviour, *i.e.*, diffusional regime (ii) (Figure 6.3), known as diffusional independence. In such cases, each nanoelectrode behaves as an individual electrode such that the obtained responses are sigmoidal, displaying steady-state and time-independent currents that are proportional to the number of electrodes. To ensure diffusional independence, the required inter-electrode distance may be estimated based on the thickness of the diffusion layer at a single nanoelectrode. This may be determined using Einstein's relation of $\delta = (2Dt)^{1/2}$, where D is the diffusion coefficient of the redox species and t is the duration of the experiment.[127,129] Hence, inter-electrode distances, d, greater than twice the diffusion thickness ($d > 2\delta$) should ensure a diffusionally independent regime. While in many cases this simple calculation can prove very accurate, care must be taken when using this equation, especially at the nanoscale, as it will only provide an estimate, *e.g.*, it does not take into account electrode geometry.[50] Therefore, a more in-depth approach based on theoretical simulation studies is often employed to assess ideal spatial requirements. For instance, finite element analysis of diffusion profiles existing at arrays of nanowire electrodes have been undertaken to explore the effects of altering inter-electrode separations on diffusional independence for a range of scan rates.[50] Conversely, current responses at nanoelectrode arrays designed to work in diffusional regime (i) are diffusion limited and time-dependent because of individual planar diffusion layers. In addition, currents at nanoelectrode arrays exhibiting behaviour (iii) tend to tail off, giving rise to peak-shaped and time-dependent responses. Effectively the overlap of analyte diffusion profiles between neighbouring nanoelectrodes results in increased analyte competition and ultimately reduces the overall electrochemical performance of the array.[59] Finally, nanoelectrode arrays applying to the diffusional regime (iv) behave as larger electrodes of the same total area as that occupied by the entire array (electrodes and insulating area between them) due to complete overlap of diffusion layers, *i.e.*, large peak-shaped diffusion-limited responses.

In square wave and pulse voltammetry, redox species are repeatedly oxidised and reduced as the voltage is pulsed positively and negatively for a series of cycles, such that the initial condition for each cycle is the complex diffusion layer that has evolved from prior pulses. At single nanoelectrodes exhibiting radial diffusion behaviour, currents reach a maximum around the formal redox potential that is likely to be proportional to bulk concentration. By contrast, at nanoelectrode arrays, it is hypothesised that molecules oxidised (or reduced) at one nanoelectrode may diffuse to a neighbouring one rather than to bulk solution so that it may consequently be oxidised (or reduced) again during a subsequent voltage step up until reaching the formal potential. This should result in more oxidative (or reductive) species at the surface of each nanoelectrode between sampling intervals, and consequently in increased signal and higher sensitivity. Furthermore, nanoelectrode arrays with overlapping diffusion profiles should exhibit higher signals as electrodes are closer to each other, thereby reducing the distance for oxidised (or reduced) species to travel from one concentration profile to the neighbouring one. However, complete overlap of diffusion layers would minimise redox species movement, resulting in decreased signals. Thus, nanoelectrode arrays exhibiting diffusional behaviour (iii) would be best suited for square wave and pulse voltammetric techniques. Similarly, it is hypothesised that regime (iii) in Figure 6.3 would also be best suited for normal pulse and staircase voltammetric techniques. Simulations are still required, however, to confirm these hypotheses.

6.4.2 Emerging and Future Applications

The advent of nanoelectrodes is pushing back traditional boundaries such that electrochemical measurements that were challenging (or even impossible) with classic micro and ultra-microelectrodes are now becoming achievable. For instance, nanoelectrodes are finding increased use to improve the spatial resolution of scanning electrochemical microscopy as this technique strongly relies on the size of the scanning probe.[130,131] In this technique, higher resolution may be achieved when the probe electrode is comparable in size to the features that are being imaged. Consequently, nanoelectrodes as scanning probes are beginning to enable much greater spatial resolution. Examples of recent development include nanoscale imaging of live-cells using multifunctional nanoprobes,[132] localised chemical delivery at nanoscale interfaces[133] and surface topography and reactivity of various modified nano-sized substrates and single molecule electrochemical detection.[134] In this regard, imaging biologically or chemically modified nano-sized surfaces such as enzyme nanoarrays or proteins attached on nanoelectrodes opens up possibilities for the study of the mass transport process and the direct probing, for example, biological, biomedical, electrochemical, chemical and/or physical properties, of single (bio)molecules. These developments could lead to new advances in areas such as nanobiotechnology, nanomedicine and molecular biology.

6.5 Conclusion and Outlook

The availability of reproducible, low-cost and highly sensitive nanoelectrodes will significantly contribute to the growing area of electrochemical analysis. Fully integrated nanosensors devices will enable new advances in applications including in-line monitoring process analytical technologies and point-of-use sensing applications. The former, approved by the US FDA, are finding increased use in the food and pharmaceutical sectors, while the latter are becoming more popular in personalised health care, security monitoring and environmental analysis. Key requirements for remote sensor devices are the needs to be portable, easy-to-use, low maintenance and highly sensitive, to enable rapid detection and quantification of target analytes. Silicon chip based nanosensors devices address these requirements in that they are integrated devices compatible with very low-volume sample delivery techniques and may be fabricated at high densities (low form factors) for simultaneous multi-analyte detection and rapid data capture. They may be directly electrically addressed and interfaced with microelectronic circuitry thereby allowing real-time signal capture, signal amplification (if required), the possibility of rapid data processing and direct signal readout. They are readily fabricated in a form factor that is portable, software control allows ease of use for the non-specialist, and have little or no maintenance requirements. Finally, they provide detection limits that are comparable or better than the laboratory-based equivalent analytical instrumentation, examples of which are presented above.

Although nanoelectrodes, particularly those fabricated on silicon chips, have genuine potential in the area of diagnostics, a complete understanding of the electrochemical processes, in terms of diffusion and mass transport, electron transfer kinetics and the effects of electrode geometry and material composition is required for sensors to be optimised for real-world applications. While a large body of work has been presented in the literature to date in this regard, of particular interest going forward is the need to develop new theoretical models that will accurately describe pulse voltammetric techniques at three-dimensional electrode structures, allowing more informed decisions on electrode geometry and design.

References

1. T. S. Harrison, *J. Hist. Neurosci.*, 2000, **9**, 165–179.
2. D. H. Hubel, *Science*, 1957, **125**, 549–550.
3. R. Feeney and S. P. Kounaves, *Electroanalysis*, 2000, **12**, 677–684.
4. C. Amatore, *Spectra 2000*, 1990, **18**, 43–46.
5. M. Fleischmann and S. Pons, *Anal. Chem.*, 1987, **59**, 1391A–1399A.
6. A. Wahl, S. Barry, K. Dawson, J. MacHale, A. Quinn and A. O'Riordan, *J. Electrochem. Soc.*, 2014, **161**, B3055–B3060.
7. R. B. Morris, D. J. Franta and H. S. White, *J. Phys. Chem.*, 1987, **91**, 3559–3564.
8. P. Ugo, L. M. Moretto and F. Vezzà, *ChemPhysChem*, 2002, **3**, 917–925.
9. S. Viswanathan, C. Rani, A. V. Anand and J. A. A. Ho, *Biosens. Bioelectron.*, 2009, **24**, 1984–1989.
10. K. Dawson, M. Baudequin and A. O'Riordan, *Analyst*, 2011, **136**, 4507–4513.
11. S. Barry, K. Dawson, E. Correa, R. Goodacre and A. O'Riordan, *Faraday Discuss.*, 2013, **164**, 283–293.
12. D. W. Arrigan, *Analyst*, 2004, **129**, 1157–1165.
13. X. Duan, Y. Huang, Y. Cui, J. Wang and C. M. Lieber, *Nature*, 2001, **409**, 66–69.
14. Y.-H. Liau, N. F. Scherer and K. Rhodes, *J. Phys. Chem. B*, 2001, **105**, 3282–3288.
15. T. Miyazaki, K. Kobayashi, K. Ishida, S. Hotta, T. Horiuchi, H. Yamada and K. Matsushige, *Jpn. J. Appl. Phys.*, 2003, **42**, 4852.
16. H. Daniels, R. Brydson, A. Brown and B. Rand, *Ultramicroscopy*, 2003, **96**, 547–558.
17. J. E. Mark, *Physical Properties of Polymers Handbook*, Springer, 2007.
18. E. Tan and C. Lim, *Appl. Phys. Lett.*, 2004, **84**, 1603–1605.
19. K. J. Klabunde and R. Richards, *Nanoscale Materials in Chemistry*, John Wiley & Sons, Inc., Hoboken, 2001.
20. L. Wang, D. Fine and A. Dodabalapur, *Appl. Phys. Lett.*, 2004, **85**, 6386–6388.

21. C. Burda, X. Chen, R. Narayanan and M. A. El-Sayed, *Chem. Rev.*, 2005, **105**, 1025–1102.
22. R. Bhattacharya and P. Mukherjee, *Adv. Drug Delivery Rev.*, 2008, **60**, 1289–1306.
23. J. F. Smith, T. P. Knowles, C. M. Dobson, C. E. MacPhee and M. E. Welland, *Proc. Natl. Acad. Sci. U. S. A.*, 2006, **103**, 15806–15811.
24. K. L. Wooley, *J. Polym. Sci., Part A: Polym. Chem.*, 2000, **38**, 1397–1407.
25. P. K. Jain, X. Huang, I. H. El-Sayed and M. A. El-Sayed, *Acc. Chem. Res.*, 2008, **41**, 1578–1586.
26. C. L. Nehl, H. Liao and J. H. Hafner, *Nano Lett.*, 2006, **6**, 683–688.
27. P. Yang, H. Yan, S. Mao, R. Russo, J. Johnson, R. Saykally, N. Morris, J. Pham, R. He and H.-J. Choi, *Adv. Funct. Mater.*, 2002, **12**, 323.
28. H. Pardoe, W. Chua-Anusorn, T. G. St Pierre and J. Dobson, *J. Magn. Magn. Mater.*, 2001, **225**, 41–46.
29. D. L. Leslie-Pelecky and R. D. Rieke, *Chem. Mater.*, 1996, **8**, 1770–1783.
30. K. Ono, J. Okabayashi, M. Mizuguchi, M. Oshima, A. Fujimori and H. Akinaga, *J. Appl. Phys.*, 2002, **91**, 8088–8092.
31. T. Namazu, Y. Isono and T. Tanaka, *J. Microelectromech. Syst.*, 2000, **9**, 450–459.
32. M. Nowicki, A. Richter, B. Wolf and H. Kaczmarek, *Polymer*, 2003, **44**, 6599–6606.
33. B. Ji and H. Gao, *J. Mech. Phys. Solids*, 2004, **52**, 1963–1990.
34. A. Sanchez, S. Abbet, U. Heiz, W.-D. Schneider, H. Häkkinen, R. Barnett and U. Landman, *J. Phys. Chem. A*, 1999, **103**, 9573–9578.
35. B. R. Cuenya, *Thin Solid Films*, 2010, **518**, 3127–3150.
36. C. Rao, G. Kulkarni, P. J. Thomas and P. P. Edwards, *Chem. – Eur. J.*, 2002, **8**, 28–35.
37. R. Prasher, P. Bhattacharya and P. E. Phelan, *Phys. Rev. Lett.*, 2005, **94**, 025901.
38. A. Shakouri, *Proc. IEEE*, 2006, **94**, 1613–1638.
39. D. G. Cahill, W. K. Ford, K. E. Goodson, G. D. Mahan, A. Majumdar, H. J. Maris, R. Merlin and S. R. Phillpot, *J. Appl. Phys.*, 2003, **93**, 793–818.
40. X. Huang, J. Tu, B. Zhang, C. Zhang, Y. Li, Y. Yuan and H. Wu, *J. Power Sources*, 2006, **161**, 541–544.
41. M. S. Park, Y. M. Kang, G. X. Wang, S. X. Dou and H. K. Liu, *Adv. Funct. Mater.*, 2008, **18**, 455–461.
42. M.-C. Daniel and D. Astruc, *Chem. Rev.*, 2004, **104**, 293–346.
43. M. C. Henstridge and R. G. Compton, *Chem. Rec.*, 2012, **12**, 63–71.
44. R. G. Compton and C. E. Banks, *Understanding Voltammetry*, World Scientific, Singapore, 2007.
45. J. K. Holt, H. G. Park, Y. Wang, M. Stadermann, A. B. Artyukhin, C. P. Grigoropoulos, A. Noy and O. Bakajin, *Science*, 2006, **312**, 1034–1037.
46. K. Dawson, A. l. Wahl, R. Murphy and A. O'Riordan, *J. Phys. Chem. C*, 2012, **116**, 14665–14673.

47. *Handbook of Electrochemistry*, ed. C. G. Zoski, Elsevier, Amsterdam, 2007.
48. T. J. Davies and R. G. Compton, *J. Electroanal. Chem.*, 2005, **585**, 63–82.
49. C. Amatore, J. Saveant and D. Tessier, *J. Electroanal. Chem. Interfacial Electrochem.*, 1983, **147**, 39–51.
50. A. Wahl, K. Dawson, J. MacHale, S. Barry, A. J. Quinn and A. O'Riordan, *Faraday Discuss.*, 2013, **164**, 377–390.
51. K. Dawson, A. Wahl, A. Pescaglini, D. Iacopino and A. O'Riordan, *J. Electrochem. Soc.*, 2014, **161**, B3049–B3054.
52. E. J. Dickinson and R. G. Compton, *J. Phys. Chem. C*, 2009, **113**, 17585–17589.
53. R. J. Forster, *Chem. Soc. Rev.*, 1994, **23**, 289–297.
54. C. G. Zoski, *Electroanalysis*, 2002, **14**, 1041–1051.
55. I. GrabecáŠvegl, *Analyst*, 1996, **121**, 1817–1822.
56. K. Dawson, J. r. Strutwolf, K. P. Rodgers, G. Herzog, D. W. M. Arrigan, A. J. Quinn and A. O'Riordan, *Anal. Chem.*, 2011, **83**, 5535–5540.
57. J. T. Cox and B. Zhang, *Annu. Rev. Anal. Chem.*, 2012, **5**, 253–272.
58. R. G. Compton, G. G. Wildgoose, N. V. Rees, I. Streeter and R. Baron, *Chem. Phys. Lett.*, 2008, **459**, 1–17.
59. K. Dawson, M. Baudequin, N. Sassiat, A. J. Quinn and A. O'Riordan, *Electrochim. Acta*, 2013, **101**, 169–176.
60. J. Wang, X. Rongrong, T. Baomin, J. Wang, C. L. Renschler and C. A. White, *Anal. Chim. Acta*, 1994, **293**, 43–48.
61. Y. Hou, Y. Cheng, T. Hobson and J. Liu, *Nano Lett.*, 2010, **10**, 2727–2733.
62. H. Li and N. Wu, *Nanotechnology*, 2008, **19**, 275301.
63. T. Ito, A. A. Audi and G. P. Dible, *Anal. Chem.*, 2006, **78**, 7048–7053.
64. Y. H. Lanyon, G. De Marzi, Y. E. Watson, A. J. Quinn, J. P. Gleeson, G. Redmond and D. W. M. Arrigan, *Anal. Chem.*, 2007, **79**, 3048–3055.
65. P. W. Bohn, *Annu. Rev. Anal. Chem.*, 2009, **2**, 279–296.
66. J. G. Osteryoung and R. A. Osteryoung, *Anal. Chem.*, 1985, **57**, 101A–110A.
67. J. J. O'Dea, J. Osteryoung and R. A. Osteryoung, *Anal. Chem.*, 1981, **53**, 695–701.
68. J. Zipper and S. Perone, *Anal. Chem.*, 1973, **45**, 452–458.
69. A. J. Bard and L. R. Faulkner, *Electrochemical Methods: Fundamentals and Applications*, John Wiley & Sons, Inc., New York, 1980.
70. N. Godino, X. Borrise, F. X. Munoz, F. J. del Campo and R. G. Compton, *J. Phys. Chem. C*, 2009, **113**, 11119–11125.
71. C. Ma, N. M. Contento, L. R. Gibson and P. W. Bohn, *Anal. Chem.*, 2013, **85**, 9882–9888.
72. Y. H. Lanyon and D. W. M. Arrigan, *Sens. Actuators, B*, 2007, **121**, 341–347.
73. M. Pal and V. Ganesan, *Analyst*, 2010, **135**, 2711–2716.
74. A. S. Aricò, P. Bruce, B. Scrosati, J.-M. Tarascon and W. Van Schalkwijk, *Nat. Mater.*, 2005, **4**, 366–377.

75. A. W. Hassel, B. Bello-Rodriguez, S. Milenkovic and A. Schneider, *Electrochim. Acta*, 2005, **50**, 3033–3039.
76. F. Vullum and D. Teeters, *J. Power Sources*, 2005, **146**, 804–808.
77. Z. C. Lin, C. Xie, Y. Osakada, Y. Cui and B. Cui, *Nat. Commun.*, 2014, **5**, 1–10.
78. S. Eustis and M. A. El-Sayed, *Chem. Soc. Rev.*, 2006, **35**, 209–217.
79. S. Guo and E. Wang, *Anal. Chim. Acta*, 2007, **598**, 181–192.
80. S. Hrapovic, Y. Liu, K. B. Male and J. H. Luong, *Anal. Chem.*, 2004, **76**, 1083–1088.
81. L. Dai, *Carbon Nanotechnology: Recent Developments in Chemistry, Physics, Materials Science and Device Applications*, Elsevier, 2006.
82. T. You, O. Niwa, M. Tomita and S. Hirono, *Anal. Chem.*, 2003, **75**, 2080–2085.
83. M. Yang, F. Qu, Y. Lu, Y. He, G. Shen and R. Yu, *Biomaterials*, 2006, **27**, 5944–5950.
84. Y. Lin, F. Lu, Y. Tu and Z. Ren, *Nano Lett.*, 2004, **4**, 191–195.
85. J. Wang, *Chem. Rev.*, 2008, **108**, 814–825.
86. K. M. Taylor-Pashow, J. Della Rocca, R. C. Huxford and W. Lin, *Chem. Commun.*, 2010, **46**, 5832–5849.
87. A. Mehdi, C. Reye and R. Corriu, *Chem. Soc. Rev.*, 2011, **40**, 563–574.
88. J. Zhang, X. Liu, X. Guo, S. Wu and S. Wang, *Chem. - Eur. J.*, 2010, **16**, 8108–8116.
89. Q. Li, L. Zeng, J. Wang, D. Tang, B. Liu, G. Chen and M. Wei, *ACS Appl. Mater. Interfaces*, 2011, **3**, 1366–1373.
90. S. Guo and S. Dong, *TrAC, Trends Anal. Chem.*, 2009, **28**, 96–109.
91. Y. Jin, Y. Shen and S. Dong, *J. Phys. Chem. B*, 2004, **108**, 8142–8147.
92. X. Kang, Z. Mai, X. Zou, P. Cai and J. Mo, *Anal. Biochem.*, 2007, **369**, 71–79.
93. R. N. Goyal, V. K. Gupta, M. Oyama and N. Bachheti, *Talanta*, 2007, **72**, 976–983.
94. J. Tashkhourian, M. Nezhad, J. Khodavesi and S. Javadi, *J. Electroanal. Chem.*, 2009, **633**, 85–91.
95. L. Lin, P. Qiu, X. Cao and L. Jin, *Electrochim. Acta*, 2008, **53**, 5368–5372.
96. A. Safavi, N. Maleki and E. Farjami, *Electroanalysis*, 2009, **21**, 1533–1538.
97. K. B. Male, S. Hrapovic, Y. Liu, D. Wang and J. H. Luong, *Anal. Chim. Acta*, 2004, **516**, 35–41.
98. Q. Xu, Y. Zhao, J. Z. Xu and J.-J. Zhu, *Sens. Actuators, B*, 2006, **114**, 379–386.
99. X. Kang, Z. Mai, X. Zou, P. Cai and J. Mo, *Anal. Biochem.*, 2007, **363**, 143–150.
100. X. Luo, A. Morrin, A. J. Killard and M. R. Smyth, *Electroanalysis*, 2006, **18**, 319–326.
101. S. Bharathi, M. Nogami and S. Ikeda, *Langmuir*, 2001, **17**, 1–4.
102. M. Hasan, S. B. Newcomb, J. F. Rohan and K. M. Razeeb, *J. Power Sources*, 2012, **218**, 148–156.
103. Y. Hu, J. Jin, P. Wu, H. Zhang and C. Cai, *Electrochim. Acta*, 2010, **56**, 491–500.

104. S. Patchkovskii, S. T. John, S. N. Yurchenko, L. Zhechkov, T. Heine and G. Seifert, *Proc. Natl. Acad. Sci. U. S. A.*, 2005, **102**, 10439–10444.
105. B. Özyilmaz, P. Jarillo-Herrero, D. Efetov and P. Kim, *Appl. Phys. Lett.*, 2007, **91**, 192107.
106. A. C. Neto, F. Guinea, N. Peres, K. S. Novoselov and A. K. Geim, *Rev. Mod. Phys.*, 2009, **81**, 109.
107. C. Lee, X. Wei, J. W. Kysar and J. Hone, *Science*, 2008, **321**, 385–388.
108. Y. Gogotsi, *J. Phys. Chem. Lett.*, 2011, **2**, 2509–2510.
109. D. A. Brownson and C. E. Banks, *Analyst*, 2010, **135**, 2768–2778.
110. D. A. Brownson, D. K. Kampouris and C. E. Banks, *Chem. Soc. Rev.*, 2012, **41**, 6944–6976.
111. Y. Shao, J. Wang, H. Wu, J. Liu, I. A. Aksay and Y. Lin, *Electroanalysis*, 2010, **22**, 1027–1036.
112. S. M. Oja, M. Wood and B. Zhang, *Anal. Chem.*, 2012, **85**, 473–486.
113. K. Dawson and A. O'Riordan, *Annu. Rev. Anal. Chem.*, 2014, **7**, 163–181.
114. L. Rassaei, P. S. Singh and S. G. Lemay, *Anal. Chem.*, 2011, **83**, 3974–3980.
115. R. W. Murray, *Chem. Rev.*, 2008, **108**, 2688–2720.
116. V. Gubala, L. F. Harris, A. J. Ricco, M. X. Tan and D. E. Williams, *Anal. Chem.*, 2011, **84**, 487–515.
117. S. Trasatti and O. Petrii, *Pure Appl. Chem.*, 1991, **63**, 711–734.
118. R. Woods, *J. Electroanal. Chem. Interfacial Electrochem.*, 1974, **49**, 217–226.
119. J. M. Doña Rodríguez, J. A. Herrera Melián and J. Pérez Peña, *J. Chem. Educ.*, 2000, **77**, 1195.
120. C. Zuliani, D. A. Walsh, T. E. Keyes and R. J. Forster, *Anal. Chem.*, 2010, **82**, 7135–7140.
121. T. Biegler, D. Rand and R. Woods, *J. Electroanal. Chem. Interfacial Electrochem.*, 1971, **29**, 269–277.
122. K. Dawson, A. Wahl, S. Barry, C. Barrett, N. Sassiat, A. J. Quinn and A. O'Riordan, *Electrochim. Acta*, 2014, **115**, 239–246.
123. *Laboratory Techniques in Electroanalytical Chemistry, Revised and Expanded*, ed. P. Kissinger and W. R. Heineman, CRC Press, Boca Raton, 1996.
124. T. R. Brumleve, J. J. O'Dea, R. A. Osteryoung and J. Osteryoung, *Anal. Chem.*, 1981, **53**, 702–706.
125. T. J. Davies, S. Ward-Jones, C. E. Banks, J. del Campo, R. Mas, F. X. Muñoz and R. G. Compton, *J. Electroanal. Chem.*, 2005, **585**, 51–62.
126. A. J. Bard, M. V. Mirkin, P. R. Unwin and D. O. Wipf, *J. Phys. Chem.*, 1992, **96**, 1861–1868.
127. A. Molina, J. Gonzalez, E. Laborda and R. G. Compton, *Russ. J. Electrochem.*, 2012, **48**, 600–609.
128. R. Antiochia, I. Lavagnini, F. Magno, F. Valentini and G. Palleschi, *Electroanalysis*, 2004, **16**, 1451–1458.
129. R. M. Wightman and D. O. Wipf, *Electroanal. Chem.*, 1989, **15**, 267–353.

130. I. Dumitrescu, P. V. Dudin, J. P. Edgeworth, J. V. Macpherson and P. R. Unwin, *J. Phys. Chem. C*, 2010, **114**, 2633–2639.
131. M. V. Mirkin, F. R. F. Fan and A. J. Bard, *J. Electroanal. Chem.*, 1992, **328**, 47–62.
132. P. Novak, C. Li, A. I. Shevchuk, R. Stepanyan, M. Caldwell, S. Hughes, T. G. Smart, J. Gorelik, V. P. Ostanin and M. J. Lab, *Nat. Methods*, 2009, **6**, 279–281.
133. Y. Takahashi, A. I. Shevchuk, P. Novak, Y. Zhang, N. Ebejer, J. V. Macpherson, P. R. Unwin, A. J. Pollard, D. Roy and C. A. Clifford, *Angew. Chem., Int. Ed.*, 2011, **50**, 9638–9642.
134. F. O. Laforge, J. Velmurugan, Y. Wang and M. V. Mirkin, *Anal. Chem.*, 2009, **81**, 3143–3150.

CHAPTER 7

Carbon Nanomaterials in Electrochemical Detection

JONATHAN P. METTERS AND CRAIG E. BANKS*

Faculty of Science and Engineering, School of Science and the Environment, Division of Chemistry and Environmental Science, Manchester Metropolitan University, Chester Street, Manchester M15 GD, UK
*Email: c.banks@mmu.ac.uk

7.1 Introduction

Electroanalysts are always searching for that next best electrode material. The electroanalytical characteristics looked for are significant improvements in the voltammetric peak height/analytical signal and improvements in the analytical sensitivity (the gradient arising from a plot of signal *versus* concentration) as well improvements (reductions) in the overpotential. Carbon nanomaterials are being extensively explored to see if they meet these criteria and are usually immobilised onto solid electrodes with the improvements usually judged compared to these underlying electrode substrates, which act as a support for the chosen carbon nanomaterials.

Carbon is, without doubt, the most commonly utilised electrode material in electroanalysis and, according to the degree of graphitization, carbon is morphologically diverse and a variety of structures exist in a plethora of micro- and nano-structures: glassy carbon, carbon fibre, nanotubes, buckminsterfullerenes, graphene, amorphous powders and diamond to name just a few. In fact the range is quite extensive and Figure 7.1 provides a

RSC Detection Science Series No. 6
Electrochemical Strategies in Detection Science
Edited by Damien W. M. Arrigan

Published by the Royal Society of Chemistry, www.rsc.org

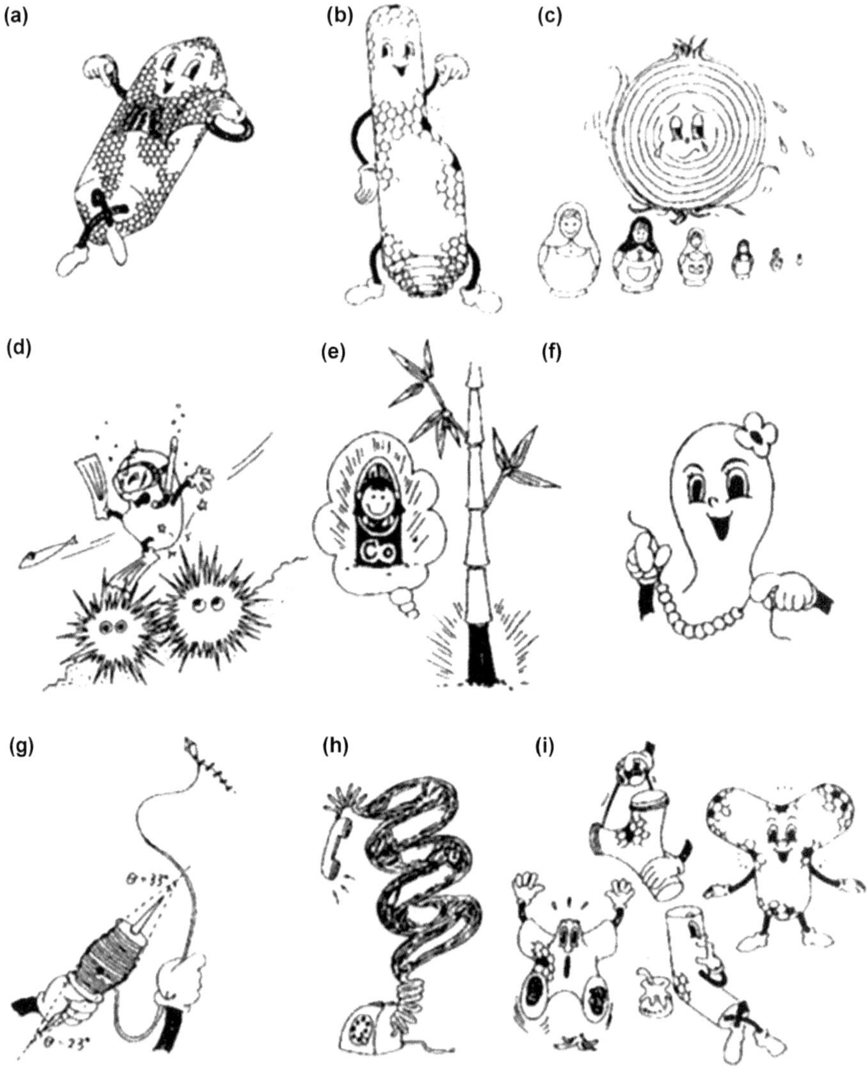

Figure 7.1 Comic style representation of various forms of nanocarbons: (a) tube with both ends closed, (b) tube with differential diameters, (c) buckyonion, (d) sea urchins, (e) bamboo-like nanotubes, (f) necklaces with buckyonion beads, (g) spinning cones, (h) helical telephone coils and (i) connectors and tripods.
(Reprinted from ref. 1 with permission from the Materials Research Society.)

summary of some of the exciting nanocarbon structures that have been reported.

The field of electroanalysis has reported the use of a range of electrode substrates such as mercury, platinum, gold, silver and carbon. Mercury electrodes were originally utilised with its accessible potential window highlighted in Figure 7.2; the obvious replacement of mercury with solid

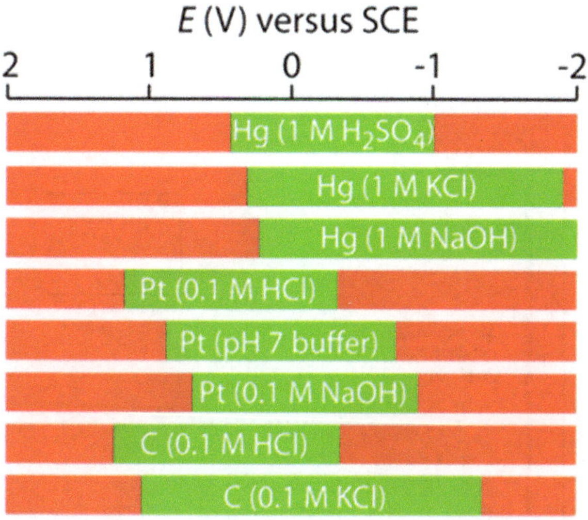

Figure 7.2 Approximate potential windows (relative to a saturated calomel reference electrode) for mercury, platinum and carbon (graphite) electrodes in acidic, neutral and basic aqueous solvents. The useful potential windows are shown in green; potentials in red result in the oxidation or reduction of the solvent or the electrode.
Compiled from ref. 2 and 3.

electrodes is driven by legislation banning its use in many countries due to its toxicity. Carbon electrodes are becoming more widely used over metal electrodes due to their useful potential windows (Figure 7.2). Carbon electrodes exhibit low background currents, high electrical conductivity, rich surface chemistry and comparative chemical inertness as well as a wide potential window in aqueous solutions. A key driver is also that the cost of carbon materials is very low.

Carbon electrodes are useful since they possess different microstructures and surface chemistries depending on the type and the way each was fabricated/processed, which can be used for the benefit of the intended electrochemical application. Consequently, the electrochemical properties vary from material to material and as a result can be challenging when applied to electroanalysis. All these electrodes have the basic structure of a six-membered aromatic ring and sp^2 bonding. The structural characteristics of carbon materials refer to the parameters L_a and L_c, which are the intra-planar and inter-planar sizes of the material, respectively. Figure 7.3 shows a schematic overview of some encountered carbon materials.

One key structural difference between these carbon electrodes that dramatically affects their electrochemical performance is the relative proportion of edge plane over that of basal plane upon the electrode surface. This affects its electrochemical reactivity and it is well established that electron transfer processes occur predominantly at edge plane sites while the basal plane is effectively electrochemically inert;[4–6] a carbon electrode or carbon nanomaterials consisting of a high density of edge plane sites will

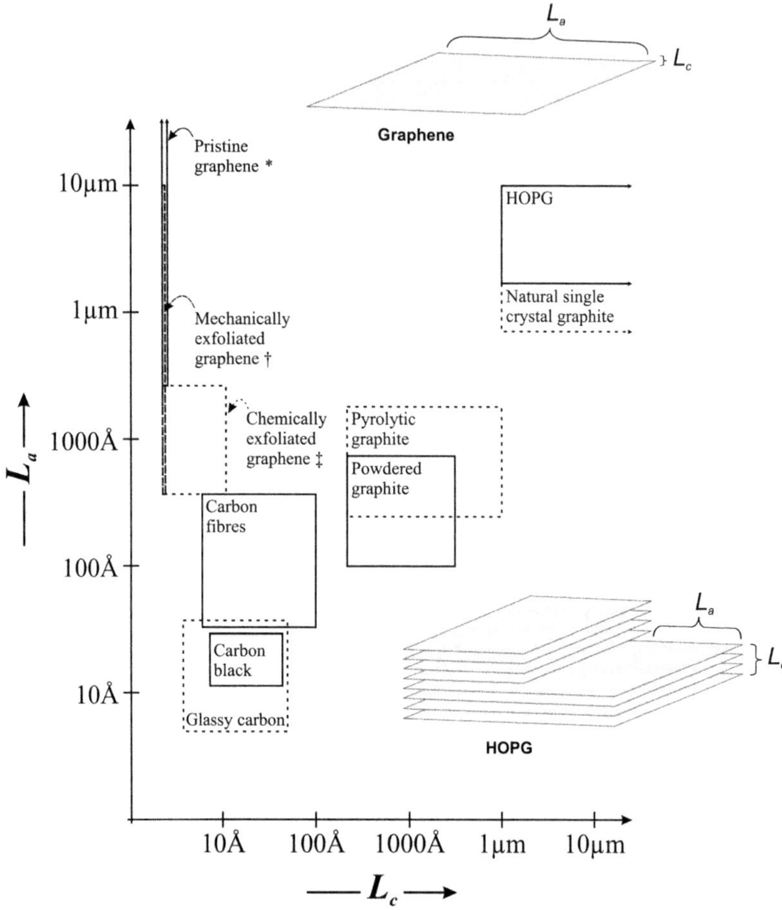

Figure 7.3 Approximate ranges of L_a and L_c values for various sp^2 carbon materials. Note, there is large variation of L_a and L_c with sample history and thus the values shown should be considered representative, yet approximate. (*) Pristine graphene; commercially available from "Graphene Supermarket", produced *via* a substrate-free gas-phase synthesis method.[7,8] (‡) Chemically exfoliated graphene; commercially available from "NanoIntegris", produced *via* a surfactant intercalation process—note that this range is also representative of graphene produced through other chemical exfoliation routes such as the reduction of GO.[9,10] (†) Mechanically exfoliated graphene was fabricated through the so-called "Scotch tape method". Note that graphene synthesised *via* CVD has been excluded given that crystal size and quality are large variables through this route; however, single graphene crystals with dimensions of up to 0.5 mm have been reported.[11,12] A schematic representation of the L_a and L_c microcrystalline characteristics of graphene and of HOPG is also shown.
(Reproduced from ref. 13 with permission from The Royal Society of Chemistry.)

show a near reversible voltammetric signature while if it consists mainly of basal plane sites it will show irreversible behaviour. Clearly this is an important parameter when choosing carbon nanomaterials as electrode materials/electrode modifiers.

In this chapter, we overview pertinent examples of carbon nanomaterials utilised in electroanalysis by researchers with the sole aim to improve the electroanalytical output. Due to the sheer diversity and range of available carbon nanomaterials (Figure 7.1) it is impossible to overview every single material without writing a dedicated book on the subject; consequently, we present pertinent interesting examples.

7.2 Carbon Nanotubes

Carbon nanotubes are separated into two general classifications: single-walled (SWCNTs) and multi-walled carbon nanotubes (MWCNTs) and usually have one dimension in the nano range, most often the diameter of the tube, with its length in the nano to micron range. Figure 7.4 depicts images of both SWCNTs and MWCNTs. There are several morphological

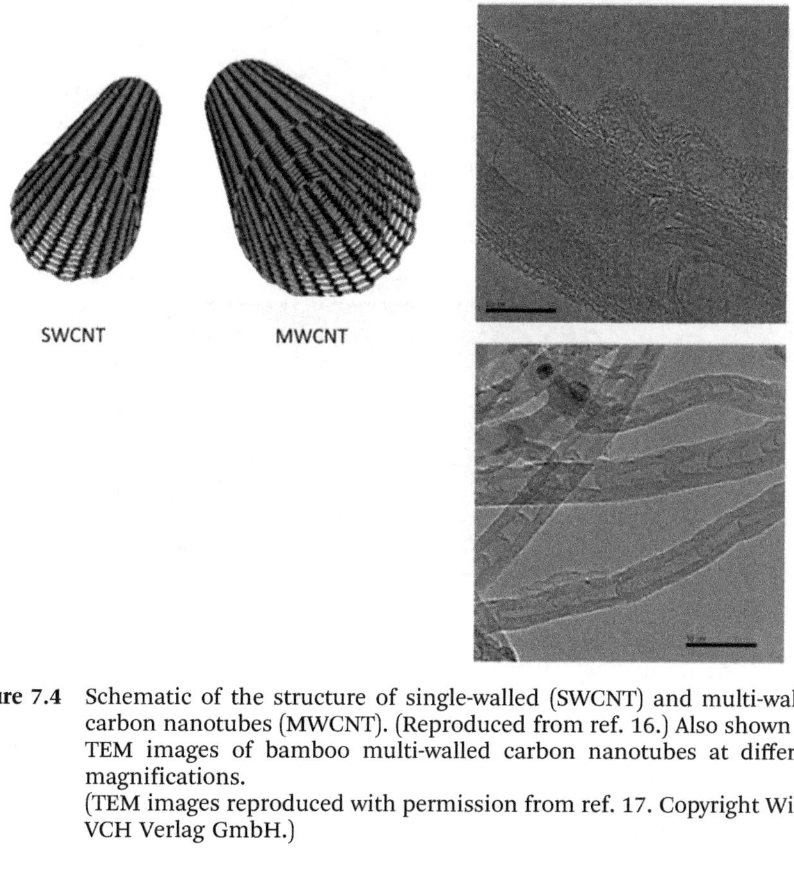

Figure 7.4 Schematic of the structure of single-walled (SWCNT) and multi-walled carbon nanotubes (MWCNT). (Reproduced from ref. 16.) Also shown are TEM images of bamboo multi-walled carbon nanotubes at differing magnifications.
(TEM images reproduced with permission from ref. 17. Copyright Wiley-VCH Verlag GmbH.)

variations of MWCNTs including hollow-tube, herringbone or bamboo-like MWCNTs.[14,15] From inspection of TEM images as shown in Figure 7.4 of bamboo multi-walled carbon nanotubes it is evident that the nanotubes are periodically closed along the length of the nanotube consisting of approximately four graphite layers dividing the nanotubes into compartments giving the nanotube its unique "bamboo" name. A proportionally higher number of edge plane like defect sites occur on herringbone and bamboo MWCNTs because, in these two cases, the plane of the graphite sheets is at an angle to the axis of the tube, resulting in a high proportion of the graphite sheets terminating at the surface of the tube. The different structures of MWCNTs give different properties compared with those of SWCNTs; the former are generally more electrochemically reactive than the latter due to differing edge plane contents (Figure 7.5).

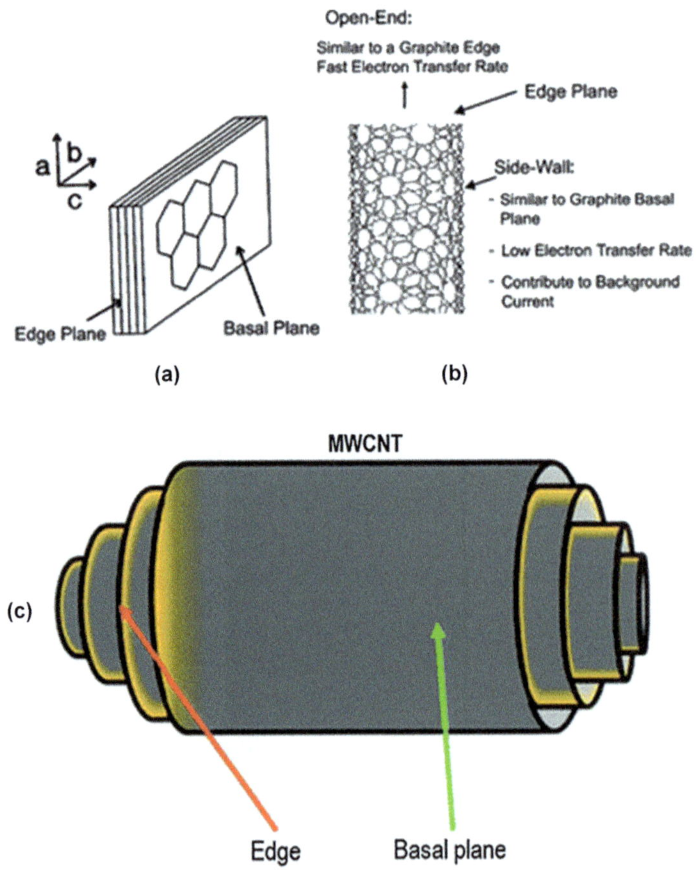

Figure 7.5 Structures of highly ordered pyrolytic graphite (a) and open-end SWCNT (b) and a MWCNT (c).
(Reproduced from ref. 28 and 29 with permission from The Royal Society of Chemistry.)

Carbon nanotubes have either metallic or semi-conducting electronic properties, and have a variety of density of electronic states distributions which are dependent on their diameter. The lower density of electronic states results in a lower conductivity over that of metals; due to their surface chemistry the electrochemistry at carbon nanotubes is highly fascinating. Carbon nanotubes continue to gain impetus in electroanalytical sensing and there are many papers that interested readers are directed to. Table 7.1 shows a non-exhaustive list of applications of carbon nanotube modified electrodes.[18] Consequently, we overview pertinent areas that matter to electroanalysts who are using, or thinking of using, carbon nanotubes.[19–27,30,31]

Compton *et al.* were also the first to unambiguously demonstrate that metallic impurities are inherent in multi-walled carbon nanotubes fabricated *via* chemical vapour deposition (CVD) which can dominate their electrochemical response.[41,42] Such work was the first to demonstrate that the "electrocatalytic" response of MWCNTs is due to metallic impurities rather than the nanotubes themselves; the exact metallic impurity depends on the metallic catalyst used in the CVD fabrication process. Follow-on work has demonstrated that this is also the case for SWCNTs, which have been shown to contain residual iron oxide impurities, which can dominate their electrochemical activity,[43] and that extensive washing, in the form of acid washing/pre-treatment, as commonly employed in the field to "clean" the nanotubes prior to use, does not leave them metallic impurity free as was once routinely claimed.[44]

In fact it has been demonstrated that metallic impurities dominate a whole range of electrochemical processes and electroanalytical targets such as glucose,[45] organic peroxides,[46,47] hydrazine,[48–50] hydrogen peroxide[49,51] and glutathione.[52]

More seminal work from Compton has shown that carbon nanotubes immobilised upon an electrode surface can give rise to a porous layer.[53–55] Originally, researchers would compare a carbon nanotube modified electrode with that of the underlying bare electrode and report that the voltammetric wave reduced in overpotential was attributed to electrocatalysis of the carbon nanotube material itself. Compton and co-workers debunked this and demonstrated that a porous nanotube layer upon the electrode surface creates a thin-layer model where the electroactive species is trapped in "pockets" between the nanotubes.[56–59]

Other elegant work has shown that graphitic impurities, again rather than the CNTs, can dominate the electrochemical response.[60] For example, the electrochemical oxidation of carbamazepine has been shown to be caused by the presence of nanographitic impurities within the carbon nanotubes rather than the electrocatalysis of the nanotubes. Such work has been extended to the cases of *p*-aminophenol[61] and acetaminophen.[62]

Jones *et al.* were the first to highlight the heterogeneity and electrocatalysis of multi-walled carbon nanotubes, which is shown to be dependent on batch-to-batch variation; this variation is often an overlooked parameter and needs to be considered, especially when carbon nanotubes are used

Table 7.1 A non-exhaustive list of applications of carbon nanotube modified electrodes.

Modification	Underlying electrode	Analyte	Linear range	Limit of detection	Comments	Ref.
SWCNT	GCE	Hydrogen peroxide	¡	¡	Electrochemical performance highly dependent on the dispersing agent and dispersion procedure used to disintegrate SWCNT aggregates for electrode modification	32
MWCNT	GCE	Guanine (G) & acid denatured DNA (A)	¡	(G): 7.5 nM (A): 150 $ng\,mL^{-1}$ (DNA)	Excellent reproducibility and selectivity	33
MWCNT	GCE	Lead(II) & cadmium(II)	Cd(II): 0.2–1 μM	51 nM	Sodium dodecyl sulfate (SDS), Triton X-100 (TX-100), dodecyl pyridinium chloride (DPC), and bovine serum albumin (BSA) were examined as four typical interferents	34
SWCNT & phytic acid	PtE	Dopamine	0.2–10 μM	80 nM	No interference from ascorbic or uric acid while ensuring good sensitivity, selectivity, and reproducibility	35
SWCNT integrated to cobalt(II)tetra-aminophthalocyanine	AuE	Epinephrine	Up to 130 μM	6 μM	Results obtained may not be the same for all metal phthalocyanine complexes integrated with CNTs for all analytes, not even for this neurotransmitter and its related complexes	36

MWCNT, Nafion and cobalt(II)-5-nitrosalophen	Carbon paste electrode	Uric acid & ascorbic acid	Uric acid: 1×10^{-7} to 1×10^{-4} M; ascorbic acid: 5×10^{-7} to 1×10^{-4} M	Uric acid: 6×10^{-8} M; ascorbic acid: 1×10^{-7} M	The voltammetric detection system was very stable and the reproducibility of the electrode response, based on the six measurements during one month, was less than 3.5% for the slope of the calibration curves of uric acid and ascorbic acid	37
Thionine–Nafion supported on MWCNT	Carbon paste electrode	Ascorbic acid & dopamine	1–80 µM	0.4 µM	The prepared modified electrode does not show any considerable response toward the electro-oxidation of sulfhydryl compounds, such as, cysteine, penicillamine and glutathione	38
MWCNT grafted polydiphenylamine	GCE	Carbon monoxide	10–200 ppm	0.01 ppm	Long-term stability for ten days with no significant change in the sensing current in comparison to the freshly prepared sensor	39
Galvanostatic electropolymerization of aniline on MWCNTs	AuE	Nitrite	5 µM–15 mM	1 µM		40

Key: MWCNTs = multi-walled carbon nanotubes, SWCNTs = single-walled carbon nanotubes, GCE = glassy carbon electrode, AuE = gold electrode, ¡ = information not provided.

commercially and typically academics only ever study small batches/same batch.[63] Such work also pointed out that the exact location of the catalyst nanoparticle will likely never be unambiguously confirmed.[63]

Ge and co-workers[64] have explored neutron activation analysis (NAA) as a non-destructive standard quantification method and inductively coupled plasma mass spectrometry (ICP-MS) as a practical approach to quantify metallic impurities within CNTs. Because metallic impurities are usually trapped within the graphene walls of the CNTs or inside the tube, special destructive pre-treatments are required before ICPMS can be utilised. Ge *et al.* investigated different sample pre-treatment procedures including dry ashing coupled with acid extraction, wet digestion, and a combination of dry ashing with acid digestion. Obviously, the quantitative determination of such metal impurities in CNTs is highly dependent on the sample pre-treatment. This is a key area of research with many reports on the topic, with notable work reporting the use of high-pressure microwave-assisted wet digestion with analysis performed *via* ICP-OES and ICP-MS and compared with direct solid sampling electrothermal atomic absorption spectrometry (DSS-ET AAS, deuterium lamp background corrector) with the proposed method, allowing the determination of all elements in CNTs with lower limits of detection in comparison with other techniques.[65]

Important work has proposed a process that allows high-quality and high-purity SWNT samples to be prepared based on the thermal heat of SWCNT powder in the presence of chlorine.[66] The process results in high chlorine diffusion into metal impurities embedded within carbon shells inducing the process of metal chloride formation and sublimation.[66] The authors demonstrated that the process combines selective elimination of catalytic impurities and high yields and importantly preserves the quality of the resulting purified nanotubes.[66] Of course, the use of high-purity metal-catalyst-free MWCNTs can be used to avoid potential experimental misinterpretations.[67–69]

Other useful approaches with carbon nanotubes involve wiring of the redox centre of enzymes using nanotubes as the electrical wires/connectors;[70] this is summarised in Figure 7.6. It was elegantly demonstrated that aligned reconstituted glucose oxidase on the edge of carbon nanotubes could be linked to an electrode surface using the carbon nanotubes as a "nano-connector" that electrically contacts (or wires) the active site of the enzyme with the electrode. Such an approach permits the molecular wiring of enzymes, allowing electron transport distances to be extended where the rate of electron transport is controlled by the length of the SWCNTs.[71] Such pioneering work highlights the benefit of using carbon nanotubes due to their unique structure and ability to be functionalised rather than just simply using them as electrode modifiers. Consequently, this area is ripe for expansion since CNT molecular wires offer efficient electron transfer to wire enzymes to electrode surfaces, which has, in turn, the potential for realisation of efficient electrical sensing devices.[72]

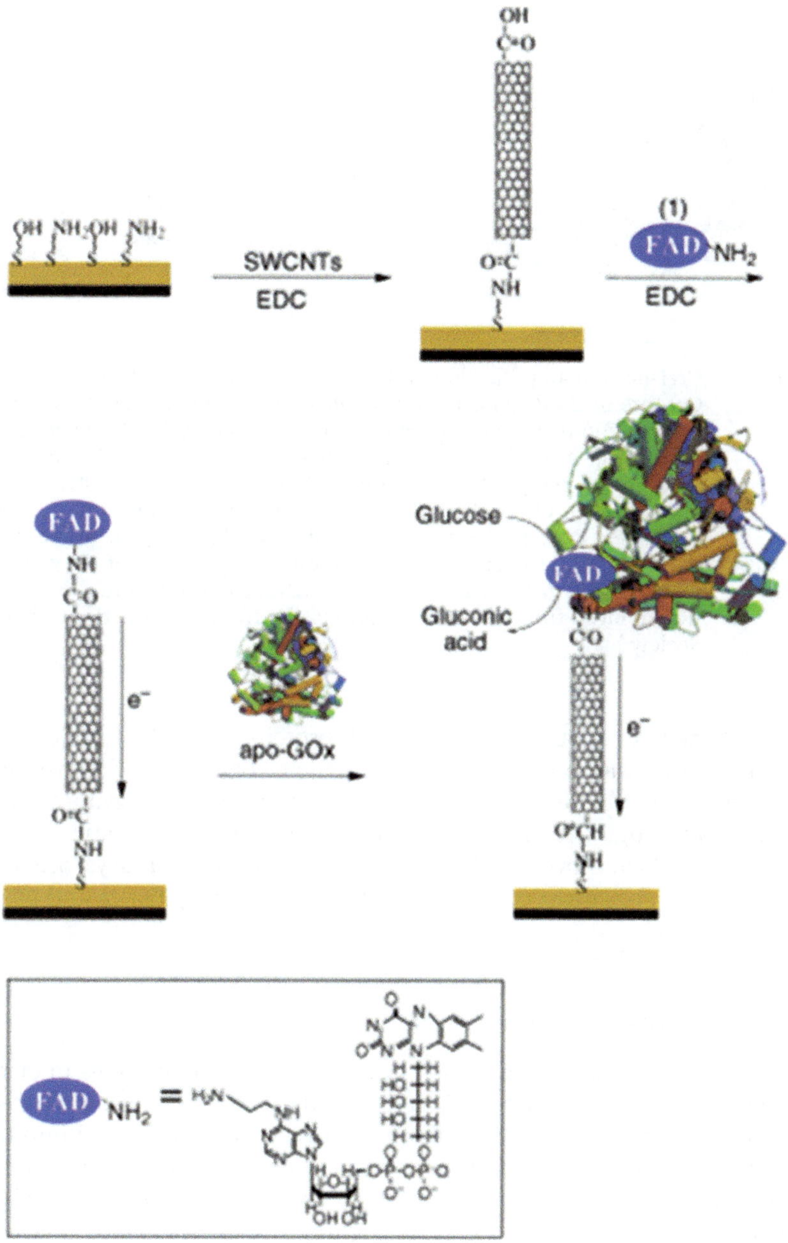

Figure 7.6 Assembly of the SWCNT electrically contacted glucose oxidase electrode. (Figure reproduced from ref. 70 with permission from Wiley-VCH Verlag GmbH.)

Lastly, another useful embodiment of CNTs is their deployment in amplifying enzyme-linked electrical detection of proteins and DNA (Figure 7.7). Wang and co-workers demonstrated a CNT-based dual amplification route

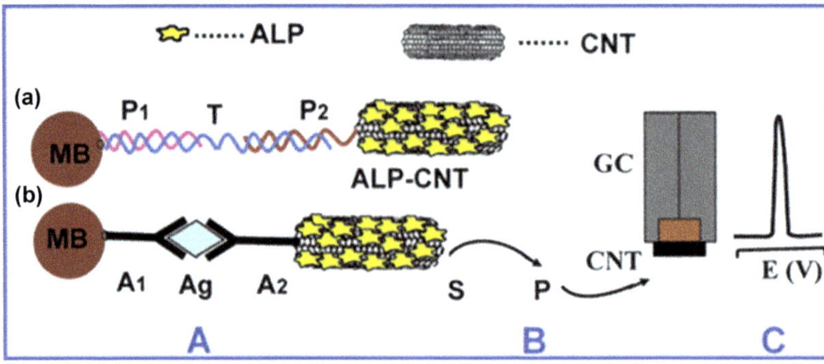

Figure 7.7 CNT-derived amplification of the recognition and transduction events. (A) Capture of the ALP-loaded (ALP = alkaline phosphatase) CNT tags to the streptavidin-modified magnetic beads by a sandwich DNA hybridization (a) or A–Ag–A interaction (b). (B) Enzymatic reaction. (C) Electrochemical detection of the product of the enzymatic reaction at the CNT-modified glassy carbon electrode. MB, magnetic beads; P_1, DNA probe 1; T, DNA target; P_2, DNA probe 2; A_1, first antibody; Ag, antigen; A_2, secondary antibody; S and P, substrate and product, respectively, of the enzymatic reaction; GC, glassy carbon electrode; CNT, carbon nanotube layer. (Reproduced from ref. 73 with permission from The American Chemical Society.)

for ultrasensitive electrical bioassays of proteins and DNA. The use of CNT plays a dual amplification role in both the recognition and transduction events, which in this case was as carriers for numerous enzyme tags and for accumulating the resulting product of the enzymatic reaction (Figure 7.7). Such novel CNT-derived amplification bioassays are being explored since they open up new vistas for medical diagnostics and protein analysis.[73]

7.2.1 Carbon Nanohorns

Related to the above section, which overviews carbon nanotubes in electroanalysis, are carbon nanohorns. Single-walled carbon nanohorns (SWCNHs) are a relatively new material, similar to SWCNTs, and are classed into three variants: "*dahlia*-", "*bud*-" and "*seed*-2 like. Figure 7.8 depicts a schematic image of SWCNHs where the structure of a single nanohorn consists of cone shaped wrapped graphene sheets. Figure 7.9 shows TEM images of SWCNHs.

SWCNHs belong to the nanotube family and as the name suggests have a horn-shaped tip which is closed, are 2–5 nm in diameter at an angle of 18–22° and have a tubule length of 30–50 nm depending upon the fabrication method. A key difference between the SWCNH and CNTs is that the former is free from metallic catalysts, which can plague the latter (see earlier for a discussion on the problems occurring in CNTs); SWCNHs can be prepared

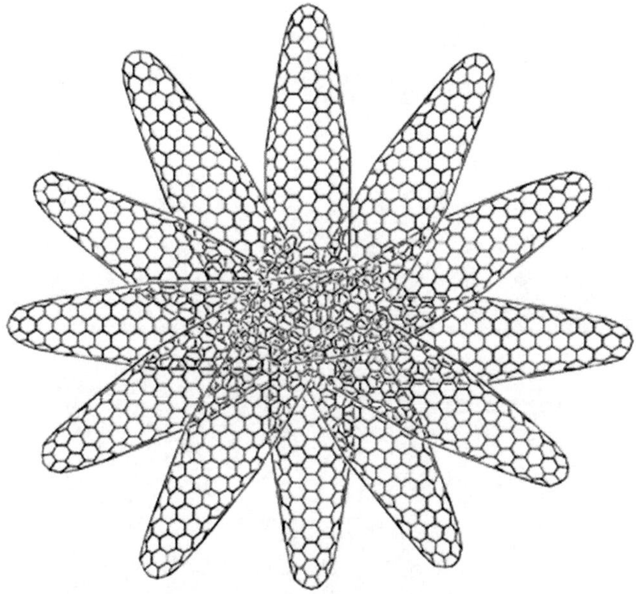

Figure 7.8 Schematic representation of multiple carbon nanohorns forming an aggregate.
(Reproduced from ref. 74 with permission from The Royal Society of Chemistry.)

through CO_2 laser ablation of a graphite target. The three different classes of SWCNHs can be obtained through differing fabrication approaches.[76] SWCNHs have not been extensively used in electroanalysis, unlike their CNT counterparts, but due to the similar structural properties insights from CNTs may be applied to this potential electrode material. Table 7.2 provides an overview of where SWCNHs have been applied in the field of electroanalysis, indicating that electroanalysts are yet to fully utilise SWCNHs.

From the structure of SWCNHs it is clear that the proportion of edge plane sites is low with a high proportion of basal plane sites, suggesting that an electrode modified with SWCNHs will give a poor voltammetric response.

Nevertheless, one of the first examples of their uses was reported towards the electroanalytical determination of uric acid, dopamine and ascorbic acid.[77]

Figure 7.10 depicts the linear sweep voltammetric response obtained at a bare GCE where no useful signals were obtained, but when the electrode is modified with SWCNHs three distinct and well-resolved peaks are observed. Such a change in response was attributed to the increase in edge-plane density and adsorption ability of the SWCNHs,[77] which is quite surprising given their structure. The authors reported large peak separations between uric acid and dopamine, and dopamine and ascorbic acid, up to 152 and

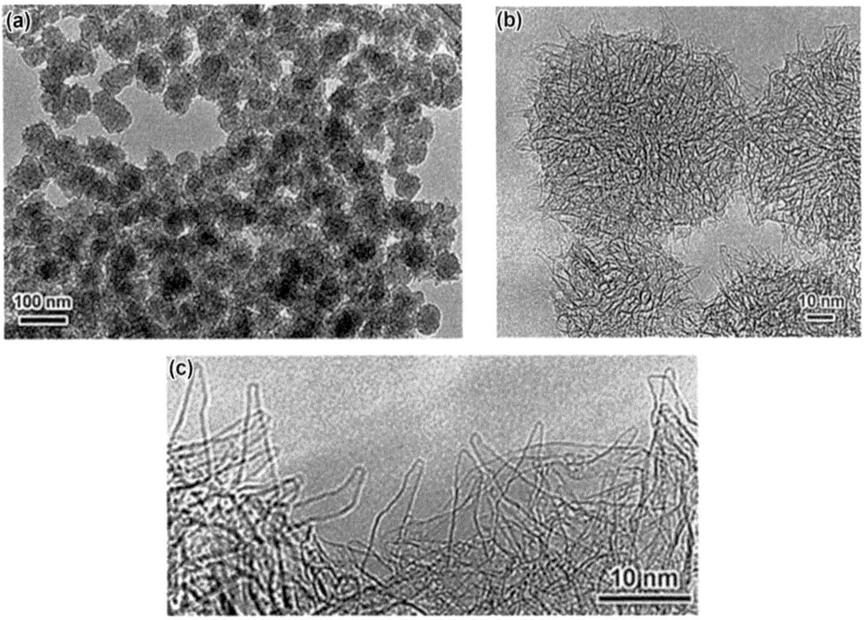

Figure 7.9 Experimental example of SWCNHs: (a) TEM micrograph of a graphitic carbon product that was generated abundantly by CO_2 laser ablation at room temperature. The product consisted of nearly uniform sized spherical particles with a diameter of 80 nm. (b) A magnified TEM micrograph of the graphitic carbon particles shows aggregations of tubule-like structures sticking out of the particle surface. (c) A highly magnified TEM micrograph of the edge regions of the graphitic particles shows conical horn-like protrusions of up to 20 nm long on the particle surface with some modified shapes. Each of these carbon nanohorns was made of a single graphene sheet with closed caps whose diameter was similar to that of the fullerene molecules.
(Reproduced from ref. 75 with permission from Elsevier.)

221 mV, respectively. The calibration curves for uric acid, dopamine and ascorbic acid were obtained in the ranges of 0.06–10, 0.2–3.8 and 30–400 μM, respectively, with detection limits (S/N = 3) of 20 nM, 60 nM and 5 μM respectively. The method was applied to the determination of uric acid, dopamine and ascorbic acid in urine samples.[77] Table 7.2 overviews other reports where SWCNHs have been utilised in electroanalysis. Mostly, Zhao *et al.* used SWCNHs modified with nano-gold to provide a disposable immunosensor and used α-fetoprotein as a model target analyte and demonstrated its applicability for detection in serum samples.[81]

Valentini *et al.*[82] utilised oxidised single-walled carbon nanohorns that were immobilised onto screen-printed electrodes. These electrodes were explored for the electrochemical detection of epinephrine, in the presence of serotonine-5-HT, dopamine, norepinephrine, ascorbic acid, acetaminophen and uric acid. A linear range of 2 to 2500 μM was found for epinephrine and a limit of detection of 0.1 μM deduced.[82]

Table 7.2 Applications of single-walled carbon nanohorn modified electrodes.

Modification	Underlying electrode	Analyte	Linear range	Limit of detection	Comments	Ref.
Single-walled carbon nanohorn	GCE	Uric acid, dopamine & ascorbic acid	Uric acid: 0.06–10 μM, dopamine: 0.2–3.8 μM & ascorbic acid: 30–400 μM	Uric acid: 20 nM; dopamine: 60 nM & ascorbic acid: 5 μM	The analytes were also explored toward their determination within urine samples	77
Single-walled carbon nanohorn	Interdigitated platinum electrode	Ozone gas	¡	50 ppb	The sensor showed a rapid response as well as a fast recovery	78
A sandwich nanohybrid of SWNHs–TiO_2–porphyrin	GCE	Chloramphenicol	Up to 135.7 μM	0.9 nM	The sandwich nanostructure of SWNHs–TiO_2–porphyrin provides a functional electrocatalyst to construct a sensitive biosensor and can be extended for application in photovoltaics and photo-catalysis in solar cells	79
Nafion–glucose oxidase–SWCNH	GCE	Glucose	0.6–6 mM	6 μM	L-Lactate, glutathione, L-cysteine, *p*-aminophenol and ascorbate acid did not interfere with the detection	80

Key: GCE = glassy carbon electrode, SWCNH = single-walled carbon nanohorn, ¡ = information not provided.

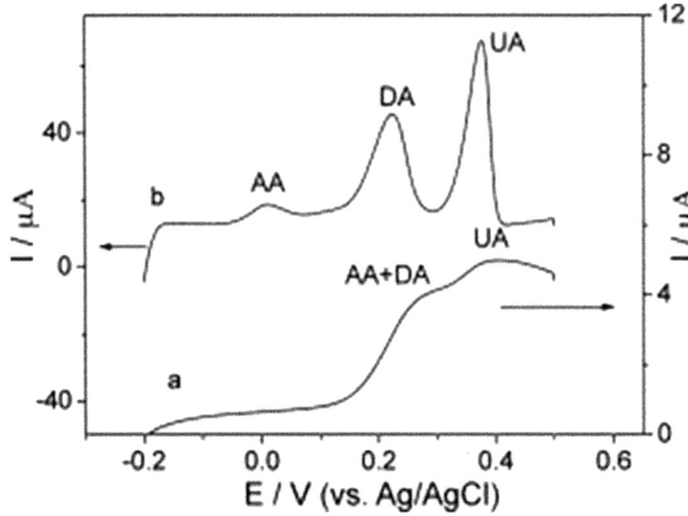

Figure 7.10 Linear sweep voltammograms at the bare GCE (a) and SWCNH-modified GCE (b) in 0.1 M pH 7.0 PBS containing 100 μM uric acid, dopamine and ascorbic acid. Scan rate: 100 mV s^{-1}.
(Reproduced from ref. 77 with permission from The Royal Society of Chemistry.)

7.3 C_{60}

Another carbon nanomaterial that has received interest is C_{60}, known as buckminsterfullerene or buckyball, which has a truncated icosahedral structure, a polygon with 60 vertices and 32 faces, 12 of which are pentagonal and 20 hexagonal.[83,84] C_{60} appears to be attractive to researchers for the modification of electrode surfaces since it is chemically stable, is free from metallic impurities and has been reported to be electrocatalytic towards a range of target analytes in aqueous media.

The review by Griese and co-workers published in 2008[84] charts the history of C_{60} modified electrodes, which includes reports as early as 1995 discussing the electrocatalytic oxidation of ascorbic acid.[85] From our earlier discussions on carbon materials, it is well known that edge plane sites are the origin of electron transfer and due to the unique structure of C_{60}, and particularly its lack of such edge plane sites, *how is this material "electrocatalytic"*?

Szucs and co-workers initially reported that solution-cast C_{60} films upon gold and glassy carbon substrates can be reduced electrochemically in aqueous solutions, producing an electron-conducting cation-exchanger material.[86] In this approach, a two-step pre-treatment in alkaline and neutral buffer solutions is undertaken and applied to the modified C_{60} electrodes to induce a partially reduced, conducting C_{60} film. This is reported to result in the partial reduction of the C_{60} to form conductive films, which have been attributed to be the origin of the electrocatalytic property of C_{60} modified electrodes.[85]

Tan and co-workers[83] reported the electrocatalytic oxidation of L-cysteine in aqueous solutions at a C_{60} modified glassy carbon electrode where a reduction in overpotential coupled with an increment in the magnitude of the voltammetric peak was observed compared to the bare glassy carbon electrode; this electrocatalysis was reported to be due to the presence of C_{60} acting as a mediator for the electrochemical oxidation of cysteine, where the presence of C_{60} provided an alternative reaction sites to the underlying electrode surface for electron transfer processes to occur.[83] This work inspired many other researchers to utilise the reported electrocatalysis of C_{60} modified electrodes.[87] For example, Goyal *et al.* reported the electrocatalytic detection of uric acid[88] with detection limits as low as 0.12 μM, stating that the use of C_{60} modified electrodes is superior to prior literature reports and suffers from no surface fouling. Goyal *et al.* have also reported the electrochemical detection of *N*-(4-hydroxyphenyl)ethanamide[89] and atenolol.[87] Figure 7.11 shows the electrochemical signatures, which are absent at a bare unmodified electrode (glassy carbon) but with the introduction of C_{60} a well-defined oxidation wave appears; atenolol could be detected over the concentration range of 0.25–1.5 mM with a respectable detection limit of 0.16 mM at physiological pH.[87]

In these excellent examples, note that no pre-treatment of the C_{60} was applied and, rather, the modified electrode was used as is (see above). Compton and co-workers have questioned the apparent reported electrocatalysis of C_{60} modified electrodes and have demonstrated that graphitic impurities in the C_{60} sample are actually the origin of the beneficial electrochemical responses.[90] Using the example provided by Tan *et al.*,[83] the electrochemical oxidation of L-cysteine, they elegantly demonstrated that with highly pure C_{60} the electrocatalytic oxidation of L-cysteine is not observed but rather the bare glassy carbon electrode gives rise to useful voltammetric profiles, with casting increasing amounts of C_{60} found to block the electrode surface. Additionally the authors went on to show that an

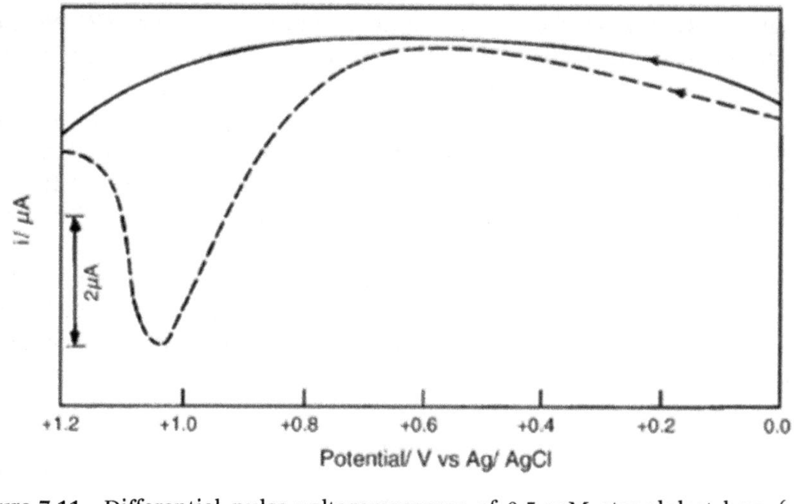

Figure 7.11 Differential pulse voltammograms of 0.5 mM atenolol at bare (solid line) and C_{60} modified glassy carbon electrode (dashed line) at pH 7.2. (Reprinted from ref. 87 with permission from Elsevier.)

improvement in the bare electrode response (glassy carbon) was possible using graphite modified electrodes, giving rise to similar magnitudes in the peak height and reduction in the overpotential as reported by Tan *et al.*[83] Compton and co-workers pointed out that the C_{60} used in the work of Tan *et al.* was only 98% pure, adding weight that C_{60} itself is not electroactive but instead the graphitic impurities are the origin of the improvements observed by Tan *et al.* Such work did not stop the momentum of C_{60} modified electrodes, with work reporting the "electrocatalytic" detection of adenine and guanine,[91] nandrolone,[92] adenosine and guanosine[93] and most recently salbutamol[94] and methyl prednisolone[95] and the determination of dopamine in the presence of ascorbic acid.[96]

The choice of whether to electrochemically pre-treat the C_{60} modified electrode is up to the researchers' discretion. Following the work of Compton *et al.*, researchers choose to consistently pre-treat their electrodes before use, thus claiming this is the origin of the electrocatalysis of C_{60} modified electrodes.

Griese *et al.*[97] explored the role of pre-treatment upon the electrocatalysis of C_{60} modified electrodes. In a common way to induce a partially reduced, conductive C_{60} film supported on a glassy carbon electrode, the modified surface is pre-treated in 1 M KOH in the range 0.0 to −1.5 V at a scan rate of 10 mV s^{-1} followed by cycling the potential from +0.55 to −0.05 V for 20 min at a scan rate of 20 mV s^{-1} in a pH 7.2 phosphate buffer solution. However, given that all of the analytes explored are electrochemical oxidations, it is likely the largely oxidative potentials (and repetitive scans) required to electrochemically oxidise the analyte of interest return the reduced film to its neutral state.[84] To explore the effect of the "pre-treatment" the electrochemical oxidation of *N*-(4-hydroxyphenyl)ethanamide was utilised with a glassy carbon electrode as the supporting electrode, which is used as the underlying electrode in most C_{60} work.

Figure 7.12 shows the voltammetric response at a bare GC electrode (no pre-treatment) resulting from the electrochemical oxidation of *N*-(4-hydroxyphenyl)ethanamide. A control experiment was undertaken in which the bare GC electrode was subjected to the pre-treatment and then used for the electrochemical oxidation of *N*-(4-hydroxyphenyl)ethanamide where, as is evident from inspection of Figure 7.12, a decrease in the peak potential is observed (∼20 mV), indicating an "electro-catalytic" response, as commonly reported in the literature.[97]

The authors noted that the response, in terms of the decrease of the oxidation potentials, is identical to that reported previously.[89] Exchanging the underlying glassy carbon electrode with that of a bare boron-doped diamond electrode indicated no difference in the voltammetric peak observed between the pre-treated and not pre-treated electrode. It is inferred that the pre-treatment in 1 M KOH introduces surface oxygenated functionalities on the surface of the glassy carbon electrode, which either act as catalytic sites promoting a means for binding the absorbing material or through other mechanisms, facilitating the improved electrochemical

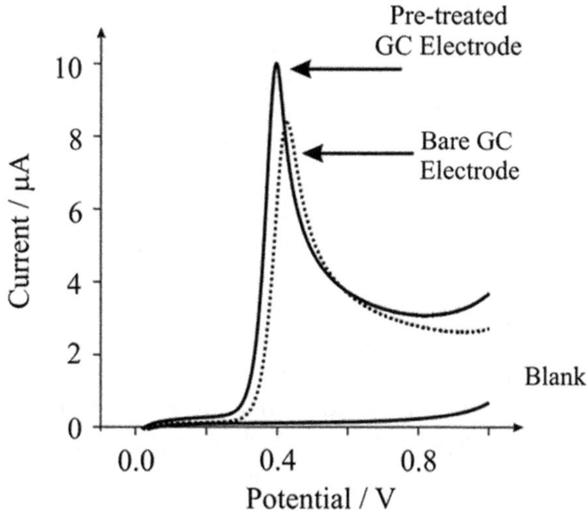

Figure 7.12 Typical linear sweep voltammograms for the electrochemical oxidation of *N*-(4-hydroxyphenyl)ethanamide at a bare glassy carbon electrode (not pre-treated, dotted line) compared with that of a bare glassy carbon electrode (solid line) that has been subjected to a pre-treatment. All scans recorded at 20 mV s^{-1} *vs* SCE.
(Reprinted from ref. 97 with permission from Elsevier.)

oxidation of *N*-(4-hydroxyphenyl)ethanamide. The same observations were also made for the electrochemical oxidation of salbutamol.

In summary, the reported electro-catalytic responses for *N*-(4-hydroxyphenyl)ethanamide and salbutamol[89,94] are due to the underlying electrode, where it is likely that surface oxygenated species are introduced, promoting "electro-catalysis" rather than due to the C_{60}[84] and as such C_{60} itself, which owing to its structure (lack of edge plane sites) provides no electrocatalysis in aqueous solutions.

Following on from this, Xiao *et al.* reported that a porous C_{60} film results that exhibits some degree of thin-layer behaviour; the applied pre-treatment results in restructuring of the C_{60} upon electrode surfaces that leads to the observed changes in the voltammetric behaviour. These changes in porosity, combined with a degree of thin-layer effects contribute to the observed voltammetric wave shape that in the past has been mistaken for signs of the electrocatalysis of C_{60}. Thus, in summary, the arguments that C_{60} is effectively inert and is not electrocatalytic within the potential window of aqueous electrolytes is convincing; however, authors continue to utilise such "electrocatalytic" effects of C_{60}.[98]

7.3.1 Carbon Onions

As mentioned above, an important family of graphitic nano-sized materials is the fullerenes (C_{60}, C_{70}, *etc.*), discovered in 1985 by Kroto *et al.*[99] and related "onions" discovered by Iijima in 1980.[100] Kroto and McKay later proposed that such carbon onions consist of nested giant fullerene cages.[101] The term carbon onion, or sometimes carbon nano-onions or onion-like carbon, is used to describe both giant fullerene-like nested structures, which are spherical cages, and also small polyhedral nanostructures. These unique 0D structures are explored in the electrochemistry field due to their reported high surface area, high mesoporosity, facile electrical conductivity, and electrochemical stability. They are widely explored in energy storage applications but there are hardly any reports in the field of electroanalysis.

Henstridge and co-workers explored the electrochemical characteristics of nanometre sized polyhedral graphite onions dispersed amongst arc-MWCNTs; these are formed during the electric arc discharge method of producing ultrapure MWCNTs (arc-MWCNTs). Figure 7.13 shows HRTEM images of the MWCNTs, which reveal that the closed tube ends along with the carbon onions are randomly dispersed amongst the arc-MWCNTs where the latter have very little amorphous carbon deposits or other unwanted

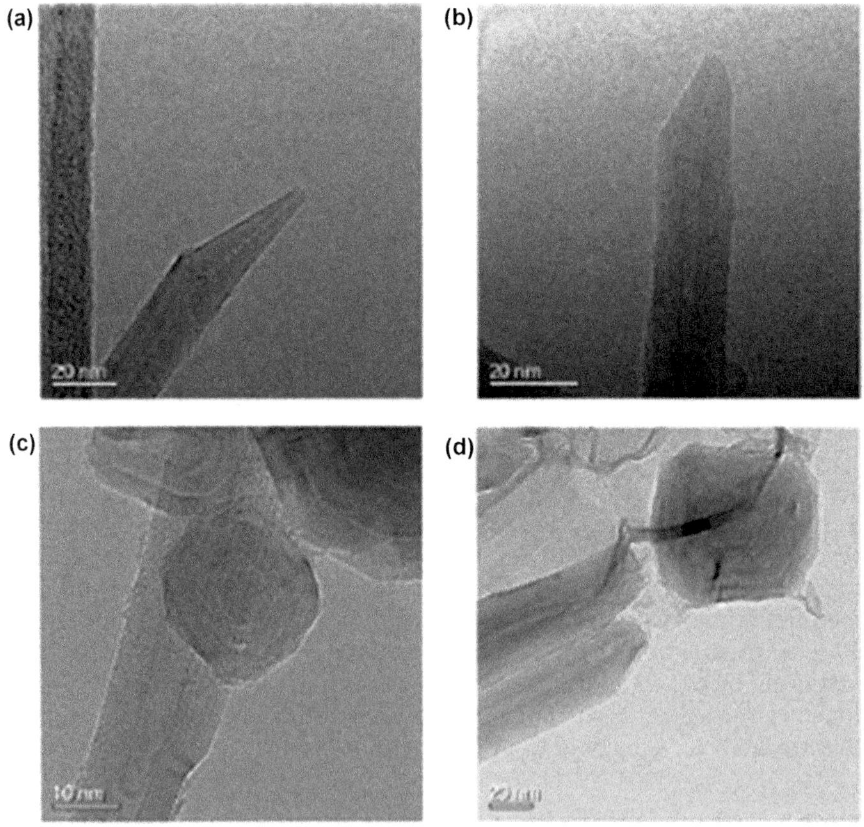

Figure 7.13 HRTEM images of (a) and (b) arc-MWCNTs showing their closed-ended nature and lack of significant amounts of amorphous carbon; (c) Swiss-roll-like and closed shell giant fullerene-like nano-onions; (d) a polyhedral graphite onion.
(Reproduced from ref. 102 with permission from John Wiley & Sons.)

impurities and are formed as closed-ended tubes. Through critical comparison with commercially available open-ended hollow-tube multi-walled carbon nanotubes made using the chemical vapour deposition method (cvd-MWCNTs), glassy carbon, edge plane pyrolytic graphite (EPPG) and basal plane pyrolytic graphite (BPPG) electrodes, it has been demonstrated that the carbon onions dispersed amongst the arc-MWCNTs have incomplete graphite shells or rolled-up Swiss-roll like structures that possess edge-plane-like defect sites that are responsible for the observed voltammetric responses; the arc-MWCNTs are fabricated as closed-ended tubes with very few, if any, edge-plane-like defect sites evident in their HRTEM characterization. Such work demonstrates that carbon onions are no more electroactive than open-ended MWCNTs, which in turn are no more electrocatalytic than an EPPG electrode.[102]

Clearly, based on such work one can see why carbon onions/nano-onions are not being investigated in electroanalysis. One other study in this area has reported arc plasma fabricated carbon nano-onions (25–35 nm) that were explored for a range of redox probes; the electrodes modified with the carbon nano-onions reportedly gave rise to beneficial electrochemical activity compared to a bare (unmodified) glassy carbon electrode.[103]

7.3.2 Nanocarbons

Related to carbon onions are nanocarbons. The electrochemical performance of a carbon paste electrode produced from nanocarbon nanoparticles (diameter 27 ± 1 nm) has been critically compared with that of traditionally used graphite (diameter less than 20 μm) mixed with mineral or paraffin oil towards a range of model analytes.[104] The results indicated that generally these electrodes behaved similarly to glassy carbon and demonstrate that the nanocarbon can be used as a replacement for graphite, especially given that, at the time of writing, the former is 25 times cheaper than the latter. Such work builds upon prior work from this group, who reported that nanocarbon black (diameter 14 ± 10 nm) can be used as a much cheaper alternative to multi-walled carbon nanotubes as an electrode modifier for use in adsorptive stripping voltammetry.[105]

7.4 Graphene

Related to the above materials is graphene, which is a two-dimensional crystal consisting of a monolayer of carbon atoms arranged in a honeycombed network. As discussed in the introduction, carbon materials are characterised by L_a and L_c parameters, which correspond to an L_c of 0.35 nm for true (monolayer) graphene with L_a values in the range from below 50 nm up to 3000 nm and greater; as such, at least one dimension of this material is in the nanometre range and will be considered in the present chapter. These values are dictated by the methodology used to fabricate the graphene, which is vast and greatly expanding. Table 7.3 provides an overview of different fabrication strategies.

Table 7.3 Comparison of different graphene preparation methods. (Reproduced from ref. 134 with permission from The Royal Society of Chemistry.)

Preparation methods	Starting materials	Operating techniques	Advantages	Disadvantages	Implications	Ref.
Mechanical exfoliation	HOPG	Scotch-tape	Simplicity, high structural and electronic quality	Delicate and time-consuming, low yields	Fundamental research	106–108
Epitaxial growth on SiC	4H-/6H-SiC wafer	Thermal desorption of Si from the SiC substrate under high temperature (>1000 °C) and UHV	Large-scale production, high qualities	High temperature requirement, high cost, non-uniform, low yields	Basic research and graphene electronics	109–112
CVD epitaxial growth	Hydrocarbon gas (such as CH_4)	Chemical vapour deposition (CVD) under high temperature	Large-scale production, high qualities, uniform, high compatibility with current CMOS technology	High temperature requirement, high cost, complicated process, low yields	Basic research and graphene electronics	113–117
Chemical reduction of graphite oxide	Graphite	Graphite exfoliation and oxidation, subsequent reduction of exfoliated graphite oxide	Facile scalability, high yields, low cost, excellent processability	Structural defects, disruption of the electronic structure of graphene	Composite materials, electronics, optoelectronics and potential technologically viable devices	118–120
Organic synthetic protocols	PAHs	A controlled organic chemical reaction	Precisely defined structures, tunable solubility and good processability	Limited size range, complicated synthetic process	Building blocks for electronics and optoelectronics	121–123
In situ electron beam-irradiation of PMMA nanofibres	PMMA	*In situ* electron beam-irradiation	One-step synthesis process in a fine-tuned way, no high temperature	Time-consuming, low yields, non-uniform	Building blocks for electronic and optoelectronics	124

Arc discharge of graphite	Graphite	Direct current arc discharge	One-step synthesis process, low cost, easy doping, large-scale production	Non-uniform, impure	Novel composite materials	125, 126
Thermal fusion of PAHs	PAHs	Precursor-controlled solid-state pyrolysis	Large-area production, simplicity, low cost	Low quality, non-uniform, high temperature	Building blocks for electronic and optoelectronics	127
Liquid-phase exfoliation	Graphite	Dispersion and exfoliation of graphite in organic solvents	Direct, simplicity, benign, large-scale production, low-cost, practicability	Time-consuming, impure	Electronic devices, transparent electrodes and conductive composites	128, 129
Conversion of nanodiamond	Nanodiamond	Annealing nanodiamond at high temperature	Simple, direct	High temperature, high cost	Composite materials	130
Sodium reduction of ethanol	Ethanol and sodium	Reduction of ethanol by sodium metal, followed by pyrolysis of the ethoxide product	Non-graphitic precursors, a gram-scale production, low-cost, practicability	Impure, the danger of violent reaction, low quality	Composite materials	131
Unzipping of carbon nanotubes	Carbon nanotubes	Solution-based oxidative action of potassium permanganate and sulfuric acid, or plasma etching	Direct, simple, large-scale production, low-cost, high quality (plasma etching)	Time-consuming, complicated process	Electronics and composite materials	132, 133

HOPG: highly oriented pyrolytic graphite; UHV: ultrahigh vacuum; CMOS: complementary metal-oxide semiconductor; PAHs: polyacyclic hydrocarbons; PMMA: poly(methyl methacrylate).

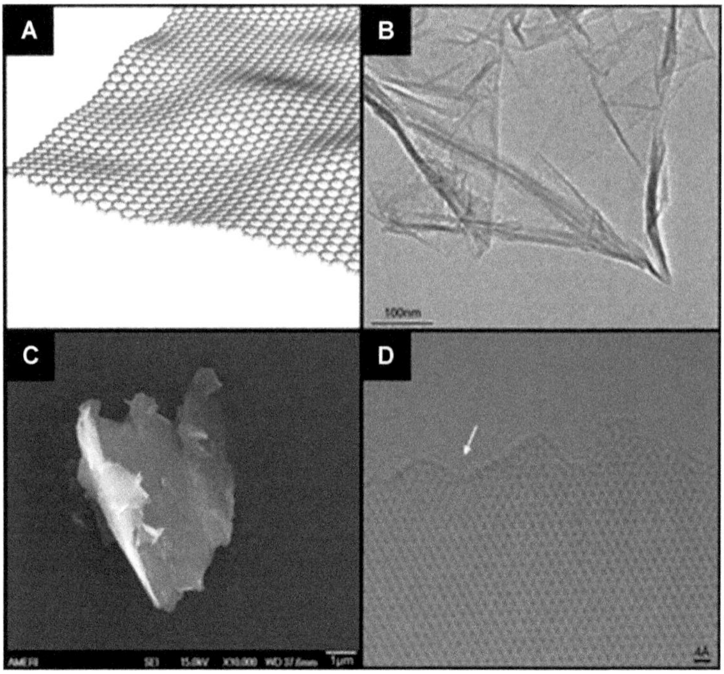

Figure 7.14 Conceptual model depicting the structure of graphene (A) and TEM (B) and SEM (C) images of a single atomic layer of graphite, known as graphene. A high-resolution TEM (D) image is also shown, where the white arrow indicates the edge of the graphene sheet. Note, in reality the graphene utilised in most work is 1–4 + layers thick.
(Parts (B) and (D) are reproduced from ref. 136 with permission from The Royal Society of Chemistry and (C) is reproduced from ref. 137 with permission from Elsevier.)

Figure 7.14 shows a conceptual image of graphene, the world's thinnest electrode material along with TEM images of fabricated graphene. Graphene is being utilised in a plethora of areas[135] with some limited success whilst many hurdles are still required to be overcome before graphene will truly reach consumerism.[135] In this chapter's area of interest, and when applied as an electrode material, graphene is the new, hot material. There are extensive reports of graphene and graphene related materials reported for the sensing of target analytes. Researchers have deviated from true monolayer graphene, such that graphene oxide, reduced graphene oxide and partially reduced graphene oxide have all been explored and reported; many other graphene variants have been fabricated and explored in a range of applications.

Figure 7.15 depicts a slab of highly ordered pyrolytic graphite (A) and if viewed closer consists of edge plane band sites and basal plane sites[5] (see earlier in the chapter) along with a side on view (C), which effectively shows that HOPG consists of multi-layers of graphene.

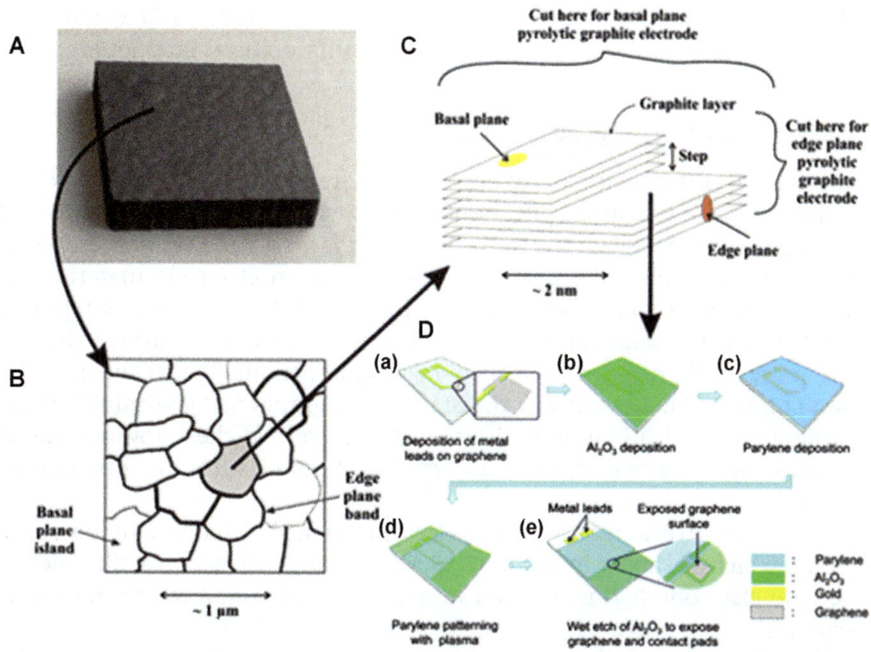

Figure 7.15 (A) Image of a commercially available slab of HOPG. (B) Schematic representation of the side on view of a HOPG surface, highlighting its basal plane and edge plane like-sites/defects, which exhibit contrasting behaviours in terms of electrochemical activity—the electron transfer kinetics of the latter are overwhelmingly dominant over that of the former, which in comparison are relatively (electrochemically) inert. (C) Schematic representation of a HOPG surface showing the discrete basal plane and edge plane islands. (D) Example procedure for fabricating monolayer graphene sheets into working electrodes for electrochemical characterisation.
(Part (D) is reprinted with permission from ref. 138. Copyright 2011 American Chemical Society.)

The Nobel Prize in Physics 2010 was awarded to Geim and Novoselov, who produced a simple yet ingenious method to fabricate graphene from HOPG, the so-called "Scotch tape" method where graphene layers are mechanically cleaved and, after careful manipulation, graphene and few-layer graphene will result.[107] Due to its unique geometry, how does one connect to this material given the size of the graphene relative to the electrical connectors of the potentiostat? As shown in Figure 7.15D, a useful approach to overcome this limitation has been presented. Although this is time-consuming, it allows one to study true graphene, that is, single layer graphene with many other ways experimental techniques reported. In the field of electroanalysis, the most common way is to adopt the approach that has been used in the case of carbon nanotubes, that is, to disperse the graphene material into a solvent, apply gentle sonication or another agitation methodology and then take aliquots and pipette these onto the desired electrode surface and allow the solvent to evaporate, leaving behind the immobilised graphene.

7.4.1 Graphene Electroanalysis

This field is a fast paced area and it is therefore not appropriate to list endlessly all the various applications of graphene electroanalysis but rather give a general overview of the most important issues. The very first examples reporting graphene to be potentially utilised as an electrode material was towards the detection of NADH and dopamine.[139,140] The authors reported that through the addition of graphene onto an electrode surface the voltammetric peak heights increased in magnitude and occurred at facile overpotentials. Such work was the start of the graphene "goldmine" where researchers frantically modified electrodes, examined these towards target analytes and reported "electrocatalysis" compared with the unmodified electrode.

Other early work reported the electroanalytical detection of paracetamol,[141] which is shown in Figure 7.16, where the poor response obtained at the bare electrode is transformed into a very large voltammetric reversible response.

Niu *et al.* constructed graphene/AuNP/chitosan nanocomposite electrodes (Figure 7.17).[142] Such electrodes demonstrated high electrocatalytic activity

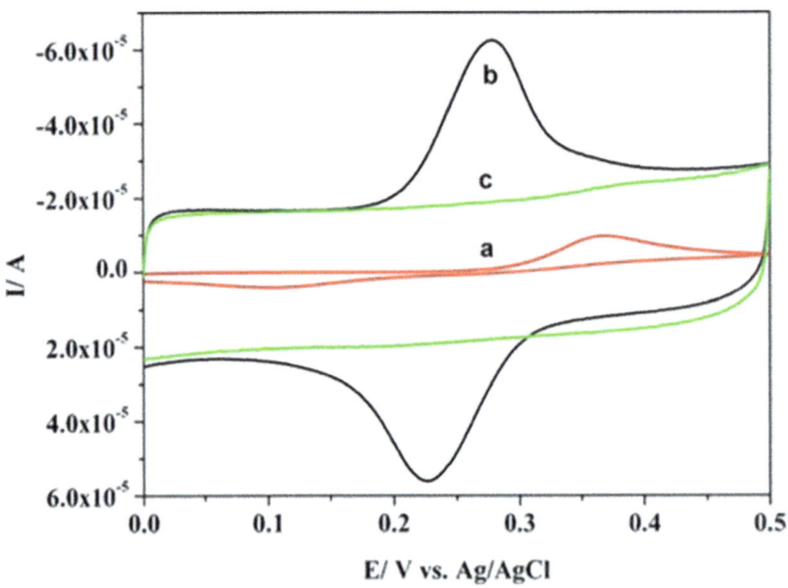

Figure 7.16 Cyclic voltammograms recorded at a bare GC electrode (a) with 100 μM paracetamol; graphene/GC electrode with (b) 20 μM paracetamol and without paracetamol (c) in a buffer of 0.1 M $NH_3 \cdot H_2O$–NH_4Cl, pH 9.3, scan rate: 50 mV s^{-1}.
(Reproduced from ref. 141 with permission from Elsevier.)

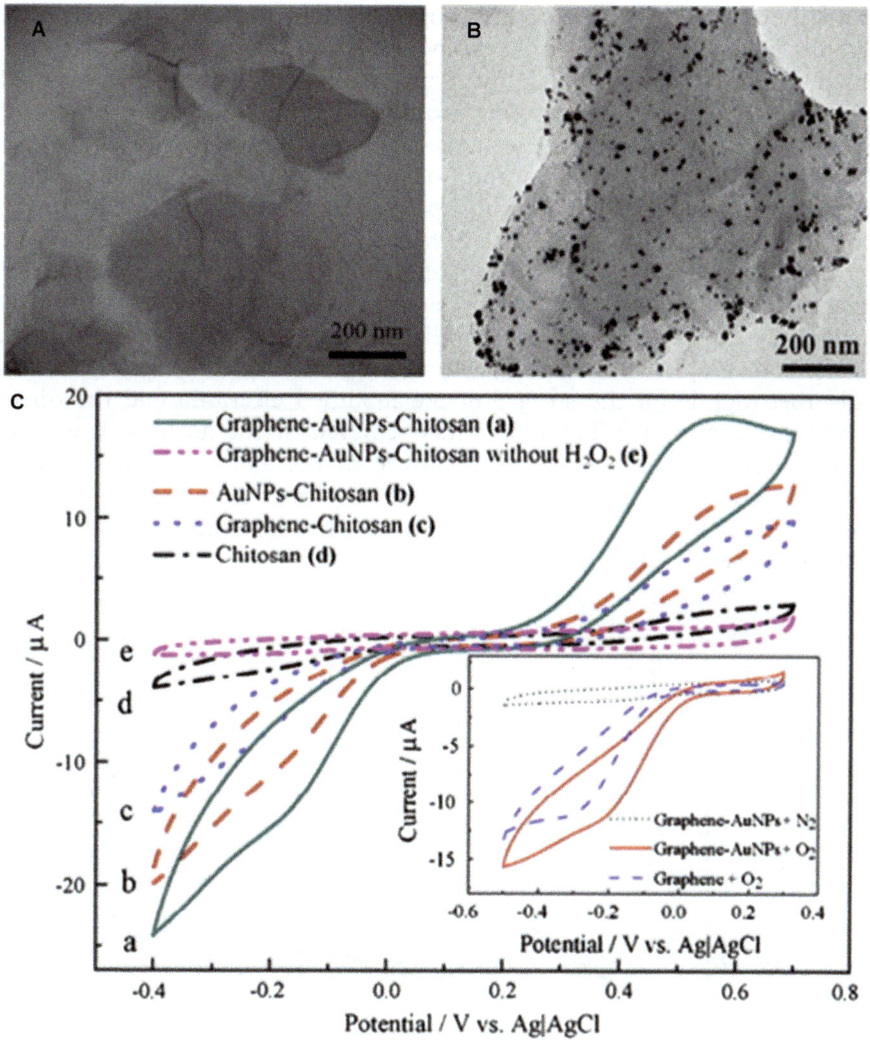

Figure 7.17 Nanoparticle-decorated graphene for enzymatic biosensing. Transmission electron microscopy images of (A) polyvinylpyrrolidone-protected graphene and (B) gold nanoparticle (AuNP)-decorated graphene. (C) Cyclic voltammograms of (a) graphene/AuNPs/chitosan, (b) AuNPs/chitosan, (c) graphene/chitosan and (d) chitosan-modified electrodes in N_2-saturated phosphate buffer (0.05 M, pH 7.4) containing 2.5 mM H_2O_2, and graphene/AuNPs/chitosan-modified electrode (e) in N_2-saturated phosphate buffer at a scan rate of 0.05 V s^{-1}. Inset: cyclic voltammograms of graphene/chitosan (dashed, blue), graphene/AuNPs/chitosan-modified electrodes (solid, red) in phosphate buffer saturated with O_2, and graphene/AuNPs/chitosan-modified electrodes in phosphate buffer saturated with N_2 (dotted, black).
(Reprinted from ref. 142 with permission from Elsevier.)

toward H_2O_2 and O_2. The authors suggested that the synergistic effect of graphene and the AuNPs might promote electrocatalysis toward H_2O_2. The high sensitivity and good stability of such a modified electrode contributed to the construction of a practical glucose biosensor. The resulting biosensor exhibited good amperometric response to glucose with a linear range from 2 to 10 mM ($R = 0.999$) at −0.2 V and from 2 to 14 mM ($R = 0.999$) at 0.5 V and a detection limit of 180 μM. Vitally, preliminary work was also carried out into the determination of glucose in human blood samples.[142]

An additional application of graphene has focused on chitosan-dispersed graphene nano-flakes prepared by a chemical route to reduce graphene oxide; the nano-flakes were immobilized on a glassy carbon electrode to construct a graphene modified electrode (Figure 7.18).[143] Cytochrome *c* was adsorbed tightly on the surface of the modified electrode and the direct electron transfer of cytochrome *c* was achieved by the authors. It was reported that cytochrome *c* on the surface of the electrode maintains its bioactivity and shows an enzyme-like activity for the reduction of nitric oxide, displaying a potential application for the fabrication of novel biosensors to sense nitric oxide.[143]

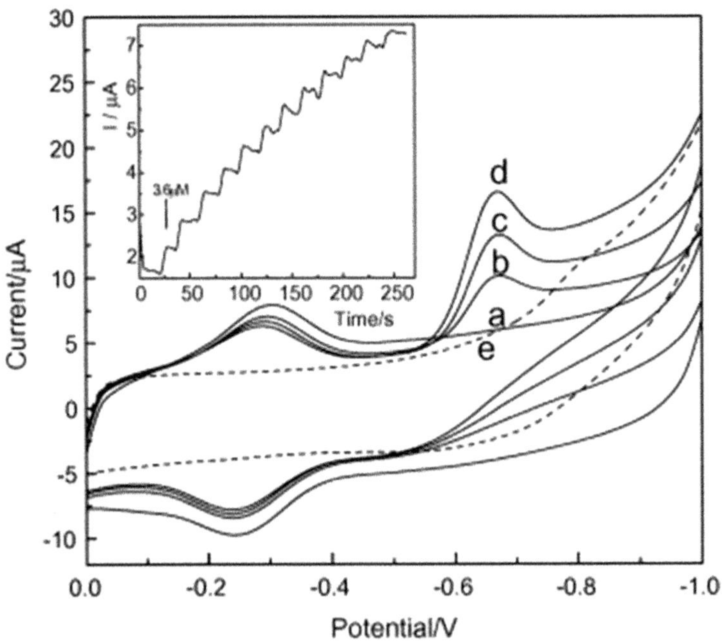

Figure 7.18 Cyclic voltammograms of cytochrome-*c*/chitosan–graphene/GC electrode in 0.1 M pH 7.0 phosphate buffer solution containing (a) no NO and (b) 0.1, (c) 0.2, (d) 0.3 and (e) 0.3 mM NO at a chitosan–graphene/GC electrode (dash) at a scan rate of 100 mV s^{-1}. Inset: recorded current–time curves with successive addition of 3.6 μM NO at the potential of −0.67 V.
(Reproduced from ref. 143 with permission from Elsevier.)

Many other papers have followed this trend, similar to that obtained for carbon nanotubes, as described earlier in this chapter. The field is complex to analyse as differing types of graphene are used that exhibit different % functionalities, different sizes giving rise to different edge-plane and basal-plane characteristics as well as the number of layers. For graphene electroanalysis to be successfully exploited, fundamental research has been undertaken to answer the questions: *Why is graphene reported to be such a good electrode material?*

There are some key reviews that the reader is directed to, which cover fundamental insights into graphene electrochemistry[13,144–146] and here we cover pertinent examples.

Works by Li[147] and Dryfe[148] have tried to study the electrochemistry of graphene with differing degrees of success. Work from decades of carbon electrochemistry can be usefully applied to the area of graphene, as has been the case for carbon nanotubes. Recently, Brownson[149] has explored and critically compared the electrochemistry of mono-, double- and few-layer CVD grown graphene. The grown graphene was transferred using PMMA onto an insulating substrate (silicon dioxide wafers). Characterisation has been performed by Raman spectroscopy, optical spectroscopy, atomic force microscopy and X-ray photoelectron spectroscopy. The graphene electrodes were electrochemically characterised using both inner-sphere and outer-sphere redox probes with electrochemical performances of the graphene electrodes compared to other available graphitic electrodes, namely, that of basal- and edge-plane pyrolytic graphite electrodes constructed from highly ordered pyrolytic graphite (HOPG), with information on heterogeneous rate constants (k^0) obtained. Monolayer graphene was found to exhibit slow heterogeneous electron transfer (HET) kinetics towards the redox probes studied, with HET rates *ca.* 2 and 8 times faster at quasi-graphene and HOPG, respectively, relative to that of the monolayer graphene electrode. Critically contrasting the performance of monolayer graphene to quasi-graphene and HOPG electrodes reveals that increasing the number of graphene layers results in improved electrochemical properties, where in terms of the electrochemical reversibility of the probes studied: monolayer-graphene < quasi-graphene < HOPG, as governed by the respective HET electrochemical rate constants. Given that edge plane sites are the pre-dominant origin of fast electron transfer kinetics at graphitic materials, the slow HET rates at pristine single-layer graphene electrodes are likely due to graphene's fundamental geometry, which consists of a small edge plane and large basal plane contribution. In the case of quasi-graphene and HOPG, they possess increasing global coverage of electrochemically reactive edge plane sites (respectively) and thus exhibit superior electrochemical performances over that of monolayer graphene. Last, the case of a double-layer graphene electrode is considered, which as a result of its fabrication possesses a large global coverage of edge plane like-sites/defects. In agreement with the former conclusions, the double-layered defect-graphene electrode is found to exhibit fast/favourable electrochemical properties, which is attributed to its large edge plane content (*i.e.* defect abundant graphene) and thus is further evidence that the electrochemical response is dependent on the density of edge plane sites at graphene based electrodes (influenced by the coverage of graphene-defects and the number of graphene layers).

Figure 7.19 summarizes the observations by Brownson[149] where the electrochemical signature of differing graphene surfaces is evident as

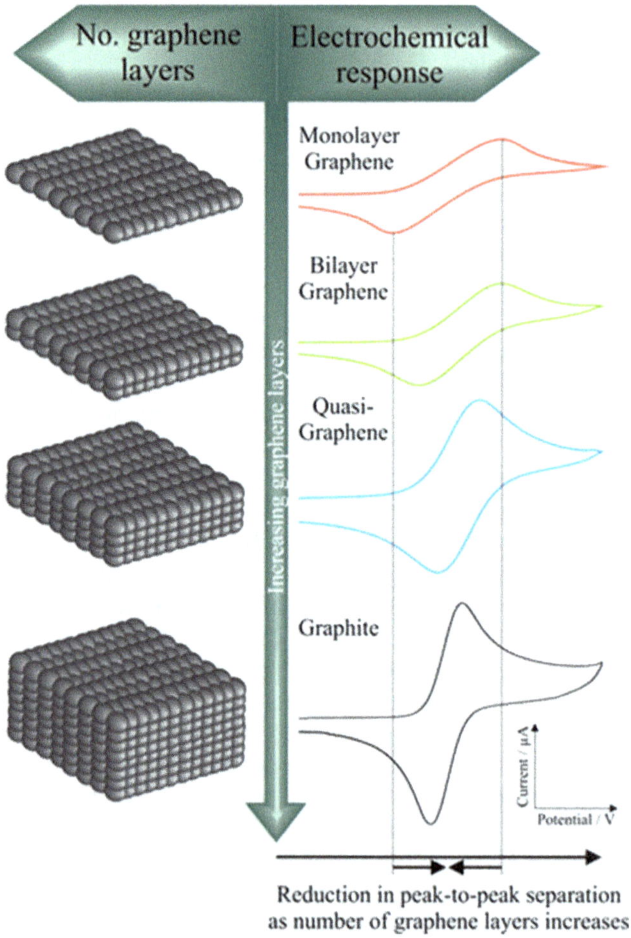

Figure 7.19 Schematic diagram depicting the relative electrochemical reactivities of *n*-layered pristine graphene towards a typical redox probe such as hexaammine–ruthenium(III) chloride. The peak-to-peak separation in the voltammetric waves decreases as the number of graphene layers increases, indicating an increase in the heterogeneous electron transfer rate and thus better electrochemical performance. Note that this will change if defects (holes, dangling bonds, *etc.*) are introduced into the pristine graphene.
(Reproduced from ref. 135.)

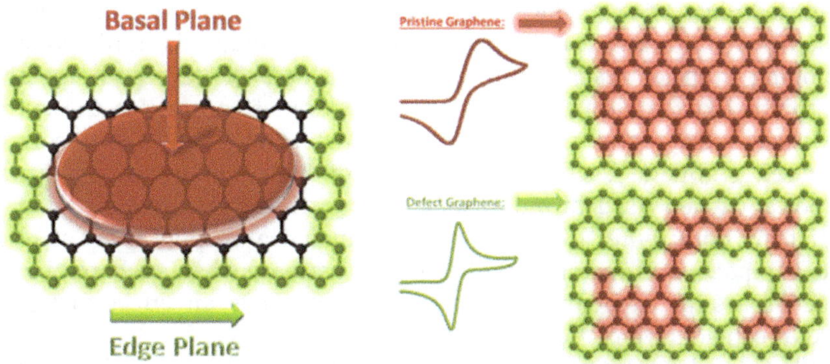

Figure 7.20 Schematic representation of graphene indicating the edge and basal sites, where the heterogeneous electron transfer rate of the former (k^0_{edge}) is anomalously faster than that of the latter (k^0_{basal}). Also shown is the effect of increasing the edge plane content (through the introduction of surface defects) of graphene on the expected electrochemical response.

evaluated using an outer-sphere redox probe. Note that the cyclic voltammetric response for a single layer is drawn out, but improves as the number of layers increase. Such work verifies for the first time that the electrochemistry of graphene is due to the number of layers and that a single pristine graphene layer (o/c: 0.05) exhibits slow electron transfer, even for the facile electron transfer probe hexamine–ruthenium.[149] Figure 7.20 overviews contemporary thinking in the field where, similar to that observed for HOPG and carbon nanotubes, edge plane sites are critical for near-reversible voltammetric responses and in the case of true graphene, a single layer with no defects and large basal plane surface, the material is relatively electrochemically inert,[13] which can be overcome with the introduction of edge plane sites/defects; typically these are introduced in the fabrication approach, such as in the Hummer's method (and its variants).

It has been shown that surfactants that are used in the fabrication of graphene, *via* density gradient ultracentrifugation and additionally in their solubilisation with the aim of reducing the likelihood of coalescing, can dominate the electrochemistry.[150]

Brownson and co-workers demonstrated that commercially available graphene that is being widely used by researchers has a surfactant on the basal plane of the graphene, due to the unique way it has been fabricated, that is in itself electrochemically active.[150] The surfactants inherent to the graphene detrimentally affected the electrochemical oxidation of NADH while in the electrochemical oxidation of acetaminophen a change in mechanism was observed. This work was the first to demonstrate that surfactants, an intrinsic property of graphene fabricated *via* such a route, can dominate the electrochemistry of graphene, advocating that control experiments in the form of surfactant modified carbon electrodes are encouraged particularly in de-convoluting the origin of the electrochemical response of graphene modified electrodes.[150] Such work was extended to demonstrate that the sensing of cadmium, *via* anodic stripping voltammetry, is inhibited at the stripping step corresponding to the transition of cadmium metal to cadmium ions; such an observation is in distinct contrast to the current electrochemical literature of graphene for metal analysis.[151] Other work has explored graphene used in the sensing of hydrogen peroxide, critically important when used in biosensors, where it has been demonstrated the surfactant contributes towards the electroanalytical response.[152] Lastly, the wider impact of surfactants has been demonstrated when such graphene is used as a supercapacitor where the surfactant has an effect that needs careful consideration.[153]

Other fundamental work has shown that graphene fabricated from graphite (*via* chemical oxidation of natural graphite followed by thermal exfoliation/reduction) contains metallic impurities, namely, cobalt, copper, iron, molybdenum and nickel oxide particles which can influence the electrochemistry of graphene towards specific analytes. Thus, these impurities could lead to inaccurate claims of the electro-catalytic effect of graphene. Note that while thorough physical characterisation of the graphite and fabricated graphene is clearly missing from this work, it nicely highlights the importance of metallic impurities and it is clear that a pure source of graphite (or as high a purity as can be obtained) should be used to alleviate such problems; alternatively, commercially available pure graphene should be purchased from reputable suppliers. Nonetheless, this valuable work highlights the importance of sufficient control experimentation when exploring graphene within electrochemistry.[47]

Other work has shown that both natural and synthetic graphite contain a large amount of metallic impurities that persist in the samples of graphite oxide after the oxidative treatment, and chemically reduced graphene after chemical reduction.[48] A method for the purification of graphenes based on thermal treatment at 1000 °C in a chlorine atmosphere to reduce the effect of such impurities on the electrochemical properties has been proposed. Other notable work has reported that carbonaceous debris in graphene oxide/reduced graphene oxide can also affect/dominate the electrochemical response.[154]

While work is, in the main, focused upon exploring the electroanalytical response of graphene or reduced graphene oxide, other researchers have explored graphene oxide modified electrodes. For example, Prabakaran and Pandian[155] have reported on a silver nanoparticle decorated graphene oxide modified electrode towards the detection of Sudan I in red chilli powder samples; such work is important in the field of food adulteration where Sudan dyes are illegally used as food additives. The authors demonstrated a limit of detection of 11.4×10^{-7} M with an assessable linear range from 3.9×10^{-6} to 3.2×10^{-5} M.[155]

Other work reports polypyrrole/graphene nanocomposites for the detection of adenine and guanine. Previous work on graphene modified electrodes indicates that these analytes strongly adhere to basal-plane sites.[156]

Gao *et al.*[157] have reported porous films of polypyrrole/graphene supported upon glassy carbon electrode with a linear range of 0.06–100 μM and 0.04–100 μM for adenine and guanine, respectively, with limits of detections of 0.02 and 0.01 μM, respectively. The new electrode configurations were applied to the quantification of adenine and guanine in calf thymus DNA.

Other strategies involve the modification of graphene with ZnO nanoparticles with carbon nanotubes, exploring this hybrid electrode for the sensing of glucose—a linear range was exhibited of 10 μM to 6.5 mM with a limit of detection of 4.5 μM. This was further applied to glucose sensing in human serum with results independently verified with a photometric method.[158]

7.4.2 Graphene Screen-printed Electrodes

When using graphene, or any carbon nanomaterials, one problem is how to electrically wire to the chosen material (see above). Usual approaches involve taking the chosen carbon nanomaterial in powder form and making a suspension in a suitable solvent. From this aliquots are taken and dropped onto a chosen surface, usually an electrode, where the solvent evaporates to leave the carbon nanomaterial immobilised. If the conditions for drying are carefully controlled, potentially a graphene modified electrode is produced with an even coverage of graphene. However, generally this is not the case and capillary forces present as a result of solvent evaporation can push graphene platelets to the edges of the underlying electrode (Figure 7.21), which can potentially leave concentrated zones of graphene at the edges of the electrode surface in addition to areas where there is little or no graphene coverage, which is akin to the so-called coffee-ring effect; Figure 7.21 shows a schematic overview of this process. The resultant uneven graphene

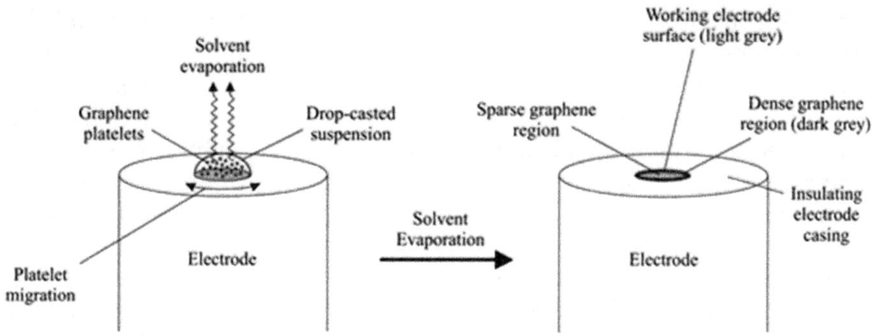

Figure 7.21 Typical coverage of graphene resulting from drop-casting a dispersion of the graphene of interest (suspended in a suitable solvent) onto a supporting electrode surface. Such an approach is extensively utilised within the academic literature to "connect to" and electrically wire the graphene.
(Reproduced from ref. 159 with permission from The Royal Society of Chemistry.)

distribution effectively leaves areas of both fast (multilayer graphene) and slow (single layer graphene) electron transfer and, thus, an electrochemically heterogeneous surface.[159]

Another obvious approach is to take insights from graphitic paste electrodes, which have been used as early as 1958 due to their ease of fabrication and modification, where the electrode surface can be easily renewed by rubbing on a filter paper or equivalent.[160,161] In such electrodes the chosen graphitic material is mixed with an organic liquid to form a paste with various organic liquids or binders used to hold the graphitic material together into a useful structure. Connection is usually made *via* inserting a copper wire into the paste, which allows an easy connection to the potentiostat. Other important work in this area has reported graphitic paste electrodes that utilise an organic material both as the binder and as the reactive material. In an example by Tomcik *et al.*[162] *trans*-1,2-dibromocyclohexane was used as a binder and as the reactive material facilitating the electrocatalytic determination of vitamin B_{12}.

As such, a logical extension is to replace the graphite with graphene and utilise many years of knowledge that allows one to overcome the identified problems met when trying to utilise graphene as an electrode material. To this end, Parvin[163] demonstrated a graphene paste electrode utilising functionalised graphene and explored its use towards the electroanalytical detection of chlorpromazine; it was reported to better than a carbon alternative.[163,164] Interestingly, a report by Li *et al.*[164] emerged, at around the same time as the former work, that concerned the fabrication of a graphene doped carbon paste electrode and demonstrated its superior electrochemical performance over that of the standard/unmodified carbon paste electrode, towards the sensing of ascorbic acid. Following these pioneering reports, further work has emerged on the analytical detection of β-nicotinamide adenine dinucleotide (NADH),[165] paracetamol,[166] cadmium/lead[167] and the simultaneous determination of hydroquinone, catechol and resorcinol[168] at graphene doped carbon paste electrodes. Moreover, modified graphene based paste electrodes have recently been reported with constituents such as Prussian blue, platinum and organic solvents.[169–171] Figueiredo-Filho *et al.* have reported upon the fabrication of a composite graphene paste (*i.e.* no graphite) electrode for the first time.[172] The critical comparison was made with graphitic paste electrodes towards a range of redox probes and other important electrochemical targets. The graphene nano-powder utilised was commercially obtained and consists of 12 nm flakes that have not been oxidised, reduced or chemically modified in any way. The graphene platelets have an average particle (lateral) size reported to be *ca.* 4500 nm (1500–10 000 nm).[172] Through the implementation of careful control and comparison experiments with a graphite paste alternative the authors were able to fully assess the benefits of employing a graphene paste electrode and *demonstrated* that there appear to be no advantages or benefits of using graphene as a paste electrode (in place of graphite), with both materials/electrodes exhibiting similar electrochemical responses and electron transfer kinetics towards various redox probes. In terms of electroanalysis the graphite paste alternative was superior to the graphene electrode given that capacitative currents at the graphene paste precluded low level sensing of the target analytes, with the graphite electrode additionally possessing economic benefits.[172] A limitation, however, is that graphene paste electrodes still need to be mass produced, a formidable feat. An alternative is to turn to the mass production offered by screen-printing technology.

Screen-printed electrodes offer a plethora of benefits over other traditional, bulk electrode configurations. In short, screen-printed electrodes can combine ease of use and portability with simple, inexpensive fabrication techniques. The modest cost of screen-printed configurations has further enhanced their desirability because it allows the devices to become disposable. This ease of fabrication, low-cost nature and robust reproducibility has led to their implementation for the monitoring of an almost inexhaustible list of analytes.[173–178] Additionally, the potentially flexible nature further expands the potential of these highly useful sensors; this is elegantly highlighted through the fabrication of paper-based screen printed sensors,[179–181] sensors fabricated on ultra-thin polyester substrates[180] and on clothing garments such as swimwear made from neoprene.[182] Furthermore, the facile nature of screen-printed electrode modification through incorporation of a mediator, for example in the bulk material of the electrode,[183] offers a significantly more uniform and robust method of electrode modification than other techniques often reported such as drop-casting, while expensive bulk materials can also be utilised to form noble metal sensors consisting of single- and multi-walled carbon nanotubes,[184] gold,[185] platinum[186] and palladium[187] for example.

Using the benefits of screen-printed electrodes described above, Randviir *et al.* reported upon true screen-printed graphene electrodes.[159] The only prior work was reported by Ping *et al.*[188] who demonstrated the fabrication and characterisation of graphene SPEs utilising an in-house synthesised graphene ink towards the beneficial 'electrocatalytic' detection of ascorbic acid, dopamine and uric acid in real samples.[188] The developed ink was not a true screen-printable ink, defined as an ink that can be used on a screen for many hours (typically a screen life of >2 or 3 h) and as such only a handful of electrodes were produced.[188] Contrastingly, in their work, Randviir *et al.*[159] have explored two commercially available graphene inks, each with differing graphene contents and have found different electrochemical signatures. Interestingly, the authors demonstrated that these inks have a long screen life (>3 h) and can be utilised to reproducibly print thousands of electrodes without the ink drying out, which is a parameter not considered before. All inks, however, contain some form of graphitic additive, namely carbon black, to provide a conductive ink. It was found that the fabrication of graphene inks with solely organic solvents and binder afforded a non-conducting graphene electrode, suggesting either that graphene itself is not electrochemically useful (as discussed above) or the orientation of the graphene within the ink/formed electrode is not favourable.[159] Furthermore, these inks can be printed into various shapes, diameters, bands and arrays with little effort required to modify the electrode design.[189,190]

In the above examples researchers generally seek to demonstrate that graphene is beneficial towards their target analytes and usually modify solid electrode substrates. In translating the proven electrochemical protocol from the laboratory to the field, screen-printed electrochemical sensors have been found to be useful. To this end, graphene screen-printed electrodes have been developed. Importantly, generally, the field utilise screen-printed electrodes as the supporting substrate rather than a true graphene screen-printed electrode (see later). Of note is the development of an electrochemical immunosensor for miRNAs based upon reduced graphene oxide/carbon nanotube modified screen printed gold electrodes with a limit of detection of 10 fm for prostate and lung cancer markers.[191]

Li *et al.* reported upon a sensitive HIV-1 sensor using an electrochemical molecular beacon combined with a Nafion–graphene composite film modified screen-printed electrode. Figure 7.22 shows how the electrode was constructed. The sensor was explored towards the determination of a HIV-1 target in real serum samples.[191] A further exciting example demonstrating the potentially expansive combination of graphene modification with screen-printed electrodes was highlighted by Westh and co-workers who reported on a screen-printed carbon electrode modified with graphene for the real-time measurement of unoccupied active sites in cellulose.[192] The research demonstrated how the temporal development in the concentrations of the three enzyme forms can be derived from a combination of the electrochemical data and adsorption measurements.[192]

7.5 3D Graphene Foam

Following on from the above, graphene foams have been developed, again to overcome the experimental problems of how to electrically wire graphene and manipulate it into useful structures. To this end material scientists have developed freestanding 3D graphene foam. The approach was produced by Dong *et al.*[193] who took a 3D nickel foam and grew graphene *via* chemical vapour deposition where the nickel skeleton defines the final graphene structure. The nickel is then chemically etched to leave behind the freestanding graphene macro-structure with graphene micro-structure characteristics. The graphene foam exhibits a well-defined 3D macro-porous structure with an average pore diameter of *ca.* 200 μm, with closer inspection revealing ripples/wrinkles at the grain boundaries of the graphene flakes, evidently an inherent property of the interconnected 3D scaffold structure and CVD grown graphene;[194] it is the differing thermal expansion coefficients of nickel and graphene that give rise to these ripples and wrinkles. SEM images of the graphene structure are depicted in Figure 7.23.

Note that there are broken parts of the graphene structure, which is due to the etching process required to remove the nickel skeleton. Raman analysis reveals that the graphene comprising the 3D freestanding foam consists of

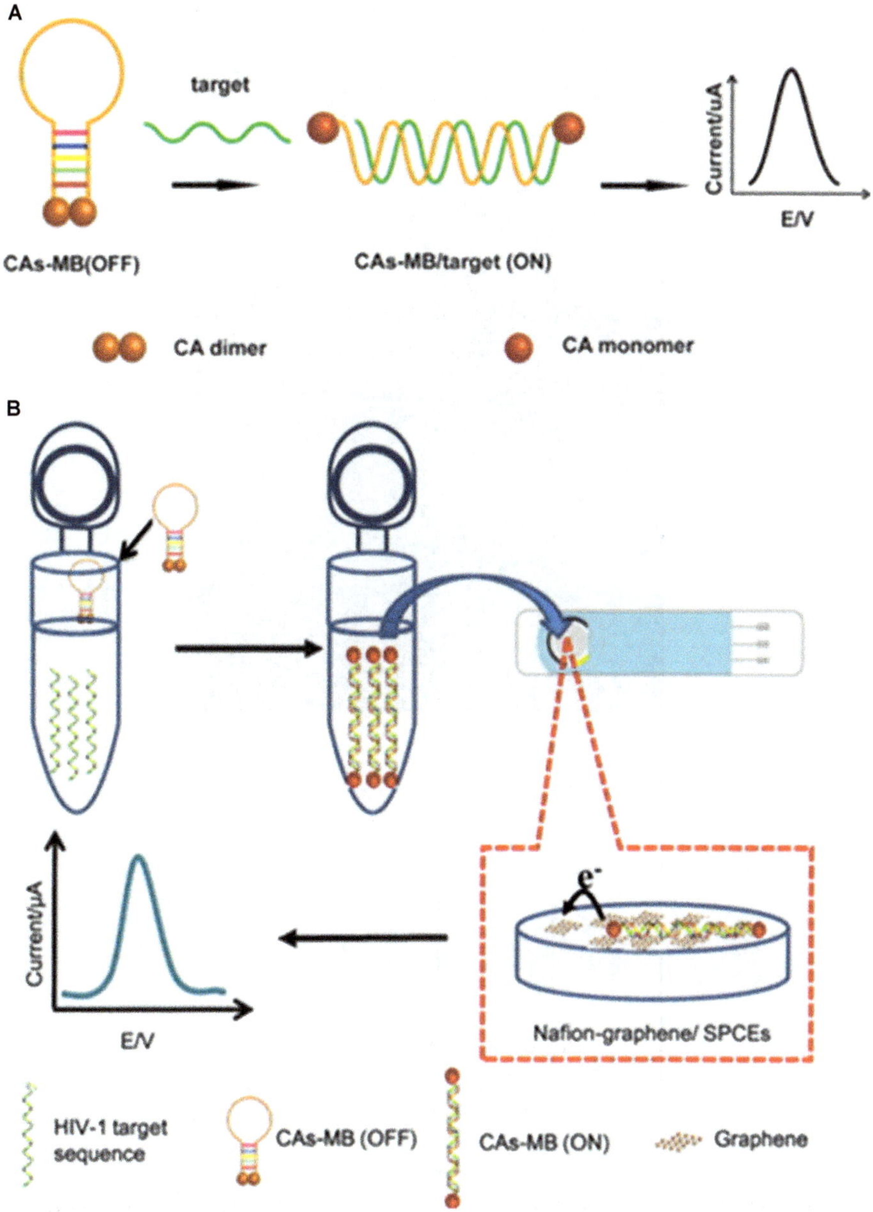

Figure 7.22 Representative diagram of electrochemical molecular beacon sensing assay for HIV-1 oligonucleotide in a homogeneous solution. (A) Hybridization of electrochemically active–inactive switching molecular beacon with the HIV-1 target oligonucleotide. (B) Detection procedures of the electrochemical HIV-1 molecular beacon sensing assay. CAs-MB = special electrochemical molecular beacon.
(Reproduced from ref. 191 with permission from Elsevier.)

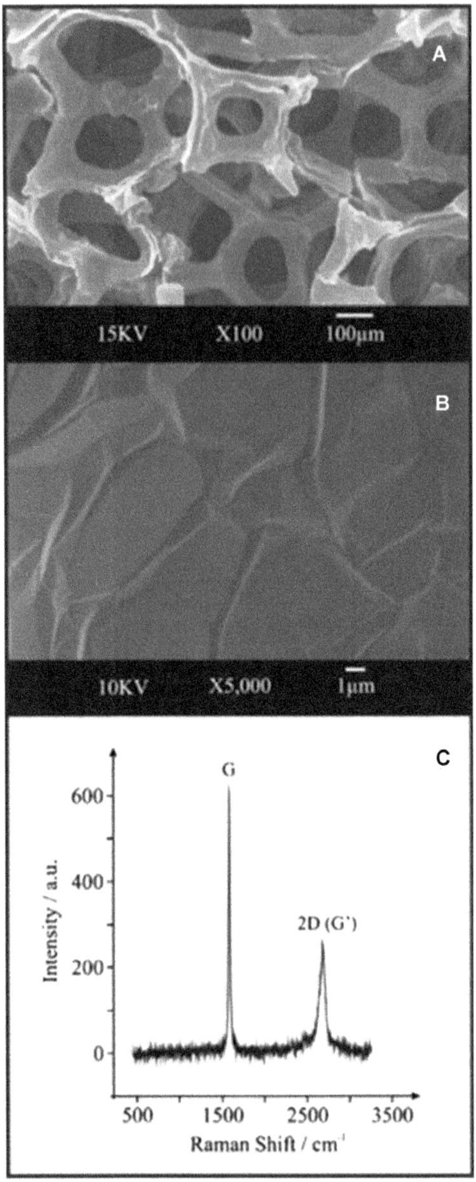

Figure 7.23 SEM image of 3D-graphene foam (A) and at a higher magnification (B). Raman spectrum of 3D-graphene foam (C).
(Reproduced from ref. 197. Copyright Wiley-VCH Verlag GmbH.)

mono- to few-layer graphenes that were of high quality and defect free. This unique 3D freestanding graphene foam provides a highly conductive material with a porous nature offering benefits in supercapacitors due to its large specific surface area with pores able to be filled with nanoparticles, polymers and other functional materials. To this, end Dong *et al.* modified

3D freestanding graphene foam with ZnO for supercapacitor applications and also for the sensing of dopamine,[193] with other work exploring a cobalt oxide modified version for glucose sensing.[195] Table 7.4 shows an overview of electrochemical applications of 3D graphene foams; generally uses involve energy storage applications.

Brownson *et al.*[196] explored an unmodified 3D freestanding graphene foam in aqueous solutions, finding poor voltammetric profiles when evaluated with model redox probes. The authors found that the graphene exhibits quasi-superhydrophobicity, corresponding to a contact angle of 120°, which is less than that expected for true superhydrophobicity (~160°). Consequently, the authors went on to demonstrate that the 3D freestanding graphene foam can give rise to beneficial electrochemical signatures in non-aqueous media, namely ionic liquids. Figure 7.24 depicts the voltammetric profile of the graphene foam, which exhibits faster electron transfer and larger currents over a graphite foam electrode, even after normalising for electrode area.

As such, graphene foam provides beneficial signatures in non-aqueous media over that of a 3D graphitic foam and holds promise in this area. Further work by Figueiredo-Filho went on to show that the limitations observed above in an aqueous media can be overcome by employing a washing pre-treatment. The authors critically compared and contrasted the electroanalytical performance of a 3D graphene foam and 3D graphitic foam electrodes in aqueous based solutions where the hydrophobic nature of the 3D graphene foam was partially overcome through the use of a pre-treatment in the form of washing with acetone. Through critical comparison of the 3D graphene foam with that of a 3D graphite foam electrode, towards the target analytes studied here, the work reported found that there are no significant benefits of using the former over that of the later in aqueous-based electroanalysis; the low level sensing of target analytes is inhibited by large capacitive currents. Future work was identified to require exploration of how the wettability can be improved/hydrophobicity be reduced at the graphene foam, in order to overcome the imposed electroanalytical limitation from the background current, which might yield significant electroanalytical improvements.[197]

Other work by Xiao[198] reported a different fabrication approach to realise a 3D graphene electrode that did not suffer from any hydrophobicity as reported by Figueiredo-Filho.[196] Yue *et al.*[199] used 3D graphene foams modified with ZnO nanowires to explore the simultaneous sensing of dopamine, uric acid and ascorbic acid, with 1 nM detection limits reported. This system was further applied to the sensing of uric acid in the serum of patients with Parkinson's disease. The benefits of the ZnO modified graphene foam are: a large surface area with mesoporous 3D graphene structures to facilitate ion diffusion easily, high conductivity from 3D graphene foam, and active sites of ZnO surface for high selectivity. An additional work[200] has reported $Ni(OH)_2$ modified graphene foam for the enzymatic sensing of glucose with a linear range of 1 μM to 1.17 mM, while Ma[201] reported CuO nanoflower

Table 7.4 Overview of the key reports detailing the use of 3D graphene foam.

Application	Configuration	Comments	Ref.
High-performance supercapacitor and enzyme-less glucose detection	3D Graphene/Co_3O_4 composite	The bare graphene foam has a high specific surface area of 850 m^2 g^{-1}, and this is further increased greatly by the Co_3O_4 nanomesh surrounding the graphene scaffold. The 3D support offers a large capacity to uniformly anchor metal oxides with well-defined size, shapes and crystallinity	195
Supercapacitor and electrochemical sensor	Zinc oxide (ZnO) on three-dimensional (3D) graphene foam	The graphene/ZnO hybrids display superior capacitive performance (400 F g^{-1}) as well as excellent cycle life. They exhibit high sensitivity for detection of $[Fe(CN)_6]^{3+}$ and dopamine, with extrapolated lower detection limits of 1.0 μM and 10.0 nM, respectively.	193
A flexible and freestanding biosensor for non-enzymatic determination of glucose and H_2O_2	Mn_3O_4 grown on 3D freestanding graphene foam	High sensitivity, large linear range and low detection limit for the detection of glucose and H_2O_2; the sensor exhibited excellent performance for real-time detection of glucose and H_2O_2 in serum and food samples	203
Highly compression-tolerant supercapacitor	Polypyrrole-mediated graphene foam electrodes	Achieves superb compression tolerance without significant variation of capacitances under long-term compressive loading and unloading processes	204
Supercapacitor	MnO_2–graphene foam hybrid	Morphology of the MnO_2 nanostructures can be readily controlled by the solution acidity. Serving as a free-standing supercapacitor electrode, this hybrid gives a remarkable specific capacitance (560 F g^{-1} at the current density of 0.2 A g^{-1}) and excellent cycling stability	205

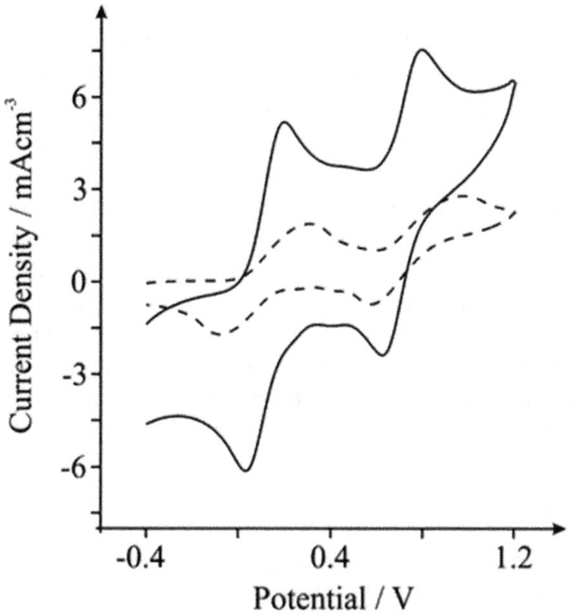

Figure 7.24 Typical cyclic voltammetric profiles recorded towards 1 mM TMPD $[C_4MIM][NTf_2]$ utilising a scan rate of 100 mV s^{-1}, *vs.* Ag. Obtained at the freestanding 3D graphene foam (solid line) and carbon (dashed line) foam electrodes.
(Reproduced from ref. 196 with permission from The Royal Society of Chemistry.)

modified graphene foam for the electroanalytical detection of ascorbic acid, providing a linear range of 0.43–200 μM and a limit of detection of 0.43 μM with a very high sensitivity compared to literature reports.

Clearly, there will be no end of researchers modifying 3D graphene electrodes with metallic structures and applying them to a range of target analytes. Given the range of 3D foam structures that can be obtained with differing porosity and functionality, the use of 3D graphene electrodes as either an unmodified or modified substrate provides a highly useful electrode geometry. Further innovative approaches to the production of 3D graphene foam involve the report of biocompatible mesocellular graphene foam, which was reported to be beneficial towards the sensing of glucose *via* mediation with glucose oxidase. A linear range of 1–12 mM and a limit of detection of 0.25 mM were reported.[202]

7.6 Conclusions

Clearly, there is a wide diversity of carbon nanostructures that are at the disposal of the electroanalyst and as material scientists progress their research, even more exotic structures will arise. As we can see from this chapter, researchers have been focusing on the "main stream" carbon nanomaterials such as carbon nanotubes and graphene and other forms of these materials, as detailed above (see Figure 7.1 for example) are yet to fully realise their potential and are ripe for further investigation. A key structural consideration for all carbon nanomaterials is that they should possess, if near-reversible voltammetric profiles are required, a high proportion of edge plane sites. Additionally, while we have seen some unique benefits for some carbon nanocarbons, such as CNTs as molecule wires, the general literature simply takes a carbon nanomaterial and immobilises it upon an existing carbon macroelectrode. In these cases the benefits should be clearly identified and any voltammetric behaviour needs to be compared to edge plane pyrolytic graphite electrodes and equivalents in order to succinctly identify the benefits of their use. The sheer range and diversity of these carbon nanomaterials in terms of their structure makes the field of electroanalysis hugely exciting, with some likely unique benefits that are yet to be realised.

References

1. E. Osawa, M. Yoshida and M. Fujita, *MRS Bull.*, 1994, **19**, 33.
2. R. N. Adams, *Electrochemistry at Solid Electrodes*, Marcel Dekker, Inc., New York, 1969.
3. Voltammetric Methods, http://chemwiki.ucdavis.edu/Analytical_Chemistry/Analytical_Chemistry_2.0/11_Electrochemical_Methods/11D_Voltammetric_Methods, Accessed August 2014.
4. T. J. Davies, C. E. Banks and R. G. Compton, *J. Solid State Electrochem.*, 2005, **9**, 797.
5. T. J. Davies, R. R. Moore, C. E. Banks and R. G. Compton, *J. Electroanal. Chem.*, 2004, **574**, 123.
6. C. E. Banks and R. G. Compton, *Analyst*, 2005, **130**, 1232.
7. A. Dato, V. Radmilovic, Z. Lee, J. Phillips and M. Frenklach, *Nano Lett.*, 2008, **8**, 2012.
8. Graphene Supermarket, www.graphene-supermarket.com, Accessed August 2014.
9. M. Lotya, P. J. King, U. Khan, S. De and J. N. Coleman, *ACS Nano*, 2010, **4**, 3155.
10. NanoIntegris, www.nanointegris.com, Accessed August 2014.
11. D. A. C. Brownson and C. E. Banks, *Phys. Chem. Chem. Phys.*, 2012, **14**, 8264.
12. X. Li, C. W. Magnuson, A. Venugopal, R. M. Tromp, J. B. Hannon, E. M. Vogel, L. Colombo and R. S. Ruoff, *J. Am. Chem. Soc.*, 2011, **133**, 2816.
13. D. A. C. Brownson, D. K. Kampouris and C. E. Banks, *Chem. Soc. Rev.*, 2012, **41**, 6944.
14. *Biosensing Using Nanomaterials*, ed. A. Merkoci, Wiley Series in Nanoscience and Nanotechnology, John Wiley & Sons, Inc., Hoboken, New Jersey, 2008.
15. J. P. Metters and C. E. Banks, *Vacuum*, 2012, **86**, 507.

16. S. Y. Madani, A. Mandel and A. M. Seifalian, *Nano Rev.*, 2013, **4**, 1.
17. C. E. Banks, X. Ji, A. Crossley and R. G. Compton, *Electroanalysis*, 2006, **18**, 2137.
18. X. Ji, R. O. Kadara, J. Krussma, Q. Chen and C. E. Banks, *Electroanalysis*, 2010, **22**, 7.
19. M. Trojanowicz, *TrAC, Trends Anal. Chem.*, 2006, **25**, 480.
20. C. E. Banks and R. G. Compton, *Analyst*, 2006, **131**, 15.
21. L. Aqui, P. Yanez-Sedeno and J. M. Pingarron, *Anal. Chim. Acta*, 2008, **622**, 11.
22. G. A. Rivas, M. D. Rubianes, M. C. Rodriguez, N. F. Ferreyra, G. L. Luque, M. L. Pedano, S. A. Miscoria and C. Parado, *Talanta*, 2007, **74**, 291.
23. C. B. Jacobs, M. J. Peairs and B. J. Venton, *Anal. Chim. Acta*, 2010, **662**, 105.
24. G. G. Wildgoose, C. E. Banks, H. C. Leventis and R. G. Compton, *Microchim. Acta*, 2006, **152**, 187.
25. G. Lota, K. Fic and E. Frackowiak, *Energy Environ. Sci.*, 2011, **4**, 1592.
26. B. S. Sherigara, W. Kutner and F. D'Souza, *Electroanalysis*, 2003, **15**, 753.
27. K. Gong, Y. Yan, M. Zhang, L. Su, S. Xiong and L. Mao, *Anal. Sci.*, 2005, **21**, 1383.
28. G. Liu, Y. Lin, Y. Tu and Z. Ren, *Analyst*, 2005, **130**, 1098.
29. A. Ambrosi and M. Pumera, *Phys. Chem. Chem. Phys.*, 2010, **12**, 8943.
30. C. E. Banks, T. J. Davies, G. G. Wildgoose and R. G. Compton, *Chem. Commun.*, 2005, 829.
31. C. E. Banks, R. R. Moore, T. J. Davies and R. G. Compton, *Chem. Commun.*, 2004, 1804.
32. J. Tkac and T. Ruzgas, *Electrochem. Commun.*, 2006, **8**, 899.
33. R. Zhang, X. Wang and C. Chen, *Electroanalysis*, 2007, **19**, 1623.
34. Y.-C. Tsai, J.-M. Chen and F. Marken, *Microchim. Acta*, 2005, **150**, 269.
35. S. Jo, H. Jeong, S. R. Bae and S. Jeon, *Microchem. J.*, 2008, **88**, 1.
36. K. I. Ozoemena, D. Nkosi and J. Pillay, *Electrochim. Acta*, 2008, **53**, 2844.
37. S. Shahrokhian and H. R. Zare-Mehrjadi, *Electroanalysis*, 2007, **19**, 2234.
38. S. Shahrokhian and H. R. Zare-Mehrjardi, *Electrochim. Acta*, 2007, **52**, 6310.
39. P. Santhosh, K. M. Manesh, A. Gopalan and K.-P. Lee, *Sens. Actuators, B*, 2007, **125**, 92.
40. M. Guo, J. Chen, J. Li, B. Tao and S. Yao, *Anal. Chim. Acta*, 2005, **532**, 71.
41. C. E. Banks, A. Crossley, C. Salter, S. J. Wilkins and R. G. Compton, *Angew. Chem., Int. Ed.*, 2006, **45**, 2533.
42. B. Sljukic, C. E. Banks and R. G. Compton, *Nano Lett.*, 2006, **6**, 1556.
43. K. Kruusma, N. Mold, K. Jurkschat, A. Crossley and C. E. Banks, *Electrochem. Commun.*, 2007, **9**, 2330.
44. K. Jurkschat, X. Ji, A. Crossley, R. G. Compton and C. E. Banks, *Analyst*, 2007, **132**, 21.

45. L. Siegert, D. K. Kampouris, J. Kruusma, V. Sammelselg and C. E. Banks, *Electroanalysis*, 2009, **21**, 48.
46. E. J. Stuart and M. Pumera, *J. Phys. Chem.*, 2010, **114**, 21296.
47. S. Y. Chee and M. Pumera, *Analyst*, 2012, **137**, 2039.
48. A. Ambrosi, C. K. Chua, B. Khezri, Z. Sofer, R. D. Webster and M. Pumera, *Proc. Natl. Acad. Sci. U. S. A.*, 2012, **109**, 12899.
49. M. Pumera and Y. Miyahara, *Nanoscale*, 2009, **1**, 260.
50. M. Pumera and H. Iwai, *J. Phys. Chem. C*, 2009, **113**, 4401.
51. M. Pumera, *Langmuir*, 2007, **23**, 6453.
52. A. Ambrosi and M. Pumera, *Chem. – Eur. J.*, 2010, **16**, 1786.
53. I. Streeter, G. G. Wildgoose, L. Shao and R. G. Compton, *Sens. Actuators, B*, 2008, **133**, 462.
54. M. C. Henstridge, E. J. F. Dickinson, M. Aslangolu, C. Batchelor-McAuley and R. G. Compton, *Sens. Actuators, B*, 2010, **145**, 417.
55. M. J. Sims, N. V. Rees, E. J. F. Dickinson and R. G. Compton, *Sens. Actuators, B*, 2010, **144**, 153.
56. T. W. B. Lo, L. Aldous and R. G. Compton, *Sens. Actuators, B*, 2012, **162**, 361.
57. H. Zanin, P. W. May, D. J. Fermin, D. Plana, S. M. C. Viera, W. I. Milne and E. J. Corat, *Appl. Mater. Interfaces*, 2014, **6**, 990.
58. M. C. Henstridge, E. J. F. Dickinson and R. G. Compton, *Russ. J. Electrochem.*, 2012, **48**, 629.
59. E. Laborda, A. Molina, F. Martinez-Ortiz and R. G. Compton, *Electrochim. Acta*, 2012, **73**, 3.
60. E. L. K. Chng and M. Pumera, *Electrochem. Commun.*, 2011, **13**, 781.
61. X. Chia, A. Ambrosi and M. Pumera, *Electrochem. Commun.*, 2014, **38**, 1.
62. L. Wang, A. Ambrosi and M. Pumera, *Anal. Chem.*, 2013, **85**, 6195.
63. C. P. Jones, K. Jurkschat, A. Crossley and C. E. Banks, *J. Iran. Chem. Soc.*, 2008, **5**, 279.
64. C. Ge, F. Lao, W. Li, Y. Li, C. Chen, Y. Qui, X. Mao, B. Li, Z. Chai and Y. Zhao, *Anal. Chem.*, 2008, **82**, 9426.
65. P. A. Mello, L. F. Rodrigues, M. A. G. Nunes, J. C. P. Mattos, E. I. Muller, V. L. Dressler and E. M. M. Flores, *J. Braz. Chem. Soc.*, 2011, **22**, 1040.
66. G. Mercier, C. Herold, J.-F. Mareche, S. Cahen, J. Glieze, J. Ghanbaja, G. Lamura, C. Bellouard and B. Vigolo, *New J. Chem.*, 2013, **37**, 790.
67. M. Merisalu, J. Kruusma and C. E. Banks, *Electrochem. Commun.*, 2010, **12**, 144.
68. L. Highton, R. O. Kadara, N. Jenkinson, B. L. Riehl and C. E. Banks, *Electroanalysis*, 2009, **21**, 2387.
69. C. P. Jones, K. Jerkschat, A. Crossley, R. G. Compton, B. L. Riehl and C. E. Banks, *Langmuir*, 2007, **23**, 9501.
70. F. Patolsky, Y. Weizmann and I. Willner, *Angew. Chem., Int. Ed.*, 2004, **43**, 2113.
71. J. Wang, *Electroanalysis*, 2005, **17**, 7.
72. C. Dekker, *Phys. Today*, May, 1999, 22.

73. J. Wang, G. Liu and M. R. Jan, *J. Am. Chem. Soc.*, 2004, **126**, 3010.
74. D. M. Guldi, *Phys. Chem. Chem. Phys.*, 2007, **9**, 1400.
75. S. Lijma, M. Yudasak, R. Yamada, S. Bandow, K. Suenaga, F. Kokai and K. Takahashi, *Chem. Phys. Lett.*, 1999, **309**, 165.
76. S. Zhu and G. Xu, *Nanoscale*, 2010, **2**, 2538.
77. S. Zhu, H. Li, W. Niu and G. Xu, *Biosens. Bioelectron.*, 2009, **25**, 940.
78. Y. Park, K.-Y. Dong, J. Lee, J. Choi, G.-N. Bae and B.-K. Ju, *Sens. Actuators, B*, 2009, **140**, 407.
79. W. Tu, J. Lei, L. Ding and H. Ju, *Chem. Commun.*, 2009, 4227.
80. X. Liu, L. Shi, W. Niu, H. Li and G. Xu, *Biosens. Bioelectron.*, 2008, **23**, 1887.
81. C. Zhao, D. Lin, J. Wu, L. Ding, H. Ju and F. Yan, *Electroanalysis*, 2013, **25**, 1044.
82. F. Valentini, E. C. Conte, L. Sabatini, N. Ditaranto, F. Cataldo, G. Palleschi, M. Bonchio, F. Giacalone, Z. Syrgiannis and M. Prato, *Biosens. Bioelectron.*, 2014, **59**, 94.
83. W. T. Tan, A. M. Bond, S. W. Ngooi, E. B. Lim and J. K. Goh, *Anal. Chim. Acta*, 2003, **491**, 181.
84. S. Griese, D. K. Kampouris, R. O. Kadara and C. E. Banks, *Electroanalysis*, 2008, **20**, 1507.
85. A. Szucs, A. Loix, J. B. Nagy and L. Lamberts, *J. Electroanal. Chem.*, 1995, **397**, 191.
86. A. Szucs, A. Loix, J. B. Nagy and L. Lamberts, *Synth. Met.*, 1996, 77, 227.
87. R. N. Goyal and S. P. Singh, *Talanta*, 2006, **69**, 932.
88. R. N. Goyal, V. K. Gupta, A. Sangal and N. Bachheti, *Electroanalysis*, 2005, **17**, 2217.
89. R. N. Goyal and S. P. Singh, *Electrochim. Acta*, 2006, **51**, 3008.
90. R. T. Kachoosangi, C. E. Banks and R. G. Compton, *Anal. Chim. Acta*, 2006, **566**, 1.
91. R. N. Goyal and S. P. Singh, *J. Nanosci. Nanotechnol.*, 2006, **6**, 3699.
92. R. N. Goyal, V. K. Gupta and N. Bachheti, *Anal. Chim. Acta*, 2007, **597**, 82.
93. R. N. Goyal, V. K. Gupta, M. Oyama and N. Bachheti, *Talanta*, 2007, **71**, 1110.
94. R. N. Goyal, D. Kaur, S. P. Singh and A. K. Pandey, *Talanta*, 2008, **75**, 63.
95. R. N. Goyal, N. Bachheti, A. Tyagi and A. K. Pandey, *Anal. Chim. Acta*, 2007, **605**, 34.
96. R. N. Goyal, V. K. Gupta, N. Bachheti and R. A. Sharma, *Electroanalysis*, 2008, **20**, 757.
97. S. Griese, D. K. Kampouris, R. O. Kadara and C. E. Banks, *Electrochim. Acta*, 2008, **53**, 5885.
98. L. Xiao, G. G. Wildgoose and R. G. Compton, *Sens Act. B*, 2009, **138**, 524.
99. H. W. Kroto, J. R. Heath, S. C. O'Brien, R. F. Curl and R. E. Smalley, *Nature*, 1985, **318**, 162.
100. S. Iijima, *J. Cryst. Growth*, 1980, **50**, 675.

101. H. W. Kroto and K. G. McKay, *Nature*, 1988, **331**, 328.
102. M. C. Henstridge, L. Shao, G. G. Wildgoose, R. G. Compton, G. Tobias and M. L. H. Green, *Electroanalysis*, 2008, **20**, 498.
103. R. Borgohain, J. Yang, J. P. Selegue and D. Y. Kim, *Carbon*, 2014, **66**, 272.
104. D. Lowinsohn, P. Gan, K. Tschulik, J. S. Foord and R. G. Compton, *Electroanalysis*, 2013, **25**, 2435.
105. T. W. B. Lo, L. Aldous and R. G. Compton, *Sens. Actuators, B*, 2012, **162**, 361.
106. K. S. Novoselov, A. K. Geim, S. V. Morozov, D. Jiang, Y. Zhang, S. V. Dubonos, I. V. Grigorieva and A. A. Firsov, *Science*, 2004, **306**, 666.
107. K. S. Novoselov, D. Jiang, F. Schedin, T. J. Booth, V. V. Khotkevich, S. V. Morozov and A. K. Geim, *Proc. Natl. Acad. Sci. U. S. A.*, 2005, **102**, 10451.
108. K. A. Ritter and J. W. Lyding, *Nanotechnology*, 2008, **19**, 015704.
109. E. Rollings, G.-H. Gweon, S. Y. Zhou, B. S. Mun, J. L. McChesney, B. S. Hussain, A. V. Fedorov, P. N. First, W. A. d. Heer and A. Lanzara, *J. Phys. Chem. Solids*, 2006, **67**, 2172.
110. W. A. d. Heer, C. Berger, X. Wu, P. N. First, E. H. Conrad, X. Li, T. Li, M. Sprinkle, K. Hass, M. L. Sadowski, M. Potemski and G. Martinez, *Solid State Commun.*, 2007, **143**, 92.
111. T. Ohta, A. Bostwick, T. Seyller, K. Horn and E. Rotenberg, *Science*, 2006, **313**, 951.
112. C. Berger, Z. Song, X. Li, X. Wu, N. Brown, C. Naud, D. Mayou, T. Li, J. Hass, A. N. Marchenkov, E. H. Conrad, P. N. First and W. A. d. Heer, *Science*, 2006, **312**, 1191.
113. J. Coraux, A. T. N'Diaye, C. Busse and M. Thomas, *Nano Lett.*, 2008, **8**, 565.
114. A. Reina, X. Jia, J. Ho, D. Nezich, H. Son, V. Bulovic, M. S. Dresselhaus and J. Kong, *Nano Lett.*, 2009, **9**, 30.
115. K. S. Kim, Y. Zhao, H. Jang, S. Y. Lee, J. M. Kim, K. S. Kim, J.-H. Ahn, P. Kim, J.-Y. Choi and B. H. Hong, *Nature*, 2009, **457**, 706.
116. Y. S. Dedkov, M. Fonin, U. Rudiger and C. Laubschat, *Phys. Rev. Lett.*, 2008, **100**, 107602.
117. P. W. Sutter, J.-I. Flege and E. A. Sutter, *Nat. Mater.*, 2008, 7, 406.
118. X. Li, G. Zhang, X. Bai, X. Sun, X. Wang, E. Wang and H. Dai, *Nat. Nanoetch.*, 2008, **3**, 538.
119. V. Lee, L. Whittaker, C. Jaye, K. M. Baroudi, D. A. Fischer and S. Banerjee, *Chem. Mater.*, 2009, **21**, 3905.
120. D. A. Dikin, S. Stankovich, E. J. Zimney, R. D. Piner, G. H. B. Dommett, G. Evmeneko, T. S. Nguyen and R. S. Ruoff, *Nature*, 2007, **448**, 457.
121. A. Rouhanipour, M. Roy, X. Feng, H. J. Rader and K. Mullen, *Angew. Chem., Int. Ed.*, 2009, **48**, 4602.
122. L. Zhi and K. Mullen, *J. Mater. Chem.*, 2008, **18**, 1472.
123. J. Wu, W. Pisula and K. Muellen, *Chem. Rev.*, 2007, **107**, 718.
124. H. Duan, E. Xie, L. Han and Z. Xu, *Adv. Mater.*, 2008, **20**, 3284.

125. C. Wu, G. Dong and L. Guan, *Phys. E*, 2010, **42**, 1267.
126. Z. Wang, N. Li, Z. Shi and Z. Gu, *Nanotechnology*, 2010, **21**, 175602.
127. X. Wang, L. Zhi, N. Tsao, Z. Tomovic, J. Li and K. Mullen, *Angew. Chem., Int. Ed.*, 2008, **47**, 2990.
128. A. B. Bourlinos, V. Georgakila, Z. R. Vasilios, T. A. Steriotis and A. K. Stubos, *Small*, 2009, **5**, 1841.
129. Y. Hernandez, V. Nicolosi, M. Lotya, F. M. Blighe, Z. Sun, S. De, I. T. McGovern, B. Holland, M. Byrne, Y. K. Gun'ko, J. J. Boland, P. Niraj, G. Duesberg, S. Krishnamurthy, R. Goodhue, J. Hutchison, V. Scardaci, A. C. Ferrari and J. N. Coleman, *Nat. Nanoetchnol.*, 2008, **3**, 563.
130. K. S. Subrahmanyam, S. R. C. Vivekchand, A. Govindaraj and C. N. R. Rao, *J. Mater. Chem.*, 2008, **18**, 1517.
131. M. Choucair, P. Thordarson and J. A. Stride, *Nat. Nanoetchnol.*, 2009, **4**, 30.
132. D. V. Kosynkin, A. L. Higginbotham, A. Sinitskii, J. R. Lomeda, A. Dimiev, B. K. Price and J. M. Tour, *Nature*, 2009, **458**, 872.
133. L. Jiao, L. Zhang, X. Wang, G. Diankov and H. Dai, *Nature*, 2009, **458**, 877.
134. D. Chen, L. Tang and J. Li, *Chem. Soc. Rev.*, 2010, **39**, 3157.
135. E. P. Randviir, D. A. C. Brownson and C. E. Banks, *Mater. Today*, 2014, **17**, 426.
136. A. Dato, Z. Lee, K.-J. Jeon, R. Erni, V. Ramilovic, T. J. Richardson and M. Frenklach, *Chem. Commun.*, 2009, 6095.
137. C. Liu, S. Alwarappan, Z. Chen, X. Kong and C.-Z. Li, *Biosens. Bioelectron.*, 2010, **25**, 1829.
138. W. Li, C. Tan, M. A. Lowe, H. D. Abruna and D. C. Ralph, *ACS Nano*, 2011, **5**, 2264.
139. Y. Wang, Y. Li, L. Tang, J. Lu and J. Li, *Electrochem. Commun.*, 2009, **11**, 889.
140. W.-J. Lin, C.-S. Liao, J.-J. Jhang and Y.-C. Tsai, *Electrochem. Commun.*, 2009, **11**, 2153.
141. X. Kang, J. Wang, H. Wu, J. Liu, I. A. Aksay and Y. Lin, *Talanta*, 2010, **81**, 754.
142. C. Shan, H. Yang, D. Han, Q. Zhang, A. Ivaska and L. Niu, *Biosens. Bioelectron.*, 2010, **25**, 1070.
143. J.-F. Wu, M.-Q. Xu and G.-C. Zhao, *Electrochem. Commun.*, 2010, **12**, 175.
144. D. A. C. Brownson and C. E. Banks, *Phys. Chem. Chem. Phys.*, 2012, **14**, 8264.
145. M. Pumera, *Chem. Rec.*, 2009, **9**, 211.
146. D. A. C. Brownson and C. E. Banks, *The Handbook of Graphene Electrochemistry*, Springer, 2014.
147. Z.-S. Wu, D.-W. Wang, W. Ren, J. Zhao, G. Zhou, F. Li and H.-M. Cheng, *Adv. Funct. Mater.*, 2010, **20**, 3595.

148. A. T. Valota, I. A. Kinloch, K. S. Novoselov, C. Casiraghi, A. Eckmann, E. W. Hill and R. A. W. Dryfe, *ACS Nano*, 2011, **5**, 8809.
149. D. A. C. Brownson, S. A. Varey, F. Hussain, S. J. Haigh and C. E. Banks, *Nanoscale*, 2013, **6**, 1607.
150. D. A. C. Brownson, J. P. Metters, D. K. Kampouris and C. E. Banks, *Electroanalysis*, 2011, **23**, 894.
151. D. A. C. Brownson and C. E. Banks, *Electrochem. Commun.*, 2011, **13**, 111.
152. D. A. C. Brownson and C. E. Banks, *Analyst*, 2011, **136**, 2084.
153. D. A. C. Brownson and C. E. Banks, *Chem. Commun.*, 2012, **48**, 1425.
154. X. Li, X. Yang, L. Jia, X. Ma and L. Zhu, *Electrochem. Commun.*, 2012, **23**, 94.
155. E. Prabakaran and K. Pandian, *Food Chem.*, 2015, **166**, 198.
156. E. P. Randviir and C. E. Banks, *RSC Adv.*, 2012, **2**, 5800.
157. Y.-S. Gao, J.-K. Xu, L.-M. Lu, L.-P. Wu, K.-X. Zhang, T. Nie, X.-F. Zhu and Y. Wu, *Biosens. Bioelectron.*, 2014, **62**, 261.
158. K.-Y. Hwa and B. Subramani, *Biosens. Bioelectron.*, 2014, **62**, 127.
159. E. P. Randviir, D. A. C. Brownson, J. P. Metters, R. O. Kadara and C. E. Banks, *Phys. Chem. Chem. Phys.*, 2014, **16**, 4598.
160. R. N. Adams, *Anal. Chem.*, 1958, **30**, 1576.
161. E. S. Jacobs, *Anal. Chem.*, 1963, **35**, 2112.
162. P. Tomcik, C. E. Banks, T. J. Davies and R. G. Compton, *Anal. Chem.*, 2004, **76**, 161.
163. M. H. Parvin, *Electrochem. Commun.*, 2011, **13**, 366.
164. F. Li, J. Li, Y. Feng, L. Yang and Z. Du, *Sens. Actuators, B*, 2011, **157**, 110.
165. A. Gasnier, M. L. Pedano, M. D. Rubianes and G. A. Rivas, *Sens. Actuators, B*, 2013, **176**, 921.
166. H. Bahramipur and F. Jalali, *Afr. J. Pharm. Pharmacol.*, 2012, **6**, 1298.
167. W. Wonsawat, S. Chuanuwatanakul, W. Dungchai, E. Punrat, S. Motomizu and O. Chailapakul, *Talanta*, 2012, **100**, 282.
168. O. Stamm, U. Latscha, P. Janecek and A. Campana, *Am. J. Obstet. Gynecol.*, 1976, **124**, 193.
169. M. Saber-Tehrani, A. Pourhabib, S. W. Husain and M. Arvand, *Electroanalysis*, 2012, **24**, 2395.
170. X. Yuan, Y. Chai, R. Yuan, O. Zho and C. Yang, *Anal. Methods*, 2012, **4**, 3332.
171. Y. Bo, W. Wang, J. Qi and S. Huang, *Analyst*, 2011, **136**, 1946.
172. L. C. Figueiredo-Filho, D. A. C. Brownson, M. Gomez-Mingot, J. Iniesta, O. Fatibello-Filho and C. E. Banks, *Analyst*, 2013, **138**, 6354.
173. J. P. Smith, J. P. Metters, C. Irving, O. B. Sutcliffe and C. E. Banks, *Analyst*, 2014, **139**, 389.
174. J. P. Smith, J. P. Metters, D. K. Kampouris, C. Lledo-Fernandez, O. B. Sutcliffe and C. E. Banks, *Analyst*, 2013, **138**, 6185.
175. M. Gomez-Mingot, V. Montiel, C. E. Banks and J. Iniesta, *Analyst*, 2014, **139**, 1442.

176. A. V. Kolliopoulos, J. P. Metters and C. E. Banks, *Anal. Methods*, 2013, **5**, 851.
177. O. Ramdani, J. P. Metters, L. C. S. Figueiredo-Filho, O. Fatibello-Filho and C. E. Banks, *Analyst*, 2013, **153**, 1053.
178. A. V. Kolliopoulos, J. P. Metters and C. E. Banks, *Anal. Methods*, 2013, **5**, 3490.
179. C. W. Foster, J. P. Metters and C. E. Banks, *Electroanalysis*, 2013, **25**, 2275.
180. C. W. Foster, J. P. Mettters, D. K. Kampouris and C. E. Banks, *Electroanalysis*, 2014, **26**, 262.
181. J. P. Metters, S. Housein, D. K. Kampouris and C. E. Banks, *Anal. Methods*, 2013, **6**, 103.
182. K. Malzahn, J. R. Windmiller, G. Valdes-Ramirez, M. J. Schoning and J. Wang, *Analyst*, 2011, **136**, 2912.
183. R. O. Kadara, N. Jenkinson and C. E. Banks, *Electroanalysis*, 2009, **21**, 2410.
184. J. P. Metters, M. Gomez-Mingot, J. Iniesta, R. O. Kadara and C. E. Banks, *Sens. Actuators, B*, 2013, **177**, 1043.
185. J. P. Metters, R. O. Kadara and C. E. Banks, *Analyst*, 2012, **137**, 896.
186. J. P. Metters, F. Tan, R. O. Kadara and C. E. Banks, *Anal. Methods*, 2012, **4**, 1272.
187. J. P. Metters, F. Tan and C. E. Banks, *J. Solid State Electrochem.*, 2013, **17**, 1553.
188. J. Ping, J. Wu, Y. Wang and Y. Ying, *Biosens. Bioelectron.*, 2012, **34**, 70.
189. J. P. Metters, R. O. Kadara and C. E. Banks, *Sens. Actuators, B*, 2012, **169**, 136.
190. D. A. C. Brownson, C. W. Foster and C. E. Banks, *Analyst*, 2012, **137**, 1815.
191. B. Li, Z. Li, B. Situ, Z. Dai, Q. Liu, Q. Wang, D. Gu and L. Zheng, *Biosens. Bioelectron.*, 2014, **15**, 330.
192. N. Cruys-Bagger, H. Tatsumi, K. Borch and P. Westh, *Anal. Biochem.*, 2014, **447**, 162.
193. X. Dong, Y. Cao, J. Wang, M. B. Chan-Park, L. Wang, W. Huang and P. Chen, *RSC Adv.*, 2012, **2**, 4364.
194. O. C. Compton and S. T. Nguyen, *Small*, 2010, **6**, 711.
195. X.-C. Dong, H. Xu, X.-W. Wang, Y.-X. Huang, M. B. Chan-Park, H. Zhang, L.-H. Wang, W. Huang and P. Chen, *ACS Nano*, 2012, **6**, 3206.
196. D. A. C. Brownson, L. C. S. Figueiredo-Filho, X. Ji, M. Gomez-Mingot, J. Iniesta, O. Fatibello-Filho, D. K Kampouris and C. E. Banks, *J. Mater. Chem. A*, 2013, **1**, 5962.
197. L. C. S. Figueiredo-Filho, D. A. C. Brownson, O. Fatibello-Filho and C. E. Banks, *Electroanalysis*, 2014, **26**, 93.
198. X. Xiao, P. R. Miller, R. J. Narayan, S. M. Brozik, D. R. Wheeler, I. Brener, J. Wang, D. B. Burckel and R. Polsky, *Electroanalysis*, 2014, **26**, 52.

199. H. Y. Yue, S. Huang, J. Chang, C. Heo, Y. Yao, S. Adhikari, F. Gunes, L. C. Liu, T. H. Lee, E. S. Oh, B. Li, J. J. Zhang, T. Q. Huy, N. V. Luan and Y. H. Lee, *ACS Nano*, 2014, **8**, 1639.
200. B. Zhan, C. Liiu, H. Chen, H. Shi, L. Wang, P. Chen, W. Huang and X. Dong, *Nanoscale*, 2014, **6**, 7424.
201. Y. Ma, M. Zhao, B. Cai, W. Wang, Z. Ye and J. Huang, *Biosens. Bioelectron.*, 2014, **59**, 384.
202. Y. Wang, H. Li and J. Kong, *Sens. Actuators, B*, 2014, **193**, 708.
203. P. Si, X.-C. Dong, P. Chen and D.-H. Kim, *J. Mater. Chem. B*, 2013, **1**, 110.
204. Y. Zhao, J. Liu, Y. Hu, H. Cheng, C. Hu, C. Jiang, L. Jiang, A. Cao and L. Qu, *Adv. Mater.*, 2013, **25**, 591.
205. X. Dong, X. Wang, J. Wang, H. Song, X. Li, L. Wang, M. B. Chan-Park, C. M. Li and P. Chen, *Carbon*, 2012, **50**, 4865.

CHAPTER 8

Dispersible Electrodes: An Approach to Developing Sensing Devices that can Quickly Detect Ultralow Concentrations of Analyte

SAIMON MORAES SILVA AND J. JUSTIN GOODING*

School of Chemistry and Australian Centre for NanoMedicine, The University of New South Wales, Sydney 2052, Australia
*Email: Justin.Gooding@unsw.edu.au

8.1 Introduction

Detection systems for monitoring ultra-trace levels of analyte face a unique challenge beyond that faced by analytical systems that detect higher concentrations of analyte. This unique challenge arises from the need for the sample for analysis to be representative of the bulk. It is a challenge of volume, sample volume. The sample volume required to be representative of the bulk increases as the analyte concentration decreases according to the Poisson distribution. As an example of the impact of the Poisson distribution on reliable sampling and analysis, if the detection limit of an analytical method was required to be 1 molecule per millilitre then the sample volume required to be 95% sure the sample is representative of the bulk is around 50 mL. A volume of 50 mL may not appear large but consider a

RSC Detection Science Series No. 6
Electrochemical Strategies in Detection Science
Edited by Damien W. M. Arrigan

Published by the Royal Society of Chemistry, www.rsc.org

microfluidic device with a volume flow rate of 0.25 mL min^{-1} then it would take around 200 min for the entire sample to pass through the device. Taken to its extreme, in water quality analysis where we may need to detect 1 colony forming unit per litre, this comes to 50 L that need to be analysed. Such a volume is totally impractical to be analysed in a reasonable time frame using most analytical methods. So, when an analytical method is required to analyse ultra-trace levels of analyte, the major challenge may well be the sample volume, rather than developing a method with sufficient sensitivity and low enough background.

In portable sensing, the need for a solution to the issue of the need to analyse large sample volumes is particularly interesting. Typically a sensor is a solid-state device with a modified surface that detects the analyte.[1,2] Sensors with exquisite sensitivity have been developed. For example, nanowire field effect transistors have been shown to be able to detect a single virus particle.[3] Similarly, some of the nanopore devices can detect even a single ion passing through them.[4] In both cases the sensing element is a nanoscale device in a single position in space. Therefore, although a single entity can be detected, it samples an exceedingly small volume. Hence, as amazing as such devices are, they do not solve the challenge at hand, how to detect ultra-trace levels of analytes in large samples within a reasonable time frame. The issue here is, essentially, the sensor may be able to detect a single analyte species but will that analyte species ever find the sensing interface. The sensor being in a defined position in space is the key, because it means the distance some analyte might need to be transported, whether it is by convection or diffusion, may be large.

The strategy we have chosen to address this challenge is to turn the sensing paradigm around. That is, rather than make the analyte find the sensor, we make the sensor find the analyte. As our strategy is an electrochemical strategy, we call this 'dispersible electrodes'. In essence, the dispersible electrode concept is to subdivide an electrochemical sensor into tiny pieces and disperse this throughout the sample volume, such that diffusion path-lengths are short. The nanomaterials utilised for this are conducting gold-coated magnetic nanoparticles (Au@MNPs) as the active element in the selective capture and direct electroanalytical quantification, of analytes. The magnetic properties of the Au@MNPs allow the bulk sensor to be reformed on application of a magnetic field, while the gold coating means the particles are conducting. Further, this gold surface also allows for the Au@MNPs to be modified such that they selectively bind to the target analyte. The implication of short diffusional path-lengths is a more rapid response. An additional benefit is that more of the sample solution is interrogated by the dispersible electrodes than with conventional sensors. This means more of the analyte is collected and hence the sensitivity of the device is also enhanced. In other words, in the 'dispersible electrodes' strategy, the analyte is captured by Au@MNPs dispersed in the solution. After completing the analyte capturing process a magnetic field is applied and brings the particles back to a conventional macroelectrode, and the

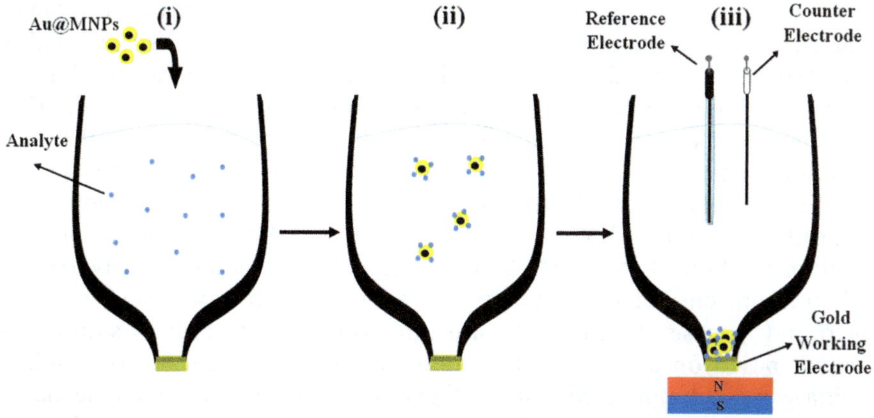

Figure 8.1 Outline of the steps necessary for the measurement of a target analyte: (i) dispersion of Au@MNPs into the sample; (ii) analytes bind selectively to Au@MNPs; (iii) captured analyte is attracted to a transducing electrode for quantification.

amount of analyte can be directly quantified by electrochemical methods (Figure 8.1).

In this chapter the dispersible electrode concept is discussed in some detail. In Section 8.2 some initial applications of Au@MNPs are overviewed to indicate the scope of the idea. Section 8.3 covers the synthetic approach used to synthesize the Au@MNPs that are employed as dispersible electrodes. The synthetic approach will be followed by the electrochemical characterization of Au@MNPs (Section 8.4). Section 8.5 discusses the functionalisation of gold-coated magnetic nanoparticles for electroanalytical applications. The detection of electroactive species is described in Section 8.6. The final section (Section 8.7) explores the detection of non-electroactive species.

8.2 The Use of Au@MNPs in Electrochemical Sensing

The application of Au@MNPs in electrochemistry is especially attractive due to the synergism between magnetic iron oxide nanoparticles and gold. The Au@MNPs present the properties of gold such as high conductivity, optical properties, chemical stability and biocompatibility, as well as its facile and predictable surface modification combined with the nontoxic nature, ease of synthesis, and high magnetic moment of the magnetic nanoparticles.[5,6] A few strategies for using these Au@MNPs in electrochemical research have been reported. None of these strategies combine the main features we exploit to create dispersible electrodes, notably the gold being conducting and compatible with modification and the ability to magnetically manipulate the particles such that particles can be dispersed throughout a sample but then collected using magnets.

In a study by Yang *et al.*, a H_2O_2 biosensor based on Au@MNPs coated with horseradish peroxidise (HRP) and graphene sheets–Nafion (GS–Nafion) film modified screen-printed carbon electrode was developed.[7] The magnetic feature is only used in the fabrication of this biosensor. That is, the Au@MNPs coated with HRP were adsorbed on the printed electrode modified with GS–Nafion by using the aid of an external magnetic field to deliver the final device. This electrochemical biosensor, under the optimized conditions, showed a linear range between 20 and 2500 μM and a detection limit of 12 μM to H_2O_2. This sensor also presented a good selectivity to H_2O_2 when in presence of some common interferents such as uric acid, ascorbic acid and glucose.[7] Similarly, Pundir and colleagues reported a sulfite biosensor based on an Au@MNPs modified electrode.[8] In this work, sulfite oxidase was immobilized onto carboxylated gold coated magnetic nanoparticles and electrodeposited onto the surface of a gold electrode through *N*-ethyl-*N′*-(3-dimethylaminopropyl)carbodiimide (EDC) and *N*-hydroxysuccinimide (NHS) chemistry. Hence the magnetic properties of the Au@MNPs were not used at all. To get the response to this analyte, amperometry was used. The proposed biosensor presented a linear range and detection limit of 0.50–1000 μM and 0.15 μM, respectively. The biosensor was applied in real samples of red and white wine and showed a good recovery.[8] In another study by Pundir and colleagues, an amperometric glutathione biosensor based on chitosan–Au@MNPs modified Pt electrode was reported.[9] Like their other study, the Au@MNPs were electrochemically deposited onto the Pt electrode. Then, chitosan was used to introduce amino groups onto the surface of Au@MNPs and finally the glutathione oxidase was immobilized covalently onto the surface of gold coated magnetic nanoparticles modified Pt electrode. With this architecture, it was possible to achieve a detection limit of 0.1 $\mu mol\ L^{-1}$ for glutathione and to detect this analyte in human hemolysated erythrocytes.[9] All these aforementioned studies exhibit good sensitivity and selectivity, but these sensing devices involve immobilising the nanoparticle on a bulk electrode before exposure to the analyte and hence the sensing interface is still in a single position in space. For this reason, the detection of species at ultra-trace levels in a reasonable time in large volume samples is not addressed in these studies, despite the use of Au@MNPs.

8.3 Synthetic Approach of Gold-shell-protected Magnetic Nanoparticles

Magnetic iron oxide nanoparticles have been extensively studied and applied in a wide range of areas including biomedicine, catalysis, environmental remediation, and sensing applications.[10–12] They are attractive due to their exceptional properties including nontoxic nature, ease of synthesis and high magnetic moment.[13] Nevertheless, magnetic nanoparticles exhibit some limitations with regards to being active elements in sensing and biosensing applications. These include low electrical conductivity and limited optical properties combined with poorly defined surface modification chemistry.[14] In addition, due to their large surface area to volume ratio and low surface charge at neutral pH, magnetic nanoparticles show low stability and tend to aggregate in solution.[15] These limitations can be overcome by modifying the magnetic nanoparticles with a gold coating because, as mentioned above, gold has inherent electrical conductivity, optical properties due to surface plasmons, chemical stability and the ability to functionalise its surface with thiolated compounds. Furthermore, gold allows facile and predictable functionalization of the particles, which can stabilize them against aggregation.[16] The synthesis of gold-coated magnetic nanoparticles involves two key processes, the magnetite nanoparticle core synthesis and the further coating with gold. There are numerous methods for the synthesis of magnetite nanoparticle cores, the selection of which synthetic method, whether organic phase, aqueous phase or otherwise, is fundamentally dependent on the desired properties and final purpose of the synthesised nanoparticles.[17] Among these synthetic approaches, the aqueous phase precipitation methodology is very attractive for the synthesis of magnetic nanoparticles due to their simple and efficient chemical pathway, as well as the ability to produce large-scale quantities of magnetic nanoparticles with nontoxic precursors chemicals and conditions.[18]

In a study by Lyon *et al.*, a route to magnetic core/shell particles, that are soluble in aqueous media, by reduction of Au^{3+} onto the surface of magnetic nanoparticles using several iterations of hydroxylamine seeding was reported.[18] In this method it was demonstrated that both γ-Fe_3O_4 and partially oxidized Fe_3O_4 can be used as cores to synthesize Au@MNPs, but the partially oxidized Fe_3O_4 particles are more resistant to Au deposition than γ-Fe_3O_4 particles. The authors showed that the morphology and the optical properties of the Au@MNPs are dependent on the amount of Au deposited, while the magnetic property is not affected by the Au shell. Based on this method, and attempting to develop Au@MNPs that present good resistance to aggregation and to verify a full gold coat, we tried to synthesize 50–150 nm Au@MNPs *via* the use of a biocompatible polyelectrolyte, polyethyleneimine (PEI), for the dual function of attaching gold seeds and preventing the formation of large aggregates.[17] The first step in the synthesis method involves the immobilization of the polycation, PEI, on the Fe_3O_4. The positively charged PEI binds to Fe_3O_4 that is negatively charged *via* electrostatic self-assembly to form a stabilizing polyelectrolyte layer (Figure 8.2). The PEI on Fe_3O_4 is of essential importance to the second step, because PEI allows the controlled attachment of negatively charged Au seeds to the Fe_3O_4 nanoparticles *via* electrostatic and weak covalent bonding with the amines on PEI. A dense layer of attached Au seeds on the Fe_3O_4 surface plays an important role in ensuring that all the initial bare Fe_3O_4 particles were coated with Au. Attempts to coat Fe_3O_4–PEI particles without, or with only a small number of attached Au seeds, results in a majority of the Fe_3O_4 remaining uncoated. Finally, gold shells are grown around the nanoparticles

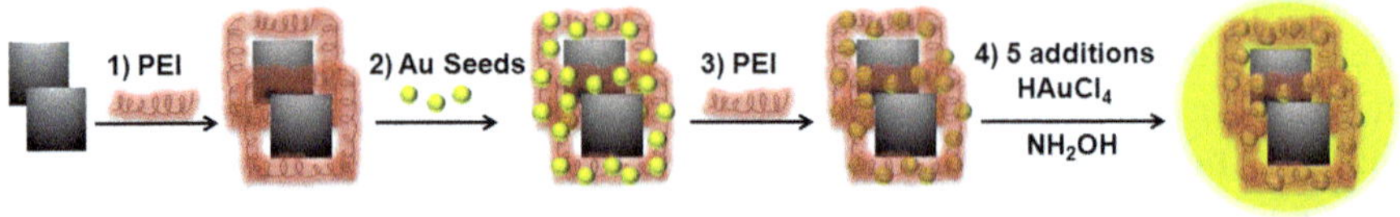

Figure 8.2 Synthetic route illustration of Au@MNPs. In the first step polyethyleneimine (PEI) is self-assembled onto magnetite (Fe_3O_4). Secondly, Fe_3O_4–PEI is mixed with Au seeds, to obtain Fe_3O_4–PEI–Au_{seeds}, after which PEI is again added to obtain Fe_3O_4–PEI–Au_{seeds}–PEI. Finally five additions of gold chloride in the presence of hydroxylamine form an even gold coating.

such that several iterations of gold coating were deposited *via* reduction of gold chloride to gold with a reducing agent, hydroxylamine, in the presence of Fe_3O_4–PEI–Au_{seed}. The completeness of the Au coating is an important criterion in the use of the gold magnetic nanoparticles for our analytical purposes and magnetic nanoparticles can be considered fully coated if all the Fe_3O_4 NPs in each sample are coated with a protective layer of Au.[17]

Altogether, with this synthesis procedure, it is possible to produce 50 to 150 nm gold-shell magnetic nanoparticles utilising magnetic nanoparticles functionalised with polyethyleneimine. This aqueous-based procedure performed at room temperature produces gold-coated magnetic nanoparticles that show the twin properties of being non-toxic magnetic nanoparticles with a high magnetic moment coupled with the inherent electrical conductivity and chemical stability of gold.[17] However, with a view to obtaining improved monodispersity, and with the stability against aggregation in mind, we are moving to a different synthetic method developed by Gao *et al.*[19] With this new method it is possible to obtain Au@MNPs with well-defined structural characteristics including size, shell thickness, and core–shell separation. In the next section the electrochemical characterization of the Au@MNPs employed as 'dispersible electrodes' will be discussed.

8.4 Electrochemical Characterization of Au@MNPs

Information about Au@MNPs behaviour in terms of electroactive surface area, electron transfer ability, and porous structure is of utmost importance to understand and evaluate the performance of Au@MNPs in electrochemistry. Such information can be ascertained using cyclic voltammetry (CV) and electrochemical impedance spectroscopy (EIS). Using CV and the well-known electrochemical probe ferricyanide (0.1 M in phosphate buffer saline (pH 7)), by increasing the number of theoretical layers of Au@MNPs on the electrode surface an increase in Faradaic current (Figure 8.3a) is observed, which can be related to a rise in the electroactive surface area.[20] This increase in the electroactive surface area can be calculated by the Randle–Sevcik equation when considering it as a roughened electrode.[21] As shown in the Figure 8.3b there is a relative increase in electrochemical

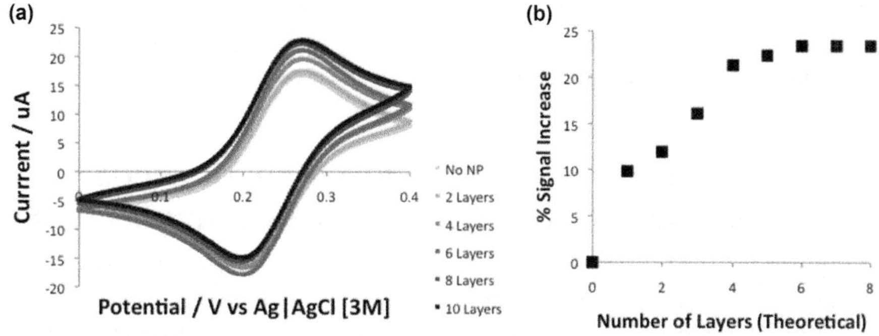

Figure 8.3 (a) Cyclic voltammograms of the ferricyanide/ferrocyanide couple at Au@MNPs on gold electrode as a function of the number of 'theoretical layers'. Electrolyte: 0.1 mM potassium ferricyanide in PBS buffer (pH 7.0). (b) Percentage electrochemical area increase as a function of the number of 'theoretical layers'.
(Reprinted from ref. 20, Copyright 2011, with permission from Elsevier.)

surface area compared with the underlying macroelectrode. The roughness reached a plateau after six theoretical layers. The 'theoretical layer' was calculated based on the number of particles required to cover the area of the polycrystalline gold macroelectrode uniformly considering the average particle size ($\sim$100 nm in our case) and mass of particles in solution.[20] The fact that no further increase in roughness can be seen suggests that the ferricyanide does not penetrate into the particle spaces; namely, a compact structure is formed during potential scanning. That is, only the top layer of particles are interacting with solution redox active species, but not with the internal surfaces within the pores due to the pore size being less than the diffusion radius of ferricyanide. This conclusion is supported by EIS of a solution of potassium ferricyanide or of a redox couple with much slower electron transfer kinetics, where no evidence of the double semicircle Nyquist plot characteristic of a porous electrode is observed. The Nyquist plot of Au@MNPs film in 1 mM potassium ferricyanide was virtually identical to that of a bare gold electrode where the electrochemistry is dominated by mass transport regime. The absence of a semicircle in Nyquist plot suggests very little resistance to electron transfer.[20]

Moreover, by taking the oxidation and reduction peak potential and using the Nicholson method[22] to assess the electron transfer kinetics of the system, it is possible to calculate the rate constant of electron transfer of the bare Au@MNPs. The calculated value for the electron transfer kinetics of the Au@MNPs system is estimated to be 0.015 s^{-1}, similar to macroelectrodes with the ferricyanide redox species.[21]

From the electrochemical characterization of the Au@MNPs coated surface, it can be concluded that the nanoparticles, due the gold coating, behave in an identical manner to bulk gold electrodes. Additionally, the resistance observed is of similar value compared to bulk gold electrode.

Having understood the behaviour of Au@MNPs to be similar to a bulk electrode, the investigation into the effect that modification has on the nanoparticles films will be discussed in the next section.

8.5 Functionalisation of Gold-coated Magnetic Nanoparticles for Electroanalytical Applications

Gold magnetic nanoparticles are promising for analytical applications because, in addition to the ability to spatially control the Au@MNPs *via* an applied magnetic field, the gold coating provides a platform for the controllable functionalization of the nanoparticles surface. The capability provided by the attachment of functional molecules, both organic and inorganic, is particularly important for the use of Au@MNPs in analytical applications because they enable selective binding with target analytes, further attachment of analytical reporter tags, such as fluorophores, and can also ensure an optimal stability of particles against aggregation, which is of importance for sensing applications.[20] This section will show that, as well as conventional gold flat surfaces, the Au@MNPs also can be functionalized in a specific manner by using thiol-containing molecules (RS-H).

In a study by Lai *et. al.*, the Au@MNPs were modified with 3-mercaptopropionic acid (MPA) and thiocholine.[20] These two molecules were chosen to simply focus on generic molecules that can be used to functionalize the surface of the particles during sensing interface construction. An additional benefit of using MPA and thiocholine is that they provide electrostatic repulsion between particles and hence reduce particle aggregation. So, the MPA and thiocholine alkanethiol functional groups chemisorb onto the surface of the Au@MNPs, *via* gold-thiolate bonds, leading to the formation of a functional SAM (self-assembled monolayer) of the thiolated molecules on the gold surface.[20]

An attempt was made to estimate the possible detection limits of the dispersible electrode concept by determining how few redox centres per nanoparticle were required to give a discernible electrochemical signal above the background current.[20] This was achieved, after modification of the Au@MNPs with a mixed SAM of MPA and thiocholine, by attaching a redox active Os complex, $[Os(bpy)_2(4\text{-AMP})Cl]PF_6$, to the carboxylic acid moieties of the SAM on the Au@MNPs *via* carbodiimide coupling. The Os complex essentially served as a model of a captured redox active analyte (Figure 8.4). The attachment of the Os complex to the Au@MNPs can be electrochemically quantified using cyclic voltammetry and the charged passed under the Faradaic peaks (Figure 8.5).[20] By varying the ratio of the two components in the SAM, such that the MPA was diluted out by the thiocholine, the number of Os complexes per particle was reduced. Using this strategy, it was shown that as little as 0.8 redox species per particle could be detected.

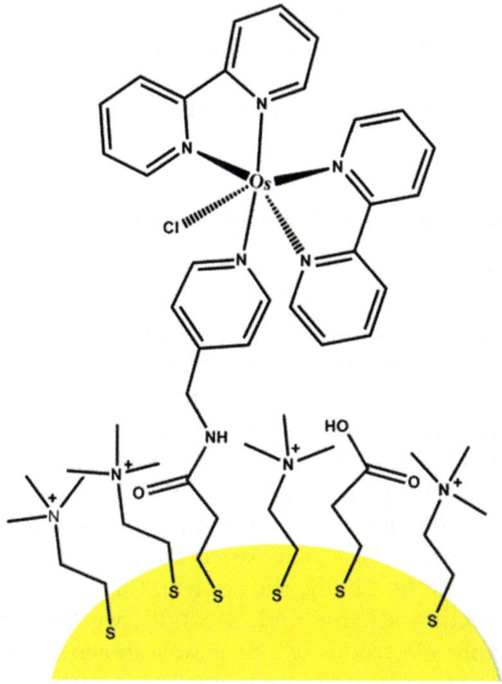

Figure 8.4 Schematic diagram showing the attachment of $[Os(bpy)_2(4\text{-AMP})Cl]PF_6$ onto the Au@MNPs.

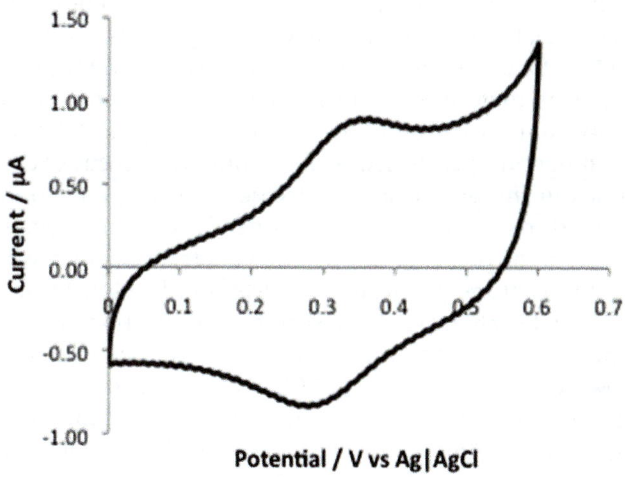

Figure 8.5 Representative cyclic voltammogram of $[Os(bpy)_2(4\text{-AMP})Cl]PF_6$ attached to an Au@MNP-on gold electrode.
(Reprinted from ref. 20, Copyright 2011, with permission from Elsevier.)

From the above, it can be seen that the Au@MNPs can be modified using a simple method by which different alkanethiol molecules form SAMs on gold surfaces and it shows that these SAMs behave as potential linkers for the attachment of functional molecules for electroanalytical applications. The Au@MNPs surface modification also improves the aggregation stability. These results open the door for the tailoring of the binding properties of the Au@MNPs *via* the use of different thiol molecules, allowing the detection of different analytes.

8.6 Detection of Electroactive Species

Thus far in this chapter we have covered the basic synthesis and electrochemical properties of the Au@MNPs as 'dispersible electrodes'. In this section we demonstrate the first use of these 'dispersible electrodes' as sensors for the detection of ultralow concentrations of analyte. It is prudent to remind the reader here that the features we are looking for with using these dispersible electrodes is much more rapid response time and increased sensitivity than a similar sensing system on a planar electrode.

To test the concept the 'dispersible electrodes' were initially configured as sensors for the detection of Cu^{2+} by employing the tripeptide ligand glycine–glycine–histidine (GlyGlyHis).[23] The key reasons for choosing this system to start with were twofold. Firstly, the analyte, Cu^{2+}, is electroactive, which simplifies the detection scheme and, secondly, we have used this system previously on planar electrodes where it was shown to have very low detection limits (0.3 nM) with good selectivity.[24] This latter point is exceedingly important, as to test the concept, a comparison with something that can already detect very low concentrations of analyte is required.

In this surface modification, MPA was first immobilized onto the Au@MNPs surface, followed by attachment of GlyGlyHis using EDC coupling including NHS. Once bound onto the Au@MNPs, GlyGlyHis molecules form complexes in the presence of Cu^{2+} *via* the deprotonation of the amide nitrogen atoms and formation of a 4 N square complex.[23] Figure 8.6 shows cyclic voltammograms performed to determine the electrochemistry of the nanoelectrodes in the presence and absence of Cu^{2+}, once brought to the surface of a gold macroelectrode using a neodymium magnet. The distinct peaks observed in the presence of Cu^{2+} demonstrate clearly the ability of the nanoparticles to capture the desired analyte and bring it back to the macroelectrode to be detected. With this surface modification, and using square wave voltammetry, a calibration curve was obtained that showed an extremely low detection limit of 2 pM (0.13 ppt). This detection limit represents a more than 100-fold decrease in the lowest detection limit relative to a planar GlyGlyHis modified gold electrode and planar gold electrode with Au@MNPs-GGH immobilized on its surface.[24] The reason the detection limit is so much lower than with the planar electrodes for the same affinity ligand is because virtually all the analytes are collected and detected.[23]

The other important observation was that the response time was much more rapid and the sensitivity much greater (Figure 8.7) for the dispersible electrode sensors relative to planar surfaces. Note in Figure 8.7, the faster response time and higher current for the dispersible sensors despite the

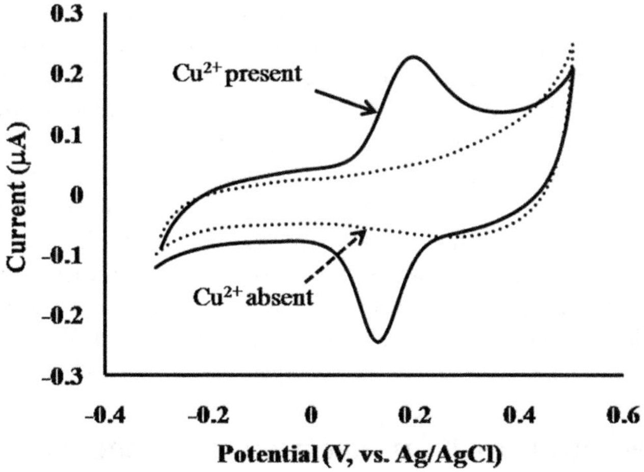

Figure 8.6 Cyclic voltammograms of nanoelectrodes performed in the absence and presence of Cu^{2+}.
(Reproduced from ref. 23.)

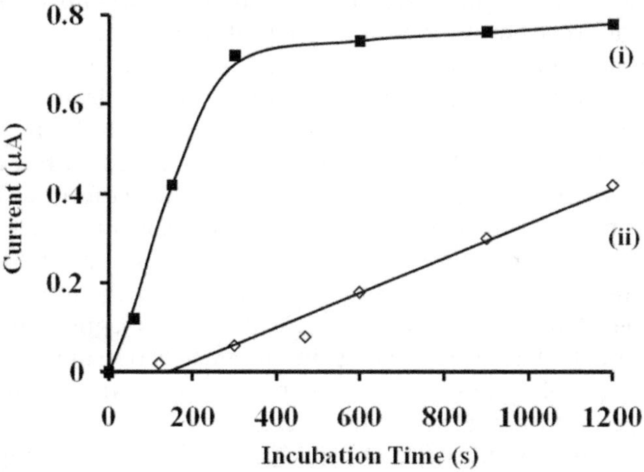

Figure 8.7 Measured current as a function of incubation time of (i) nanoelectrodes system in 0.1 nM Cu^{2+} compared to (ii) static GlyGlyHis modified gold disc electrode in 50 nM Cu^{2+}.
(Reproduced from ref. 23.)

concentration of Cu^{2+} being 500 times lower than for the planar electrodes and actually below the detection limit of the planar electrode sensor.[23]

As the Au@MNPs basically collect all the analyte molecules present in the sample solution, the volume of sample considerably affects the electrochemical response. The dependence on sample volume is a result of more Cu^{2+} being available in larger volumes.[23] That is, unlike most sensors that detect the concentration of an analyte, the dispersible electrode system detects the amount of analyte. What this means practically is the dispersible electrode system is expected to be sensitive to the loss of a few nanoparticles. However, we see the same response for a given low concentration of Cu^{2+} solution if we change the number of nanoparticles. The insensitivity of the dispersible electrodes to the number of nanoparticles is because the particles are in excess of the amount of analyte at these low concentrations.[23] Saying that, the concentration of Au@MNPs in the bulk solution used in the dispersible electrode concept should be greater than the concentration of the analyte. Taking into account that this concept was developed for detecting low concentrations of analyte, this is not a difficult criterion to fulfil.

8.7 Detection of Non-electroactive Species

The detection of copper showed that the 'dispersible electrode' concept gives faster response times, higher sensitivities and lower detection limits as predicted. However, the vast majority of analytes, such as proteins and small molecules[25] of fundamental importance in biosensing, are not electroactive. A few studies that employ Au@MNPs in the electrochemical detection of non-electroactive species have been reported. Pham and Sim presented an electrochemical impedance immunosensor for detecting the interaction between human immunoglobulin (IgG) and protein A from *Staphylococcus aureus*[26] based on the immobilization of human IgG on the surface of modified Au@MNPs. The Au@MNPs were modified with a SAM of 11-mercaptoundecanoic acid, which is as an excellent platform to immobilize human IgG onto the Au@MNPs. The immunosensor was able to detect the minimum concentration of the target analyte as low as 20 nM.[26] In this study, a magnetic carbon paste electrode was used as the working electrode and the Au@MNPs were fixed in the electrode *via* magnetic force. In this case the magnetic force was used just in the construction of this biosensor.[26] Rodríguez and colleagues recently reported a Au@MNPs based disposable magnetic genosensors for the determination of DNA hybridization processes.[27] In this architecture, an enzymatic amplification approach was utilized for the detection of a specific hybridization process, in combination with Au@MNPs. A thiolated 19-*mer* capture probe was attached to Au@MNPs and hybridized with a biotinylated target.[27] Then, a streptavidin peroxidase was attached to the biotinylated target and the resulting modified Au@MNPs, which were collected using a magnetic field on the surface of carbon screen printed electrodes (SPEs). In this study, hydroquinone was used as mediator. It showed a detection limit for the target DNA of 31 pM.[27] The studies mentioned above presented very high sensitivity and selectivity to their non-electroactive analyte targets. However, in both cases the magnetic field was used only as aid in the construction of the sensing interface, which means that the sensor is in a single position in the space. So, the sample still needs to be representative of the bulk.

We have detected protein analytes using Au@MNPs as dispersible electrodes. In this case the Au@MNPs were selective capture vehicles for prostate-specific antigen (PSA), which is used as a biomarker for prostate cancer.[28] A promising advantage of detecting PSA at very low concentrations is the ability to monitor the recurrence of prostate cancer after treatment and to assess the quality of treatment.[29,30]

The biosensor was constructed, by modifying Au@MNPs with anti-PSA antibody. For sensing, the modified Au@MNPs were dispersed in a PSA sample solution to capture the PSA present.[28] Following the addition of an enzyme-conjugated anti-PSA antibody for a different epitope (antibody binding site) on PSA to the capture antibody, a magnetic field is applied to bring the analyte to the transducing electrode for the electrochemical measurements. As with the detection of Cu^{2+}, the system for detecting PSA had higher sensitivity, lower detection limits, and faster response times than the same chemistry on a planar electrode surface. In fact the lowest detected PSA concentration was found to be 100 $fg\,mL^{-1}$, which is orders of magnitude lower than the $pg\,mL^{-1}$ detection limits for planar electrodes. Namely, the Au@MNPs-based electrochemical immunosensor resulted in a 2500-fold decrease in the detection limit of PSA in comparison to a planar gold electrode using the same surface modification.[28] In comparison to other sensors reported in the literature— for example a study presented by Sardesai *et al.*[31] achieved a detection limit of 40 $pg\,mL^{-1}$ of PSA using an electrochemiluminescent immunosensor based on carbon nanotube with $[Ru\text{-}(bpy)_3]^{2+}$–silica–secondary antibody nanoparticles—the dispersible electrodes still present substantially greater sensitivity.

8.7.1 The Biochemiresistor: An Ultrasensitive Biosensor for Small Organic Molecules

The final capability that we have demonstrated with the dispersible electrodes is the detection of small organic molecules using a new approach called the biochemiresistor.[32] In a biochemiresistor, changes in resistance across a nanoparticle film are measured. Changes in the resistance, owing to the presence of small organic molecule, are used to transduce the presence and amount of the organic molecules. The idea of the biochemiresistor is derived from the chemiresistor that uses conventional nanoparticle films without any biological recognition molecule.[33] In the biochemiresistor Au@MNPs were modified with a self-assembled monolayer (SAM) of mercaptoethanolamine followed by covalent attachment of the veterinary antibiotic enrofloxacin to the amine-terminated SAM using carbodiimide coupling.[32] The immobilized enrofloxacin was used as an epitope to which anti-enrofloxacin IgM antibodies can selectively bind. As the Au@MNPs are coated with proteins, the resistance of a resultant Au@MNPs film is expected to be high. Placing these IgM-modified Au@MNPs in a sample solution containing enrofloxacin causes competition between the surface-bound enrofloxacin and enrofloxacin in solution for the antibodies (Figure 8.8a).

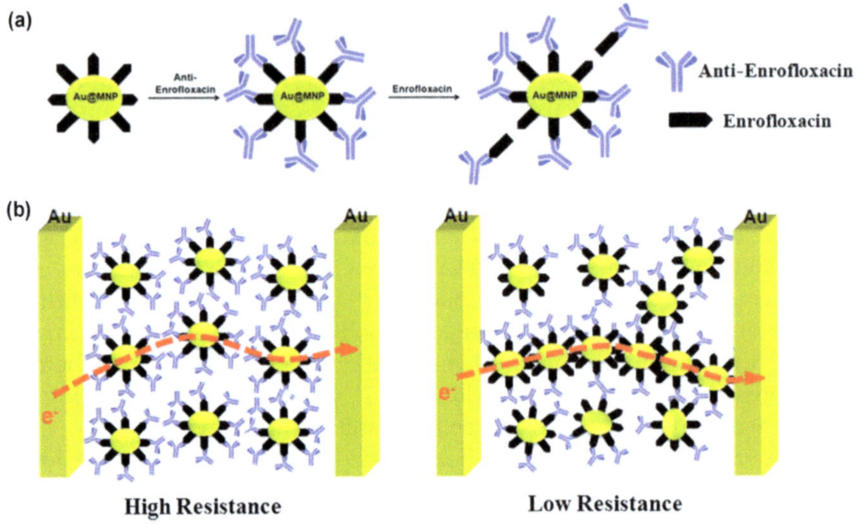

Figure 8.8 Design of the biochemiresistor: (a) Au@MNPs functionalised with enrofloxacin. The Au@MNPs bound enrofloxacin serves as selective binding sites for anti-enrofloxacin IgM antibodies Ab–Au@MNPs. When dispersed into a solution that contains enrofloxacin, some of the anti-enrofloxacin dissociates from the Ab–Au@MNPs to bind with the enrofloxacin in solution. (b) Ab–Au@MNPs magnetically assembled between two interdigitated electrodes. These films have high resistance. After exposure to the solution with enrofloxacin, particles in the nanoparticle film can approach each other more closely and the resistance of the film decreases.

If some of the IgM dissociate from the surface of the Au@MNPs, a so-called displacement assay, then the resistance of the resultant films will decrease. In other words, the modified Au@MNPs act as vehicles to capture and detect the analyte. So, the Au@MNPs are first fully coated with the target analyte (enrofloxacin) and coupled with the specific antibody (anti-enrofloxacin). Then, upon exposure to a solution that also contains the analyte (enrofloxacin), a competition for the antibodies is established between the surface bound enrofloxacin and enrofloxacin in solution. Displacement of the antibody from the surface of the modified Au@MNPs will change the contact between particles in the nanoparticle film, once formed, which can be detected as a change in resistance in the film.

Detection of the small organic molecules using the biochemiresistor is possible because the Au@MNPs modified with antibodies (Ab–Au@MNPs) allow the temporal and spatial biorecognition separation of the analyte from the measurement of the resistance; and the nanoparticle films can be magnetically assembled over interdigitated electrodes.

The biochemiresistor shows a good selectivity to anti-enrofloxacin even in the presence of different types of proteins such as BSA and anti-biotin; the proposed biochemiresistor shows significant changes in the resistance measurements just when exposed to the anti-enrofloxacin antibody. This change of resistance is due to the complexation of the specific antibodies onto the nanoparticle surface, creating a much thicker resistive layer between each particle in the film.

The example of the biochemiresistor sensor using Ab–Au@MNPs for the detection of enrofloxacin presented a detection limit of 0.28 fM. This detection limit is 106 times lower than that of a Faradaic immunosensor formed on a flat electrode using the same antibodies, surface bound epitope and assay format.[34] In addition, even at ultralow concentrations, the response time was significantly shorter than the Faradaic immunosensor.[34]

This innovative kind of sensor presented here is a general idea that can be applied to several other analytes by altering the antibodies attached onto the Au@MNPs.

8.8 Conclusions

Throughout this chapter we have presented a new detection system for monitoring ultra-trace levels of analyte in a reasonable time frame. It was achieved by using the Au@MNPs as 'dispersible electrodes'. In this strategy the particles are dispersed throughout the sample solution to capture the analyte target. After the particles' capturing process has been completed, a magnetic field is applied that brings the nanoparticles to the sensor interface for the direct electrochemical measurement. The fast response and the ultra-high sensitivity obtained can be related to the short diffusion path-lengths and also because more of the sample solution is interrogated by the Au@MNPs in comparison to conventional electrodes. In Section 8.3 we showed how we can synthesize the Au@MNPs *via* the use of the polyelectrolyte polyethyleneimine and the reduction of gold chloride with a reducing agent. The electrochemical characterization and the functionalisation of the gold coated magnetic nanoparticles showed that it presents improved or similar properties to gold flat surface. Ultra-low detection limits were achieved by employing our innovative detection system for both electroactive and non-electroactive species.

The new paradigm discussed here is really only in its infancy, and certainly various new concepts will be presented in the future. The initial results are extremely exciting with such low detection limits obtained without compromising the response time. There is still considerable work to perform to fully understand the scope of the idea as well as to provide fundamental information about what influences the electrochemical performance of the dispersible electrodes; parameters such as kinetics of electron transfer and the impact of the assembled bulk electrode, essentially being a porous electrode, need to be investigated. It is the performance of these more basic studies that are the main motivations to look for improvements in the synthesis of the Au@MNPs so that a better defined system can be investigated.

Acknowledgements

The authors thank the Australian Research Council and the University of New South Wales for financial support for different aspects of this work. SMS is a scholarship student from CNPq, Conselho Nacional de Desenvolvimento Científico e Tecnológico and INCTBio, Instituto Nacional de Ciência e Tecnologia em Bioanalítica – Brasil.

References

1. A. X. Oliveira, S. M. Silva, F. R. F. Leite, L. T. Kubota, F. S. Damos and R. D. S. Luz, *Electroanalysis*, 2013, **25**, 723–731.
2. R. A. Durst, A. J. Baumner, R. W. Murray, R. P. Buck and C. P. Andrieux, *Pure Appl. Chem.*, 1997, **69**, 1317–1323.
3. F. Patolsky, G. F. Zheng, O. Hayden, M. Lakadamyali, X. W. Zhuang and C. M. Lieber, *Proc. Natl. Acad. Sci. U. S. A.*, 2004, **101**, 14017–14022.
4. J. Sanchez-Quesada, M. R. Ghadiri, H. Bayley and O. Braha, *J. Am. Chem. Soc.*, 2000, **122**, 11757–11766.
5. Z. C. Xu, Y. L. Hou and S. H. Sun, *J. Am. Chem. Soc.*, 2007, **129**, 8698–8699.
6. C. S. Levin, C. Hofmann, T. A. Ali, A. T. Kelly, E. Morosan, P. Nordlander, K. H. Whitmire and N. J. Halas, *ACS Nano*, 2009, **3**, 1379–1388.
7. X. Yang, F. B. Xiao, H. W. Lin, F. Wu, D. Z. Chen and Z. Y. Wu, *Electrochim. Acta*, 2013, **109**, 750–755.
8. R. Rawal, S. Chawla and C. S. Pundir, *Biosens. Bioelectron.*, 2012, **31**, 144–150.
9. N. Chauhan, J. Narang and M. C. S. Pundir, *Int. J. Biol. Macromol.*, 2012, **51**, 879–886.
10. S. H. Sun, H. Zeng, D. B. Robinson, S. Raoux, P. M. Rice, S. X. Wang and G. X. Li, *J. Am. Chem. Soc.*, 2004, **126**, 273–279.
11. S. F. Hasany, N. H. Abdurahman, A. R. Sunarti and R. Jose, *Curr. Nanosci.*, 2013, **9**, 561–575.
12. W. Wu, Q. G. He and C. Z. Jiang, *Nanoscale Res. Lett.*, 2008, **3**, 397–415.
13. L. Y. Lu, L. N. Yu, X. G. Xu and Y. Jiang, *Rare Met.*, 2013, **32**, 323–331.
14. A. K. Gupta and M. Gupta, *Biomaterials*, 2005, **26**, 3995–4021.
15. K. C. de Souza, G. F. Andrade, I. Vasconcelos, I. M. de Oliveira Viana, C. Fernandes and E. M. B. de Sousa, *Mater. Sci. Eng., C*, 2014, **40**, 275–280.
16. M. Brust, M. Walker, D. Bethell, D. J. Schiffrin and R. Whyman, *J. Chem. Soc., Chem. Commun.*, 1994, 801–802.
17. I. Y. Goon, L. M. H. Lai, M. Lim, P. Munroe, J. J. Gooding and R. Amal, *Chem. Mater.*, 2009, **21**, 673–681.
18. J. L. Lyon, D. A. Fleming, M. B. Stone, P. Schiffer and M. E. Williams, *Nano Lett.*, 2004, **4**, 719–723.
19. Y. D. Jin, C. X. Jia, S. W. Huang, M. O'Donnell and X. H. Gao, *Nat. Commun.*, 2010, **1**, 41.

20. L. M. H. Lai, I. Y. Goon, M. Lim, A. B. Hibbert, R. Amal and J. J. Gooding, *J. Electroanal. Chem.*, 2011, **656**, 130–135.
21. A. J. Bard and L. R. Faulkner, *Electrochemical Methods: Fundamentals and Applications*, John Wiley & Sons, Inc., New York, 2nd edn, 2001.
22. R. S. Nicholson, *Anal. Chem.*, 1965, **37**, 1351–1355.
23. I. Y. Goon, L. M. H. Lai, M. Lim, R. Amal and J. J. Gooding, *Chem. Commun.*, 2010, **46**, 8821–8823.
24. E. Chow, E. L. S. Wong, T. Bocking, Q. T. Nguyen, D. B. Hibbert and J. J. Gooding, *Sens. Actuators, B*, 2005, **111**, 540–548.
25. L.-J. Feng, X.-H. Zhang, P. Liu, H.-Y. Xiong and S.-F. Wang, *Anal. Biochem.*, 2011, **419**, 71–75.
26. T. T. H. Pham and S. J. Sim, *J. Nanopart. Res.*, 2010, **12**, 227–235.
27. O. A. Loaiza, E. Jubete, E. Ochoteco, G. Cabanero, H. Grande and J. Rodríguez, *Biosens. Bioelectron.*, 2011, **26**, 2194–2200.
28. K. Chuah, L. M. H. Lai, I. Y. Goon, S. G. Parker, R. Amal and J. J. Gooding, *Chem. Commun.*, 2012, **48**, 3503–3505.
29. C. S. Thaxton, R. Elghanian, A. D. Thomas, S. I. Stoeva, J. S. Lee, N. D. Smith, A. J. Schaeffer, H. Klocker, W. Horninger, G. Bartsch and C. A. Mirkin, *Proc. Natl. Acad. Sci. U. S. A.*, 2009, **106**, 18437–18442.
30. J. M. Nam, C. S. Thaxton and C. A. Mirkin, *Science*, 2003, **301**, 1884–1886.
31. N. Sardesai, S. M. Pan and J. Rusling, *Chem. Commun.*, 2009, 4968–4970.
32. L. M. H. Lai, I. Y. Goon, K. Chuah, M. Lim, F. Braet, R. Amal and J. J. Gooding, *Angew. Chem., Int. Ed.*, 2012, **51**, 6456–6459.
33. H. Wohltjen and A. W. Snow, *Anal. Chem.*, 1998, **70**, 2856–2859.
34. S. M. Khor, G. Z. Liu, J. R. Peterson, S. G. Iyengar and J. J. Gooding, *Electroanalysis*, 2011, **23**, 1797–1804.

CHAPTER 9

Amperometric Ion Sensing Approaches at Liquid/Liquid Interfaces for Inorganic, Organic and Biological Ions

HYE JIN LEE,*[a] DAMIEN W. M. ARRIGAN,[b] MD. NURUL KARIM[a] AND HYERIM KIM[a]

[a] Department of Chemistry and Green-Nano Materials Research Center, Kyungpook National University, 80 Daehakro, Buk-gu, Daegu-city 702-701, Republic of Korea; [b] Nanochemistry Research Institute, Department of Chemistry, Curtin University, GPO Box U1987, Perth, Western Australia 6845, Australia
*Email: hyejinlee@knu.ac.kr

9.1 Introduction

Since the first report on a charge transfer process across a polarized interface between two immiscible electrolyte solutions (ITIES),[1] there have been tremendous efforts in both elucidating the theoretical aspects of the species transfer and utilizing these processes for a wide spectrum of applications, such as ion-selective sensors, pharmacokinetics, energy conversions and biomembrane studies. As a result, "electrochemistry at the ITIES" has matured into a distinct field over the last few decades.[2–6] This chapter focuses on the detection applications of ion transfer reactions at a polarized ITIES

RSC Detection Science Series No. 6
Electrochemical Strategies in Detection Science
Edited by Damien W. M. Arrigan

Published by the Royal Society of Chemistry, www.rsc.org

which are implemented *via* amperometric or voltammetric strategies for the detection of inorganic, organic and biological ionized species.

One of the greatest advantages of utilizing charge transfer reactions across the ITIES is that these can be used as a platform for the non-redox detection of species such as alkali and alkaline earth metal ions, as well as certain proteins and drug molecules. They can also be used to detect species whose redox activity is associated with problems in traditional electrochemical detection, such as surface fouling by reaction products or insufficient selectivity *via* the applied potential. In addition, currents associated with the ion transfer reaction can be directly correlated to the concentration of a target ion species, thus forming the basis of an excellent sensing platform. These measurements can be enhanced by the use of different voltammetric techniques already available for application at solid electrodes, which results in improvements of the analytical performance. In addition to sensitivity, the selectivity of an ITIES-based sensor can be tailored by the use of an ionophore selective to a specific target species. Accordingly, the first section of this chapter will deal with the principles of both direct transfer and ionophore-assisted transfer of species together with the commonly used voltammetric methods for ion sensing at the ITIES.

It is over the last decade or so, in particular, that there has been significant advances in developing ITIES-based sensors. This has involved creating more convenient types of interfaces for detection platforms using different combinations of microfabrication, nanotechnology, chemical and biological functionalization, and mechanical tools. In particular, recent trends can be seen in the evaluation of new materials to create an ITIES. A review of some of these advances is given in Section 9.3. In addition, there has been parallel progress on being able to use amperometric measurements to detect new target molecules such as radioactive compounds, drug molecules, pesticides, proteins, *etc.*, based on designing new ionophores and exploiting interfacial behavior. This advancement alongside some interesting examples will be discussed in Section 9.4.

9.2 Principles for Utilizing Ion Transfer Reactions across the ITIES for Sensing

9.2.1 Ion Transfer Reaction at the ITIES

An ion species can transfer across a polarized ITIES either directly or with the assistance of an ionophore present in the organic phase which can selectively complex with the ion. In addition, ions can transfer across the ITIES by coupling with an electron transfer process. To understand the transfer behavior of ions at an ITIES, numerous theoretical research studies have been performed and several excellent review papers are available.[7–9] Since the main focus here is on the sensing and detection applications of electrochemistry at the ITIES, only the basic principles of ion transfer reactions across the ITIES will be introduced.

When two immiscible electrolytes are in contact with each other, electrolyte ions will partition into either of the phases depending upon their Gibbs solvation energies, and can thus establish a Galvani potential difference at the ITIES. At thermodynamic equilibrium, when the ion species in both phases possess the same electrochemical potential, the Galvani potential difference can be written as:[8,10]

$$\Delta_{\mathrm{B}}^{\mathrm{A}}\emptyset = \frac{\Delta G^{\circ}_{\mathrm{tr},\, i\,\mathrm{A}\rightarrow\mathrm{B}}}{z_i F} + \frac{RT}{z_i F}\ln\left(\frac{a_i^{\mathrm{B}}}{a_i^{\mathrm{A}}}\right) \tag{9.1}$$

where $\Delta_{\mathrm{B}}^{\mathrm{A}}\emptyset$ is the Galvani potential difference between phases A and B, $\Delta G^{\circ}_{\mathrm{tr},\, i\,\mathrm{A}\rightarrow\mathrm{B}}$ is the Gibbs energy of transfer for species i between the phases, z_i is the ion charge, F is the Faraday constant, R is the universal gas constant, T is the temperature, and a_i^{A} and a_i^{B} are the activities of species i in phases A and B, respectively.

The term:

$$\frac{\Delta G^{\circ}_{\mathrm{tr},\, i\,\mathrm{A}\rightarrow\mathrm{B}}}{z_i F}$$

can be written as $\left(\Delta_{\mathrm{B}}^{\mathrm{A}}\emptyset_i^{\circ}\right)$, which is the standard Galvani ion-transfer potential of the ion species. Therefore, eqn (9.1) becomes:

$$\Delta_{\mathrm{B}}^{\mathrm{A}}\emptyset = \Delta_{\mathrm{B}}^{\mathrm{A}}\emptyset_i^{\circ} + \frac{RT}{z_i F}\ln\left(\frac{a_i^{\mathrm{B}}}{a_i^{\mathrm{A}}}\right) \tag{9.2}$$

Eqn (9.2) can also be written in terms of concentrations if the standard Galvani potential is replaced with the formal potential $\Delta_{\mathrm{B}}^{\mathrm{A}}\emptyset_i^{\circ\prime}$ which incorporates the activity coefficients of the ions:

$$\Delta_{\mathrm{B}}^{\mathrm{A}}\emptyset = \Delta_{\mathrm{B}}^{\mathrm{A}}\emptyset_i^{\circ\prime} + \frac{RT}{z_i F}\ln\left(\frac{c_i^{\mathrm{B}}}{c_i^{\mathrm{A}}}\right) \tag{9.3}$$

Eqn (9.3) shows that the Galvani potential difference relies on the concentration of ion species in both aqueous and organic phases and is similar to the Nernst equation developed for redox-active species.[1] This shows that the interface can be chemically polarized, by the choice of a suitable ion species and controlling its relative concentrations on each side of the interface. However, by the use of suitable electrodes and an external energy source, a potential difference can be imposed across the interface. In this case, the ions must move in response to the potential difference, according to eqn (9.3). The movement of ions produces a current, which is dependent on the applied potential. The ion transfer reaction can thus be monitored using voltammetry and the relation of voltammetric response and ion concentration follows the well-known models regularly applied at solid electrodes, such as the Randles–Sevcik equation for redox species. This means the current associated with the transfer of target ions increases as a function

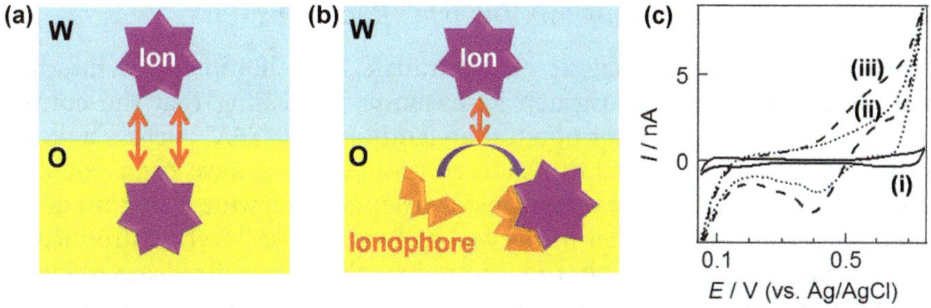

Figure 9.1 Schemes showing (a) direct and (b) assisted ion transfer processes between a water (w) and organic (o) interface. (c) Representative cyclic voltammograms for (i) direct transfer of Cd^{2+} ions in the absence of ionophore ETH 1062 and (ii) assisted transfer of Cd^{2+} ions by ETH 1062 across a single microhole water/PVC-NPOE gel interface; 10 mM BTPPATPBCl was used as organic supporting electrolyte. In (i) 10 mM $CdCl_2$ was used while the Cd^{2+} concentration was 20 μM in (ii) and 40 μM in (iii).

of the ion concentration just as in the case of a typical redox reaction at solid electrodes.[2,7,8] This is one of the most attractive points when developing ion sensors.

Figure 9.1 shows schemes of direct and assisted (or facilitated) ion transfer reactions that have been widely used for the detection of target ion species in conjunction with voltammetry. Direct ion transfer across the ITIES combined with voltammetry can be easily employed for ion sensing. This relies on the Gibbs transfer energy of the ion falling within the limits set for the available potential window by the ions of the background supporting electrolytes. Direct ion-transfer detection will be discussed further in Section 9.4.2. When hydrophilic ions, for example Cd^{2+} ions, are present in the water phase in contact with the organic phase and the interface between the phases is polarized, voltammetry can be used to monitor the transfer process of ions across the interface. In the absence of ionophores specific to the target ions, the transfer reaction of a hydrophilic ionic species such as Cd^{2+} usually limits the potential window and thus the full voltammetric cycle involving the target ion species may not be visualized [see Figure 9.1c(i)]. However, when the Cd^{2+} selective ionophore, ETH 1062,[11] is introduced into the organic phase, the Gibbs transfer energy is lowered by the complexation of Cd^{2+} with ETH 1062, and resulting in the realization of a full voltammogram for the Cd^{2+} ion transfer reaction. As the concentration of Cd^{2+} increases (Figure 9.1(c)(ii & iii), the wave at 610 mV (*versus* Ag/AgCl) proportionally increases, which can be used as the basis for Cd^{2+} detection. The major advantage of assisted ion transfer is that the selectivity for the target ion species can be easily tailored by the choice of target-specific ionophores and thus can expand the versatility of the ion transfer process for sensing applications. This will be further discussed in Section 9.2.2.

9.2.1.1 Voltammetry of Ion Transfer Reactions

Since the quantitative analysis of ion transfer at a liquid–liquid interface was first investigated by Gavach and Henry,[12] revealing that the current associated with the transfer of tetrabutylammonium (TBA^+) across a water/nitrobenzene interface was diffusion controlled, there have been extensive research reports demonstrating transfer reactions of a wide spectrum of ion species. One of the seminal works was by Samec *et al.*,[13] who introduced a four-electrode potentiostat in order to reduce the ohmic drop caused by the resistive organic solutions. They were able to obtain cyclic voltammograms (CVs) for tetramethylammonium (TMA^+) ion transfer and to determine quantitative information including the diffusion coefficient of TMA^+ in both phases.[13] Most voltammetric techniques, such as cyclic, differential pulse, differential pulse stripping, square wave and hydrodynamic voltammetries have been employed for the investigation and application of ion transfer processes at the polarized ITIES.

Voltammetric methods are usually realized by monitoring current changes associated with target ionic species diffusing to the ITIES and transferring from the water to the organic phases, or *vice versa,* when an appropriate potential corresponding to the Gibbs transfer energy of the target species is applied. Figure 9.2 compares different voltammetric techniques applied to

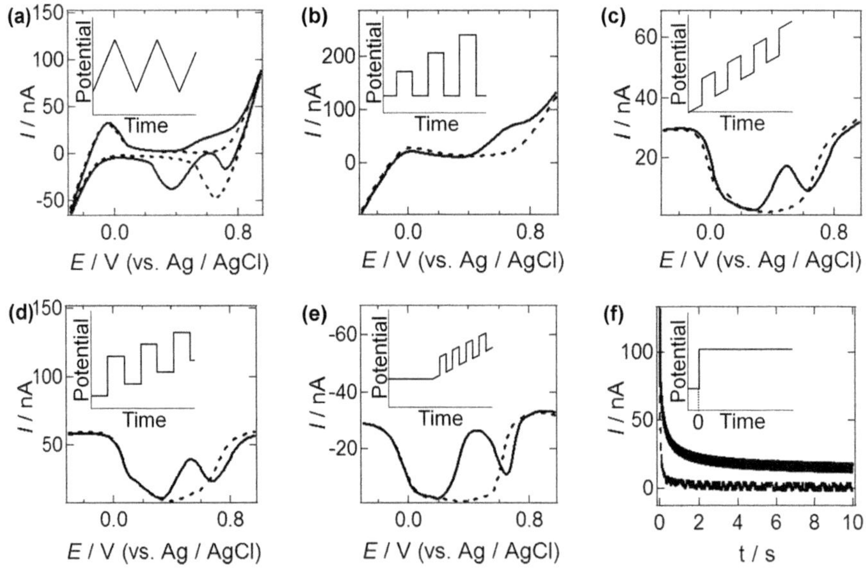

Figure 9.2 Comparison of different voltammogram analyses including (a) cyclic, (b) normal pulse, (c) differential pulse, (d) square wave, and (e) differential pulse stripping as well as (f) chronoamperometry analysis for 100 μM TMA^+ ion transfer (solid line) across a microhole array supported water/PVC-NPOE gel interface. The dashed curves represent measurements with 10 mM LiCl in the absence of TMA^+ ions. Inset in each figure shows schematic of the corresponding potential–time waveform.

the analysis of TMA^+ transfer reactions at water/poly(vinyl chloride)–nitrophenyl octyl ether (PVC-NPOE) gel interfaces patterned in a micro-interface array format. In Figure 9.2(a), the CV method is applied to this micro-interface array. Generally, the positive and negative ends of the potential window are set by the transfer of ions of the aqueous and organic supporting electrolytes in the absence of TMA^+ ions; in Figure 9.2(a), the positive end is limited by transfer of Li^+ (from aqueous to organic) or tetraphenylborate (TB^-) ions (from organic to aqueous), while the negative end is limited by the transfer of tetraoctylammonium (TOA^+) (from organic to aqueous) or Cl^- ions (from aqueous to organic) across the polarized interface. With the addition of TMA^+ ions, a steady-state voltammogram on the forward scan was observed because a hemispherical diffusion flux governs the transfer from aqueous to organic phase, just like the case of microdisc electrodes. However, a peak-shaped voltammogram was obtained on the reverse scan (Figure 9.2(a)), due to a linear diffusion flux within the organogel filled in the microhole used to form the micro-interfaces.

The use of CV allows one to rapidly establish considerable information on the thermodynamics and kinetics of ion transfer processes. Even though a CV represents a complicated, time-dependent function of different physical and chemical parameters,[14,15] measurements of half-wave transfer potentials from CV data can be used to evaluate the Gibbs transfer energy of the ion species. The reversibility of the ion transfer and the stoichiometry of metal ion–ionophore complexes can also be assessed, using, for example, a series of theoretical approaches for elucidating thermodynamic information from experimental data.[16,17]

Normal pulse amperometry (NPV) has been often used to improve the sensitivity of amperometric ion sensors while reducing the impact of constant polarization of the ITIES.[18,19] Figure 9.2(b) shows a normal pulse voltammogram for 0.1 mM TMA^+ ion transfer across the water/PVC-NPOE gel micro-interface. The interfacial potential is pulsed between a base potential (E_1), where faradaic current due to ion transfer is negligible, and a pulse potential (E_2), which increases in magnitude during the course of the experiment and encompasses the value at which the target ion transfer is observed. The faradaic current proportional to the target ion concentration is sampled at the end of each pulse, at which point the charging current has decreased to a minimum value and hence has minor impact on the total current measured. Returning the applied potential to the base potential leads to a negative faradaic current for the reverse transfer of the ion from the gelled organic phase to the water phase so that before the next pulse is applied the interfacial conditions are returned to the same initial state. The technique can provide a stable reproducible current for amperometric ion detection. For example, Lee *et al.* developed an amperometric detector for the detection of Na^+, $NH_4{}^+$ and K^+ using NPV that was integrated with ion chromatography and used it to determine concentrations of these ions in Lake Geneva.[19]

Figure 9.2(c) shows the technique of differential-pulse voltammetry (DPV), where fixed magnitude pulses are superimposed on a linear potential ramp and applied to the TMA^+ ion transfer reaction. This technique was introduced because charging current is still present, although with a minor impact, when NPV is used. Further lowering of the detection limit can be achieved by additional discrimination against this charging current. In DPV, the current is sampled twice, just before and at the end of each pulse, and the difference in these currents produces a peak-shaped voltammogram for TMA^+ ion transfer. This differential current helps to further minimize the influence of the charging current on the total current signal. The height of the peak (peak current) is also directly proportional to the concentration of the transferring analyte and the peak voltammogram is often easier to measure at lower concentrations than a wave-shape voltammogram like that obtained in NPV. Another useful pulse technique is square wave voltammetry (SWV) where a symmetric square wave, superimposed on a base staircase potential, is applied (see Figure 9.2(d)). The current is sampled twice during each square wave, at the end of the forward and reverse pulses, respectively. The difference between the forward and reverse currents is the output, plotted *versus* the applied potential. This differential current again discriminates against the charging current and enables detection of lower concentrations. At the transfer potential of the target ion, the peak current is linearly dependent on the ion concentration.

Stripping voltammetry approaches can generally be used to improve the sensitivity of the analytical method and can be applied usually with DPV or SWV, but also with linear sweep voltammetry. For example, differential pulse stripping voltammetry (DPSV) involves a two-step procedure. The first step is used to pre-concentrate the target species at the interface using a fixed potential. At the ITIES, the pre-concentration step involves selective extraction of the target ions from the aqueous phase into the organic phase by holding the applied potential at a value where ion transfer occurs rapidly. The second step involves application of the voltammetric method, DPV in this case, by which the pre-concentrated ions in the organic phase are stripped from the organic phase back to the aqueous phase (Figure 9.2(e)). The peak current resulting from the stripping process can be used as the analytical sensing signal since the DPSV response is proportional to the concentration of target ions.

Chronoamperometry involves the stepping of the applied potential at the ITIES from a value at which no faradaic (ion transfer) reaction occurs to a potential at which the interfacial concentration of the transferring ion species is effectively zero (Figure 9.2(f)). Since diffusion dominates the mass transport under such conditions, the current–time transients reflect the change in the concentration gradient at the interface due to the transfer of target ionic species. This technique can be used widely for the characterization of ion-transfer processes and also ITIES of different geometries.[20–23]

All the above voltammetric techniques have advanced the analytical performance of ion sensors based on ITIES for inorganic, organic and biological ions. In particular, the key parameters in developing sensors are the detection limit, the sensitivity for target analytes and the selectivity over other species. Pulsed techniques such as NPV, DPV, DPSV and SWV have enabled great improvement in the detection sensitivity of the sensors. These techniques have also proved to be more suitable for quantitative analysis of analytes, providing well-defined peaks compared to that of CV. For example, Pereira *et al.* used DPV to determine the transfer characteristics of the anticancer drug daunorubicin at the ITIES, which was not possible with CV.[24] It has been found that SWV offers faster analysis, better response resolution and lower detection limit; for example, SWV was used to detect lower concentrations of NH_4^+ ion.[25] Because DPSV consists of a pre-concentration step and a fast voltammetric detection step, it further improves the detection limit of an analyte compared to that of DPV. For example, Collins *et al.* employed DPSV to show a *ca.* 80-fold enhancement in the detection limit of propranolol, relative to DPV.[26]

9.2.2 Assisted Ion Transfer

One of the key aspects when designing ion sensors is the selectivity that can be achieved towards the specific target ions. Incorporation of hydrophobic ionophores that selectively complex with a particular ion species in an organic phase, which has been highly developed in the ion selective electrode (ISE) research area, is a commonly used approach. The great advances (*e.g.*, discovery of new ionophores) made in the development of potentiometric ISEs have also enhanced the investigations of amperometric ion sensors utilizing electrochemistry at the ITIES. The main difference in the ITIES compared to an ISE is use of a non-equilibrium method *versus* an equilibrium method, because the current changes due to ion transfer reactions facilitated by ionophores are measured. Table 9.1 summarizes a wide range of ionophores that have been used for the selective detection of target ion species using the ITIES methodologies; some of these were also used in ISEs. The two areas of ISEs and electrochemistry at the ITIES have, however, come increasingly close in recent years, with an overlap in methods and materials used to design new sensors.[27]

When ionophores are present in the organic phase (*e.g.*, see Figure 9.1(b)), hydrophilic cations or anions that can selectively complex with the ionophore can transfer from water to organic phase, and *vice versa,* because the ionophore lowers the solvation energy of the complexed ion in the organic phase, thus reducing the Gibbs energy of transfer of the target ion. Different mechanisms have been proposed for these interfacial transfer processes, including aqueous complexation followed by transfer (ACT), transfer to the organic phase followed by complexation (TOC), transfer by interfacial complexation (TIC) and transfer by interfacial dissociation (TID).[28]

Facilitated ion transfer was first realized by Koryta,[29] who studied the transfer of potassium and sodium ions from the aqueous phase to the organic phase by the formation of complexes in the organic phase with dibenzo-18-crown-6 (DB18C6) and valinomycin, respectively. This pioneering work opened a new route for developing ion selective sensors utilizing

Table 9.1 Summary of different target ion and ionophore combinations used to create ion selective sensors based on facilitated ion transfer across the ITIES. Examples of systems developed for ISE sensing are also included for comparison.

Ionophores	Target ions	ITIES or ISE	Ref.
Aliquat 336	Cr(VI)	ISE	127
		ITIES	84
Benzo-15-crown-5	K^+	ISE	30
	Dopamine	ITIES	99
Benzo-12-crown-4	Na^+	ITIES	128
Bis(dibenzoylmethanato)Ni(II)	ClO_4^-	ISE	129
		ITIES	130
Dibenzo-18-crown-6	K^+	ISE	30, 131, 132
		ITIES	133, 134, 31, 135, 136, 19, 128, 137, 138, 139
	Rb^+, Cs^+	ITIES	71
	Aspartame, acesulfame K	ITIES	100
	Noradrenaline	ITIES	96
	Cu^{2+}	ITIES	140
	Catecholamine	ITIES	141
	Dopamine	ITIES	97
	Decylammonium	ITIES	142
		ISE	143
	Na^+	ITIES	19, 144–146
	NH_4^+	ISE	147–149
		ITIES	19
	Ca^{2+}, Mg^{2+}	ITIES	136
Dibenzo-24-crown-8	Dopamine	ITIES	99
Dimethyldibutyltetradecyl-1,3-malonamide	Am^{3+}	ITIES	150
ETH 1062	Cd^{2+}	ISE	151
		ITIES	11, 152
ETH 1810	Li^+	ISE	153
	H^+, Li^+	ITIES	154
ETH 1778	H^+	ISE	155
		ITIES	46

Hexakis(2,3,6-tri-*O*-acetyl)-α-cyclodextrin	NH_4^+	ITIES	25
Herbicide Prometrine	H^+, Al^{3+}	ITIES	156
Heptakis(2,3,6-tri-*O*-acetyl)-α-cyclodextrin	(+), (−) ephedrinium	ITIES	41
N-Methyl-4-(4-phenoxyphenyl)thiazol-2-amine	H^+	ITIES	157
N-{2-[bis[2-(4-*tert*-butylbenzoyl)aminoethyl]-amino]ethyl}4-*tert*-butylbenzamide	NO_3^-	ITIES	158
Tris-[2-(4-*tert*-butylbenzoyl)aminoethyl]amine		ISE	159
Napthol-15-crown-5	Ca^{2+}, Sr^{2+}, Ba^{2+}	ITIES	160
Naphthalene-2-boronic acid	Glycolic, lactic, mandelic and gluconic acid	ITIES	161
2,8-Dithia [9],(2,9)-1,10-phenanthrolinophane	Cu^{2+}	ITIES	162
5-Oxo-2,8-dithia [9],-(2,9)-1,10-phenanthrolinophane			
Picolinamide–phenylene-vinylene	Cu^{2+}	ITIES	35
Phospholipid interfacial films	H^+	ITIES	163
Procaine	Dy^{3+}	ISE	164
Propylene carbonate	Li^+	ITIES	44
Phosphatidylcholine	Acetylcholine	ITIES	78
Tetraethyl-*p*-*tert*-butyl calix[4]arene tetraacetate	Na^+, K^+	ITIES	165
O,*O*″-bis[2-(Methylthio)ethyl]-*tert*-butylcalix[4]arene	Ag^+	ISE	166
		ITIES	88, 92
Tributyl phosphate, octyl(phenyl)-*N*,*N*-diisobutylcarbamoylmethyl-phosphine oxide	UO_2^{2+}	ITIES	87
tert-Butyl calix[4]arene tetramethylester	Na^+	ISE	153, 167
Urea-functionalized calix[4]arene compound	$H_2PO_4^-$, HPO_4^{2-}	ITIES	168
	HPO_4^{2-}	ISE	169
Valinomycin	K^+	ISE	170–174
	K^+, Na^+, H^+, NH_4^+	ITIES	19
	K^+, Na^+	ISE	175
		ITIES	135, 176

Table 9.1 (*Continued*)

Ionophores	Target ions	ITIES or ISE	Ref.
5,11,17,23-Tetra-tertbutyl-25,27-bis(2′amino-methylpyridine)-26,28-dihydroxy calix[4]arene	K^+	ITIES	40
2-Acetylpyridine-4-phenyl-3-thiosemicarbazone	H^+	ITIES	177
3,3′,6,6′-Tetrakis[*p*-(*n*-Dodecyloxy) benzoyloxymethyl]dibenzo-18-crown-6	K^+		
3,6′-bis[3″,4″,5″-tris[*p*-(*n*-Dodecyloxy)benzyloxy]-benzoyloxymethyl]benzo-15-crown-5	Na^+	ITIES	178
6,7-Dimethyl-2,3-di(2-pyridyl)quinoxaline	Cu^{2+}	ITIES	179
25,27-Di[(2-ethoxy)benzaldehyde]-26,28-dihydroxy-5,11,17,23-tetra(*tert*-butyl) calix[4]arene	Na^+	ITIES	180
5,11,17,23-Tetra-*tert*-butyl-25,27-diethoxycarbonylmethoxy-26,28-dimethoxy calix[4]arene	Li^+, Na^+, K^+, Rb^+, Ca^{2+}	ITIES	181
2,5,8,11,14-Pentaaza[15]-16,29-phenanthrolinophane	Cr(VI)	ITIES	33
2,8-Dithia[9],(2,9)-1,10-phenanthrolinophane	Cu^{2+}, Cu^+, Cd^{2+}, Co^{2+}, Ni^{2+}, Pb^{2+}, Zn^{2+}	ITIES	182
5,6-Diphenyl-3-(2-pyridyl)-1,2,4-triazine	Pb^{2+}, Cd^{2+}, Zn^{2+}, Hg^{2+}, Cu^{2+}, Co^{2+}, Fe^{2+}, Cr^{3+}	ITIES	183
β-Octafluoro-meso-octamethylcalix[4]pyrrole	Cl^-, Br^-	ITIES	32
3-(2-Pyridyl)-5,6-diphenyl-1,2,4-triazine	Glucose	ITIES	184
9-Ethyl-3-carbazolecarboxaldehyde-thiosemicarbazone	Pb^{2+}, Cd^{2+}	ITIES	185
1,4,7,10,13,16-Hexathiacyclooctadecane	Hg^{2+}, Pb^{2+}	ITIES	186
4-Morpholinoacetophenone-4-ethyl-3-thiosemicarbazone	Cd^{2+}	ITIES	187

the ITIES as a detection platform and which brought about a tremendous number of applications to environmental and biological sciences and industrial technologies. Table 9.1 provides a summary of ion selective sensing platforms based on such ionophore-facilitated ion transfer processes. For example, Osakai and co-workers reported K^+ ion-selective sensors based on K^+ ion transfer facilitated by DB18C6 at a water/PVC membrane micro-interface.[30] This ionophore has also been used by Shao and colleagues to study facilitated ion transfer (FIT) of K^+ at the nanoscopic interface between water and 1,2-dichloroethane (nano-W/DCE) supported at nanopipettes.[31]

An additional example is the development of sensor strips utilizing a single micro-interface composed of water and a PVC-NPOE organogel for the selective detection of Cd^{2+} ion facilitated by the ionophore ETH 1062, which was present in the gel matrix.[11] In this case, one electrode is integrated in the disposable strip, while the second electrode is off-strip and immersed in the drop of sample placed onto the strip. The use of differential pulse stripping voltammetry improved the sensitivity for Cd^{2+} ions by pre-concentrating Cd^{2+} ions in the organogel phase by holding the applied potential for 30 s at a value where the Cd^{2+} ion transfer occurred, followed by rapid stripping of the pre-concentrated Cd^{2+} ions from the organogel back to the aqueous phase.

Although many of the ionophores listed in the Table 9.1 were originally developed for potentiometric ISEs, there are also many reports on the development of ITIES-based sensors with newly discovered ionophores. These are also summarized in Table 9.1 together with their target ions. The first noticeable point from Table 9.1 is that most of the amperometric sensors developed with facilitated ion transfer reactions at the ITIES have been focused on cations (*i.e.*, alkali, alkaline earth and heavy metal ions, protons), which is probably due to the rich libraries of ionophores available for these species. But there are still great demands to discover new ionophores for anionic species. As a result, Shao's group investigated the thermodynamics and kinetics of the transfer behaviors of Cl^-, Br^-, NO_2^- and $CH_3CO_2^-$ (Ac^-) facilitated by β-octafluoro-*meso*-octamethylcalix[4]pyrrole at the micropipette supported water/DCE interface.[32] Further examples are the detection of hexavalent chromium facilitated by the polyamine macrocycle 2,5,8,11,14-pentaaza[15]-16,29-phenanthrolinophane[33] and transfer of halides facilitated by a cholapod, steroid-based receptor, illustrating a selectivity for chloride.[34]

The development of new ionophores for a particular target can be supported by the application of electrochemistry at the ITIES to investigate kinetic and thermodynamic properties, such as diffusion coefficients, partition coefficients of ionophores, as well as association constants and stoichiometries of complexes formed. One example is illustrated in Figure 9.3(a) where a lipophilic cyclodextrin (hexakis(2,3,6-tri-*O*-acetyl)-α-cyclodextrin (HexαCD)) was used for the amperometric detection of ammonium ions at the water/1,6-dichlorohexane (DCH) micro-interface.[25] HexαCD

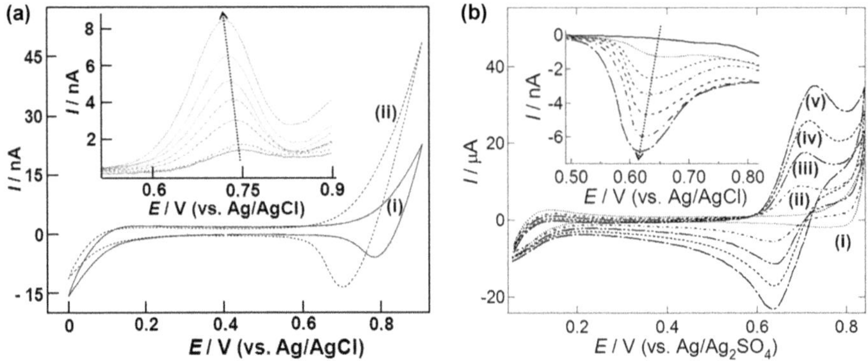

Figure 9.3 (a) Cyclic voltammograms of ammonium ion transfer facilitated by 10 mM hexakis(2,3,6-tri-*O*-acetyl)-α-cyclodextrin (HexαCD) at a microhole water/DCH interface in the (i) absence and (ii) presence of 0.64 mM ammonium ions. Inset in (a): background subtracted square wave voltammograms for assisted ammonium ion transfer by 10 mM HexαCD. Ammonium concentrations were varied from 4.2, 8.4, 17, 33, 66, 106, 183 to 327 μM. (b) Cyclic voltammograms for the assisted transfer of Cu(II) ions using picolinamide-PV across a water/1,2-DCE interface; 0.25 mM picolinamide-PV was present in the 1,2-DCE phase and 10 mM Li_2SO_4 was used in the aqueous phase. The Cu^{2+} ion concentrations were (i) 0, (ii) 15, (iii) 30, (iv) 45 to (v) 70 μM. Scan rate = 50 mV s^{-1}. Inset in (b): differential pulse stripping voltammograms for the detection of Cu^{2+} cations using a microhole water–PVC-DCH gel interface. Solid line is for 10 mM LiCl in the absence of Cu^{2+} cations. A deposition potential of 800 mV (*versus* Ag/AgCl) for 20 s was applied before stripping and other pulse conditions are as follows: potential increment = 10 mV, pulse duration = 50 ms and pulse potential = 50 mV.
(Both (a) and (b) are reproduced with permission from refs 25 and 35, respectively. Elsevier copyright.)

formed stable complexes with ammonium ions, with an association constant (log β) of 7.9. Figure 9.3(a) shows CVs for ammonium ion transfer assisted by HexαCD, where the increase in current at the positive potential limit observed when ammonium ions were present in the aqueous phase was attributed to their assisted transfer from water to DCH. This was more clearly seen when background-subtracted SWV was used, where the peak due to the assisted transfer of ammonium ions appeared at *ca.* 0.737 V. Using SWV, the ammonium ion concentration was detected in the range 4.2–327 μM (Figure 9.3(a) inset), with a limit of detection (LOD) of 0.12 μM.[25]

Another example is shown in Figure 9.3(b) where a Cu^{2+} ion selective sensor was developed based on the assisted transfer reaction of Cu^{2+} by picolinamide-(phenylene vinylene) (picolinamide-PV) at a liquid/organogel micro-interface composed of water and poly(vinyl chloride)–1,6-dichlorohexane (PVC-DCH).[35] Since this ionophore was not previously used with either an ISE or at the ITIES for Cu^{2+} ion sensing, voltammetric characterization of the picolinamide-PV-facilitated Cu^{2+} ion transfer process at a macroscale – ITIES was undertaken to determine the complexation ratio of

Cu^{2+} ion with the ligand, which was 1 : 4 (Figure 9.3(b)). Following characterization of the ion–ionophore complex, this was then implemented as an ion selective sensing platform. In this case, a single microhole-supported interface was used, but replacing the organic liquid electrolyte with the PVC-DCH organogel, to create the Cu^{2+} ion selective sensor. Combined with DPSV to improve the detection capability (see Figure 9.3(b) inset), a 20 s pre-concentration step at 800 mV (*versus* Ag/AgCl), which is above the transfer potential of Cu^{2+} ion from the water to the viscous organic phase, followed by the stripping of the ions from the organogel back to the water, achieved a linear response to Cu(II) concentration in the range 1–50 μM.[35] Readers are also referred to Table 9.1, which summarizes the extensive research on developing a wide range of ionophore-based sensing platforms for ion species. Assisted ion transfer reactions have also been proposed as the basis of protein detection at the ITIES. As will be discussed in Section 9.4.3, facilitated transfer of organic phase anions by cationic proteins enables protein detection.

9.2.3 Geometric Platforms for the ITIES

In the same way that microelectrode and nanoelectrode behavior differs from that of larger (millimeter and centimeter sized) electrodes, the size of the interface between aqueous and organic phases can greatly affect the voltammetric behavior and analytical performance of ion transfer reactions. As a result, various size classifications of the ITIES have emerged. Generally, three size scales are considered: (i) large scale, meaning millimeter or centimeter scale, (ii) microscale and (iii) nanoscale.[5]

9.2.3.1 Large Scale

For a large scale ITIES, the interface can be considered to be planar and ions are subjected to semi-infinite diffusion processes. This is probably the most widely used ITIES platform for understanding the thermodynamic and kinetic properties of ion transfer reactions and extensive reviews on this topic are available.[6,8,36] One important point to mention here is that for a diffusion-controlled reversible ion transfer process at a polarized large-scale ITIES, the Randles–Sevcik equation,[18] originally developed for reactions at solid electrodes, can be used to correlate the relationship between the current and target ion characteristics (such as concentration, charge, diffusion coefficient). From an analytical sensing application point of view, even though a linear relationship exists between the current response and the ion concentration, the mechanical instability of two liquids together with the inconvenience of system assembly, *e.g.* requiring a four-electrode configuration and careful handling of two liquids, could limit the development of practical sensing devices. Hence, these large interface systems are predominantly used for characterization of ion-transfer processes, often in preliminary experiments prior to sensor development, but do not form the physical basis of the sensor themselves.

9.2.3.2 Micro-scale

Since the introduction of microscale ITIES, featuring a glass pipette tip (of radius 25 μm) as an interface-supporting orifice,[37] for the study of ion transfer reactions, there have been numerous efforts made on the development of different types of micro-ITIES.[5,9,38,39] The use of microscale interfaces offers distinct advantages over larger-scale interfaces, including reduced ohmic drop and charging current. This allows the use of a two-electrode cell. The enhanced mass transport resulting from radial diffusion provides an improvement in sensitivity. When combined with the gelification of one or both phases, the micro-ITIES has greatly promoted the facile fabrication of sensing device. Generally, the micro-ITIES can be classified into (i) micropipette supported ITIES, (ii) microhole-supported ITIES in single or array format, based on polymer composite or silicon membranes, and (iii) micro-cavity-supported ITIES formed in polymer substrates. Table 9.2 lists some of the recent developments in creating ion selective sensors with different types of micro-ITIES organized according to target ions and the limit of detection. The characteristics of ion transfer processes depend sensitively on the geometry of the interface. The typical geometry of each microscale-ITIES with the corresponding diffusion regimes and CVs for a simple ion-transfer process (such as for tetraethylammonium (TEA^+) ion transfer) are shown in Figure 9.4. Micropipette type ITIES have also been extended to nano and dual pipette supported ITIES, where the water phase is usually located inside the pipette due to the hydrophilic nature of glass pipettes.

Micropipettes (Figure 9.4(a)(i)) provide asymmetric diffusion fields; hemispherical diffusion dominates the ingress ion transfer process, like to a microdisk interface, while linear diffusion governs the egress process of ions out of the pipette (Figure 9.4(a)(ii)), although this depends on the angle of the pipette walls. This produces asymmetric voltammograms. As can be seen in Figure 9.4(a)(iii), for the transfer of TEA^+ ions at the water/DCE micro-interface supported at the tip of a glass micropipette, the voltammogram shows that linear diffusion leads to a peak formation (egress) while hemispherical diffusion leads to a steady-state current (ingress). The steady state current (I_{SS}) due to ion ingress increases linearly with the concentration of target ions, according to the expression:[9]

$$I_{SS} = Kz_iF\,D_i^W\,C_i^W r \tag{9.4}$$

where K is a geometric constant dependent on the pipette tip size, z_i is the ion charge, F is the Faraday constant, r is the radius, D_i^W is the diffusion coefficient, and C_i^W is the concentration of ions i in the water phase.

Micropipettes for formation of micro-ITIES can be prepared by various methods,[5] such as a vertical glass puller, *e.g.*, a Kopf 720 puller (Tujunga, CA), or a laser glass puller (*e.g.*, P-2000) with borosilicate glass tubing. The inner wall of the pipette can also be silanized so that the organic phase can be placed inside the pipette for the convenient measurements of target ionic

Table 9.2 Comparison of various ITIES ion detection platforms for the detection of a wide range of inorganic and organic molecules.

	Target	Platform	Direct (D)/ assisted (A) ion transfer	Method	LOD	Ref.
Organic	Acetylcholine	Micropipette	D	CV, stripping voltammetry, chronoamperometry, NPV	100 μM	188
		Functionalized ITIES	A	CV, SWV, impedance, Faradaic resistance	—	78
	Aspartame, acesulfame K	Large-ITIES	A	CV, DPV	30 μM	100
	β-Blocker propranolol	Microhole array (silicon)	D	CV, DPV, DPSV	0.02 μM	189
		Microhole array (silicon)	D	CV, DPSV	0.05 μM	26
		Functionalized ITIES (large-ITIES)	D	CV, hydrodynamic voltammetry	—	190
	Catecholamines	Microporous (PET film)	A	CV, SWV	0.35 μM	141
	Choline	Microhole on polymer substrate	D	CV, stripping voltammetry	3 μM	191
	Clomipramine, imipramine, tacrine, verapamil	Polymer Membrane (Large-ITIES)	D	DPV	—	102
	$CH_3(CH_2)_9NH_3^+$	Large-ITIES	A	CV, SWV	—	142
	Dopamine	Large-ITIES	A	CV, DPV, SWV	2 μM	97
		Micropipette	A	CV, DPV	~0.05 μM	99
		Microhole on polymer substrate	A	CV, DPV	—	96
		Microhole array (silicon)	A	CV, DPV, SWV	0.5 μM	98
	(+), (−) Ephedrinium	Micropipette	A	CV, DPV	—	41
	Glycolic, gluconic, lactic, mandelic acid	Micro droplet	A	CV	—	161
	Imipramine	Large-ITIES	D	PA (pulse amperometry)	1 μM	103
	Paraoxon	Microhole on polymer substrate	A	CV, DPSV	0.5 μM	107
	Parathion	Microhole on polymer substrate	A	CV, DPSV	0.5 μM	108
	Polylysine dendrimers	Large-ITIES	D	CV	~1 μM	192
	Pyridine	Micropipette	D	CV	—	42
	Vitamin B1	Micropipette	D	CV, chronoamperometry, stripping voltammetry	4.6 μM	193

Table 9.2 (*Continued*)

	Target	Platform	Direct (D)/ assisted (A) ion transfer	Method	LOD	Ref.
	Vitamin C	Microhole on polymer substrate	A	CV, DPSV	1 μM	194
	Warfarin	Functionalized ITIES (Large-ITIES)	D	CV, hydrodynamic voltammetry	—	190
	Tetrabutylammonium	Micropipette (IL)	D	CV	—	65
	Tetraethylammonium	Micropipette	D	CV	—	195
		Nanopipette	D	CV, DPV	—	52, 144
		Nanopipette	D	CV	—	31
		Silica films-micro interface	D	CV	—	81
		Silicon nanoporous array	D	CV	—	55, 57
		Lipid adsorbed ITIES	D	CV	—	79
	Tetramethylammonium	Nanopipette	D	CV, DPV	—	144, 196
	Tetrapropylammonium	Silicon nanoporous array	D	CV	—	23
Inorganic	H^+	Large-ITIES	A	CV	—	157
		Micropipette	A	CV	—	177
		Microhole on polymer substrate	A	CV, DPSV	—	46
		Phospholipid interfacial films	A	CV	—	163
	H^+, Li^+	Large-ITIES	A	CV	—	154
	H^+, Al^{3+}	Large-ITIES	A	CV	—	156
	Li^+	Micropipette	A	CV	—	44
	Li^+, Na^+, K^+	Large-ITIES	A	Polarography	—	128
	Li^+, Na^+, Cs^+	Nanopipette	D	CV	—	196
	Li^+, Na^+, K^+, Rb^+, Cs^+	Micropipette	D	CV, DPV	—	43, 197
		Micropipette	A	CV	—	71
		Micropipette	A	CV, DPV		181
	Na^+	Micropipette	A	CV, DPV	25 μM	180
		Micropipette	A	CV, potential step chronoamperometry, A.C. polarography	—	146

	Microhole on polymer-PET, alumina	A	CV	—	145
	Nanopipette	A	CV, DPV	—	144
Na^+, K^+	Large-ITIES	A	CV	—	165
	Micropipette	A	CV, chronoamperometry	—	178
	Microhole on polymer substrate	A	CV	—	135
Na^+, K^+, NH_4^+	Microhole on polymer substrate	A	CV, pulse amperometry	~50 ppb	19
Na^+, K^+, Cl^-, ClO_4^-, SCN^-, NO_3^-, $(CH_3)N^+$, $(C_4H_9)_4N^+$	Thin film-modified electrodes	A (redox probe)	SWV	—	198
K^+	Micropipette	A (SECM)	Voltammetry	0.1 mM	133
	Micropipette	A (SECM)	CV	0.05 mM	176
	Micropipette	A	CV	—	134, 137, 138
	Micropipette	A	CV, DPV	50 μM	40
	Micropipette	A	Impedance	—	139
	Ionic liquid membrane	A	CV	0.05 mM	69
K^+, ClO_4^-	Nanopipette	D/A	CV	—	31
NH_4^+	Large-ITIES	D	CV	—	199
	Microhole on polymer substrate	A	CV, DPV	0.12 μM	25
Ag^+	Large-ITIES	A	CV	0.02 mM	90
	Large-ITIES	A	CV, DPV	0.2 μM	88
	Micro pipette	A (SECM)	CV	0.2 μM	92
Ca^{2+}, Sr^{2+}, Ba^{2+}	Large-ITIES	A (ESI-MS)	CV, ESI-MS	—	160
Cu^{2+}	Large-ITIES	A	CV	—	179
	Large-ITIES	A	CV, electrochemical impedance spectroscopy	—	162
	Microhole on polymer substrate	A	CV, DPSV	1 μM	35
	Functionalized ITIES with size selective membrane	A	CV	—	140

Table 9.2 *(Continued)*

Target	Platform	Direct (D)/ assisted (A) ion transfer	Method	LOD	Ref.
Cu^{1+}, Cu^{2+}, Cd^{2+}, Co^{2+}, Ni^{2+}, Pb^{2+}, Zn^{2+}	Large-ITIES	A	CV	—	182
Cd^{2+}	Large-ITIES	A	CV	—	187
	Microhole on polymer substrate	A	CV, DPSV	~20 ppb	11
Cd^{2+}, Pb^{2+}	Large-ITIES	A	CV	—	185
Cd^{2+}, Pb^{2+}, Zn^{+}	Large-ITIES	A	CV, NPV	—	183
Cd^{2+}, Pb^{2+}, Tl^{+}, Zn^{+}	Microhole on polymer substrate	A	CV, SWV	—	152
Hg^{2+}, Pb^{2+}	Large-ITIES	A	CV, stripping voltammetry	7.9 nM	186
UO_2^{2+}	Micropipette	A	CV	—	87
Cr^{6+}	Large-ITIES	A	CV, SWV	0.25 ppm	33
	Microhole on polymer substrate	A	CV, DPV	0.5 µM	84
H_2O_2	Droplet interface	A (SECM)	CV, linear sweep voltammetry	—	200
F^-, Cl^-, Br^-, NO_3^-, SO_4^{2-}	Microhole on polymer substrate	D	DPV	~1 ppt	201
Cl^-, Br^-, NO_2^-, $CH_3CO_2^-$	Micro pipette	A	CV	—	32
NO_3^-	Microhole on polymer substrate	A	CV, DPV	0.1 mM	158
BF_4^-, ClO_4^-	Micropipette (IL)	D	CV	—	202
SCN^-, ClO_4^-	Micropipette (IL)	D	CV	—	65
ClO_4^-	Microhole on polymer substrate	A	CV, DPSV	10 ppb	130
PO_4^{3-}	Large-ITIES	A	CV	—	168

species in the aqueous phase.[5] For example, a micropipette-supported water/DCE interface was developed to study a novel ionophore, the calix[4]arene derivative 5,11,17,23-tetra-tertbutyl-25,27-bis(2-aminomethylpyridine)-26,28-dihydroxy calix[4]arene (APHC4) for the facilitated transfer across of alkali and alkaline-earth metal ions. The voltammetric investigation revealed the excellent selectivity of APHC4 toward K^+ ions amongst alkali and alkaline earth metal ions.[40] This can be easily applied as a K^+ selective ion sensing platform. Extensive works on utilizing micropipette-based ITIES for ion transfer can be found in the literature.[5,41–44]

Another way to prepare a micro-ITIES is to use a microhole or an array of microholes fabricated in a thin membrane (polymer or silicon), as pioneered by Girault for polymer membranes.[45] This strategy was initially demonstrated by drilling a single hole (20 μm diameter) in 12 μm thick poly(ethylene terephthalate) (PET) films using UV excimer laser photo-ablation.[45] A single microhole-supported ITIES was also created by punching a single hole in a 12 μm thick PVC film with a sharp needle.[46] An elliptic single microhole was formed in this way, depending upon the mechanical punching force used. These perforated membranes can then be placed between the aqueous and organic phases, to produce a single micro-ITIES (Figure 9.4.(b)(i)). For such microhole-patterned micro-ITIES, the ion ingress process is governed by hemispherical diffusion (Figure 9.4.(b)(ii)) and results in steady-state voltammetric currents (Figure 9.4(b)(iii)). If the organic phase filling the microhole is gelled with PVC, the egress ion transfer can be limited by mixed linear and hemispherical diffusion regimes. Thus the voltammogram can also be asymmetric and composed of steady state currents, for the water to organic phase ion transfer, and peak currents for the organogel to water ion transfer (see Figure 9.4(b)(iii) for the transfer of TMA^+ ions across a microhole-supported interface between the aqueous and PVC-NPOE gel).

One distinctive feature of using laser photo-ablation to produce microholes in a polymer membrane is the production of microholes with a conical shape. This results in the diameter of the holes on the side where the laser directly impacts the membrane being larger than the holes on the laser exit side. Figure 9.4(c)(i) shows the geometry of a micro-array ITIES on a PET film (12 μm thick) fabricated using a UV excimer laser.[47] If the film is thin enough and the property of the organic phase chosen is similar to the aqueous phase, combined with a sufficient separation distance between the microholes so that each microhole has an individual diffusion regime, the diffusion flux of ion transfer from water to the organic phase, and *vice versa,* will be dominated by a hemispherical flux profile. This will result in steady-state voltammograms for both directions of the ion transfer process. In this situation, also, the total current is the sum of the currents from each micro-interface, because each micro-interface experiences individual diffusion regimes that are independent of those of its neighbors in the array.

Figure 9.4(c) shows an example of an array of microhole-supported ITIES for the study of TEA^+ ion transfer. Due to the use of a 12 μm thick polymer

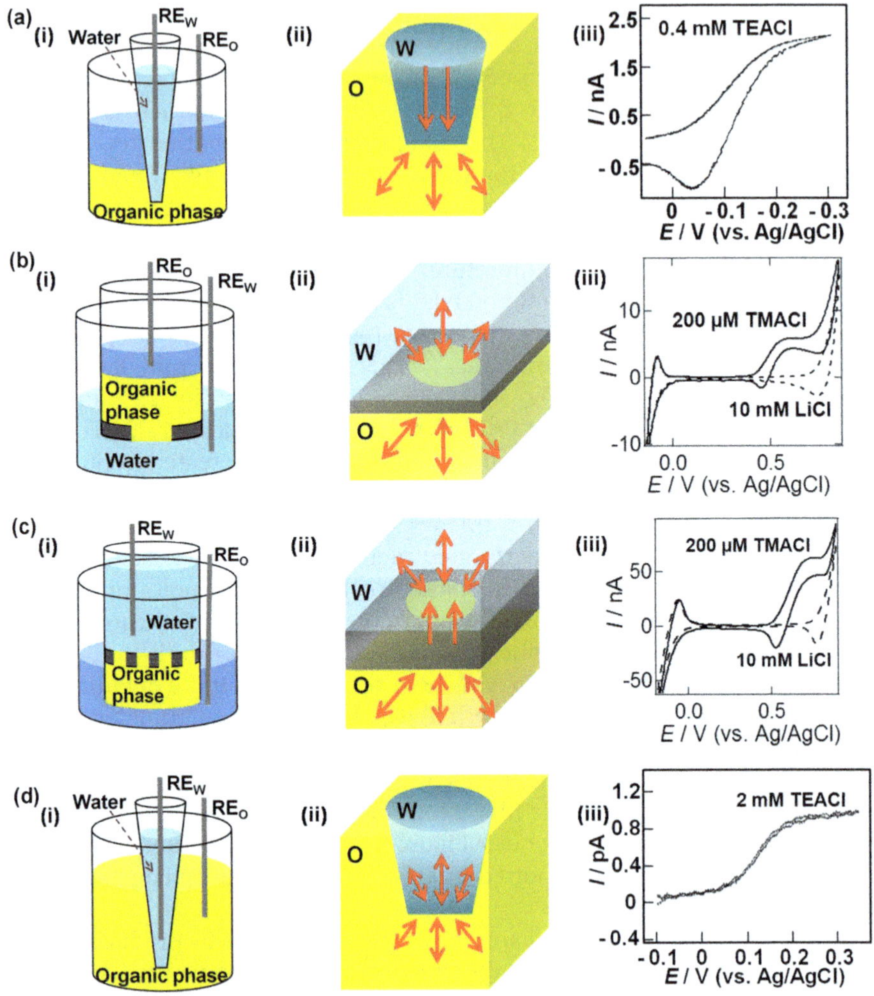

Figure 9.4 Schemes depicting various (i) electrochemical cell set-ups and (ii) diffusion profiles for ion transfer reactions across different ITIES platforms. (a) A micropipette tip, (b) a microhole, (c) a microhole array and (d) nanopipette tip. (iii) Cyclic voltammograms for TEA^+ ion transfer processes for each of the corresponding configurations.
(Both (a) and (d) are reproduced with permission from ref. 126 and 31, respectively. Copyright (1998) American Chemical Society (a) and John Wiley & Sons (d).)

film and a gelled viscous organic phase, three different diffusion regimes can be obtained (Figure 9.4(c)(ii)). The gelled organic phase usually fills up the microhole when using a hot-casting procedure—thus when the ion transfers from water to the organogel a hemispherical flux will dominate and result in a steady state voltammogram. For the ion transfer from the organogel to water, two different diffusion regimes occur, including a linear flux due to the viscous gel in the microhole and a hemispherical flux from the gel layer outside the microhole. This usually results in a semi-peak-shaped voltammogram, as can be seen in Figure 9.4(c)(iii). The steady-state current can be calculated by modification of eqn (9.4), with the geometric constant = 4 and taking into consideration the total number of diffusionally-independent microhole-interfaces in the array, as follows:

$$I_{ss} = 4Nz_iFDCr \tag{9.5}$$

where N is the number of microholes and the other parameters have been previously defined.

Building on the use of laser-ablated polymer films to create microhole-supported micro-ITIES arrays, silicon membranes containing micromachined holes were later developed.[48] These microporous silicon membranes were fabricated in clean room conditions using various etching, deposition and lithographic patterning methods that are standard processes in semiconductor chip fabrication. The essential fabrication steps were the thinning of sections of silicon wafers to 100 μm, using potassium hydroxide wet etching, followed by formation of microholes in these thinned sections using the method of deep reactive ion etching (DRIE), which uses a gas plasma to etch the silicon and open up vertical channels in the silicon. The key advantage of this fabrication process is that the vertical walls of the microholes are coated with a fluorocarbon polymer during the fabrication, so that the walls of the microholes are hydrophobic. This helps to fill the microholes with organic phase when used for electrochemistry at the ITIES in microarray format. A typical set of microhole arrays in these silicon membranes are shown in Figure 9.5 together with typical CVs for simple ion-transfer reactions.[49] Using the photolithographic patterning process, it is relatively simple to control both the hole–hole separation as well as the hole radius, thus giving control over the electrochemical signal obtained for diffusion-controlled ion transfer reactions.

Comparison of experimental and simulated CV data provided an excellent characterisation of the location of the ITIES when the organic phase was placed inside the silicon microholes.[48,49] The steady-state CVs for the ion transfer from aqueous to organic phase were in excellent agreement with simulations and consistent with the formation of an inlaid ITIES at the mouth of each microhole. As a result the current can be calculated according to eqn (9.5), just as for laser-ablated microhole-based ITIES. Interestingly, the reverse scan CV current was peak shaped, which can be attributed to the depth of the microholes (100 μm), which ensures that only linear diffusion occurs during the time scale of the reverse scan at all reasonable scan rates. This reverse peak voltammogram was also confirmed by computer simulations.[49]

These various types of microarray-supported ITIES have been employed to create a wide spectrum of ion selective sensors. For example, Lee and Girault[19] developed alkali metal ion selective amperometric sensors featuring microarray-ITIES with ion selective ionophores incorporated in the organogel. These were applied successfully as detectors for a commercial ion chromatography system.[19] In addition, a disposable type of microarray-ITIES

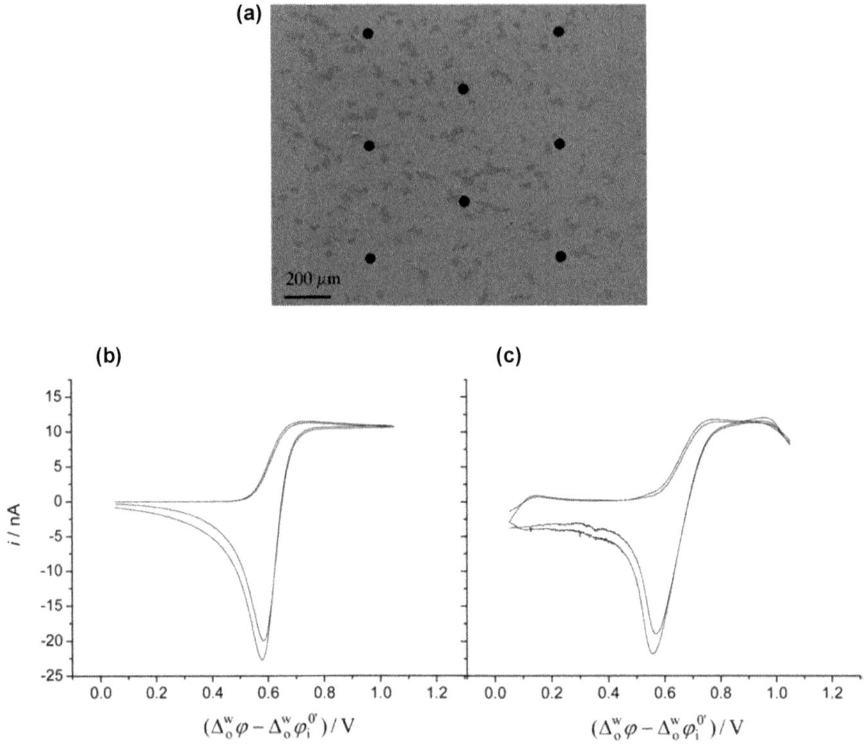

Figure 9.5 Silicon membranes to pattern micro-ITIES. (a) Scanning electron micrograph of a silicon membrane containing eight micropores (radius 26 μm, pore–pore separation *ca.* 490 μm), (b) simulated CV and (c) experimental CV for tetramethylammonium transfer at the water/gelled DCH micro-ITIES array. CV scan rates in (b) and (c) are 10 and 15 mV s^{-1}. (Adapted from ref. 49. Royal Society of Chemistry copyright.)

was prepared for the selective detection of Cd^{2+} ions using the ionophore ETH 1062.[11] There have been extensive efforts made on the use of microhole-supported ITIES for developing ion selective and sensitive sensing platforms for a wide range of ionic species and some examples are listed in Table 9.2.

9.2.3.3 *Nano-scale ITIES*

One of the recent highlights in electrochemistry at the ITIES has been the development of nano-scale ITIES.[5,50] These can offer benefits similar to those of solid nanoelectrodes, including the enhanced mass transport due to radial diffusion, decreased charging currents and a lower ohmic drop, resulting in an improvement in sensing performance. Additionally, nano-scale ITIES can also be used as a model for understanding biological nanopore function. Nano-scale ITIES are prepared by either of two ways: (a) locating the ITIES at the mouth of a single (or dual) nanopipette prepared by pulling a quartz capillary or (b) placing nanoporous materials containing geometrically irregular or regular arrays of pores at the ITIES.

Figure 9.4(d) shows a nanopipette-ITIES formed by using a pulled quartz capillary, filling it with an aqueous solution and interfacing it with the organic phase. In this case, the nano-interface formed at the mouth of the pipette possesses a fast radial diffusion flux (Figure 9.4(d)(ii)), resulting in steady-state voltammograms.[5] As shown in Figure 9.4(d)(iii), steady-state currents are observed in both scan directions, for transfer processes from water to the organic and *vice versa,* for TEA^+ ion transfer across the nanopipette-supported water/DCE interface.[31] The initial work of Shao and Mirkin[51] revealed that uncompensated ohmic potential drop (*iR* drop) and capacitive currents can be significantly decreased by the use of nano-scale ITIES. They also demonstrated an improved nanopipette-ITIES methodology by silanization of the outer pipette wall and were able to investigate the kinetics of the transfer of TEA^+ across the nano-ITIES.[52] Shao *et al.* quite recently reported the kinetics of the direct transfer of TEA^+ and ClO_4^-, as well as the facilitated transfer of K^+ by DB18C6 at nanopipette ($r = 1$ nm) supported interfaces between water and DCE.[31]

The second way to create the nano-ITIES is to place a nanoporous membrane at the interface. If either phase fills the nanopores and makes contact with the other liquid phase, then an array of nano-ITIES is formed, dependent on the size and distribution of the nanopores in the membrane used. This strategy was introduced by Dryfe and colleagues[53,54] using nanoporous filtration membranes, such as track-etched membranes. This approach demonstrated the practical utility of CV at the nano-ITIES arrays as a method to characterise the porosity of such membranes. However, the big disadvantage to modification of the ITIES with these nanoporous membranes was the difficulty in controlling both pore size and pore distribution and hence the size and density of the resultant nano-ITIES in the array. To overcome this issue, Arrigan and co-workers developed nanoporous silicon nitride membranes[55–58] in which the nanopore size and geometric pattern were controlled by electron beam lithography (EBL). In this way, pore size and pore–pore spacings were controllable using the high-precision fabrication methods already in use in nanoelectronics. The electrochemical behavior of the simple ion transfer of species such as TEA^+ and tetrapropylammonium ($TPrA^+$) as model ions at these geometrically-regular nanopore array-supported ITIES was investigated to elucidate the effect of nano-interface size on the improvement in ion sensing sensitivity[56,57] and also the fast response behavior coupled with the radial diffusion.[23] Interestingly, it was found that while analytical sensitivity, expressed as the slope of the current density *versus* concentration plots, increased with decreasing nano-ITIES size (smaller nanopores), in fact the nano-ITIES were not diffusionally independent despite being *ca.* 20-times the pore radius separated from their nearest neighbors. This observation, which was in agreement with a recent simulation study of nanoelectrode arrays,[59] indicates that greater sensitivity may be achieved by further optimising the pore–pore separation distances in such arrays. Such an enhancement in the sensitivity and responses when using nano-ITIES could definitely have great potential for near-future sensing applications.

More recently, the formation of nanoporous membranes for preparation of nano-ITIES arrays was achieved by focused ion beam (FIB) milling.[60] This is a rapid prototyping approach that uses a high energy gallium ion beam to physically mill nanopores.[61] Hence, this offers an attractive alternative to EBL and associated etching methods to prepare prototype nano-ITIES arrays with controlled pore size, density and geometry.

9.3 New Materials for ITIES-based Ion Sensing

9.3.1 Room Temperature Ionic Liquid as an Alternative Organic Phase

To date, many studies on ion transfer reactions at the ITIES have used organic phase solvents such as 1,2-DCE, hexane, nitrobenzene and octanol, all solvents possessing toxicity as an organic phase, which can limit the application of ITIES for the development of sensing platforms. Thus there has been a drive to develop alternative organic phases; the use of less toxic organic solvents such as nitrophenyl octyl ether or trifluorotoluene[62] is one option. Aprotic room temperature ionic liquids (RTILs) may be an important alternative since they can be used as a replacement for both the solvent and the electrolytes. Other advantages of RTILs[63] are their relatively low volatility, good conductivity and unique solvation properties; however, the high electrical resistance and viscosity could prohibit fast mass transport and contribute to poor long-term stability. A wide range of RTILs in contact with water have been studied for ion transfer reactions which can potentially be used as a new ITIES sensing platform.

Most of initial studies on the water/RTIL interface were focused on understanding the ion transfer properties when using different types of viscous aprotic RTILs. For example, Kakiuchi and Mirkin reported the kinetics of ion transfer at a nanopipette-based interface between water and the RTIL trihexyltetradecylphosphonium bis(1,1,2,2,3,3,4,4,4-nonafluoro-1-butanesulfonyl)imide, [$THTDP^+$][$C_4C_4N^-$]. Voltammetric characterization of tetrabutylammonium (TBA^+) ion transfer from water to the RTIL revealed a low background current and a good potential window, although very low diffusion currents was obtained due to the slow diffusion process.[64] The width of the potential window can be extended by selection of appropriate RTILs; for example, the use of tetraheptylammonium tetrakis[3,5-bis(trifluoromethyl)phenyl]borate ([$THpA^+$][$TFPB^-$]) interfaced with a $MgSO_4$ aqueous electrolyte showed a potential window of 1.1 V, which enabled the interfacial transfer of hydrophilic anions SCN^- and ClO_4^- as well as hydrophobic TBA^+.[65] Another effort was devoted to improve the stability of the water/RTIL interface; for instance, Dryfe *et al.* synthesized various hydrophobic RTILs by exchanging halide ions to form a stable RTIL/water interface for the study of TMA^+ ion transfer.[66]

Water/RTIL micro-interface arrays have been the subject of a couple of studies. For example, the voltammetric behavior of tetraethylammonium ion (TEA^+) transfer at the microarray water/RTIL interface (*e.g.*, where the RTIL is tetrahexylammonium tetrakis(pentafluorophenyl)borate) is shown in Figure 9.6(a). A steady state current on the forward scan representing a spherical diffusion flux was obtained whilst a peak formed on the reverse scan due to a linear diffusion flux in the RTIL phase. The steady state limiting current increases as a function of TEA^+ ion concentration and the obtained experimental value (4.25 nA) was in good agreement with the theoretical value (3.55 nA) from eqn (9.5). Additional water/RTIL microarrays were studied by Silvester *et al.*, using the hydrophobic RTIL trihexyltetradecylphosphonium tris(pentafluoroethyl)trifluorophosphate ($[P_{14,6,6,6}][FAP]$) held within the pores of silicon microporous membranes.[67]

One of the interesting features of ion transfer processes at the water/RTIL interface is that diffusion can be significantly slower in the RTIL phase due to the high viscosity of many RTILs. Ding *et al.* showed,[43] when using micro-ITIES, that the usual asymmetric voltammograms were not observed; instead, symmetrical diffusion profiles were established due to slow diffusion of ions transferring from RTIL to water, which resulted in peak-shaped voltammograms for both the RTIL-to-water and the water-to-RTIL transfers, as shown in Figure 9.6(b) for the transfers of tetraphenylarsonium ($TPAs^+$)

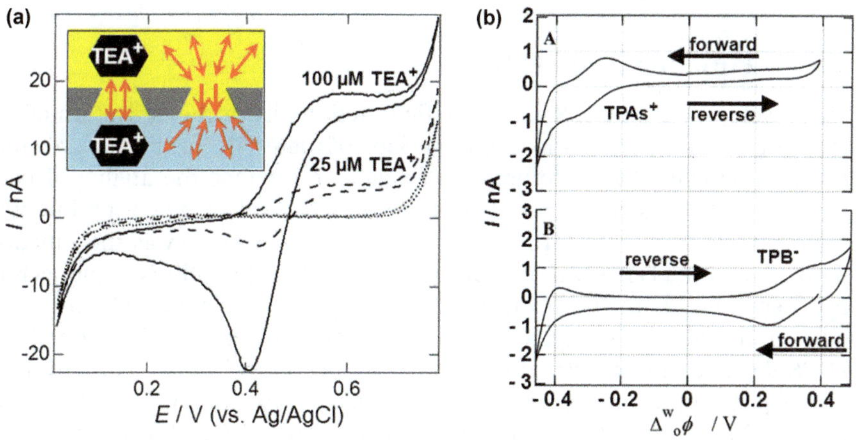

Figure 9.6 (a) Cyclic voltammograms for TEA^+ ion transfer across a microhole array supported water/RTIL interface. The RTIL was tetrahexylammonium tetrakis(pentafluorophenyl)borate etherate with 10 mM LiCl also used. Inset: diffusion profile of TEA^+ ions across the ITIES. (b) Cyclic voltammograms for tetraphenylarsonium ($TPAs^+$) (A) and tetraphenylborate (TPB^-) (B) ion transfer across the water/P_{66614}TB micro-interface; scan rate = 20 mV s^{-1}.
(Part (b) is reproduced with permission from ref. 68. Copyright Elsevier.)

and tetraphenylborate (TPB^-) ions across the water/trihexyltetradecylphosphonium tetrakis(pentafluorophenyl)borate ($P_{66614}TB$) interface.[68]

An understanding of the ion transfer characteristics at water/RTIL interfaces can of course aid the development of these interfaces as the basis for ion sensors. This was explored by Samec's group, who proposed an amperometric ISE for alkali metal cations using a thin film RTIL membrane system. For example, the interface between water containing a hydrophilic ionophore crown ether (*e.g.*, 18-crown-6) and RTIL [*e.g.*, tridodecylmethylammonium tetrakis[3,5-bis-(trifluoromethyl)phenyl]borate (TDMATFPB)] was employed as a K^+ selective sensor based on the facilitated K^+ ion transfer by the crown ether.[69] Another interesting sensing example by Samec's group was the transfer of alkali-metal cations (K^+, Rb^+, Cs^+, Na^+, Li^+), protons and ammonium ions assisted by valinomycin across the interface between water and an RTIL [*e.g.* TDMATFPB]. Selective voltammetric sensing of K^+ and Na^+ by valinomycin in table or tap water samples was highlighted.[70] Another example of facilitated ion transfer at the water/RTIL interface for the selective sensing of target ions was realized by Kakiuchi *et al.*[71] In this work, using a micropipette-supported interface between *N*-octadecylisoquinolinium tetrakis[3,5-bis(trifluoromethyl)phenyl]-borate ([C_{18}Iq][TFPB]) and water, transfers of Li^+, Na^+, K^+ and Rb^+ ions were facilitated by DB18C6. Interestingly, a high selectivity of DB18C6 towards K^+ over Na^+ in [C_{18}Iq][TFPB] was observed, in marked contrast to most observations at ITIES formed with organic solvents, which suggests that RTILs could provide a unique solvation environment and hence a unique ion sensing platform.

9.3.2 Functionalized ITIES

The development of ITIES with specific functionality or the derivation of an ITIES device from alternative combinations of materials has been a growing research activity. Such alternate materials can enhance the analytical performance of ion sensors based on the ITIES or lead to new experimental platforms derived from ITIES behavior. One such advance was the functionalization of the ITIES with metallic (*e.g.*, Au,[72] Pd,[73] Pt[74]) films as pioneered by Cheng and Schiffrin.[75] Controlled assembly of Au nanoparticles at ITIES was employed to investigate optical properties of metallic functionalized ITIES such as surface plasmon resonance and optical mirror formation.[76] The deposition technologies for creating metallic functionalized ITIES have advanced but not yet been applied to sensing applications. Indeed, recently, the processes of metal nanoparticle assembly at the ITIES was used in the fabrication of nanoelectrodes,[77] indicating the way in which ITIES behavior can be utilised to prepare alternative platforms for detection.

Another strategy to create functionalized ITIES is the formation of lipid monolayers for the study of ion transfer reactions. This can provide a model lipid monolayer system for study of ion and other interactions which may be beneficial to biological activity assessments. For example, Gulabowski *et al.*

created phosphatidylcholine-modified water/DCE interfaces in order to mimic cell membranes and to study the interactions between acetylcholine (Ach^+) and the phospholipids. Voltammetric characterisation showed that the phospholipids play an important role in assisting the transfer of Ach^+ across biomembranes.[78] In addition, Marecek *et al.* reported voltammetric studies of TEA^+ ion transfer across an ITIES modified with a layer of dipalmitoyl phosphatidic acid (DPPA), demonstrating that the transfer was based on ion exchange with the DPPA's dissociable proton.[79]

The achievement of mesoporous silica films *via* an electrodeposition process has also been reported by several groups, including Marecek[80] and more recently Herzog and co-workers.[81–83] In the latter process, surfactant transfer from the organic to aqueous phase is used to template the hydrolysis of silicate reagents such as triethoxysilane and to lead to the formation of mesoporous films at the ITIES. These mesoporous silica films were capable of modulating ion transfer and thus can have a future role in sensor development, perhaps providing improved selectivity or stability for some applications.

Although many of these functionalized ITIES studies have not been utilized for sensing, such developments are expected to contribute to advanced detection strategies in the future.

9.4 ITIES Sensing for Pioneering New Target Ions

Ion transfer reactions at ITIES have proven to be promising detection platforms aimed at a wide range of ionised species. To extend the applicability of electrochemistry at the ITIES for different target species, there have been efforts on developing new strategies to detect small organic molecules. In addition, there have been extensive efforts on designing new ionophores to improve selectivities of established target ions with ITIES as well as new target ions, together with an expansion of activity in the study of biomacromolecule behavior at the ITIES as a strategy for label-free detection. This section will highlight recent efforts on developing ITIES based sensing platforms for targeting new ions, organic molecules and proteins.

9.4.1 Inorganic Ions

Over recent decades, most of the common alkali and alkaline earth metal cations as well as many inorganic anions have been detected in environmental water samples using the ITIES methodology. These employed either direct or facilitated ion transfer reactions. Tables 9.1 and 9.2 summarise this information. Recent new examples of analyte ions detected using this approach include $HCrO_4^-$ by Lee *et al.*, based on Aliquat 336-facilitated transfer of $HCrO_4^-$ in pH 4 conditions across a microhole array-supported water/PVC-NPOE interface (see Figure 9.7(a)).[84] Chromium(VI) is highly toxic and simple and sensitive methods for its detection are required for industrial and environmental applications. The CVs of Figure 9.7(a) clearly

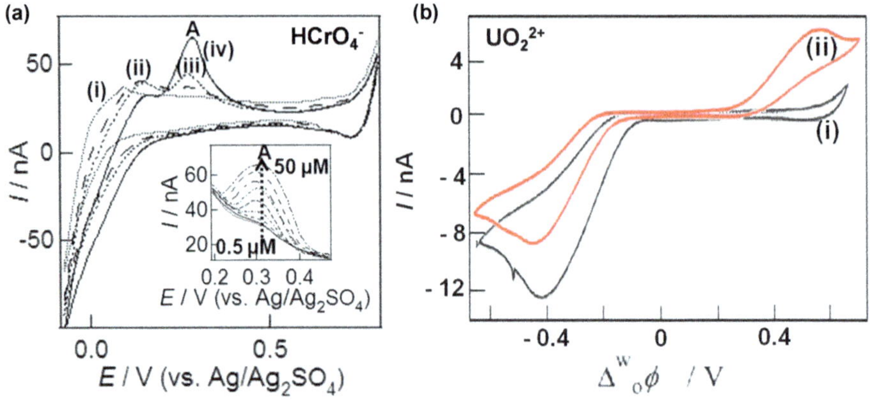

Figure 9.7 (a) Cyclic voltammograms for the assisted transfer of Cr(VI) ions by the ionophore Aliquat 336 present in the PVC-NPOE gel phase. The aqueous phase had a pH of 4 and a scan rate scan rate of 20 mV s^{-1} was used. (i) In the absence of Cr(VI) ions, and in the presence (ii) 25, (iii) 50 and (iv) 100 μM Cr(VI) ions in 10 mM K_2SO_4. Inset: differential pulse stripping voltammograms for sensing Cr(VI) ion concentrations ranging from 0, 0.5, 2, 5, 10, 20, 30, 40 to 50 μM. A sweep direction from low to high positive potentials was used and the deposition potential was 300 mV applied for 40 s. Potential increment = 10 mV, pulse potential = 50 mV and pulse duration = 50 ms. (b) Cyclic voltammograms for the transfer of 5 mM UO_2^{2+} ions in the absence (i) and presence (ii) of 36.8 mM tri-*n*-butylphosphate (TBP) across a micro-ITIES. Scan rate = 20 mV s^{-1}. (Both (a) and (b) are reproduced with permission from ref. 84 and 87, respectively. Copyright (2012) Elsevier (a) and (2011) American Chemical Society (b).)

demonstrate the back-transfer of Cr(VI) ions from the organogel to aqueous phase, with a concentration-dependent peak. The peak labeled A, at 300 mV, increased with the concentration across the range from 25 to 100 μM (Figure 9.7(a) (iii) and (iv)). This peak was assigned to the reverse-transfer of the Aliquat 336-assisted transfer of $HCrO_4^-$. An excellent detection limit of 0.5 μM and linear range of 0.5–10 μM was reported when using DPSV, which employs a pre-concentration step to achieve lower concentration detection (inset, Figure 9.7(a)).[84] O'Mahony *et al.* also previously reported on the amperometric detection of hexavalent chromium based on Cr(VI) ion transfer across the ITIES facilitated by the penta-protonated form of the polyamine macrocycle 2,5,8,11,14-pentaaza[15]-16,29-phenanthrolinophane (NeoTT).[33] Recent work on detection of perchlorate ClO_4^- was reported by Amemiya's group. Employing a stripping voltammetry approach at a gelled organic phase electrode, the detection of nanomolar concentrations was achieved.[85] A similar strategy for the detection of Ca^{2+} ions was reported by the same group, who employed ionophore-facilitated ion transfer for sub-nanomolar detection.[86]

An ITIES-based sensing platform for radioactive ions was initiated by Ding *et al.*, who investigated the tributyl phosphate-facilitated transfer of uranyl cations (UO_2^{2+}) across a water/DCE micro-interface.[87] Figure 9.7(b) shows CVs for the facilitated transfer of 5 mM UO_2^{2+} ion. The clear changes in current at positive potentials upon the addition of UO_2^{2+} show that this might be useful as a UO_2^{2+} ion sensor. Thermodynamic parameters including the Gibbs transfer energy of UO_2^{2+} ions at the water/DCE interface and the association constant for the UO_2^{2+}–tributyl phosphate complex were also determined using CV.[87]

The detection of Ag^+ ions was demonstrated based on the facilitated ion transfer of Ag^+ ions by *O*,*O*″-bis[2-(methylthio)ethyl]-*tert*-butylcalix[4]arene at both micro- and macro-liquid/liquid interfaces.[88] In this work, using DPV, a LOD of 3.5×10^{-6} M (at S/N ratio = 3) for Ag^+ was obtained, with a linear relationship in the range of $2–10\times10^{-5}$ M Ag^+ ions. Indeed, previous studies reported the transfer of Ag^+ cations based on electrolyte-facilitated transfer: complexation of Ag^+ with tetraphenylborate electrolyte species was possible and enabled detection of silver cations at trace concentrations.[89–91] Another interesting work based on Ag^+ transfer facilitated by a calixarene was developed to explore Ag^+ toxicity in living cells using scanning electrochemical microscopy (SECM). An LOD of 0.2 μM for Ag^+ ions was also reported.[92] In addition to ion interaction with live cell imaging applications based on an ion transfer process at the ITIES, the Li^+ ion transfer reaction was applied to the understanding of the chemical mechanisms of lithium-based batteries. The Li^+ ion transfer across the water/1,2-DCE interface facilitated by solvation effects was studied using propylene carbonate (PC) as an ionophore, which is one of the most commonly used solvents in lithium-based batteries.[44]

Another highlight in this area is the design of an ITIES detection platform that can be readily used in-the-field (out of the laboratory), in addition to real sample analysis. For example, Lee *et al.* reported a disposable tape ion sensor for Cd^{2+} ions using microarrays fabricated on a thin strip device. The assisted transfer reaction of Cd^{2+} ions by the ionophore ETH 1062 across the water/PVC-NPOE organogel micro-interface was used to detect Cd^{2+} ions in aqueous solution, with DPSV offering detection limits as low as 20 ppb in a 20 μL sample volume.[11]

9.4.2 Small Organic Ions

Compared to the inorganic ion sensing, there have been fewer reports on the use of ITIES targeted towards small ionized organic molecules,[93] besides the widespread study of tetraalkylammonium cation transfers as model reactions at the ITIES. Nevertheless, there have been studies of organic ion transfer over previous decades, such as picrate[15] transfer and protonated amine transfers.[94,95] Recent examples include the work of Pereira *et al.*, who demonstrated a simple method for quantification of the neurotransmitters protonated noradrenaline (NAH^+) and protonated dopamine (DAH^+) *via* voltammetric measurements of the ion transfer assisted by dibenzo-18-crown-6 at a water/DCH micro-interface.[96] LODs of 0.35 and 1.7 mM for DAH^+ and NAH^+, respectively, were achieved.[96] The same group also reported the direct transfer of the lipophilic anticancer drug daunorubicin (DNR) at the ITIES using voltammetric techniques, and elucidated the solvation mechanisms and explored potential analytical applications. Using DPV, a LOD of 0.80 μM DNR was established, together with a linear response in the range of 12–82 μM.[24]

Other groups have also studied the detection of neurotransmitters. The Arrigan group has initially studied the transfer of protonated dopamine across the ITIES facilitated by dibenzo-18-crown-6, which was selective over ascorbate. However, the lowest concentration detectable was *ca.* 0.1 mM using CV at a millimeter-sized ITIES.[97] A key advantage for protonated dopamine detection at the ITIES is non-interference from vitamin C (ascorbic acid) electro-oxidation and no fouling of the interface by dopamine oxidation products. Protonated dopamine was also detected using this ionophore in conjunction with microporous silicon membrane-supported micro-interfaces, which greatly enhanced the detection of lower concentrations of dopamine, down to 0.5 μM as compared to 2 μM at large interfaces.[98] Substantial improvement was reported by Shao's group, who achieved a LOD of 50 nM for facilitated dopamine detection at a micropipette-supported ITIES, illustrating the power of this approach.[99] Detection of dopamine remains of interest with this approach. The detection of food additives, such as the artificial sweeteners aspartame and acesulfame K, was also demonstrated by the same research group based on the direct and facilitated ion transfer using DB18C6 at the ITIES; LODs of *ca.* 30 μM were reported.[100]

Other recent highlights for the detection of organic ions include the detection of protonated forms of heptylamine, octylamine, decylamine, procaine and procainamide, facilitated by dibenzo-18-crown-6, using a water/solvent polymeric membrane interface and yielding LODs of 2×10^{-6} M.[101] Various drug substances have also been detected with this approach, such as verapamil, clomipramine, tacrine and imipramine.[102] In particular, imipramine, an antidepressant agent, was determined using a flow-injection pulse amperometric method based on the ion transfer across the water/PVC-NPOE organogel interface, with a detection limit of 0.42 μM and a linear response range of 1–80 μM.[103]

Figure 9.8(a) shows the voltammetric detection of anticancer drugs at the ITIES. Direct transfer of irinotecan across a large-scale water/DCE interface was investigated and the current responsible for the irinotecan transfer was found to be concentration dependent, illustrating its possible use in drug detection platforms.[104] Related recent studies of drug transfer at ITIES include the detection of ractopamine, which transferred at the upper edge of the available potential window,[105] at a water/DCE micro-interface array and the demonstration of hollow micro-needles as substrates to form a micro-ITIES array, enabling the possible detection of drugs with *in situ* sampling.[39]

Enantioselective detection of drugs is often a requirement in development and production operations and electrochemistry at ITIES has proved useful

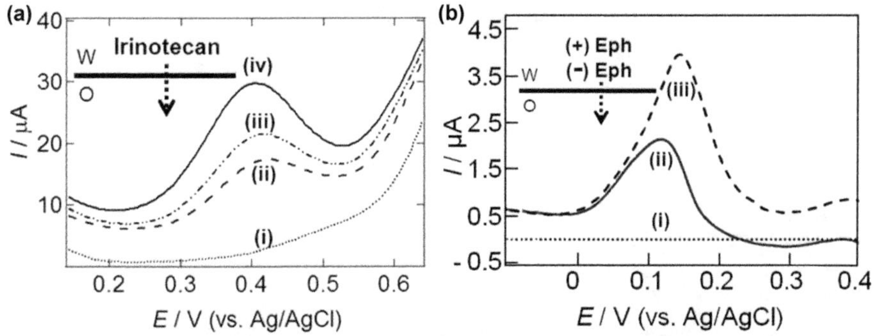

Figure 9.8 (a) A series of differential pulse voltammograms for the sensing of different concentrations of the anti-cancer drug irinotecan, using a large-scale water/1,2-DCE interface. Plot (i) is for tris acetate buffer of (pH 7.5) in the absence of irinotecan and in the presence of (ii) 50, (iii) 100 and (iii) 150 μM irinotecan. (b) Differential pulse voltammograms for the detection of the chiral ionic compound, ephedrinium (Eph^+) ion, transfer ringed across a micropipette supported water (w)/DCE (o) interface *via* selective complexation with the ionophore, cyclodextrin (CD): (i) is in the absence of chiral selector (Eph^+), while (ii) and (iii) correspond to the presence of (−)Eph^+ and (+)Eph^+ enantiomers, respectively. Inset: simplified diffusion profile of (−)Eph^+ and (+)Eph^+ enantiomers across the ITIES.
(Part (a) is reproduced with permission from ref. 104. Part (b) is reproduced with permission from ref. 41, copyright Royal Society of Chemistry.)

for this analysis. For example, the chiral transfer of ephedrinium ion enantiomers (Eph^+) facilitated by heptakis(2,3,6-tri-*O*-acetyl)-α-cyclodextrin (AcαCD) was reported at the micropipette-supported ITIES.[41] Figure 9.8(b) shows background-subtracted DPVs for the facilitated chiral transfer of Eph^+ using AcαCD. In the absence of the chiral selector (AcαCD), no transfer across the liquid/liquid interface was observed, within the available potential window (Figure 9.8(b)(i)). Different transfer potentials of the enantiomers, (−)Eph^+ and (+)Eph^+, were observed (Figure 9.8(b)(iii)) which suggests that the facilitated transfer reaction can differentiate the ephedrinium enantiomers using DPV.[41]

Most of the above examples of organic ion transfer are cationic species. However, anion detection is also possible, as illustrated by the recent example from Amemiya's group on detection of perfluoroalkanesulfonates and perfluoroalkanecarboxylates, with picomolar sensitivity.[106]

As a final example, an ITIES-based sensing strategy targeting organic molecules was suggested, by linking enzymatic hydrolysis with an ion-transfer reaction. The selective detection of organophosphate compounds (*e.g.*, paraoxon) was proposed by the use of a disposable amperometric proton selective sensor. This was prepared using polydimethylsiloxane (PDMS) technology and a microhole-supported water/PVC-NPOE organogel interface.[107] The detection scheme is outlined in Figure 9.9(a);

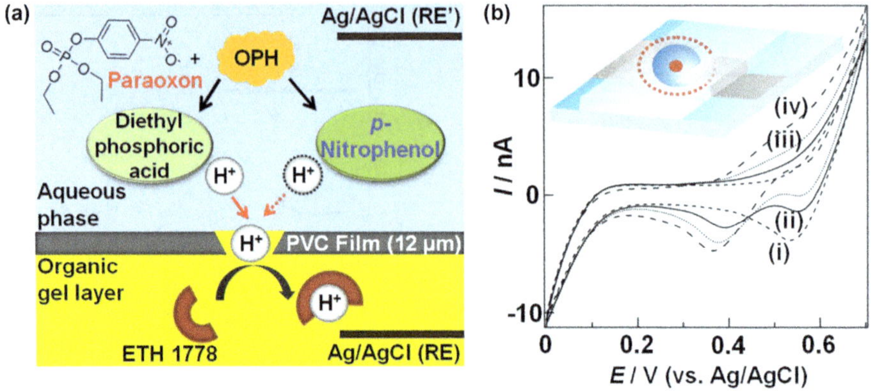

Figure 9.9 (a) Schematic diagram of a paraoxon sensing platform using a proton selective ion sensor in conjunction with a single microhole-water/PVC-NPOE gel interface. (b) Cyclic voltammograms for the assisted transfer of protons released from the hydrolysis of paraoxon by 40 nM organophosphorus hydrolase. The aqueous supporting electrolyte was 10 mM LiCl. Plots: (i) corresponds to no paraoxon, with (ii) 25, (iii) 50 and (iv) 100 μM paraoxon concentrations. The aqueous and organic supporting electrolytes were 10 mM LiCl and 10 mM TBATPBCl, respectively. Sweep rate = 20 mV s^{-1}. Inset: schematic of a strip sensor for paraoxon detection.
(Both (a) and (b) are reproduced with permission from ref. 107. Elsevier)

organophosphorus hydrolase (OPH) selectively hydrolyzes paraoxon, producing diethyl phosphoric acid and *p*-nitrophenol, which can then release two protons at pH 6.5. The transfer of protons across the micro-interface, assisted by the proton-selective ionophore ETH 1778, produces a current that increases in proportion to the paraoxon concentration. The micro-ITIES can be also configured as a portable paraoxon-selective strip-sensor on a PDMS substrate (inset, Figure 9.9(b)). As can be seen in the CVs of Figure 9.9(b) recorded at different paraoxon concentrations and at a fixed OPH concentration, no change in current response was observed in the absence of paraoxon. In the presence of both paraoxon and OPH, steady-state and peak currents were observed on the forward and the reverse scans, respectively, due to the fact that the protons, produced by the enzyme reaction, were transferred across the micro-interface. A LOD of 0.5 μM and a linear response in the range 0.5–100 μM paraoxon was achieved. The same device was also employed for the multiplexed detection of both parathion and methyl parathion based on their different catalytic responses with OPH, using time-resolved DPSV.[108] Such a proton selective sensor can also be utilized for the selective detection of glucose; for example, protons released by the glucose oxidase-catalyzed oxidation of glucose can then transfer from water to organic phase assisted by the ionophore ETH 1778. The proton selective sensor with DPSV was used to measure glucose concentrations up to 18 mM.[46]

9.4.3 Biomacromolecules

The ability of electrochemistry at the ITIES to detect biological macromolecules[3,93] was evident a couple of decades ago, when Vanysek and coworkers[109] described the adsorption of proteins at water/nitrobenzene interfaces. This adsorption was detected by methods such as impedance spectroscopy and CV of simple ion-transfer reactions that were distorted by the adsorbed protein film. In subsequent years, the behavior of proteins at the ITIES was studied as a function of the pH of the aqueous phase electrolyte[110–115] and it was found that current peaks associated with an ion transfer process were present when the aqueous phase pH was lower than the protein's isoelectric point (pI). Figure 9.10(a) and (b) shows CV and current–pH data for the protein lysozyme from hen egg white.[115] The CV shows clearly that the protein can be detected by voltammetry at this water/DCE interface, but also that important mechanistic information is obtained.

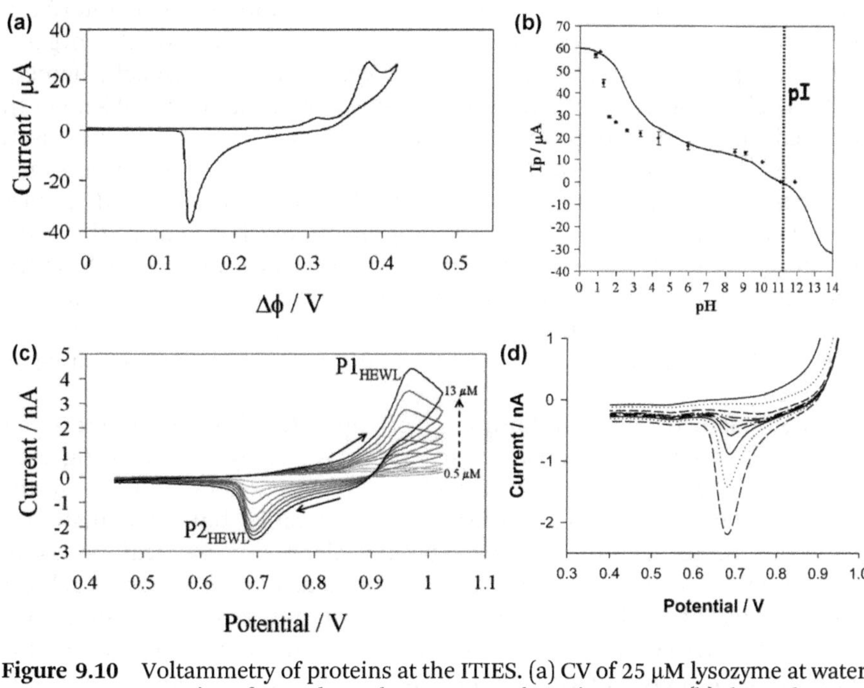

Figure 9.10 Voltammetry of proteins at the ITIES. (a) CV of 25 μM lysozyme at water/DCE interface, where the aqueous phase is pH 1.9, (b) dependence of the forward peak current (points) for lysozyme voltammetry on the aqueous phase pH; the solid line is a theoretical response based on acid–base titration of lysozyme; (c) background-subtracted CVs of lysozyme (0.5–13 μM) at the micro-ITIES array between water and gelled DCH; (d) AdSV of 0.1 μM lysozyme at gelled water/DCH micro-ITIES array, following adsorption times of 0–1800 s.
(Parts (a) and (b) from ref. 115 (2009); part (c) from ref. 118 (2010); part (d) from ref. 119. Copyright Royal Society of Chemistry (a)–(c) and (2012) the American Chemical Society (d).)

First, there is a main peak associated with a diffusion-controlled ion transfer process (*ca.* 0.38 V, on the forward sweep) which is also concentration dependent and enabled detection of lysozyme in solution down to micromolar concentrations. There is a pre-peak associated with this (*ca.* 0.30 V) which is indicative of the adsorption of a reaction product on the interface. Finally, on the reverse scan (in the negative direction), there is a single large current peak (*ca.* 0.14 V), which displays the features of a desorption process. Previous studies on cytochrome *c* ion-transfer voltammetry indicated that this colored protein did not transfer to the organic phase[110,111] and hence it was suggested that electrolyte species interacting with the protein are responsible for the peak currents detected. Since the protein, lysozyme for example, is detectable when present in aqueous solutions at pH values below its pI (Figure 9.10(b)) it can be suggested that the cationic protein interacts with the anions of the organic phase electrolyte, namely, tetraphenylborate and its chloro- and fluoro- derivatives. Indeed, studies of the effect of changing this organic phase electrolyte anion reveal shifts in the position of the peak potential, which is consistent with the formation of interfacial complexes between the cationic protein and the anions of the organic electrolyte phase. Using an on-line mass spectrometry approach, namely, biphasic electrospray mass spectrometry, such complexes were subsequently detected only under conditions in which the flowing ITIES was positively polarized,[116] indicating that the application of a sufficient potential difference was necessary to form these interfacial complexes at the ITIES. However, detailed mechanistic understanding of protein behavior at the ITIES remains a topic of active consideration. For instance, factors such as the formation of protein–anion complexes at the interface and protein unfolding[117] when subjected to adsorption/complexation remain to be fully elucidated. Nevertheless, the detection of protein remains possible at the ITIES.

Protein detection *via* voltammetry at the ITIES (*e.g.* millimeter-scale interfaces) was demonstrated for several target analytes, including cytochrome *c*, hemoglobin, lysozyme and insulin.[3,93] Generally, it was found that low-micromolar concentrations were detectable, but for realistic applications to protein detection in biological and biomedical problems substantially lower concentrations must be detected. One way to achieve this is the use of smaller interfaces, such as micro-interfaces or their arrays. Such miniaturized interfaces offer improved mass transport rates and minimized background capacitance, which may contribute to improved LODs for proteins. For example, Amemiya[21] employed glass pipette-based micro-ITIES for the study of the behavior of protamine, a polypeptide rich in arginine residues and which is hence highly positively charged in physiological pH media. Using CV at this micro-ITIES, and the addition of the specific reagent dinonylnaphthalenedisulfonate, $DNNS^{2-}$ (as its tetradodecylammonium salt),[21] to the organic phase it was found that protamine was detected by transfer to the organic phase as a $DNNS^{2-}$ complex.

Subsequent to studies on lysozyme behavior at larger ITIES,[115] Scanlon *et al.*[118] explored the behavior and detection possibilities of lysozyme voltammetry at micro-ITIES arrays formed using silicon microporous membranes (Figure 9.10(c)). It was noteworthy that radial diffusion control for lysozyme detection was not evident: rather, linear diffusion control was present, which was indicative of the influence of organic anions in the interfacial reaction. These anions, in the gelled organic phase electrolyte, were held within the silicon membrane's micropores and thus were subjected to linear diffusion processes if involved in any interfacial reaction. From a detection point-of-view, lysozyme at 0.5 μM was detected, representing an order of magnitude improvement on the performance obtained at millimeter-sized ITIES. It was also noted that lysozyme adsorption at the micro-ITIES occurred.

Analytical exploitation of protein adsorption was investigated as a function of potential applied during the adsorption process, for lysozyme,[119] hemoglobin[120] and insulin.[121] It was found that a positive applied potential favored the adsorption of a protein layer at the interface between water and organogel phases, which is consistent with the formation of protein–anion complexes and their adsorption at the micro-ITIES. Voltammetric desorption of the adsorbed layer subsequent to a controlled potential adsorption enabled measurement *via* the approach of adsorptive stripping voltammetry (Figure 9.10(d)) and produced LODs for lysozyme, hemoglobin and insulin of 30,[119] 48,[120] and 10 nM,[121] respectively.

Detection of proteins at nanomolar concentration is a step toward the targets needed for viable biomedical applications, but lower concentrations still need to be achieved. Another issue that needs to be addressed is selectivity – the ability to detect one substance in the presence of others. Although the ability to detect proteins at the ITIES is based on interfacial complexation and adsorption processes, some aspect of these behaviors may be tuned for selectivity achievement. For example, the potential-controlled adsorption might have different optimum potentials for different proteins and hence the applied potential during this stage of the AdSV experiment can be tuned to enable preferential adsorption of one protein relative to another. This strategy was examined for the detection of insulin in the presence of bovine serum albumin, and, indeed, selective detection of insulin was possible.[121] Another possible strategy to achieve selectivity is manipulation of the acid–base chemistry of proteins in a mixture. Since proteins need to be cationic to be detected by the method of voltammetry at the ITIES, provided different proteins in a mixture have different acid–base properties then the pH of the aqueous phase can be tuned to achieve a cationic state for the target analyte while other proteins present are neutral or anionic. This was demonstrated by the selective detection of amylin in a protein mixture.[122] Amylin is a polypeptide, also known as islet amyloid polypeptide, which is implicated in fibril formation and diabetes. Amylin from rat was studied at the micro-ITIES array, as this form of amylin does not aggregate and form fibrils. While amylin transfer to the organogel phase was possible, leading to a steady-state voltammogram, the reverse peak was consistent with stripping from a thin layer of amylin collected on the organic side of the ITIES. In a mixture with myoglobin and hemoglobin, amylin was detectable at physiological pH, where it retained a cationic charge whereas the proteins myoglobin and hemoglobin were neutral or anionic, and hence did not interfere in the detection.[122] Thus, some simple strategies, like exploiting selective electro-adsorption and manipulating aqueous phase pH, can deliver some aspects of selectivity for biomacromolecule detection. However, additional selectivity that may be achieved *via* biomolecular recognition remains to be integrated into ITIES-based protein detection schemes.

Besides proteins, and polypeptides, discussed above, other biological macromolecules have also been subjected to detection *via* electrochemistry at the ITIES. These include carbohydrates, such as heparin (a sulfonated polysaccharide)[123] and deoxyribonucleic acid,[124,125] which may be detected by voltammetry at the ITIES.

9.5 Summary and Outlook

Ion transfer reactions at a polarized ITIES hold great promise for a wide range of ion detection platforms. The simplicity with which ions may be pushed across the interface by virtue of the applied interfacial potential difference enables a current to be generated that is proportional to the concentration of the transferring ion. The use of micro-interfaces together with gelification of one of the phases (usually the organic phase) has significantly improved the mechanical stability and, hence, usability of this system. The improvement in mass transport properties from the use of miniaturized interfaces (microscale and nanoscale) offers scope for more sensitive detection of target ion species. Novel phase materials, such as new ionic liquids and interfacial modification with reagent layers, can offer alternative reactivities that lead to enhanced detection capabilities. In addition, the development and discovery of new ionophores that selectively interact with different target ionic species can also be noted to have continued and to expand the applicability and selectivities of ion sensors. The fabrication of ITIES on disposable platforms is also a highlight that advances the ITIES--based analytical sensors. The detection of biological molecules in a label-free manner has been demonstrated with this form of electrochemistry and further developments will lead to bio-assays and biosensors that can deliver detection performances required in biomedical applications. All of these development areas will translate ion transfer processes at the ITIES towards more practical analytical tools for environmental and biological applications.

Acknowledgements

Aspects of this research were supported by the National Research Foundation (NRF) of Korea funded by the Ministry of Science, ICT and Future Planning (NRF-2013R1A2A2A03068833), and by the Australian Research Council (DP130102040).

References

1. W. Nernst and E. H. Riesenfeld, *Ann. Phys.*, 1902, **8**, 600.
2. H. H. Girault, *Electrochim. Acta*, 1987, **32**, 383.
3. D. W. M. Arrigan, *Annu. Rep. Prog. Chem., Sect. C: Phys. Chem.*, 2013, **109**, 167.
4. Z. Samec, *Electrochim. Acta*, 2012, **84**, 21.
5. S. Liu, Q. Li and Y. Shao, *Chem. Soc. Rev.*, 2011, **40**, 2236.
6. Z. Samec, E. Samcova and H. H. Girault, *Talanta*, 2004, **63**, 21.
7. H. H. Girault, Charge transfer across liquid-liquid interfaces, in *Modern Aspect of Electrochemistry*, ed. J. O. M. Bockris, New York, Plenum Press, 1993, vol. 25.
8. P. Peljo and H. H. Girault, Liquid/liquid interfaces, electrochemistry at, *Encyclopedia of Analytical Chemistry*, Hoboken, John Wiley & Sons, Inc., 2012.
9. B. Liu and M. V. Mirkin, *Electroanalysis*, 2000, **12**, 1433.
10. P. Vanysek and L. B. Ramirez, *J. Chil. Chem. Soc.*, 2008, **53**, 1455.
11. H. J. Lee, G. Lagger, C. M. Pereira, A. F. Silva and H. H. Girault, *Talanta*, 2009, **78**, 66.
12. C. Gavach and F. Henry, *J. Electoanal. Chem.*, 1974, **54**, 361.
13. Z. Samec, V. Marecek, J. Koryta and M. W. Khalil, *J. Electroanal. Chem.*, 1977, **83**, 393.
14. J. Wang, *Analytical Electrochemistry*, Hoboken, John Wiley & Sons, Inc., 3rd edn, 2006.
15. D. Homolka, L. Q. Hung, A. Hofmanova, M. W. Khalil, J. Koryta, V. Marecek, Z. Samec, S. K. Sen, P. Vanysek, J. Weber and M. Brezina, *Anal. Chem.*, 1980, **52**, 1606.
16. F. Reymond, P. A. Carrupt and H. H. Girault, *J. Electroanal. Chem.*, 1998, **449**, 49.
17. F. Reymond, G. Lagger, P. A. Carrupt and H. H. Girault, *J. Electroanal. Chem.*, 1998, **451**, 59.
18. A. J. Bard and L. R. Faulkner, *Electrochemical Methods: Fundamentals and Applications*, New York, John Wiley & Sons, Inc., 2nd edn, 2001.
19. H. J. Lee and H. H. Girault, *Anal. Chem.*, 1998, **70**, 4280.
20. S. Wilke, M. D. Osborne and H. H. Girault, *J. Electroanal. Chem.*, 1997, **436**, 53.
21. Y. Yuan and S. Amemiya, *Anal. Chem.*, 2004, **76**, 6877.
22. Y. Yuan, L. Wang and S. Amemiya, *Anal. Chem.*, 2004, **76**, 5570.
23. M. Sairi, J. Strutwolf, R. A. Mitchell, D. S. Silvester and D. W. M. Arrigan, *Electrochim. Acta*, 2013, **101**, 177.
24. J. A. Ribeiro, F. Silva and C. M. Pereira, *Anal. Chem.*, 2013, **85**, 1582.
25. J. A. Ribeiro, F. Silva and C. M. Pereira, *Talanta*, 2012, **88**, 54.
26. C. J. Collins, C. Lyons, J. Strutwolf and D. W. M. Arrigan, *Talanta*, 2010, **80**, 1993.
27. G. A. Crespo and E. Bakker, *RSC Adv.*, 2013, **3**, 25461.

28. Y. Shao, M. D. Osborne and H. H. Girault, *J. Electroanal. Chem.*, 1991, **318**, 101.
29. J. Koryta, *Electrochim. Acta*, 1979, **24**, 293.
30. M. Imoto, T. Sakaki and T. Osakai, *Anal. Chem.*, 2013, **85**, 4753.
31. Q. Li, S. Xie, Z. Liang, X. Meng, S. Liu, H. H. Girault and Y. Shao, *Angew. Chem., Int. Ed.*, 2009, **48**, 8010.
32. R. Cui, Q. Li, D. E. Gross, X. Meng, B. Li, M. Marquez, R. Yang, J. L. Sessler and Y. Shao, *J. Am. Chem. Soc.*, 2008, **130**, 14364.
33. A. M. O'Mahony, M. D. Scanlon, A. Berduque, V. Beni, D. W. M. Arrigan, E. Faggi and A. Bencini, *Electrochem. Commun.*, 2005, **7**, 976.
34. R. A. W. Dryfe, S. S. Hill, A. P. Davis, J. B. Joos and E. P. L. Roberts, *Org. Biomol. Chem.*, 2004, **2**, 2716.
35. S. H. Lee, J. Sumranjit, P. Tongkate, B. H. Chung and H. J. Lee, *Electrochim. Acta*, 2014, **123**, 198.
36. Z. Samec, *Chem. Rev.*, 1988, **88**, 617.
37. G. Taylor and H. H. Girault, *J. Electroanal. Chem.*, 1986, **208**, 179.
38. G. Herzog and V. Beni, *Anal. Chim. Acta*, 2013, **769**, 10.
39. P. Vazquez, G. Herzog, C. O'Mahony, J. O'Brien, J. Scully, A. Blake, C. O'Mathuna and P. Galvin, *Sens. Actuators, B*, 2014, **201**, 572.
40. M. Durmaz, E. Zor, E. Kocabas, H. Bingol and E. G. Akgemci, *Electrochim. Acta*, 2011, **56**, 5316.
41. R. Kataky and P. Lopes, *Chem. Commun.*, 2009, **12**, 1490.
42. X. Meng, Z. Liang, B. Li, X. Xu, Q. Li, W. Zhao, S. Xie and Y. Shao, *J. Electroanal. Chem.*, 2011, **656**, 125.
43. T. J. Stockmann and Z. Ding, *Phys. Chem. Chem. Phys.*, 2012, **14**, 13949.
44. U. Nestor, H. Wen, G. Girma, Z. Mei, W. Fei, Y. Yang, C. Zhang and D. Zhan, *Chem. Commun.*, 2014, **50**, 1015.
45. J. A. Campbell and H. H. Girault, *J. Electroanal. Chem.*, 1989, **266**, 465.
46. S. N. Faisal, C. M. Pereira, S. Rho and H. J. Lee, *Phys. Chem. Chem. Phys.*, 2010, **12**, 15184.
47. H. J. Lee, P. D. Beattie, B. J. Seddon, M. D. Osborne and H. H. Girault, *J. Electroanal. Chem.*, 1997, **440**, 73.
48. R. Zazpe, C. Hibert, J. O'Brien, Y. H. Lanyon and D. W. M. Arrigan, *Lab Chip*, 2007, **7**, 1732.
49. J. Strutwolf, M. D. Scanlon and D. W. M. Arrigan, *Analyst*, 2009, **134**, 148.
50. S. Amemiya, Y. Wang and M. V. Mirkin, Nanoelectrochemistry at the liquid/liquid interfaces, in *Specialist Periodical Reports – Electrochemistry*, ed. R. G. Compton and J. D. Wadhawan, Royal Society of Chemistry, Cambridge, 2014, pp. 1–43.
51. Y. Shao and M. V. Mirkin, *J. Am. Chem. Soc.*, 1997, **119**, 8103.
52. Y. Wang, J. Velmurugan and M. V. Mirkin, *Anal. Chem.*, 2010, **82**, 77.
53. B. Kralj and R. A. W. Dryfe, *Phys. Chem. Chem. Phys.*, 2001, **3**, 5274.
54. R. A. W. Dryfe and B. Kralj, *Electrochem. Commun.*, 1999, **1**, 128.

55. M. D. Scanlon, J. Strutwolf, A. Blake, D. Iacopino, A. J. Quinn and D. W. M. Arrigan, *Anal. Chem.*, 2010, **82**, 6115.
56. M. D. Scanlon and D. W. M. Arrigan, *Electroanalysis*, 2011, **23**, 1023.
57. M. Rimboud, R. D. Hart, T. Becker and D. W. M. Arrigan, *Analyst*, 2011, **136**, 4674.
58. J. S. Ellis, G. Herzog, B. Glynn and D. W. M. Arrigan, *ECS Trans.*, 2011, **35**, 29.
59. N. Godino, X. Borrise, F. X. Munoz, F. J. del Campo and R. G. Compton, *J. Phys. Chem. C*, 2009, **113**, 11119.
60. M. Sairi, N. Chen-Tan, G. Neusser, C. Kranz and D. W. M. Arrigan, *ChemElectroChem*, 2015, **2**, 98.
61. C. Dekker, *Nat. Nanotechnol.*, 2007, **2**, 209.
62. A. J. Olaya, P. Y. Ge and H. H. Girault, *Electrochem. Commun.*, 2012, **19**, 101.
63. L. E. Barrosse-Antle, A. M. Bond, R. G. Compton, A. M. O'Mahony, E. I. Rogers and D. S. Silvester, *Chem. – Asian J.*, 2010, **5**, 202.
64. Y. Wang, T. Kakiuchi, Y. Yasui and M. V. Mirkin, *J. Am. Chem. Soc.*, 2010, **132**, 16945.
65. R. Ishimatsu, N. Nishi and T. Kakiuchi, *Chem. Lett.*, 2007, **36**, 1166.
66. A. D. Ballantyne, A. K. Brisdon and R. A. W. Dryfe, *Chem. Commun.*, 2008, **40**, 4980.
67. D. S. Silvester and D. W. M. Arrigan, *Electrochem. Commun.*, 2011, **13**, 477.
68. T. J. Stockmann, J. Zhang, J. C. Wren and Z. Ding, *Electrochim. Acta*, 2012, **62**, 8.
69. J. Langmaier, A. Trojanek and Z. Samec, *Electroanalysis*, 2009, **21**, 1977.
70. J. Langmaier and Z. Samec, *Anal. Chem.*, 2009, **81**, 6382.
71. N. Nishi, H. Murakami, S. Imakura and T. Kakiuchi, *Anal. Chem.*, 2006, **78**, 5805.
72. Y. Grunder, H. L. T. Ho, J. F. W. Mosselmans, S. L. M. Schroeder and R. A. W. Dryfe, *Phys. Chem. Chem. Phys.*, 2011, **13**, 15681.
73. M. Platt and R. A. W. Dryfe, *J. Electroanal. Chem.*, 2007, **599**, 323.
74. A. Trojanek, J. Langmaier and Z. Samec, *J. Electroanal. Chem.*, 2007, **599**, 160.
75. Y. Cheng and D. J. Schiffrin, *J. Chem. Soc., Faraday Trans.*, 1996, **92**, 3865.
76. P.-P. Fang, S. Chen, H. Deng, M. D. Scanlon, F. Gumy, H. J. Lee, D. Momotenko, V. Amstutz, F. Cortes-Salazar, C. M. Pereira, Z. Yang and H. H. Girault, *ACS Nano*, 2013, **7**, 9241.
77. X. Zhu, Y. Qiao, X. Zhang, S. Zhang, X. Yin, J. Gu, Y. Chen, Z. Zhu, M. Li and Y. Shao, *Anal. Chem.*, 2014, **86**, 7001.
78. R. Gulaboski, C. M. Pereira, M. N. D. S. Cordeiro, I. Bogeski, E. Ferreira, D. Ribeiro, M. Chirea and A. F. Silva, *J. Phys. Chem. B*, 2005, **109**, 12549.
79. H. Janchenova, K. Stulik, K. Maeda and V. Marecek, *J. Electroanal. Chem.*, 2009, **632**, 55.
80. V. Marecek and H. Janchenova, *J. Electroanal. Chem.*, 2003, **558**, 119.

81. L. Poltorak, G. Herzog and A. Walcarius, *Electrochem. Commun.*, 2013, **37**, 76.
82. L. Poltorak, G. Herzog and A. Walcarius, *Langmuir*, 2014, **30**, 11453.
83. L. Poltorak, M. Dossot, G. Herzog and A. Walcarius, *Phys. Chem. Chem. Phys.*, 2014, **16**, 26955.
84. M. M. Hossain, S. H. Lee, H. H. Girault, V. Devaud and H. J. Lee, *Electrochim. Acta*, 2012, **82**, 12.
85. Y. Kim and S. Amemiya, *Anal. Chem.*, 2008, **80**, 6056.
86. B. Kabagambe, M. B. Garada, R. Ishimatsu and S. Amemiya, *Anal. Chem.*, 2014, **86**, 7939.
87. T. J. Stockmann and Z. Ding, *Anal. Chem.*, 2011, **83**, 7542.
88. P. O'Dwyer and V. Cunnane, *J. Electroanal. Chem.*, 2005, **581**, 16.
89. M. H. M. Cacote, C. M. Pereira, L. Tomaszewski, H. H. Girault and F. Silva, *Electrochim. Acta*, 2004, **49**, 263.
90. A. Sherburn, M. Platt, D. W. M. Arrigan, N. M. Boag and R. A. W. Dryfe, *Analyst*, 2003, **128**, 1187.
91. A. Sherburn, D. W. M. Arrigan, R. A. W. Dryfe and N. M. Boag, *Electroanalysis*, 2004, **16**, 1227.
92. D. Zhan, X. Li, W. Zhan, F.-R. F. Fan and A. J. Bard, *Anal. Chem.*, 2007, **79**, 5225.
93. D. W. M. Arrigan, G. Herzog, M. D. Scanlon and J. Strutwolf, Bioanalytical applications of electrochemistry at liquid-liquid micro-interfaces, in *Electroanalytical Chemistry, A Series of Advances*, ed. A. J. Bard and C. G. Zoski, CRC Press, Boca Raton, 2013, pp. 105–178.
94. D. Homolka, V. Marecek, Z. Samec, K. Base and H. Wendt, *J. Electroanal. Chem.*, 1984, **163**, 159.
95. O. Dvorak, V. Marecek and Z. Samec, *J. Electroanal. Chem.*, 1991, **300**, 407.
96. J. A. Ribeiro, I. M. Miranda, F. Silva and C. M. Pereira, *Phys. Chem. Chem. Phys.*, 2010, **12**, 15190.
97. V. Beni, M. Ghita and D. W. M. Arrigan, *Biosens. Bioelectron.*, 2005, **20**, 2097.
98. A. Berduque, R. Zazpe and D. W. M. Arrigan, *Anal. Chim. Acta*, 2008, **611**, 156.
99. D. Zhan, S. Mao, Q. Zhao, Z. Chen, H. Hu, P. Jing, M. Zhang, Z. Zhu and Y. Shao, *Anal. Chem.*, 2004, **76**, 4128.
100. G. Herzog, V. Kam, A. Berduque and D. W. M. Arrigan, *J. Agric. Food Chem.*, 2008, **56**, 4304.
101. E. Torralba, J. A. Ortuno, A. Molina, C. Serna and F. Karimian, *Anal. Chim. Acta*, 2014, **826**, 12.
102. A. Molina, C. Serna, J. A. Ortuno, J. Gonzalez, E. Torralba and A. Gil, *Anal. Chem.*, 2009, **81**, 4220.
103. J. A. Ortuno, A. Gil and C. Sanchez-Pedreno, *Sens. Actuators, B*, 2007, **122**, 369.
104. H. R. Kim, Detection of irinotecan using an interface between two immiscible electrolyte solutions, MSc Degree Thesis (Kyungpook National University), 2015, pp. 1–25.

105. M. Sairi and D. W. M. Arrigan, *Talanta*, 2015, **132**, 205.
106. M. B. Garada, B. Kabagambe, Y. Kim and S. Amemiya, *Anal. Chem.*, 2014, **86**, 11230.
107. M. M. Hossain, S. N. Faisal, C. S. Kim, H. J. Cha, S. C. Nam and H. J. Lee, *Electrochem. Commun.*, 2011, **13**, 611.
108. M. M. Hossain, C. S. Kim, H. J. Cha and H. J. Lee, *Electroanalysis*, 2011, **23**, 2049.
109. P. Vanysek, J. D. Reid, M. A. Craven and R. P. Buck, *J. Electrochem. Soc.*, 1984, **131**, 1788.
110. M. Shinshi, T. Sugihara, T. Osakai and M. Goto, *Langmuir*, 2006, **22**, 5937.
111. M. Shinshi, T. Sugihara, T. Osakai and M. Goto, *Langmuir*, 2006, **22**, 8614.
112. T. Osakai, Y. Yuguchi, E. Gohara and H. Katano, *Langmuir*, 2010, **26**, 11530.
113. G. Herzog, V. Kam and D. W. M. Arrigan, *Electrochim. Acta*, 2008, **53**, 7204.
114. F. Kivlehan, Y. H. Lanyon and D. W. M. Arrigan, *Langmuir*, 2008, **24**, 9876.
115. M. D. Scanlon, E. Jennings and D. W. M. Arrigan, *Phys. Chem. Chem. Phys.*, 2009, **11**, 2272.
116. R. A. Hartvig, M. A. Mendez, M. van de Weert, L. Jorgensen, J. Ostergaard, H. H. Girault and H. Jensen, *Anal. Chem.*, 2010, **82**, 7699.
117. E. Alvarez de Eulate, L. Qiao, M. D. Scanlon, H. H. Girault and D. W. M. Arrigan, *Chem. Commun.*, 2014, **50**, 11829.
118. M. D. Scanlon, J. Strutwolf and D. W. M. Arrigan, *Phys. Chem. Chem. Phys.*, 2010, **12**, 10040.
119. E. Alvarez de Eulate and D. W. M. Arrigan, *Anal. Chem.*, 2012, **84**, 2505.
120. E. Alvarez de Eulate, L. Serls and D. W. M. Arrigan, *Anal. Bioanal. Chem.*, 2013, **405**, 3801.
121. S. O'Sullivan, E. A. de Eulate, Y. H. Yuen, E. Helmerhorst and D. W. M. Arrigan, *Analyst*, 2013, **138**, 6192.
122. E. Alvarez de Eulate, S. O'Sullivan, S. Fletcher, P. Newsholme and D. W. M. Arrigan, *Chem. - Asian J.*, 2013, **8**, 2096.
123. S. Amemiya, Y. Kim, R. Ishimatsu and B. Kabagambe, *Anal. Bioanal. Chem.*, 2011, **399**, 571.
124. B. R. Horrocks and M. V. Mirkin, *Anal. Chem.*, 1998, **70**, 4653.
125. T. Osakai, H. Komatsu and M. Goto, *J. Phys.: Condens. Matter*, 2007, **19**, 375103.
126. Y. Shao and M. V. Mirkin, *Anal. Chem.*, 1998, **70**, 3155.
127. Y. W. Choi and S. H. Moon, *Environ. Monit. Assess.*, 2004, **92**, 163.
128. Y. Kudo, Y. Takeda and H. Matsuda, *J. Electroanal. Chem.*, 1995, **396**, 333.
129. B. Rezaei, S. Meghdadi and V. Nafisi, *Sens. Actuators, B*, 2007, **121**, 600.

130. S. H. Lee, H. Kim, H. H. Girault and H. J. Lee, *Bull. Korean Chem. Soc.*, 2013, **34**, 2577.
131. W. S. Han, Y. H. Lee, K. J. Jung, S. Y. Ly, T. K. Hong and M. H. Kim, *J. Anal. Chem.*, 2008, **63**, 987.
132. P. C. Pandey and R. Prakash, *Sens. Actuators, B*, 1998, **46**, 61.
133. N. J. Evans, M. Gonsalves, N. J. Gray, A. L. Barker, J. V. Macpherson and P. R. Unwin, *Electrochem. Commun.*, 2000, **2**, 201.
134. B. Liu, Y. Shao and M. V. Mirkin, *Anal. Chem.*, 2000, **72**, 510.
135. O. Shirai, S. Kihara, Y. Yoshida and M. Matsui, *J. Electroanal. Chem.*, 1995, **389**, 61.
136. A. Yoshizumi, A. Uehara, M. Kasuno, Y. Kitatsuji, Z. Yoshida and S. Kihara, *J. Electroanal. Chem.*, 2005, **581**, 275.
137. P. D. Beattie, A. Delay and H. H. Girault, *J. Electroanal. Chem.*, 1995, **380**, 167.
138. J. A. Campbell, A. A. Stewart and H. H. Girault, *J. Chem. Soc., Faraday Trans.*, 1989, **85**, 843.
139. P. D. Beattie, A. Delay and H. H. Girault, *Electrochim. Acta*, 1995, **40**, 2961.
140. M. J. Stephenson, S. M. Holmes and R. A. W. Dryfe, *Electrochem. Commun.*, 2004, **6**, 294.
141. J. A. Ribeiro, F. Silva and C. M. Pereira, *Electroanalysis*, 2013, **25**, 2331.
142. A. Molina, E. Torralba, C. Serna and J. A. Ortuno, *Electrochim. Acta*, 2013, **106**, 244.
143. Y. Yoshida, M. Matsui, K. Maeda and S. Kihara, *Anal. Chim. Acta*, 1998, **374**, 269.
144. P. Sun, F. O. Laforge and M. V. Mirkin, *J. Am. Chem. Soc.*, 2005, **127**, 8596.
145. B. Kralj and R. A. W. Dryfe, *J. Electroanal. Chem.*, 2003, **560**, 127.
146. T. Kakutani, Y. Nishiwaki, T. Osakai and M. Senda, *Bull. Chem. Soc. Jpn.*, 1986, **59**, 781.
147. V. V. Egorov and A. A. Bolotin, *Talanta*, 2006, **70**, 1107.
148. T. Osakai, T. Kakutani and M. Senda, *Anal. Sci.*, 1987, **3**, 521.
149. T. Osakai, T. Kakutani and M. Senda, *Anal. Sci.*, 1988, **4**, 529.
150. S. Sriram, P. K. Mohapata, A. K. Pandey, V. K. Manchanda and L. P. Badheka, *J. Membr. Sci.*, 2000, **177**, 163.
151. H. A. Arida, J. P. Kloock and M. J. Schoning, *Sensors*, 2006, **6**, 435.
152. S. Wilke and H. Wang, *J. Electroanal. Chem.*, 1999, **475**, 9.
153. S. Jadhav and E. Bakker, *J. Anal. Chem.*, 2001, **73**, 80.
154. Y. Shao, S. N. Tan, V. Devaud and H. H. Girault, *J. Chem. Soc., Faraday Trans.*, 1993, **89**, 4307.
155. U. Oesch, Z. Brzozka, A. Xu and W. Simon, *Med. Biol. Eng. Comput.*, 1987, **25**, 414.
156. A. V. Juarez and L. M. Yudi, *Electrochim. Acta*, 2008, **54**, 530.
157. H. Bingol and A. Coskun, *J. Solution Chem.*, 2011, **40**, 1505.
158. Q. Qian, G. S. Wilson, K. Bowman-James and H. H. Girault, *Anal. Chem.*, 2001, 73, 497.

159. J. A. Ortuno, R. Exposito, C. Sanchez-pedreno, M. I. Albero and A. Espinosa, *Anal. Chim. Acta*, 2004, **525**, 231.
160. M. A. Rahman, H. Doe, N. Sakurada and R. Arakawa, *Electrochim. Acta*, 2001, **47**, 623.
161. N. Katif, R. A. Harries, A. M. Kelly, J. S. Fossey, T. D. James and F. Marken, *J. Solid State Electrochem.*, 2009, **13**, 1475.
162. R. Gulaboski, E. S. Ferreira, C. M. Pereira, M. N. D. S. Cordeiro, A. Garau, V. Lippolis and A. F. Silva, *J. Phys. Chem. C*, 2008, **112**, 153.
163. K. Holub, H. Janchenova, K. Stulik and V. Marecek, *J. Electroanal. Chem.*, 2009, **632**, 8.
164. M. R. Ganjali, P. Norouzi, F. Faridbod, N. Hajiabdollah, B. Larijani and Y. Hanifepour, *Anal. Lett.*, 2007, **40**, 2544.
165. J. Wickens, R. A. W. Dryfe, F. S. Mair, R. G. Pritchard, R. Hayes and D. W. M. Arrigan, *New J. Chem.*, 2000, **24**, 149.
166. A. Ceresa, A. Radu, S. Peper, E. Bakker and E. Pretsch, *Anal. Chem.*, 2002, **74**, 4027.
167. A. Shvarev and E. Bakker, *Anal. Chem.*, 2003, **75**, 4541.
168. F. Kivlehan, W. J. Mace, H. A. Moynihan and D. W. M. Arrigan, *Electrochim. Acta*, 2009, **54**, 1919.
169. F. Kivlehan, W. J. Mace, H. M. Moynihan and D. W. M. Arrigan, *Anal. Chim. Acta*, 2007, **585**, 154.
170. A. J. Michalska, C. Appaih-kusi, Y. H. Lee, S. Walkiewicz and E. A. H. Hall, *Anal. Chem.*, 2004, **76**, 2031.
171. B. Kabagambe, A. Izadyar and S. Amemiya, *Anal. Chem.*, 2012, **84**, 7979.
172. E. Totu, A. M. Josceanu and A. K. Covington, *Mater. Sci. Eng. C*, 2001, **18**, 87.
173. Z. Mousavi, J. Bobacka, A. Lewenstam and A. Ivaska, *J. Electoanal. Chem.*, 2009, **633**, 246.
174. G. A. Crespo, S. Macho and F. X. Rius, *Anal. Chem.*, 2008, **80**, 1316.
175. J. Lu, Q. Chen, D. Diamond and J. Wang, *Analyst*, 1993, **118**, 1131.
176. S. Amemiya and A. J. Bard, *Anal. Chem.*, 2000, **72**, 4940.
177. A. Benvidi, S. N. Lanjwani and Z. Ding, *J. Electroanal. Chem.*, 2010, **641**, 99.
178. Y. Qiao, B. Zhang, X. Zhu, T. Ji, B. Li, Q. Li, E. Chen and Y. Shao, *Electroanalysis*, 2013, **25**, 1080.
179. L. Tomaszewski, Z. Ding, D. J. Fermin, H. M. Cacote, C. M. Pereira, F. Silva and H. H. Girault, *J. Electroanal. Chem.*, 1998, **453**, 171.
180. F. Kaykal, E. Kocabas, H. Bingol, E. G. Akgemci and A. Coskun, *J. Electroanal. Chem.*, 2011, **654**, 96.
181. H. Bingol, F. Kaykal, E. G. Akgemci and A. Sirit, *Electroanalysis*, 2010, **22**, 2825.
182. E. S. Ferreira, A. Garau, V. Lippolis, C. M. Pereira and A. F. Silva, *J. Electroanal. Chem.*, 2006, **587**, 155.
183. H. Katano, H. Kuboyama and M. Senda, *J. Electroanal. Chem.*, 2000, **483**, 117.

184. C. M. Pereira, J. M. Oliveira, R. M. Silva and F. Silva, *Anal. Chem.*, 2004, **76**, 5547.
185. H. Bingol, E. G. Akgemci, M. Ersoz and T. Atalay, *Electroanalysis*, 2007, **19**, 1327.
186. H. Katano and M. Senda, *Anal. Sci.*, 1998, **14**, 63.
187. C. Baslak and T. Atalay, *Russ. J. Electrochem.*, 2014, **50**, 180.
188. T. Ohkouchi, T. Kakutani, T. Osakai and M. Senda, *Anal. Sci.*, 1991, 7, 371.
189. C. J. Collins and D. W. M. Arrigan, *Anal. Chem.*, 2009, **81**, 2344.
190. M. Velicky, K. Y. Tam and R. A. W. Dryfe, *J. Electroanal. Chem.*, 2012, **683**, 94.
191. H. J. Lee, C. Beriet and H. H. Girault, *Anal. Sci.*, 1998, **14**, 71.
192. G. Herzog, S. Flynn and D. W. M. Arrigan, **307**, *J. Phys. Conf. Ser.*, 2011, 012055.
193. B. Huang, B. Yu, P. Li, M. Jiang, Y. Bi and S. Wu, *Anal. Chim. Acta*, 1995, **312**, 329.
194. S. N. Faisal, M. M. Hossain and H. J. Lee, *J. Electrochem. Sci. Technol.*, 2010, **1**, 121.
195. S. E. C. Dale and P. R. Unwin, *Electrochem. Commun.*, 2008, **10**, 723.
196. P. Sun, F. O. Laforge and M. V. Mirkin, *J. Am. Chem. Soc.*, 2007, **129**, 12410.
197. T. J. Stockmann, A.-M. Montgomery and Z. Ding, *J. Electroanal. Chem.*, 2012, **684**, 6.
198. V. Mirceski, F. Quentel, M. L'Her and A. Pondaven, *Electrochem. Commun.*, 2005, **7**, 1122.
199. G. Bouchard, P.-A. Carrupt, B. Testa, V. Gobry and H. H. Girault, *Pharm. Res.*, 2001, **18**, 702.
200. F. Li, B. Su, F. C. Salazar, R. P. Nia and H. H. Girault, *Electrochem. Commun.*, 2009, **11**, 473.
201. H. J. Lee, C. M. Pereira, A. F. Silva and H. H. Girault, *Anal. Chem.*, 2000, **72**, 5562.
202. N. Tsujioka, S. Imakura, N. Nishi and T. Kakiuchi, *Anal. Sci.*, 2006, **22**, 667.

CHAPTER 10

Electrochemical Detection Using Ionic Liquids

DEBBIE S. SILVESTER*[a] AND LEIGH ALDOUS*[b]

[a] Nanochemistry Research Institute, Department of Chemistry, Curtin University, Perth, WA, Australia; [b] School of Chemistry, UNSW Australia, Sydney, NSW, Australia
*Email: d.silvester-dean@curtin.edu.au; l.aldous@unsw.edu.au

10.1 Introduction

The purpose of this chapter is to review the advances in electrochemical detection strategies using ionic liquids. Ionic liquids are a relatively new type of solvent and are being widely investigated as (potentially superior) replacements for conventional solvent/electrolyte combinations traditionally used in electrochemical sensing applications. As will be seen from this chapter, this area is very popular and rapidly developing, due to the many advantageous properties of ionic liquids. This chapter will focus on introducing ionic liquids and their electrochemical properties, followed by discussing the many ways in which they can be employed in electrochemical sensing experiments. It will be made clear that, on top of many demonstrated applications of ionic liquids, they still have much more to offer for the next generation of chemical sensors.

10.1.1 What are Ionic Liquids?

Ionic liquids (ILs) are generally defined as salts that melt at temperatures below 100 °C, with room temperature ionic liquids (RTILs) existing in the

RSC Detection Science Series No. 6
Electrochemical Strategies in Detection Science
Edited by Damien W. M. Arrigan

Published by the Royal Society of Chemistry, www.rsc.org

liquid state at 25 °C. Although there are so-called "first generation" haloaluminate ILs, this chapter will focus only on "second generation" non-haloaluminate ILs,[1] usually containing a bulky organic cation and an inorganic anion (although other ion combinations are also possible). They possess several archetypal properties such as low-volatility, chemical and thermal stability, intrinsic conductivity, high polarity, high viscosity, wide electrochemical windows and the ability to dissolve a wide range of species. RTILs were originally explored as electrochemical solvents,[1–3] but have now found use in several applications such as catalysis,[4–6] "green" chemistry,[7–9] organic reactivity,[10,11] analytical uses,[12,13] biocatalysis and enzymes[14,15] and applications in the chemical industry.[16] Some review papers are also available in relation to ILs and electrochemical sensing.[17–20]

The structures, abbreviations and nomenclature of all the ionic liquid cations and anions discussed in this chapter are given in Scheme 10.1.

The main attraction of ILs/RTILs is that they are highly tunable for various purposes, and different functional groups/carbon chain lengths can be added to the structure. It is estimated that up to 10^{18} different combinations of ILs are possible![21] As will become apparent in the next sections, imidazolium is by far the most common cation used for most IL studies (and thus by extension sensing applications), with tetraalkylammonium, tetraalkylphosphonium, pyrrolidinium and pyridinium cations also popular. Widely used anions include bis(trifluoromethylsulfonyl)imide, tetrafluoroborate and hexafluorophosphate, and the anion is the primary ion dictating hydrophobicity and hydrophilicity.[1] RTILs also have controllable miscibility, and those that are immiscible with water have even been used in liquid/liquid experiments for detecting chemical species.[22]

Since each ionic liquid has unique properties (viscosity, conductivity, solvation ability, electrochemical window, reactivity, *etc.*) and there are so many combinations, many studies rely on some "trial-and-error" before suitable characteristics are achieved. Due to gaps in our knowledge, optimum cation/anion combinations cannot be intelligently chosen at present.

As a result, research into ILs/RTILs is a highly active field, covering a vast number of ILs and applications. In particular, there has been a huge increase in publications since around 1999, coinciding with the publication of a comprehensive review by Welton,[5] which is the most highly cited IL paper to date (>8000 citations in the first 16 years).

10.1.2 Inherent Electrochemical Properties of Ionic Liquids

ILs are composed entirely of ions. To have a low melting point, and thus qualify as an IL or RTIL, they typically consist of a bulky, asymmetric organic cation and a weakly-coordinated inorganic (or organic) anion. As a direct result, they are inherently conductive and do not require additional supporting electrolyte in electrochemical experiments (*i.e.* they act as both the solvent and supporting electrolyte). Their conductivity is often "on a par"

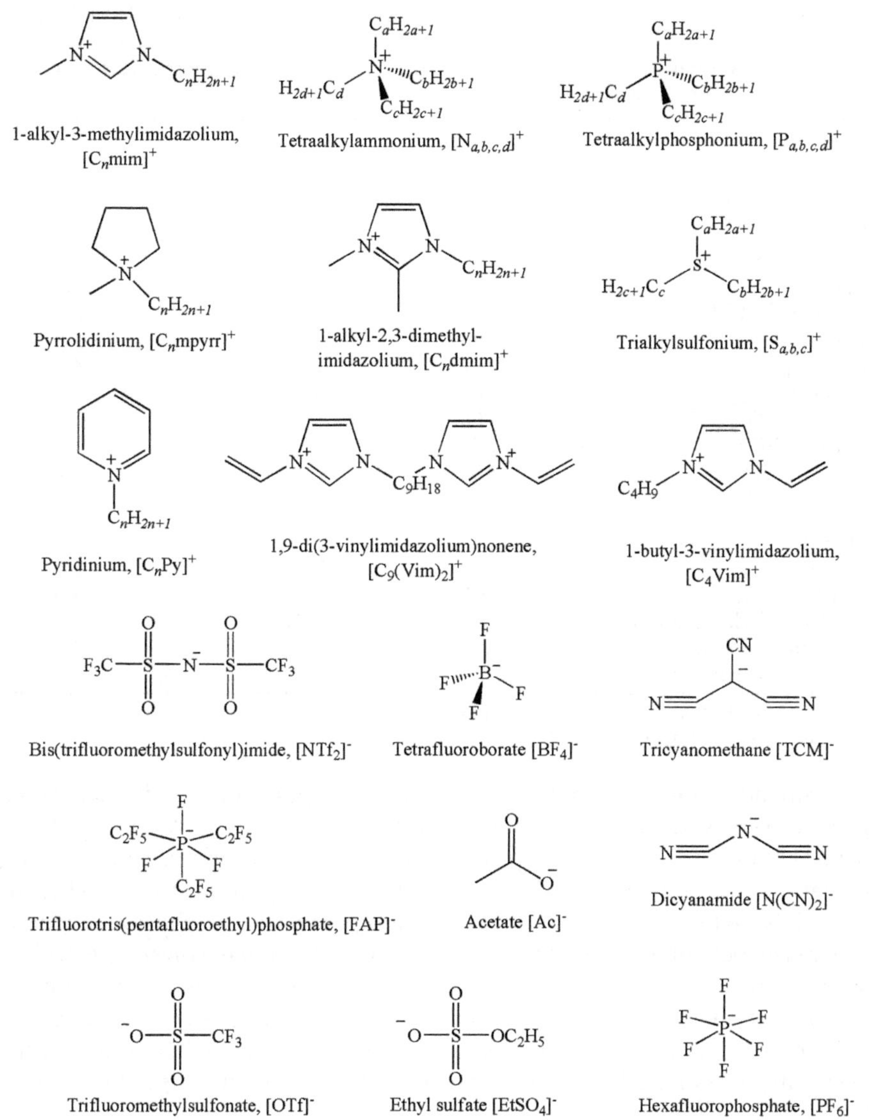

Scheme 10.1 Chemical structures, full names and abbreviations of the ionic liquid cations and anions described in this chapter.

with traditional solvents containing supporting electrolyte (*e.g.* acetonitrile with 0.1 M tetrabutylammonium perchlorate).[23] This allows them to be easily employed in electrochemical experiments without the need to add additional electrolyte, simplifying the experimental set-up, minimizing waste and providing the potential to recycle the solvent (*e.g.* after an electrosynthetic reaction).[10] Their extremely low to negligible volatility supports miniaturisation of sensor devices, and the application of thin films open to the atmosphere, *e.g.* as electrochemical gas sensors.[24]

Table 10.1 Viscosity (η), conductivity (κ) and electrochemical window (EW) data for various commonly-used aprotic RTILs and conventional solvents. References to the original papers can be found in the review paper by Barrosse-Antle *et al.*[26]

Solvent	η (cP)	κ (mS cm^{-1})	EW[a] (V)
RTILs			
[C_2mim][NTf_2]	34	8.8	4.3
[C_4mim][NTf_2]	52	3.9	4.8
[C_4dmim][NTf_2]	105	2.0	5.2
[C_6mim][FAP]	74	1.3	5.3
[C_4mpyrr][NTf_2]	89	2.2	5.2
[C_4mim][OTf]	90	3.7	4.9
[C_4mim][BF_4]	112	1.7	4.7
[$N_{6.2.2.2}$][NTf_2]	167	0.67	5.4
[C_4mim][NO_3]	266	—	3.7
[C_4mim][PF_6]	371	1.5	4.7
[$P_{14.6.6.6}$][NTf_2]	450	—	5.0
[$P_{14.6.6.6}$][FAP]	464	—	5.6
Organic			
Acetonitrile	0.34	7.6[b]	5.0[b]
Dichloromethane	0.44	—	3.5[b]
N,N-Dimethylformamide	0.92	4.07[b]	4.3[b]
Dimethyl sulfoxide	1.99	2.7[b]	4.4[b]
Propylene carbonate	2.5	—	4.7[b]

[a]Obtained (for RTILs) at 10 μm diameter Pt electrode.
[b]Containing 0.1 M Bu_4NClO_4 at 295 K.

A very advantageous property from a sensing point of view is that RTILs possess very wide electrochemical windows (up to *ca.* 7 V).[25] The large windows are a result of the high stability of the anions and cations towards oxidation and reduction. The wide windows may allow the sensing of materials that oxidise or reduce at very high potentials, *i.e.* those that would otherwise be out of the potential range in traditional solvent/electrolyte systems. ILs can also be mixed with other solvents to act as "conventional" supporting electrolytes, while also contributing some advantages to the system, *i.e.* enlarged anodic electrochemical windows when mixed with acetonitrile, especially when employing the highly hydrophobic, stable [FAP]$^-$ anion.[23]

Table 10.1 summarizes the viscosity, conductivity and electrochemical window for some selected RTILs. The references to the original data shown in Table 10.1 can be found in the review paper by Barrosse-Antle *et al.*[26]

10.1.3 Task Specific Ionic Liquids

Ionic liquids are almost invariably composed of large (ionic) molecules with low melting points. As a direct result, minor chemical changes to these large molecules can result in altered physical properties, opening up the possibility of "fine-tuning" the system. The introduction of functional groups can also be achieved, resulting in liquids that have inherent functionality. Such a concept is encompassed in the term "task specific ionic liquids".[27] Many electroanalytical studies have employed conventional ILs to investigate their effect upon the electroanalytical performance of a certain system. In some cases (*cf.* carbon paste electrodes in particular, Section 10.2.3), the results were extremely beneficial. Recently, there is an increasing trend to select or "design" the IL to have a specific role in the analytical system, such as task-specific ILs for extracting metals from aqueous systems into the distinct hydrophobic IL phase,[28,29] capturing gas molecules from the gas phase,[30] dissolving insoluble pollutants such as heavy metal oxide (nano)particles in ILs to facilitate their direct electrochemistry,[31] digesting food samples,[32] *etc.*

The remainder of this chapter will explore various applications of ILs as, and in, electroanalytical sensors. The specific ILs discussed in this chapter, and their full names, are given in Table 10.2, along with the applications in which they are employed. Note that, in many cases, the particular IL may have been chosen arbitrarily (*e.g.* availability in the laboratory), and often the IL can be changed for another and the sensing technique will still work. There are, however, several examples where the IL (and the IL structure) play a key role in the analytical technique, *e.g.* the use of an anion with a basic acetate anion for the detection acidic vapors (Section 10.2.1.2), the use of an inherently acidic IL for the dissolution and quantification of heavy metals (Section 10.2.5) and various TSILs that have been used for bioanalysis applications (Section 10.2.4).

The use of ILs in electroanalytical detection is still a growing field, with many possible future refinements and novel applications. Readers are encouraged to identify what specific properties they would like in their own systems. The diversity and flexibility of ILs will almost certainly be able to assist.

10.2 Electrochemical Detection Using Ionic Liquids

10.2.1 Gases

Ionic liquids have many advantages as electrolytes in gas sensors. They tend to solubilise a wide range of gases easily, can work in gas sensor devices at room temperature (in contrast to many gas sensors employing solid materials, *e.g.* metal oxides, that only work at high temperatures), and they have high intrinsic conductivity, meaning they can be used by personnel straight "out of the bottle" without having to make up solvent/electrolyte solutions. In particular, their extremely low volatility means that they will not evaporate, and hence they have the potential to extend the lifetime of sensors that traditionally use aqueous/organic solvents. RTIL-based gas sensing research typically uses a voltammetric/amperometric approach, where two or three electrodes are connected through the RTIL solvent.[33] In these devices, gas detection is achieved when gas partitions into the RTIL from the gas phase, then diffuses through the RTIL and is detected at the working electrode surface. More recently, the wide availability of

Table 10.2 Summary of the ionic liquids discussed throughout this chapter and their corresponding application(s).[a]

Abbreviation	Full name	Application(s)	Section(s)[b]
$[C_2mim][NTf_2]$	1-Ethyl-3-methylimidazolium bis(trifluoromethylsulfonyl)imide	Gas sensing	10.2.1.1 10.2.1.3 (×3)
$[C_2mim][BF_4]$	1-Ethyl-3-methylimidazolium tetrafluoroborate	Gas sensing Carbon paste Heavy metals Other analytes	10.2.1.1 10.2.1.3 10.2.3 10.2.5 10.2.6
$[C_4mim][NTf_2]$	1-Butyl-3-methylimidazolium bis(trifluoromethylsulfonyl)imide	Gas sensing Explosives/CWAs Bioanalysis Heavy metals Other analytes	10.2.1.1 10.2.1.2 (×5) 10.2.2 (×2) 10.2.4 10.2.5 10.2.6 (×3)
$[C_4mpyrr][NTf_2]$	*N*-Butyl-*N*-methylpyrrolidinium bis(trifluoromethylsulfonyl)imide	Gas sensing Explosives/CWAs Heavy metals Other analytes	10.2.1.1 10.2.1.3 (×4) 10.2.2 10.2.5 (×3) 10.2.6 (×2)
$[N_{4,4,4,1}][NTf_2]$	Methyl(tributyl)ammonium bis(trifluoromethylsulfonyl)imide	Gas sensing	10.2.1.1
$[P_{14.6.6.6}][FAP]$	Trihexyltetradecylphosphonium tris(pentafluoroethyl)trifluorophosphate	Gas sensing	10.2.1.1
$[C_3mim][NTf_2]$	1-Propyl-3-methylimidazolium bis(trifluoromethylsulfonyl)imide	Gas sensing	10.2.1.1
$[C_4mim][PF_6]$	1-Butyl-3-methylimidazolium hexafluorophosphate	Gas sensing Explosives/CWAs Carbon paste Bioanalysis Heavy metals Other analytes	10.2.1.1 10.2.1.3 (×3) 10.2.2 (×3) 10.2.3 (×2) 10.2.4 (×3) 10.2.5 (×2) 10.2.6 (×3)
$[C_6mim]Cl$	1-Hexyl-3-methylimidazolium chloride	Gas sensing	10.2.1.1
$[C_6mim][NTf_2]$	1-Hexyl-3-methylimidazolium bis(trifluoromethylsulfonyl)imide	Gas sensing	10.2.1.1
$[C_6mim][TCM]$	1-Hexyl-3-methylimidazolium tricyanomethane	Gas sensing	10.2.1.1
$[C_8mim][BF_4]$	1-Octyl-3-methylimidazolium tetrafluoroborate	Gas sensing	10.2.1.2

$[C_2mim][N(CN)_2]$	1-Ethyl-3-methylimidazolium dicyanamide	Gas sensing	10.2.1.2
$[C_4mim][N(CN)_2]$	1-Butyl-3-methylimidazolium dicyanamide	Gas sensing	10.2.1.2 (×2)
$[C_4mpyrr][N(CN)_2]$	*N*-Butyl-*N*-methylpyrrolidinium dicyanamide	Gas sensing	10.2.1.2
$[C_4mim][Ac]$	1-Butyl-3-methylimidazolium acetate	Gas sensing	10.2.1.2
$[C_4mim]Br$	1-Butyl-3-methylimidazolium bromide	Gas sensing	10.2.1.2
$[C_4mim][OTf]$	1-Butyl-3-methylimidazolium trifluoromethylsulfonate	Gas sensing	10.2.1.3
		Other analytes	10.2.6 (×2)
$[C_4mim][BF_4]$	1-Butyl-3-methylimidazolium tetrafluoroborate	Gas sensing	10.2.1.3
		Explosives/CWAs	10.2.2 (×2)
		Carbon paste	10.2.3
		Bioanalysis	10.2.4
		Heavy metals	10.2.5 (×3)
		Other analytes	10.2.6 (×5)
$[C_6mim][FAP]$	1-Hexyl-3-methylimidazolium tris(pentafluoroethyl)trifluorophosphate	Gas sensing	10.2.1.3
$[C_4mim][FAP]$	1-Butyl-3-methylimidazolium tris(pentafluoroethyl)trifluorophosphate	Explosives/CWAs	10.2.2
		Heavy metals	10.2.5
$[C_4mpyrr][FAP]$	*N*-Butyl-*N*-methylpyrrolidinium tris(pentafluoroethyl)trifluorophosphate	Explosives/CWAs	10.2.2
		Other analytes	10.2.6
$[S_{2,2,2}][NTf_2]$	Triethylsulfonium bis(trifluoromethylsulfonyl)imide	Explosives/CWAs	10.2.2
$[C_8Py][PF_6]$	*N*-Octylpyridinium hexafluorophosphate	Carbon paste	10.2.3 (×2)
		Bioanalysis	10.2.4
		Heavy metals	10.2.5 (×3)
$[C_3mim][PF_6]$	1-Propyl-3-methylimidazolium hexafluorophosphate	Carbon paste	10.2.3
$[C_5mim][PF_6]$	1-Pentyl-3-methylimidazolium hexafluorophosphate	Carbon paste	10.2.3
$[C_4Py][PF_6]$	*N*-Butylpyridinium hexafluorophosphate	Carbon paste	10.2.3 (×2)
$[C_6Py][PF_6]$	*N*-Hexylpyridinium hexafluorophosphate	Carbon paste	10.2.3 (×4)
$[C_8mim][PF_6]$	1-Octyl-3-methylimidazolium hexafluorophosphate	Carbon paste	10.2.3 (×2)
$[C_6mim][NTf_2]$	1-Hexyl-3-methylimidazolium bis(trifluoromethylsulfonyl)imide	Carbon paste	10.2.3
—	1-Amyl-3-methylimidazolium bromide	Carbon paste	10.2.3
—	1,3-Dipropylimidazolium bromide	Carbon paste	10.2.3
$[C_{12}mim][PF_6]$	1-Dodecyl-3-methylimidazolium hexafluorophosphate	Carbon paste	10.2.3
$[C_4mim]Br$	1-Butyl-3-methylimidazolium bromide	Carbon paste	10.2.3
$[C_6mim][PF_6]$	1-Hexyl-3-methylimidazolium hexafluorophosphate	Carbon paste	10.2.3
$[P_{1,4,4,4}][Tos]$	Tributylmethylphosphonium tosylate	Bioanalysis	10.2.4
$[Eim][OTf]$	Ethylimidazolium trifluoromethylsulfonate	Bioanalysis	10.2.4

Table 10.2 (*Continued*)

Abbreviation	Full name	Application(s)	Section(s)[b]
—	1-Ethylamine-2,3-dimethylimidazolium bromide	Bioanalysis	10.2.4
—	1-(4-Sulfonylbutyl)-3-methylimidazolium hexafluorophosphate	Bioanalysis	10.2.4
[C_2mim][$EtSO_4$]	1-Ethyl-3-methylimidazolium ethyl sulfate	Heavy metals	10.2.5
[C_2mim]Cl	1-Ethyl-3-methylimidazolium chloride	Heavy metals	10.2.5
[C_4dmim][NTf_2]	1-Butyl-2,3-dimethylimidazolium bis(trifluoromethylsulfonyl)imide	Heavy metals	10.2.5
[$N_{2,1,1,3}$][NTf_2]	Ethyl(dimethyl)(propyl)ammonium bis(trifluoromethylsulfonyl)imide	Heavy metals	10.2.5
		Other analytes	10.2.6
[$P_{14.6.6.6}$][NTf_2]	Trihexyltetradecylphosphonium bis(trifluoromethylsulfonyl)imide	Other analytes	10.2.6
[AEIm][BF_4]	1-Allyl-3-ethylimidazolium tetrafluoroborate	Other analytes	10.2.6
[C_4mim][Cys]	1-Butyl-3-methylimidazolium 2-amino-3-mercaptoproponic acid (L-cysteine) salt	Other analytes	10.2.6
[HeMIM][NTf_2]	1-(2′-Hydroxyethyl)-3-methylimidazolium bis(trifluoromethylsulfonyl)imide	Other analytes	10.2.6
[C_4Vim][PF_6]	1-Butyl-3-vinylimidazolium hexafluorophosphate	Other analytes	10.2.6
[$C_9(Vim)_2$][PF_6]$_2$	1,9-Di(3-vinylimidazolium)nonane dihexafluorophosphate	Other analytes	10.2.6
Various TSILs	—	Bioanalysis	10.2.4 (×8)
		Other analytes	10.2.6 (×3)

[a]Notably, many of these ILs were most likely chosen arbitrarily (*e.g.* availability in the laboratory) and it is possible that these may be replaced with other ILs and the sensing technique will still work. However, there are also cases where the IL structure plays a key role in the analytical technique (as discussed in the text). This table merely represents the range of ILs that have been employed in different electroanalysis applications.

[b]The parentheses after the section number indicate that this IL has been used multiple times for a particular application. For example, "10.2.4 (×4)" means that there are four different research articles (mentioned in this chapter) that have used this RTIL for bioanalysis applications.

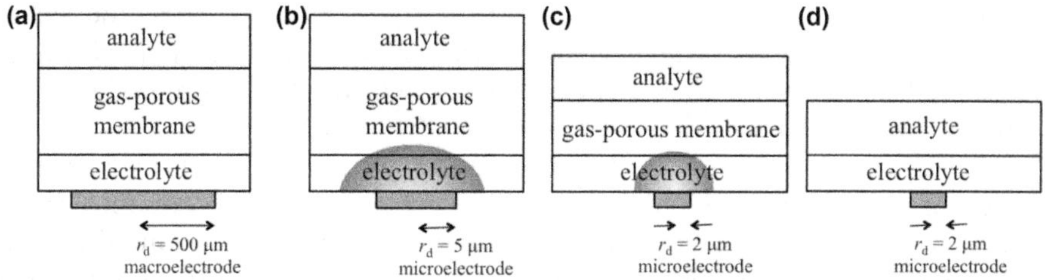

Figure 10.1 Schematic diagrams of various gas sensing devices.[24] Copyright The Electrochemical Society. Reproduced with permission.

"lab-on-a-chip"-type devices (*e.g.* screen printed electrodes) for academic research means that RTILs can be employed in microliter (or even sub microliter) quantities for gas sensing,[34] allowing for miniaturisation of devices and reduced costs compared to gas sensors using pure metal electrodes or large volumes of electrolyte.

Currently, there are no commercially-marketed gas sensors using RTILs, but aqueous/organic solvent-type sensors are based on modifications of the Clark cell, which use membranes that can aid selectivity towards particular gases. Figure 10.1 shows schematic diagrams of various gas sensing devices, with the electrode size gradually decreasing to micrometer dimensions. Figure 10.1(a) shows a traditional sensor design with a millimeter-sized working electrode, liquid supporting electrolyte (usually an aqueous buffer) and a polymer to allow gas to enter the electrolyte and avoid electrode fouling. Figure 10.1(b) shows the same configuration as (a), but where the electrode has been reduced from millimeter to micrometer size. In this case, the diffusion layer partially overlaps with the membrane, compared to complete overlap in (a). A further modification of the electrode size to smaller dimensions (Figure 10.1c) ensures that the diffusion layer does not encroach on the membrane, and that the response is dependent on the electrolyte and electrode dimensions, and not on the properties of the membrane. Finally, Figure 10.1(d) shows a sensor design using an RTIL, where the RTIL acts as both the membrane and electrolyte (see Rogers *et al.*[24] for a detailed discussion). This "membrane-free" design is possible due to the low/negligible volatility of the RTIL.

The main developments using RTILs for gas sensing in academic research will be described here in this chapter. This section is split into three parts, focusing on oxygen detection, volatile organic compound detection and the sensing of highly toxic gases.

10.2.1.1 Oxygen

By far, the highest number of published articles on RTIL-based gas sensing involves the detection of oxygen. This is probably due to the relative safety of the gas, along with the well-known chemically reversible one-electron reduction to superoxide in aprotic and RTIL solvents. Additionally, the electroreduction of oxygen is vital in various applications such as fuel cells, metal–air batteries and the electrosynthesis of reactive oxygen species (*e.g.* superoxide and hydrogen peroxide).[24] Here, we highlight some of the key studies reported for the detection of O_2 in RTILs.

As early as 2004, Wang *et al.*[35] reported an amperometric O_2 gas sensor based on a supported membrane-coated electrode. The RTIL [C_2mim][BF_4] was combined with a porous polyethylene membrane and mounted onto a three-electrode system with a glassy carbon working electrode, Ag reference and Pt ring counter electrode. The thickness of the supported RTIL membrane was estimated to be 50 μm, comparable in size to the diffusion layer thickness, meaning that fast response to the gas concentration was obtained (*ca.* 2.5 min from the introduction of gas). Linear calibration plots were obtained for 10–100% O_2 in the gas phase, with high sensitivity and good reproducibility reported. Wang *et al.*[36] used a Clark-type electrode containing a gas permeable membrane for detecting oxygen, with the RTIL replacing the conventional aqueous supporting electrolyte. Three RTILs were tested as solvents, [C_4mpyrr][NTf_2], [C_4mim][NTf_2] and [$N_{4,4,4,1}$][NTf_2], and it was found that changing the cation had an effect on the behavior of the sensor. In two of the RTILs with lower viscosities ([C_4mpyrr][NTf_2] and [C_4mim][NTf_2]) a stable analytical response over 90 days was observed with no fouling of the electrode surface, but O_2 reduction products were believed to build up near the electrode surface in the more viscous RTIL, limiting its lifetime. A linear range between 0% and 20% O_2 was observed, with a limit of detection (LOD) of 0.05 v/v% and response time of 2 min, suggesting that the sensor shows much promise for long-term detection of oxygen.

On bare metal electrode surfaces, membrane-free O_2 detection in RTILs has been demonstrated by several groups. For example, Huang *et al.*[37] used an array of 80 recessed gold microelectrodes of diameter 12 μm fabricated on a silicon chip to detect oxygen in the hydrophobic RTIL [$P_{14,6,6,6}$][FAP]. Only a very small volume of RTIL (0.2 μL) was required to connect the electrodes, resulting in a fast response to O_2 (*ca.* 20 s) over the concentration range 2–13%. Xiong *et al.*[38] constructed both Pt and Cu annular microband electrodes and investigated their suitability for the detection of oxygen in the RTIL [C_3mim][NTf_2]. A linear range from 3% to 100% was observed and a limit of detection of 0.5% was reported on the Cu microband electrode, suggesting that these cheaply constructed electrodes may be used for O_2 sensing in RTILs. Mu *et al.*[39] reported a robust flexible miniaturised O_2 gas sensor consisting of gold metal deposited as a thin film on a porous PTFE substrate. The working, counter and reference electrodes were fabricated on a planar, bendable surface and covered with a thin layer (21 μm thick) of RTIL [C_4mpyrr][NTf_2]. A linear range of 0–21%, a response time of several hundred seconds and LOD of 0.08% O_2 was reported. A second report of a flexible "paper-like" sensor was published by Hu *et al.*,[40] who used a nanoporous gold electrode array inkjet-printed onto a cellulose membrane, with 1 μL of the RTIL [C_4mim][PF_6] to connect the electrodes. Oxygen was detected in the range 0.054–0.177%, with a high sensitivity, short response time (10 s) and low detection limit of 0.0075% (or 75 ppm). This shows promising applications for cost-effect and environmentally-friendly paper-based amperometric gas sensors.

Oxygen reduction has also been studied on commercial screen-printed electrodes (SPEs) in a range of RTILs by Lee *et al.*[41] Although chemically irreversible voltammetry was observed in imidazolium RTILs (suggesting a reaction of the electrogenerated superoxide with particles in the paste of the SPEs in the presence of the mildly acidic environment of the RTIL cation), linear calibration curves were obtained on four working electrode surfaces (Pt, Au, C and Ag) and LODs were in the low percentage range. This study suggests that low-cost commercially available SPEs can be used with RTIL electrolytes for O_2 detection, although degradation of the electrode response is likely due to unfavorable chemical reactions, making these electrodes only suitable for "single-use" applications.

Other groups have employed modified electrodes to detect oxygen in RTILs. For example, Shen *et al.*[42] reported free-standing Pt–Au bimetallic membranes with a leaf-like nanostructure as the working electrode. The membrane electrode was connected to a silver wire quasi reference/counter electrode with a tiny volume (1 μL) of RTIL ([C_4mim][NTf_2], [C_4dmim][NTf_2], [C_4mpyrr][NTf_2] or [$N_{6,2,2,2}$][NTf_2]) and linear calibration graphs were observed for 0.5–36% O_2 with a fast response time (a few seconds), high current density and very small background currents (similar to that observed on microelectrodes). Xiong *et al.*[43] employed four different RTILs ([C_2mim][NTf_2], [C_4mim][NTf_2], [C_4dmim][NTf_2] and [C_6mim][FAP]) for oxygen detection on modified carbon screen-printed electrodes. A paste of the RTIL with either (a) gold nanorods or (b) so-called "quasi-platonic" gold nanoparticles was deposited on the working electrode surface as a thin film, and 30 μL of the same blank RTIL connected the working, reference and counter electrodes. Approximately linear calibration curves were observed on all surfaces in all RTILs from 20% to 100% O_2 with the quasi-platonic gold nanoparticle-modified electrodes showing a higher sensitivity than the gold nanorods, and both were much better than the bare electrode. Importantly, the electrodes modified with gold particles showed lower overpotentials for oxygen reduction, which will be beneficial in sensor applications.

While many of the above studies have focused on detecting oxygen at higher (percent) concentrations, lower limits of detection have also been reported. A LOD of 140 ppm (v/v in the gas phase) for O_2 on Pt was reported by Toniolo *et al.*[44] using a salt with a quinone moiety dissolved in the RTIL [C_4mim][NTf_2] as a catalyst. A wide linear range of 200 to 10^6 ppm was reported and the sensor showed promise for use in a mixed-gas environment, due to the lower potentials required for oxygen reduction with the mediator present. Baltes *et al.*[45] reported the lowest limit of detection for O_2 in RTILs (5 ppm) using an electrochemical membrane sensor containing working, reference and counter electrode strips (either Pt or Au) sandwiched between two alumina microfiltration membranes, filled with the RTIL [C_6mim][FAP]. Multiple-potential-step chronoamperometry was used as the technique for measuring the current response, and this technique was applied continuously to the sensor for at least one week with good results. Importantly, the authors also investigated the effect of increased temperature on the sensor, with measurements performed in the range 23–220 °C. They noted that it was still possible to detect O_2 in the ppm range, but that the sensors suffered strongly from hysteresis and irreproducible behavior, and new electrode materials may be needed.

10.2.1.2 Volatile Organic Compounds

Another increasingly popular area of research in RTIL-based gas sensing is the detection of vapors of volatile organic compounds (VOCs). Due to the use of many organic solvents in synthetic reactions for industrial processes, and the hazardous nature of VOCs to human health and the environment, it is very important to be able to detect and quantify these chemical species. Gebicki and co-workers have contributed three articles to this area,[46–48] focusing on detecting benzaldehyde, formaldehyde, methyl benzoate and acetophenone. In the first study,[47] square wave voltammetry was used to measure current responses for different concentrations of benzaldehyde in air from 10 to 100 ppm. The sensor consisted of Pt working and counter electrodes together with a salt-bridge based reference electrode, connected through the ionic liquid electrolyte ([C_6mim]Cl, [C_6mim][NTf_2] or [C_6mim][TCM]) and covered by a polydimethylsiloxane (PDMS) gas permeable membrane. Limits of quantification (LOQs) were in the range 25–53 ppm, and it was suggested that optimisation of sensor parameters was required to make the sensor achieve LOQs below the exposure limit of 2 ppm. A similar prototype sensor was used for detecting benzaldehyde and formaldehyde on Pt and Au working electrode surfaces.[46] In this study, the PDMS thickness was varied, and slightly lower LOQs were achieved with thinner layers. The third article investigated the reduction of benzaldehyde, methyl benzoate and acetophenone on commercially-available gold screen-printed electrodes using three different RTILs ([C_8mim][BF_4], [C_4mim][NTf_2] or [C_4mim][$N(CN)_2$]) as electrolytes.[48] Approximately linear calibration curves were obtained ($R^2 = 0.97$–0.99), with LODs between 1 and 6 ppm, depending on the gas and choice of RTIL cation or anion. Relatively slow responses were observed (more than 35 min to obtain a stable signal), likely due to the relatively large volume of RTIL (100 μL) used, and the authors suggested that this could be improved by using thinner RTIL layers covering the electrodes.

Dossi *et al.*[49] proposed a membrane-free amperometric gas sensor to detect 1-butanethiol vapors present in headspace samples in equilibrium with aqueous solutions. The sensor consisted of a patterned piece of filter paper printed with black wax-based ink (to define the area of the electrodes), and insulated by laminating a polyethylene layer of thickness 0.1 mm. The working, reference and counter electrodes were then screen-printed onto the top face of the sensing device using carbon ink and 1.7 μL of the RTIL

[C_4mim][NTf_2] was then allowed to soak into the paper channels, without covering the electrode surfaces. A dynamic range of 2–200 μM and LOD of 0.5 μM for 1-butanethiol (in the liquid phase) was achieved. The high electrical conductivity and negligible vapor pressure of RTILs were highlighted as key properties for making these devices quite easily and with low-cost materials. The same group later reported a similar sensing device to detect acidic phenol and 1-butanethiol vapors.[50] In this study, the RTIL of choice ([C_4mim][Ac]) contained an anion of basic character (acetate), lowering the normally high potentials required for the detection of these analytes. The sensor showed a satisfactory performance with a dynamic range of 1–200 μM, LOD of 0.3 μM (both as solution concentrations), a good repeatability and good long-term stability. The group emphasised that no permeation, diffusion or adsorption step is involved, making these inexpensive sensors better performing than conventional electrochemical gas sensors.

Most recently, Carvalho *et al.*[51] proposed "ion jelly" as a novel sensing material for gas sensors and electronic noses. They used eight interdigitated electrodes covered with a spin-coated layer of ion jelly (gelatin combined with one of the following RTILs: [C_4mim][$N(CN)_2$], [C_4mpyrr][$N(CN)_2$], [C_2mim][$N(CN)_2$] or [C_4mim]Br) to create an electronic nose to detect eight volatile compounds (ethyl acetate, acetone, chloroform, ethanol, hexane, methanol, toluene and water) at concentrations ranging from 4% to 39% in air. The detector measured conductance *versus* time and was able to detect and classify the eight chemicals with very good repeatability and a lifetime of more than three months. Although no information on the analytical utility (*e.g.* detection limits) was provided, this suggests that RTILs can be combined with polymers to provide quasi-solid materials to be used in micro-devices for gas sensing.

10.2.1.3 Toxic and Other Gases

A final area of research to be discussed is the sensing of toxic gases (*i.e.* those that are harmful to humans at low concentrations). Also included in this section are some other gases not necessarily considered "toxic" (*e.g.* CO_2), but are worthy of mentioning.

The detection of ammonia (NH_3) gas has been reported by Ji *et al.*[52] on a Pt microdisk electrode in three RTILs ([C_2mim][NTf_2], [C_4mim][OTf] and [C_4mim][BF_4]). Using cyclic voltammetry (CV), steady-state oxidation currents were measured over the concentration range 200–1000 ppm and LODs of approximately 50 ppm were reported in the three RTILs. Murugappan *et al.*[34] investigated the oxidation of NH_3 in [C_2mim][NTf_2] using CV on Pt and Au screen-printed electrodes, with linear calibration graphs in the range 240–1360 ppm, and LODs of 50 ppm on Pt and 185 ppm on Au. These values are above the permissible exposure limit of 25 ppm, but show that sensing of NH_3 in RTILs is highly feasible in RTILs, and other more sensitive methods or modified electrodes could be employed to achieve lower LODs. Murugappan *et al.*[53] have also reported the fundamental electrochemical behavior and detection of methylamine gas on a Pt microdisk electrode using CV. Good responses were observed for methylamine oxidation, with a linear calibration graph of peak current *versus* concentration in the range 104–1048 ppm and a LOD of 33 ppm reported in two RTILs, [C_2mim][NTf_2] and [C_4mpyrr][NTf_2].

Rosen *et al.*[54] reported the sensing of carbon dioxide on a microfabricated device consisting of chromium and gold layers sputtered onto a silicon wafer, covered with 2–3 μL of the RTIL [C_2mim][BF_4]. The sensing mechanism was based on the reduction of CO_2 at the cathode and measuring the current for the subsequent oxidation of that species. An approximately linear calibration graph was shown over the range of 0.4–2.7 atmospheres of CO_2 pressure; however, no LODs were reported. Wang *et al.*[55] investigated the detection of methane by electrochemical impedance spectroscopy at a gold electrode covered by a thin film of the RTIL [C_4mpyrr][NTf_2]. The working, reference and counter electrodes were coated onto a flat porous Teflon membrane (working and counter electrodes in an interdigitated set-up) with the RTIL on one side of the electrodes and the porous Teflon membrane on the other side. A concentration of 5% methane was allowed to enter the porous membrane and was detected at the working electrode, with excellent stability over 90 days. Good selectivity towards methane in the presence of other interfering gases was observed and attributed to the unique highly-ordered arrangement of ions at the RTIL–electrode interface. The same group[56] also reported a miniaturised device (with only 8% of the total sensing area compared to the previous study) for the detection of methane and sulfur dioxide (SO_2). The RTIL [C_4mpyrr][NTf_2] was deposited with a thickness of *ca.* 200 μm, and methane was studied in the range 0–5% (as 5% is the lower explosive limit) and SO_2 in the range 0–5 ppm (as 5 ppm is the permissible exposure limit). The same miniaturised device was also suggested as a wearable sensor array for real-time health and safety monitoring.[57] An operational first generation prototype was proposed, with real-time monitoring of SO_2 demonstrated. The main feature of the sensors in all of the three reports[55–57] is that it relies on the gas entering the sensor directly from the gas phase through a porous membrane. The RTIL is located on the opposite side of the working electrode, as demonstrated in Figure 10.2(b). As a result, the response time is much improved compared to conventional strategies that rely on the gas first partitioning into the RTIL and diffusing towards the working electrode (Figure 10.2a).

Various nitrogen oxides have also been detected using RTILs. Nadherna *et al.*[58] used a mixture of the RTIL [C_4mim][PF_6] with the polymer poly(ethylene glycol) methyl ether methacrylate (PEGMEMA) for detecting nitrogen dioxide (NO_2) on a fabricated sensing device by amperometry. The sensor consisted of three electrodes: a gold minigrid working electrode, and Pt counter and reference electrodes. The current *versus* concentration calibration curve was linear in the range 0.3–1.1 ppm, with a LOD of 0.01 ppm NO_2. The response was found to be reproducible and stable over

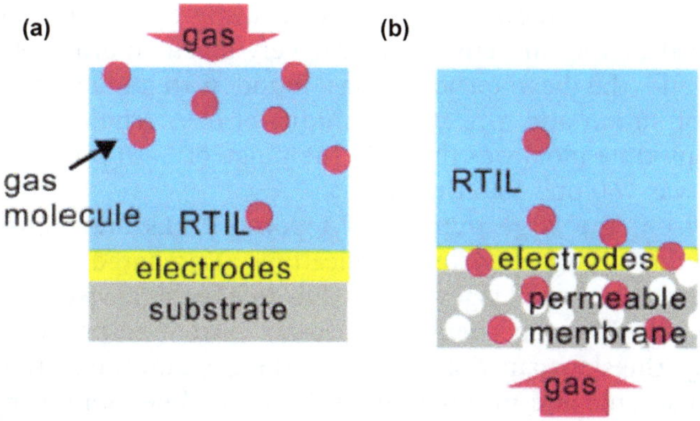

Figure 10.2 (a) Conventional sensor structure: response time is slow due to slow gas diffusion through RTIL; (b) electrodes-on-permeable-membrane structure: response time is improved due to fast gas diffusion in the permeable membrane.[56]
(Copyright IEEE. Reproduced with permission from IEEE Proceedings.)

several months. The same group[59] later reported NO_2 detection on a similar gold minigrid electrode using PEGMEMA mixed with either [C_4mim][PF_6] or [C_4mim][NTf_2], in the presence of humidity and found good responses to NO_2 over the range 0.3–1.1 ppm, a LOD of 0.04 ppm and response time of 15 s. The RTIL-polymer sensor showed the smallest dependence of the response on the relative humidity of the air compared to other NO_2 solid-state amperometric sensors reported in the literature. Nitric oxide (NO) detection was investigated by Ng *et al.*[60] using a carbon-paste electrode containing a 3D graphene material and the RTIL [C_4mim][PF_6]. NO was dissolved in phosphate buffer solution and detected in the range 1–16 μM using amperometry, with a very low LOD of 16 nM and a fast response time of less than 4 s. A membrane-free amperometric gas sensor for monitoring nitrogen oxides (NO_x, defined as mixtures of NO and NO_2 gas) was suggested by Toniolo *et al.*[61] Three Pt wires were pierced through a Teflon rod, and covered by a tiny volume (0.5 μL) of the RTIL (either [C_4mim][NTf_2] or [C_4mpyrr][NTf_2]), corresponding to an electrolyte thickness of about 70 μm. Current–time signals were recorded in the presence of different concentrations of gas and linear calibration curves were obtained over the range 0.01–103 ppm NO_x, with a detection limit of 0.96 ppb calculated at 25 °C. The authors also performed experiments at higher temperatures (100 °C), with an even lower LOD obtained (0.55 ppb), suggesting that these robust sensors can be conveniently used to detect low levels of NO_x at higher temperatures.

Finally, the sensing of ethylene gas was investigated by Zevenbergen *et al.*[62] using a thin ionic liquid layer (either [C_4mim][NTf_2] or [C_6mim][FAP]) covering a planar sensing device with a gold working electrode, Pt counter electrode and Pt quasi-reference electrode. Using CV, a pre-wave was observed before the onset of gold oxidation, which increased approximately linearly with increasing concentrations of ethylene. However, this was only seen when sufficient moisture was present in the gas stream; no response was observed in very dry environments. Very small volumes of RTIL were employed and the best response was found with an ionic liquid layer thickness of 30 μm and a relative humidity of 60%. The calibration graph from current–time plots was linear in the range of 1–10 ppm, and the LOD calculated was 760 ppb for ethylene.

Clearly, from the large number of reports, RTILs are highly favorable solvent media for gas sensing applications. Compared to conventional solvents, the main challenges occur due to the high viscosity of RTILs, leading to low diffusion coefficients (lower current responses) but, more importantly, the slow partitioning step of the gas into the RTIL results in a slow response time and may not be ideal for real-time monitoring of highly toxic gases. As many recent articles have suggested, the use of very thin RTIL films allows for miniaturisation of the sensor platform and improved response time. Alternatively, the response time can be dramatically improved if the gas to be detected is introduced on the other side of the working electrode to the RTIL, no longer requiring the partitioning and diffusion steps, which are often very slow in RTILs. These strategies may be used if such sensors are to become commercially available in the future for real-time monitoring applications.

10.2.2 Explosives and Chemical Warfare Agents

There has been increasing interest in detecting potentially dangerous analytes such as explosives and chemical warfare agents (CWAs) due to the increased threat from terrorist activity in recent years. As a result, there are several notable reports where researchers have used RTILs as the solvent of choice for their detection. The first report in 2009 by Forzani *et al.*[63] describes the detection (by reduction) of the nitroaromatic explosive vapors of 2,4,6-trinitrotoluene (TNT), picric acid (PA) and 2,4-dinitrotoluene (DNT) using a combined electrochemical and colorimetric detection technique. In contrast to standard electrochemical detection techniques, the electrochemical reactions were used to generate reaction products, which were then detected with an optical imaging device (Figure 10.3). The RTIL $[C_4mim][PF_6]$ was able to pre-concentrate the explosives, transport them quickly to the surface and promote the formation of colored reduction products. The same group also demonstrated the detection of TNT vapors using a conducting polymer nanojunction consisting of poly(ethylene dioxythiophene), PEDOT, bridged between two gold working electrodes on a silicon chip, covered with a layer of $[C_4mim][PF_6]$ RTIL.[64] The reduction current for TNT and the resulting conductance change of the polymer was simultaneously measured. Even in the presence of other redox-active interferences from the air (*e.g.* perfume, mouth wash, body spray and cigarette smoke), a linear calibration graph for TNT from 30 pM to 6 nM (in the liquid phase) was obtained. It was suggested that very low levels of TNT (ppt) could be measured in less than 2 min.

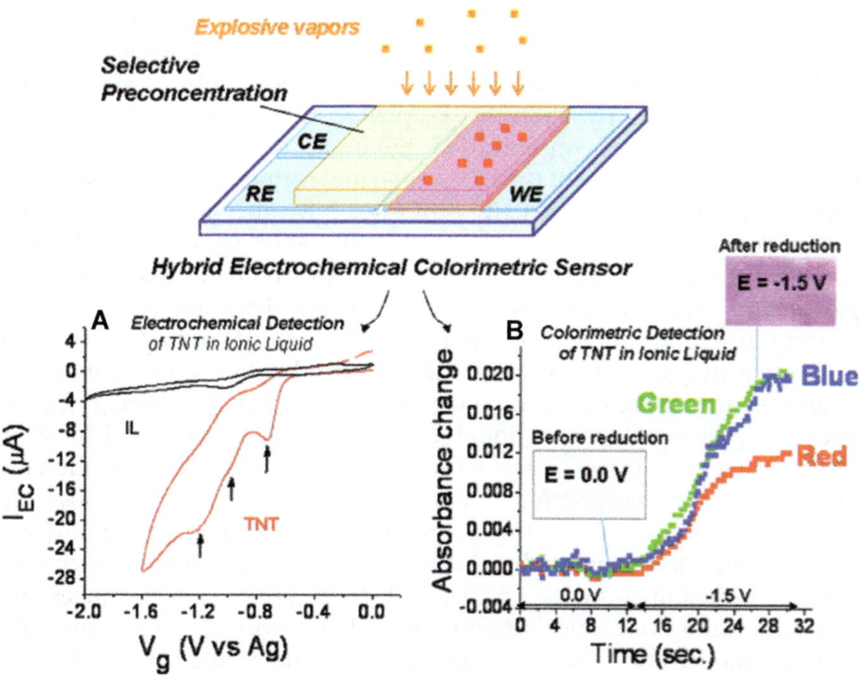

Figure 10.3 Hybrid electrochemical–colorimetric sensor with a thin layer of ionic liquid as a selective preconcentration medium. (A) Cyclic voltammograms of blank IL [C_4mim][PF_6] (black line) and 2 ppm TNT in the IL (red line) at 100 mV s^{-1}. Arrows indicate peak currents of TNT. (B) Color (absorbance) changes during the electrochemical reduction of TNT in [C_4mim][PF_6]. Inset to part (B): two images taken before (0.0 V) and after (−1.5 V) TNT reduction. The distinct color change provides a fingerprint for identification and quantification of the explosive.[63] (Copyright American Chemical Society. Reproduced with permission.)

TNT has also been detected in a 0.5 M NaCl solution using a carbon paste electrode (see definition later in Section 10.2.3) with the RTIL [C_4mim][PF_6] and three-dimensional graphene.[65] Absorptive stripping voltammetry was employed and the sensor showed low background current, high sensitivity and pronounced mesoporosity. Calibration curves were linear from 2 to 1000 ppb, with an LOD of 0.5 ppb for TNT. A separate group employed a similar hybrid material with ionic liquid-graphene nanosheets deposited on a glassy carbon electrode to detect TNT in 0.1 M phosphate buffer solution.[66] They reported a linear range of 0.03–1.5 ppm and a detection limit of 4 ppb. The electrode was used to detect TNT in samples of ground water, tap water and lake water with satisfactory results. The detection of TNT and DNT dissolved directly in four RTILs ([C_4mim][NTf_2], [C_4mpyrr][NTf_2], [C_4mim][FAP] and [C_4mpyrr][FAP]) by square wave voltammetry was also reported by Xiao *et al.*[67] They observed linear calibration curves (for reduction peak current *versus.* concentration) in the range 5–100 μM for TNT and DNT in all four RTILs, with detection limits in the liquid phase of 190 and 230 nM for TNT and DNT, respectively. Interestingly, when gas phase analysis was performed, strong redox signals were observed at 0.27 ppm (TNT) and 2.05 ppm (DNT) in the gas phase, suggesting that the RTILs do indeed act as a pre-concentration medium, with the exact behavior dependent on the nature of the RTIL anion and cation (*i.e.* tunable solvents).

Other analogues of redox-active explosives have also been detected using RTILs. For example, Yu *et al.*[68] reported an ionic liquid electrochemical quartz crystal microbalance (EQCM) sensor using both electrochemical and piezoelectric transduction mechanisms for the detection of ethyl nitrobenzene and dinitrotoluene. A thin layer of the RTIL $[C_4mim][BF_4]$ was used as the electrolyte to connect all three electrodes (working, counter and reference) and to pre-concentrate the explosive vapors by applying negative potentials. Nitromethane, a liquid explosive, has been detected by Wang *et al.*[69] using an electrochemical biosensor based on the immobilisation of a film containing one of four heme proteins (hemoglobin, myoglobin, horseradish peroxidase and cytochrome *c*), the RTIL $[C_4mim][BF_4]$ and multi-walled carbon nanotubes (MWCNTs) on a glassy carbon electrode. Using chronoamperometry on the reduction peaks, linear ranges and detection limits were calculated for each of the heme proteins, with myoglobin demonstrating the best response with a linear range of 0.01–1.36 μM and LOD of 3 nM nitromethane in 0.1 M phosphate buffer solution.

For the detection of chemical warfare agents, there appears to be only two reports, both from the same group, demonstrating the difficulty of getting access to, and handling, these dangerous compounds. Singh *et al.*[70] first reported the oxidation and reduction of the chemical warfare agent nitrogen mustard-2 (NM-2) using the RTIL triethyl sulfonium bis(trifluoromethylsulfonyl)imide ($[S_{2,2,2}][NTf_2]$). A linear calibration plot (using the reduction current *versus* concentration) was observed in the range 0.029–1.17 mM^2, with a detection limit of 0.015 mM and detection time of 20 s. The same group[71] later demonstrated the electrocatalytic oxidation of 2-chloroethyl ethyl sulfide (CEES), a chemical warfare agent simulant in phosphate buffer solution on a gold electrode coated with a copper phthalocyanine/RTIL ($[C_4mim][NTf_2]$) composite electrode. They observed linear calibration graphs (for the oxidation peak current *versus*. concentration) from 0.017 to 0.51 mM with a LOD of 1.7 μM. By using a RTIL in the modified electrode (as opposed to an organic solvent), the lifetime and performance was greatly increased. Although the LODs in both studies were significantly higher than the concentrations at which these compounds are dangerous to humans, it clearly demonstrates that RTILs can be used as alternative solvents for their detection and more work in this area would be useful.

Overall, in the field of explosive and chemical warfare agent detection, the use of RTILs has been shown to be promising, but RTILs alone do not appear to have the capability to detect sufficiently low quantities of these highly dangerous materials. As many of these reports have demonstrated, the RTIL either needs to be combined with other materials (*e.g.* graphene, carbon nanotubes, proteins) or the electrochemical technique must be combined with other detection techniques.

10.2.3 Carbon-paste Electrodes and Ionic Liquids

So-called "carbon-paste electrodes" are widely used in the field of electrochemistry, but are predominantly employed for electroanalysis. Such electrodes are traditionally made from a paste-like mixture of fine graphite powder and mineral oil.[72] This results in an often cheap, soft, easily constructed electrode with relatively large surface area. The softness means surfaces can be refreshed extremely easily, by any form of shear pressure such as rubbing on a sheet of paper. They can also be constructed to a wide variety of sizes and shapes, dictated only by the external body holding the electrode together. The ease of construction and relatively large surface area means it is easy and often effective to introduce additional functional materials into the paste, such as nanoparticles (as electrocatalysts) or chelating ligands (to introduce chemical functionality). If homogeneously dispersed throughout the paste then the introduced functionality persists despite numerous resurfacings.

The major drawback of traditional carbon-paste electrodes has always been that the introduction of insulating mineral oil renders much of the carbon surface inaccessible to the electrolyte, and increases internal resistance by blocking conduction pathways. One option is to replace the oil binder with an IL, to form a so-called "carbon ionic liquid electrode" (CILE).[17] Due to the inherent conductivity of ionic liquids, essentially all of the carbon surface becomes electrochemically accessible. The carbon particles are coated in a thin layer of IL, which possess the ability to preconcentrate some analytes at the surface of the electrode, by phase transfer, anion exchange,[73] *etc.* The clear changes in voltammetry upon switching from a paraffin oil binder to an IL binder are demonstrated in Figure 10.4 (for the quantification of rutin;[74] specific study discussed below). Certain active species can be actively incorporated into the paste during electrode fabrication, such as glucose oxidase.[75]

The early work on CILEs utilised the IL $[C_8Py][PF_6]$, with a melting point of 65 °C.[76] Maleki *et al.* demonstrated that forming a CILE with $[C_8Py][PF_6]$ resulted in an electrode which, due to the additional and more accessible surface, showed apparent electrocatalysis relative to traditional CPEs.[73] Heating and cooling the electrode before use was found to beneficially result in a reduction in the background current. The optimised CILE resulted in significant increases in the apparent rate of electron transfer for various electroactive compounds such as ferricyanide, catechol, NADH, dopamine and ascorbic acid. Current densities were improved and electrode fouling reduced. The electrolyte was noted to be significant in the interaction between the analyte and the CILE, in some cases resulting in adsorptive characteristics and ion-exchange-based accumulation.[73]

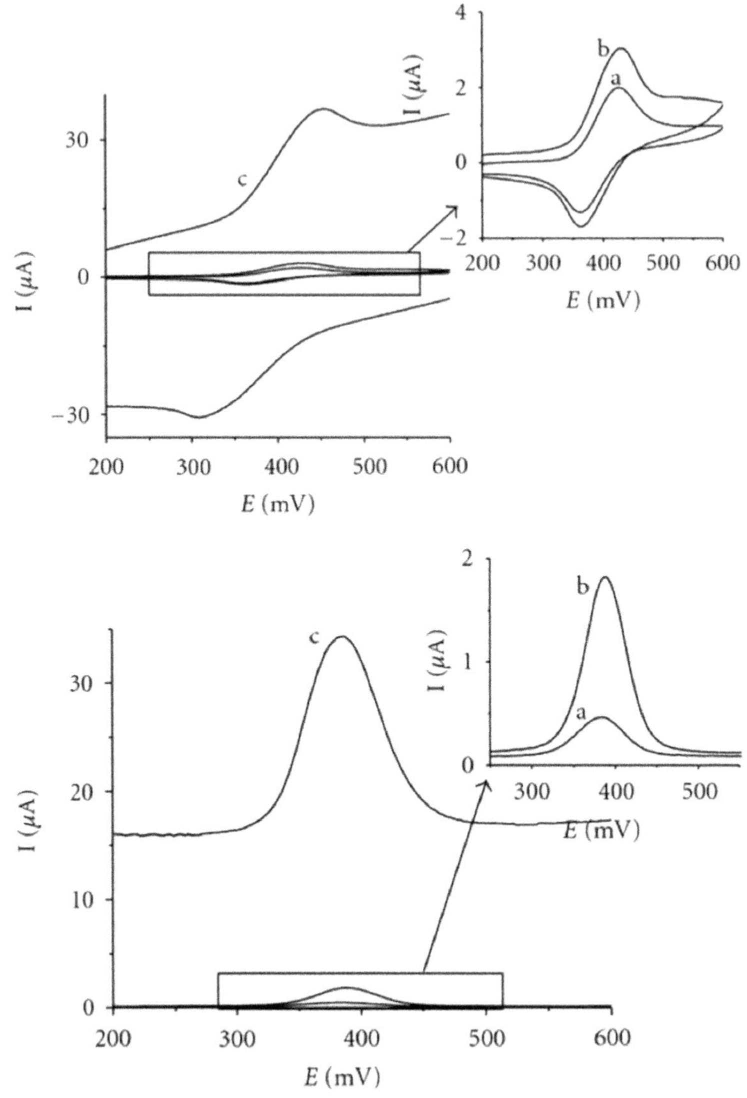

Figure 10.4 Voltammograms highlighting the difference between carbon paste electrodes prepared using (a) paraffin oil as binder and iron phthalocyanine as electrocatalyst, (b) just paraffin oil as binder and (c) the IL $[C_6mim][NTf_2]$ as binder. Top: linear sweep voltammograms for 100 μM rutin solution in acetate buffer (pH 4.0). Bottom: differential pulse voltammograms for 8 μM rutin in the acetate buffer.[74]
(Reproduced under the Creative Commons Attribution License.)

The improvement in heating was likely related to restructuring of the solid–solid interface between the IL and graphite. Lu *et al.* demonstrated an improvement in signal-to-noise by moving from $[C_8Py][PF_6]$ (melting point 65 °C) to $[C_3mim][PF_6]$ (melting point 39 °C).[76] The heating step also improved sensitivity of the CILE with respect to hydrogen peroxide. Due to the lower temperature required for the initial heat-treatment step, glucose oxidase enzymes could be easily incorporated and remain unaffected by the heating pre-treatment. A linear amperometric response could detect glucose between 2.0 and 26 mM (LOD of 0.39 mM).

The use of CILEs has also been successful in the sensitive quantification of drug molecules. For example, the use of $[C_5mim][PF_6]$ to form a CILE (instead of mineral oil) resulted in apparent faster electron transfer, higher peak currents and improved electrochemical reversibility for calcium dobesilate oxidation.[77] Calcium dobesilate displayed an adsorption-controlled oxidation at slow scan rates at the conventional CPE, but was adsorption-controlled for all investigated scan rates at the CILE.[77] Introduction of the IL $[C_4Py][PF_6]$ also resulted in an improvement in the oxidative voltammetric quantification of 3,4-dihydroxybenzoic acid, a naturally occurring polyphenol with medicinal applications,[78] and tiamulin fumarate, a commercial antibiotic.[79] A CILE was prepared using the aryl-functionalised 1-benzyl-3-methylimidazolium hexafluorophosphate (mixed with an equal weight of paraffin oil), and compared with a conventional CPE.[80] The CILE gave a significant increase in the oxidation peak current and cathodic shift for labetalol, a drug used as a clinical treatment for hypertension.[80] Interestingly, despite the introduction of the benzyl group to the IL, two benzyl groups present in the analyte and the graphitic nature of carbon, the voltammetry was diffusion-controlled. This highlights the potential role of oil content and electrolyte, which dictates whether the mechanism is adsorptive or diffusion-limited. All of the above studies utilised DPV (differential pulse voltammetry)[77–79] or SWV (square wave voltammetry)[80] to generate linear calibration curves between low μM and low mM concentrations of analytes. The reported CILE's could be successfully employed in conjunction with capsules,[77] urine,[77] juices,[78] and commercial injection[79] and tablet samples.[80]

A CILE was prepared using $[C_6Py][PF_6]$ for the determination of the nucleoside adenosine.[81] Replacement of the oil with the IL resulted in only a minor shift in oxidation potential but a significant increase in peak current, indicating that the IL was pre-concentrating or otherwise facilitating adenosine quantification at the CILE. The use of a carboxyl functionalised ionic liquid to form a CILE was successful for the sensitive and selective quantification of adenosine-5′-triphosphate.[82] Additionally, a $[C_8mim][PF_6]$-based CILE was used to quantify guanosine-5′-monophosphate.[83] In both systems, apparent increases in electrocatalytic ability were observed upon introduction of the IL, with well-defined adsorption-controlled oxidation peaks observed.[82,83] These systems used DPV, giving linear responses between low μM and low mM concentrations,[81–83] and could be employed in urine,[81] commercial injection samples[82] and chicken powder samples.[83]

An interesting study involved the quantification of Aloe-emodin (1,8-dihydroxy-3-hydroxymethylanthraquinone), an extract from *Rheum palmatum* with many potential medicinal applications.[84] Quantification was optimised using a traditional paraffin oil-based CPE, followed by the replacement of increasing quantities of oil with [C_4mim][PF_6] to form a mixed binder. Sensitivity increased until a weight ratio of carbon powder, paraffin oil and IL of 80 : 15 : 5 was reached. If more IL was added, the background capacitance rapidly increased,[84] obscuring the oxidation features of small quantities of the analyte. Incorporation of IL into the binder resulted in an enhanced apparent electron transfer rate, and an adsorption-controlled mechanism.[84] By optimising the ratio, the oxidative peak current displayed a linear relationship by DPV between 10 nM and 12.4 μM (LOD of 3.0 nM) of Aloe-emodin, an order of magnitude lower LOD than that reported in the above studies.

The detection of rutin (a bioactive flavonoid) was studied at a conventional CPE, a CPE electrode with iron phthalocyanine incorporated and a CILE prepared using [C_6mim][NTf_2] (Figure 10.4).[73] While the CPE and iron phthalocyanine/CPE performed best using regular DPV, the CILE was much more effective if a biased potential pre-accumulation/adsorption step was included before DPV. The CILE could achieve LOD values as low as 5 nM, compared to 80 nM for the iron phthalocyanine/CPE system. However, despite being less sensitive, the conventional CPE displayed the widest linear range and the smallest relative standard deviation between repeats, especially when used to quantify rutin in extracts from buckwheat seeds.[73]

Interestingly, Zhang and Zheng observed that the detection limit of rutin could be driven even lower by the introduction of a hydrophilic IL. The ratio of 1-amyl-3-methylimidazolium bromide and paraffin oil was investigated, and the optimum value was found to be 2 : 3;[85] too much hydrophilic IL and the CILE was unstable, too little IL and significant increases in rutin oxidation peak current and surface accumulation were not apparent. Using SWV, rutin could be quantified between 0.4 nM and 0.1 μM (LOD of 0.1 nM), and the CILE could be applied to quantify rutin in urine and tablet samples.[85]

Hydroquinone is a widely utilised reagent but is also a serious environmental pollutant. She *et al.* investigated the quantification of hydroquinone using a mineral oil CPE, a [C_8mim][PF_6] CILE and one prepared using a 50 : 50 ratio of oil : IL.[86] The hydroquinone electrochemistry was consistently diffusion-based, and in this case the pure CILE was found to result in the highest oxidation peak current (LOD of 81 μM by CV).[86] Sun *et al.* subsequently prepared a "carbon ionogel electrode" (where sol–gel-type chemistry is used to make a ceramic-like carbon based electrode) containing [C_4mim][PF_6].[87] Incorporation of the IL resulted in both enhanced peak current values and reversibility, improving resolution between hydroquinone and catechol features, and a LOD of 70 nM hydroquinone was achieved by DPV.[87]

The surface of a [C_4Py][PF_6]-based CILE was used as a base for the potentiostatic formation of poly(crystal violet) with the simultaneous incorporation of graphene.[83] This procedure dramatically enhanced peak resolution for hydroquinone and catechol oxidation features. Both hydroquinone and catechol could be quantified individually by DPV (LOD values of 6.2 and 13 nM, respectively) with a linear range between 0.02

and 600.0 μM.[83] When quantified simultaneously the LOD values increased by *ca.* an order of magnitude but the linear ranges remained unchanged. All features were diffusion-based, and the CILE could detect both analytes in "artificial water" samples.[83]

The above studies highlight how surface modification of a CILE can be employed to enhance its performance. Although the modification of a carbon paste electrode somewhat defeats some of the advantages inherent in these types of electrodes, they also provide easily renewed yet relatively rough surfaces for screening surface modification processes. The incorporation of an IL to make a CILE ultimately means that more of the graphite surface is available for modification. The incorporation of nanoparticles is another effective method of introducing functionality or improving electrocatalysis of CPEs, including CILEs.

For example, gold nanoparticles were electrodeposited onto a $[C_8Py][PF_6]$-based CILE, followed by underpotential deposition of palladium to produce a sensor for formaldehyde.[88] The same authors then prepared palladium nanoparticles and silver–palladium nanoparticle colloids, which could be incorporated homogeneously throughout the CILE.[89] In both systems the nanomaterials were optimised for their electrocatalytic and antifouling abilities, allowing formaldehyde to be quantified between low μM and high mM ranges (LOD for both systems was 2 μM)[88,89] even in the presence of methanol, ethanol and formic acid.[89] The precise role of the IL was not elucidated. Similar studies have incorporated nanoparticles throughout the CILE, such as the introduction of gold nanoparticles into a CILE, which resulted in a stable electrode that was effective for the oxidation and thus quantification of tryptophan.[90]

Vahedi *et al.* displayed the effectiveness of employing both ILs and nanomaterials when preparing CPEs.[91] The incorporation of the hydrophilic IL 1,3-dipropylimidazolium bromide (1 : 4 ratio with liquid paraffin) and an equivalent mass of MgO nanoparticles was far more effective for the quantification of the antihypertensive agent methyldopa than the use of just IL or nanoparticles. The IL resulted in a *ca.* 100 mV negative shift in the oxidation peak for this system, and both IL and nanoparticles cumulatively increased the active electrode surface area. A LOD of 30 nM was noted, and the composite was effective even in genuine patient human urine samples.[91]

Chen and Huang utilised coordination alteration-induced redox reactions in ILs to spontaneously form size-controlled Au, Pd and AuPd nanoparticles in a range of ILs.[92] These nanoparticles were formed without additional stabilising agents, and the resulting colloids could be easily mixed with graphite powder (without the traditional use of a mortar and pestle[92]) to form homogeneous nanoparticle-functionalised CILEs. The various permutations possible were demonstrated to be effective for the formation of hydrogen peroxide-, formic acid- or ethanol-responsive CILEs.[92]

The explosion of interest in ILs has coincided with the widespread interest in "modern" carbon nanomaterials such as carbon nanotubes and graphene. This is expressed in the electroanalytical literature, and especially in the CILE literature. It is impossible to cover all of the literature concerning the incorporation of carbon nanomaterials. Additionally, the precise role of the IL is rarely elucidated, in comparison to studies that focus upon the introduction of the IL as the sole modification to the system.

As a selective summary, rutin was electrochemically quantified using a CILE, where the carbon powder was multiwalled carbon nanotubes (MWCNT) and the IL [C_{12}mim][PF_6] (1 : 5 w/w ratio, respectively), resulting in a reported limit of detection of 10 nM.[93] When the surface of a regular [C_6Py][PF_6]-based CILE was modified with a mixture of Nafion, graphene oxide and [C_2mim][BF_4], rutin could also be quantified with a limit of detection of 16 nM.[94] Modifying the surface of a [C_6Py][PF_6]-based CILE with graphene oxide followed by cyclic voltammetry could result in the simultaneous reduction of the graphene oxide and its modification with poly(acridine orange); the resulting limit of detection for rutin using this composite was 8.3 nM.[95] When the surface of a [C_6Py][PF_6]-based CILE was modified with a graphene–manganese oxide nanocomposite, it resulted in significant apparent electrocatalysis and a reported limit of detection of 2.7 nM for rutin.[96] Such publications quote various degrees of enhanced electrocatalysis (larger peak current and/or reduced peak-to-peak separation) and "synergy" between the various additions to the CILE or its surface, but it is often difficult to elucidate the precise role the IL plays in such electrodes; it is likely they contribute in multiple ways.

Elyasi *et al.* used thermal decomposition to produce Pt nanoparticle-modified multi-walled carbon nanotubes, then used 70 : 30 paraffin oil : hydrophilic [C_4mim][Br] to prepare a paste-based electrode that demonstrated significantly enhanced peak current for Sudan I oxidation (a synthetic azo-colourant).[97] Mo *et al.* used MWCNT and a range of ILs to prepare pastes that they used to modify the surface of a GC electrode, also introducing a range mono- and geminal-surfactants.[98] The extremely hydrophobic IL [$P_{14,6,6,6}$][NTf_2] and a C_{28}-based geminal surfactant were optimum for the open-circuit accumulation then electrochemical quantification of Sudan I. All components together were stated to demonstrate good synergy, resulting in a LOD of 30 nM for Sudan I.[98]

As discussed previously (Figure 10.4), the introduction of IL results in a larger available surface area, resulting in larger Faradaic and non-Faradaic currents. Most of the above studies largely employed graphite-based electrodes that were modified with relatively smaller quantities of carbon nanomaterials. This is possibly due to the cost and relatively lower availability of these materials compared to graphite, and possibly because ILs can be *too* effective at making carbon surfaces "available". For example, Kachoosangi *et al.* noted that MWCNT/IL-based CILEs displayed extremely large background charging currents due to the very high surface area.[99] This can be extremely detrimental when the purpose is to analyze low concentrations of analytes, especially those that are diffusion-limited. By using a rotating disc electrode to increase mass transport, the authors demonstrated improved signal-to-noise ratio while also retaining the advantages (*i.e.* faster electron transfer rate) inherent to using nanomaterials and ILs.[99]

Various other analytes mentioned above have been subsequently quantified with nanocarbon-based CILEs, such as methyldopa (at an IL/MWCNT-based CILE)[100] and adenosine-5′-monophosphate (using IL-graphene-chitosan).[101] A 50 μm microelectrode was prepared from a gel formed from single walled carbon nanotubes (SWCNTs) and $[C_6mim][PF_6]$. This gel was then packed into an etched cavity in a conventional 50 μm Pt microelectrode.[102] When used for the electroanalytical oxidation of nitric oxide (NO), overlaid diffusional and thin-layer responses were observed due to penetration of the analyte into the porous electrode.

Quantification using such composite electrodes does not need to be restricted to non-equilibrium processes. For example, a potentiometric Er(III) sensor was prepared using multi-walled carbon nanotubes (MWCNTs) as the body and 5-(dimethylamino)naphthalene-1-sulfonyl 4-phenylsemicarbazide as the Er(III)-complexing reagent.[103] Comparison between oil and $[C_4mim][BF_4]$ showed that the latter as a binding agent demonstrated better sensitivity, selectivity and response time to Er(III). The optimised Er(III) sensor had a response range from 100 nM to 0.1 M, with an LOD of 50 nM.[103] Conversely, the introduction of cerium acetylacetonate into a CPE was found to result in selective interaction with monohydrogen phosphate anions ($[HPO_4]^{2-}$).[104] An optimised composition of 5% MWCNT, 65% graphite powder, 15% IL and 15% of cerium acetylacetonate resulted in a *ca.* 20 s response time to $[HPO_4]^{2-}$ and a dynamic concentration range (1 μM to 0.1 M).

10.2.4 Biosensors and Bioanalysis

A biosensor is an analytical device that employs a biological component as part of the quantification process; often this relies upon enzymes as key biochemical contributors. By extension, such biochemical quantification often relates to analytes of direct biological interest. While significant progress has been made in controlling and applying such "biological machinery",[14,15] research has also progressed on non-biosensors to investigate analytes (*cf.* biomolecules) of direct biological interest, referred to here as bioanalysis.

The significant number of publications that have come out in the general area of "biosensors and bioanalysis" cannot be covered as a subsection of a chapter. These analytical applications have also been extensively reviewed.[17,19,105–107] To emphasise the various ways ILs have been incorporated, glucose sensing will be covered in depth. A separate section will cover the breadth of the application of ILs towards various biological analytes.

10.2.4.1 Glucose Quantification

Biosensors geared towards glucose quantification have frequently employed the enzyme glucose oxidase or, less frequently, glucose dehydrogenase.[108] As such glucose can be quantified by the direct electrochemical interrogation of the enzymes, often necessitating their (controlled) immobilisation at the surface of the electrode. Alternatively, the action of glucose oxidase upon glucose (in an oxygenated solution) will result in the stoichiometric formation of hydrogen peroxide, and thus glucose can be quantified by the quantification of hydrogen peroxide. Both systems have been investigated in conjunction with ILs. Non-enzymatic sensors, which typically rely upon either direct electron transfer between the electrode and glucose (*e.g.* glucose is oxidised directly *via* an inner-sphere mechanism at the surface of an appropriate electrode) or oxidation of glucose *via* a non-enzymatic redox mediator, have also featured ILs in key roles.

Indirect biosensors for glucose must often be excellent, selective hydrogen peroxide sensors, with the incorporation of glucose oxidase being an important but also secondary issue. The combination of the IL $[C_8Py][PF_6]$ and graphite to prepare a CILE was found to be highly effective for the electroanalytical quantification of hydrogen peroxide.[75] The use of $[C_8Py][PF_6]$, which is solid at room temperature, resulted in a more homogeneous paste than those prepared ILs which were liquid at room temperature. Conversely, the electrochemical properties (peak current and potentials) for the oxidation and reduction of hydrogen peroxide were far superior at the CILE than conventional paraffin oil-based CPEs. Loading of glucose oxidase was related to the amperometric response of the biocomposite to glucose, with higher loadings resulting in a more sensitive response (higher A $mol.^{-1}$ responses) but also smaller linear ranges before the response reached a plateau (*ca.* 5–20 mM glucose); thus the dynamic range with respect to hydrogen peroxide quantification was fixed but the range for glucose could be selected by choice of the enzyme loading.

Poly(sodium 4-styrenesulfonate) was self-assembled upon MWCNT (through hydrophobic interactions).[109] Gold nanoparticles were formed by the reduction of hydrogen tetrachloroaurate in an amino-terminated ionic liquid, then the IL was mixed with the polymer-coated MWCNT in order to assemble by electrostatic interactions. This resulting assembly displayed good apparent electrocatalysis towards hydrogen peroxide. Thus, when glucose oxidase was also immobilised within the composite, the resulting bionanocomposite could be used for glucose quantification with an LOD of 25 μM.

A mixture of Prussian blue, graphite paste and ionic liquid was found to be a sensitive electrode for the quantification of hydrogen peroxide.[110] The CILE was then adapted to be a glucose biosensor by the immobilisation of glucose oxidase. Three routes towards the enzyme immobilisation were evaluated: (i) covalent crosslinking with glutaraldehyde and bovine serum albumin, (ii) physical and electrostatic entrapment within a Nafion matrix and (iii) addition of glucose oxidase to the surface followed by a layer of silica sol–gel. The system prepared by covalent crosslinking displayed the greatest sensitivity and stability. All systems were largely selective for glucose, with the exception of ascorbic acid, which acted as an interference.

An interesting example of the analytical application of ILs is in organic electrochemical transistors (OECTs). The devices generally consist of a conducting polymer in contact with an electrolyte, with the electrolyte forming an integral part of the device.[111] A biased voltage across the system causes ions from the electrolyte to enter or leave the polymer film, thus altering the drain current.[112] Figure 10.5 displays an excellent example of this. Glass modified with a monolayer of (tridecafluoro-1,1,2,2-tetrahydrooctyl)trichlorosilane resulted in the localisation of a thin layer of phosphonium-based IL ($[P_{1,4,4,4}][Tos]$, Figure 10.5a) to connect the source and drain electrodes, and leaving a gap to the gate electrode. The IL contains glucose oxidase and ferrocene; when a glucose solution comes into contact with the immiscible IL, a drop in the current drain is observed that is proportional to the glucose concentration. Given the multifaceted role of ILs (to form a thin layer of non-volatile electrolyte with controllable immiscibility, good analyte extraction properties, and the ions themselves

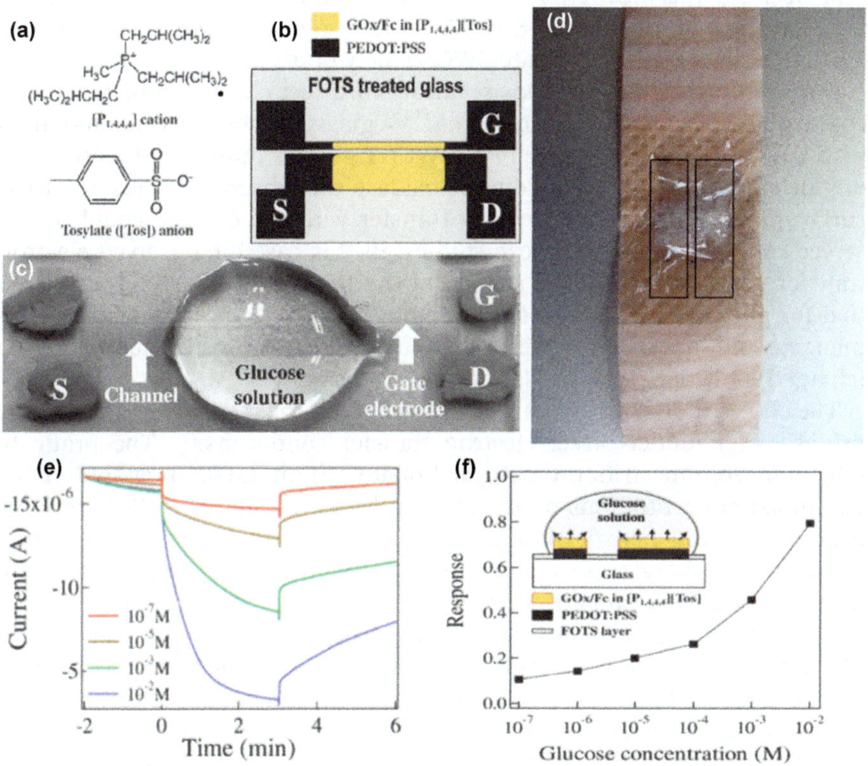

Figure 10.5 Displaying an "electrochemical transistor" sensor, based upon top, (a) the IL $[P_{1,4,4,4}][Tos]$, (b) confined within the yellow layer and containing glucose oxidase and ferrocene, (c) with a drop of glucose solution on top of the assembly, and (d) the same assembly replicated upon a plaster. (e) Transient response of the drain current in the presence of various concentrations of glucose, and (f) the normalised response. Inset in (f): concept of the device in operation.[112]
(Copyright Royal Society of Chemistry.)

are part of the transduction resulting in an analytical response), it is clear that ILs can have much more scope in this area.

The direct electrochemistry of glucose oxidase, in theory, allows the development of a simpler, more robust electroanalytical assembly. The glucose oxidase component still remains a crucial part of the biosensor, yet the glucose can now be quantified directly rather than indirectly (*cf.* extrapolating the glucose content by hydrogen peroxide measurements).

The electrostatic immobilisation of glucose oxidase was achieved on single-wall carbon nanotube (SWCNT)/poly(sodium 4-styrenesulfonate) composites, with and without various ILs.[113] Spectroscopic evaluation of the glucose oxidase highlighted no conformational changes in the enzyme upon introduction of the ILs. However, direct electrochemistry of the enzyme in the presence of glucose revealed that the kinetic rate of glucose oxidation decreased by at least an order of magnitude upon introduction of the IL. This decrease was attributed to a reduction in the conductivity and electron transfer rate constants in the SWCNT-IL networks (and upon the IL-blocked SWCNT surface) relative to bare SWCNT networks.

One of the simplest methods resulting in direct electron transfer to glucose oxidase was reported by Liu *et al.*[114] A glassy carbon electrode modified with a thin layer of a [C_4mim][PF_6]-MWCNT gel was immersed in a solution of glucose oxidase for 12 h. Glucose oxidase was observed to adsorb at the surface, where-after direct electron transfer with the enzyme could be observed. As opposed to the above studies, in this case the observed electron transfer rate adsorbed at the surface of the IL-MWCNT (9.08 s^{-1}) exceeded that for glucose oxidase adsorbed at bare MWCNT (1.7 s^{-1}). Such improvements could not be observed when the MWCNTs were modified with (charged) polymers such as Nafion and chitosan.

The choice of IL is also highly influential upon the response of the glucose oxidase, even under direct electron transfer conditions.[115] The protic IL ethylimidazolium trifluoromethylsulfonate ([Eim][OTf], prepared from equimolar concentrations of an acid and base) resulted in both high ionic conductivity and beneficial conformational changes of the microenvironment around the glucose oxidase. This resulted in high currents and unaltered biocatalytic properties. Well-defined redox peaks were observed for the incorporated glucose oxidase, and decreases in the cathodic peak current of the glucose oxidase peak could be directly related to the glucose content in solution. A LOD of 1.5 μM glucose was achieved under ideal conditions, and the assembly could be used to quantify glucose in authentic serum samples

A nanocomposite was constructed by first mixing IL SWCNT with an amine-terminated IL (1-ethylamine-2,3-dimethylimidazolium bromide), and then electrodepositing gold nanoparticles onto this composite.[116] More amine-terminated IL was self-assembled onto the exposed gold nanoparticles surfaces, and then glucose oxidase was allowed to self-assemble at the composite by electrostatic interactions. The direct electrochemistry of glucose oxidase could thus be observed, and glucose could be quantified with an LOD of 0.8 μM.

The above highlights various composites or electrodes based upon graphite or nanocarbons. Other reports include ILs in conjunction with porous carbon nanofibers,[117] fullerene,[118] graphene oxide[119] and also graphene coated in polymeric ionic liquids.[120,121]

Li *et al.* have reported a three-component film for glucose oxidase entrapment and direct electrochemistry.[122] Unlike many studies, the composite does not contain any (nano)carbon component, beyond the glassy carbon electrode the composite was supported upon. It was composed of gold nanoparticles, *N,N*-dimethylformamide and $[C_4mim][PF_6]$, and resulted in significant, reversible peaks for direct glucose oxidase electron transfer. Without the IL, very significant reductions in current were observed, and the thermal stability of the composite was markedly reduced. Without *N,N*-dimethylformamide the enzyme features were lost upon repeated scanning, and without gold nanoparticles the electron transfer was irreversible. The whole composite could be used for glucose quantification between 0.1 and 20 μM in 0.050 M pH 5 phosphate buffer, by monitoring the reduction in the glucose oxidase features and concurrent increase in hydrogen peroxide features. It could successfully quantify glucose in human plasma and beer samples.

While most studies have utilised glucose oxidase, the enzyme glucose dehydrogenase can also be effective. A homogeneous mixture of MWCNT and IL was prepared.[108] This approach allowed it to operate as an oxygen sensor between 0% and 100% O_2 content in the headspace, with an LOD of 126 $\mu g\ L^{-1}$. Subsequent entrapment of glucose dehydrogenase within a chitosan layer at the surface of the composite allowed it to operate as a glucose biosensor, with a linear range of 0.02–1 mM glucose and an LOD of 9 μM. This assembly could thus be used to quantify both glucose and oxygen levels in blood samples.

Biosensors are not suitable for all analytical roles, especially those that require measurement outside of physiological conditions such as extremes in pH or temperature. Additionally, relative to biosensors, non-enzymatic sensors are (currently) more readily manufactured, stored and can better serve certain analytical functions. For this reason the investigation and evaluation of non-enzymatic sensors has paralleled biosensor development and, as a result, ILs have also found beneficial applications in non-enzymatic glucose sensors.

A gel was prepared from a mixture of MWCNT and the IL 1-(4-sulfonylbutyl)-3-methylimidazolium hexafluorophosphate, and incorporation of gold nanoparticles (which subsequently interacted strongly with the MWCNT) resulted in electrocatalytic activities towards non-enzymatic glucose oxidation in alkaline media.[123] Under these conditions the sulfonyl group of the IL cation would have been deprotonated, and was believed to take the form of MWCNT/gold nanoparticles surrounded by a zwitterionic species. Under these conditions the gel could be successfully used for the quantification of glucose in the linear range 5–120 μM, even in the presence of physiologically relevant concentrations of chloride, uric acid and ascorbic acid. However, the zwitterionic nature of the assembly did not prevent fouling by proteins commonly found in genuine biological samples.

A much wider range of ILs has also been investigated in relation to IL–MWCNT–gold nanoparticle gels for the direct non-enzymatic oxidation of glucose.[124] In general, increasing hydrophobicity of the IL was detrimental to the glucose oxidation currents observed at the composite. For example, hydrophilic anions such as $[BF_4]^-$ were preferential over hydrophobic anions such as $[PF_6]^-$. The imidazolium cation was superior to other aromatic and non-aromatic cations; extension of the alkyl chain also reduced the observed current (and therefore sensitivity). The zwitterionic IL cation from above, 1-(butyl-4-sulfonate)-3-methylimidazolium, was the most hydrophilic IL cation investigated and also resulted in the highest observed current densities using the gold nanoparticle-containing bucky gels.

The direct electrocatalytic oxidation of glucose can be achieved using inexpensive electrocatalysts such as nickel, albeit also in alkaline media.[125] Nanoplatelets of $Ni(OH)_2$ were incorporated in a CILE; as a result glucose could be quantified between 50 μM and 23 mM, with an LOD of 6 μM glucose. This composite was relatively selective for glucose over ascorbic acid and uric acid, typical interferences in biological media.

Alloyed nanoparticles of PtRu, PtPd and PtAu were synthesised at the surface of MWCNT by electrodeposition in the presence and absence of ultrasonic irradiation, in order to introduce electrocatalysis for the non-enzymatic oxidation of glucose.[126] Ultrasonication resulted in smaller, better distributed nanoparticles. The combination of the nanoparticles, IL and MWCNT all contributed beneficially towards the largest active surface area, lowest electron transfer resistance and thus the optimum electrochemical characteristics for amperometric detection and quantification of glucose. A PtRu-MWNT-IL nanocomposite-modified GC electrode displayed a LOD of 50 μM in neutral media, with no significant interference from ascorbic acid, uric acid, *p*-acetamidophenol and fructose.

ILs have also been reported in the "upstream" production of electrodes that can be used for the direct (non-enzymatic) oxidation of glucose.[127] The electrodeposition of a PtZn alloy followed by Zn-dealloying in a zinc chloride-1-ethyl-3-methylimidazolium chloride IL resulted in a nanoporous Pt electrode. This rough, high-surface area was effective for the amperometric quantification of glucose content at pH 7.4. The roughness factor could be easily controlled by controlling the Pt-to-Zn ratio in the IL from which the electrode was initially prepared. Interestingly, as the roughness factor of the final Pt electrode increased, so did the observed selectivity for this electrode for glucose oxidation over interferences such as ascorbic acid and *p*-acetamidophenol.

10.2.4.2 Other Analytes

The above glucose system highlights a range of approaches taken towards biosensors and bioanalysis, and the various applications of ILs in these approaches. The complexity of human biology means that there will always be a vast range of analytes for bioanalysis; the incredible range, sophistication and (electro)catalytic abilities of compounds such as proteins and enzymes will also allow the preparation of a vast range of biosensors.

Other bio-analytes that ILs have assisted in the quantification of include a wide range of pharmaceuticals and food molecules, as covered elsewhere in this chapter (primarily in the carbon paste electrode section, Section 10.2.3).

Lu *et al.* immobilised the protein hemoglobin within a chitosan and $[C_4mim][BF_4]$ matrix.[128] Investigation of the composite revealed that the IL resulted in improved thermal and chemical stability, indicative of good biocompatibility for the protein. Direct electrochemistry of the hemoglobin was observed, and bio(electro)catalytic activity was demonstrated with respect to oxygen reduction and trichloroacetic acid.

Various inorganic nanomaterials have also been incorporated with ILs, nanocarbon and hemoglobin to improve the direct electrochemistry of the protein.[129–132]

Many other studies have employed carbon nanomaterials in conjunction with ILs and other enzymes or proteins, the ILs facilitating dispersion and utilisation of the nanomaterial, and in many cases the stability of the biomolecule. This includes horseradish peroxidase,[133,134] myoglobin,[135,136] cytochrome *c*,[137,138] *etc.* An almost unlimited range of analytes is possible, and ILs have been reported for the quantification of DNA,[139] *Escherichia coli*,[140] specific cancer biomarkers,[141–143] doping agents,[144,145] *etc.*

ILs have also been employed for their inherent ability to stabilise some nanoparticles, resulting in small particles and extremely stable colloids. One example is the application of iridium nanoparticles in $[C_4mim][PF_6]$, mixed with chitosan-immobilised polyphenol oxidase.[146] Chlorogenic acid was oxidised by the enzyme to the corresponding quinone, which could be electrochemically quantified at the iridium-modified electrode surface. A similar process employed pine nut peroxidase in conjunction with $[C_4mim][NTf_2]$ to quantify rosmarinic acid, *via* electrochemistry of its quinone form.[147]

It is obvious from this section, which covers only a fraction of the work performed, that biosensors and bioanalysis is a vast and promising area. Nanomaterials have had by far the largest impact upon facilitating biosensors, but ILs clearly have the ability to allow full use of nanomaterials such as graphene and carbon nanotubes. The IL clearly also has an effect upon the structure, and therefore stability and activity (for better or worse). ILs therefore clearly have an important role, particularly as greater understanding is gathered on specific IL–biomolecule interactions.

10.2.5 Heavy Metals

An increasingly popular area of research on ionic liquid-based sensing concerns the detection of heavy metals. This section describes some of the strategies that researchers have used to detect heavy metals including lead, cadmium, mercury, copper and arsenic, mainly for environmental (*e.g.* water quality monitoring) applications. Most articles focus on cadmium (Cd^{2+}) or lead (Pb^{2+}) detection in aqueous samples, typically using glassy carbon (GC) film-modified electrodes or carbon-paste electrodes as sensing materials. For example, Jia *et al.*[148] reported a glassy-carbon electrode modified with a composite of poly(sodium 4-styrenesulfonate) (PSS) with the RTIL $[C_4mim][PF_6]$ and a bismuth film for detecting Cd^{2+} and Pb^{2+} in acetate buffer solutions. Using anodic stripping voltammetry (ASV) with a 120 s deposition time, a linear range from 1 to 50 μg L^{-1} was observed for both metals, with detection limits of 0.07 μg L^{-1} for Cd^{2+} and 0.09 μg L^{-1} for Pb^{2+}. The electrode showed good analytical responses for Cd^{2+} and Pb^{2+} in real-world waste water samples. Guo *et al.*[149] also used a GC electrode, with the modifying film consisting of the RTIL $[C_4mim][BF_4]$ combined with a thiol-functionalised mesoporous molecular sieve. The electrode exhibited a linear response (using ASV) in the concentration range 29–870 nM of Cd^{2+} in acetate buffer samples, with a LOD of 1 nM obtained after a 240-s accumulation time.

Carbon ionic liquid electrodes (CILEs) have been widely used for detecting heavy metals, typically using RTILs as a replacement for the organic binder commonly used (*e.g.* paraffin oil). For example, Li *et al.*[150] reported the detection of Cd^{2+} and Pb^{2+} using a CILE containing graphite powder, hydroxyapatite and the RTIL $[C_8Py][PF_6]$. By using square-wave anodic stripping voltammetry (SWASV) with a 180 second preconcentration time, the sensor exhibited linear behavior in the range 1–100 nmol L^{-1}, with detection limits of 0.5 and 0.2 nmol L^{-1} for Cd^{2+} and Pb^{2+}, respectively. The sensor worked in both perchloric acid and waste water samples. Ping *et al.*[151] used a CILE containing graphite, the RTIL $[C_8Py][PF_6]$ and an electrochemically-deposited bismuth film for the detection of Cd^{2+} and Pb^{2+} in acetate buffer solutions. Using SWASV and a 120-s deposition time, the electrode exhibited a linear range from 1 to 100 μg L^{-1} with detection limits of 0.1 μg L^{-1} for cadmium and 0.12 μg L^{-1} for lead. The electrode was used to determine the concentrations of the two metals in soil sample extracts, with good performance. A CILE consisting of graphite powder, the RTIL $[C_4mim][PF_6]$, paraffin oil and an electrochemically deposited bismuth film was reported by Wang *et al.*[152] for the simultaneous determination of Cd^{2+} and Pb^{2+} in acetate buffer and soil extracts. Using SWASV with a 120-s deposition time, linear calibration graphs were obtained for both metal ions in the range 1–90 μg L^{-1}, and LODs of 0.12 μg L^{-1} for Cd^{2+} and 0.25 μg L^{-1} Pb^{2+}. Bagheri *et al.*[153] used a CILE with graphite powder, the RTIL $[C_4mpyrr][NTf_2]$ and triphenylphosphine-modified multi-walled carbon nanotubes (MWCNTs) for the detection of Cd^{2+}, Pb^{2+} and Hg^{2+}. Using SWASV with a 75 s preconcentration step, linear calibration plots were observed for all three metals in the range 0.1–150 nM. LODs were calculated as 0.07, 0.06 and 0.09 nM for Cd^{2+}, Pb^{2+} and Hg^{2+}, respectively, in aqueous buffer solutions. The electrode was successfully applied to determine metal concentrations in soil, gasoline, fish, tap water and waste-water samples with reasonable accuracy. Afkhami *et al.*[154] employed a CILE containing MWCNTs, the RTIL

$[C_4mpyrr][NTf_2]$ and a newly synthesised Schiff base to determine Cd^{2+} and Pb^{2+} in aqueous buffer solutions. SWASV was employed with a deposition time of 190 s, with linear calibration graphs obtained in the range 1–1100 μg L^{-1} and LODs of 0.25 μg L^{-1} for Pb^{2+} and 0.74 μg L^{-1} for Cd^{2+}. The electrode was employed in a wide range of real-world samples such as tap water, waste water, fish, shrimp, rice, tobacco, soya, sugar and hair samples. Zhai *et al.*[155] employed a CILE containing ordered mesoporous carbon, the RTIL (either $[C_2mim][EtSO_4]$ or $[C_2mim][BF_4]$) and chitosan to determine Pb^{2+} in nitric acid samples. DPASV (differential pulse anodic stripping voltammetry) was applied with a 200-s accumulation time, giving a linear range from 0.05 to 1.4 μM and a LOD of 25 nM for Pb^{2+}. This electrode was used to determine Pb^{2+} concentrations in rain water and piped water samples with good accuracy.

In contrast to the strategies employing GC film-modified electrodes or CILEs, Sbartai *et al.*[156] used a boron-doped diamond (BDD) microcell electrode to detect Pb^{2+} directly in the RTIL $[C_4mpyrr][NTf_2]$ (*i.e.* extracted from citrate buffer aqueous solution). Microcrystalline BDD was deposited on a silicon wafer and the working, reference and counter electrodes were etched out from the BDD wafer by micromachining. The electrodes were covered with the RTIL containing the complexing agent trioctylphosphine oxide (TOPO). DPASV was applied and calibration curves with and without TOPO were linear in the range 0–4 μg L^{-1} with a LOD of 0.3 μg L^{-1} for Pb^{2+} calculated. The level of extraction of Pb^{2+} for water to the RTIL was nearly double in the presence of TOPO.

In addition to Pb^{2+} and Cd^{2+}, other heavy-metal ions such as mercury (Hg^{2+}), copper (Cu^{2+}) and arsenic (As^{3+}) ions have also been detected using RTIL-based sensors. Niu *et al.*[157] reported the detection of Hg^{2+} using a disposable screen-printed antimony film electrode modified with MWCNTs and the RTIL $[C_4mim][BF_4]$. Using ASV, with a 120-s accumulation step, Hg^{2+} was detected in the range 20–140 μg L^{-1}, with a LOD of 0.36 μg L^{-1} in water/hydrochloric acid solutions. The electrode was applied to measurements of Hg^{2+} in drinking water and waste water samples with good analytical performances. The detection of Hg^{2+} was also investigated by Safavi *et al.*,[158] who employed a nanocomposite sensor containing thiolated amino acid-capped gold nanoparticles on a carbon–ionic liquid electrode (graphite powder with the RTIL $[C_8Py][PF_6]$). A linear range from 10 nM to 20 μM Hg^{2+} in phosphate buffer solutions was observed, with a detection limit of 2.3 nM. Good responses were observed for the sensor in tap and waste water samples. Zhou *et al.*[159] proposed a glassy carbon electrode modified with gold nanoparticles and $[C_4mim][BF_4]$ ionic liquid-functionalised graphene oxide (Figure 10.6). A combination of ASV and differential pulse voltammetry was employed, with a deposition time of 180 s and accumulation time of 660 s showing the best response to Hg^{2+}. The electrode exhibited good linearity in the range 0.1–100 nM, and a LOD of 0.03 nM was obtained in water/HCl solutions. The practical application of the sensor was investigated in tap water, bottled water and sea water samples.

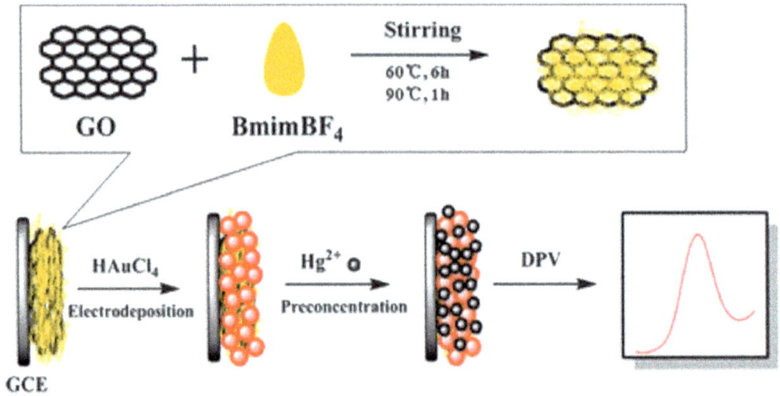

Figure 10.6 Electrochemical assay strategy based on AuNPs–GO–IL modified onto a glassy carbon electrode.[159] Note: GO, IL, AuNPs and Hg^{2+} are not drawn to scale.
(Copyright Royal Society of Chemistry.)

The detection of Cu^{2+} was demonstrated by Huang and Lin[160] who used a nanoporous gold electrode coated with a self-assembled monolayer of 3-mercaptopropylsulfonate (MPS), which was formed from dealloying zinc from a 40–60 mol.% zinc chloride–[C_2mim]Cl ionic liquid. A linear range of 0.1–5 μg L^{-1} and a LOD of 0.002 μg L^{-1} (0.031 nM) Hg^{2+} was reported. Although the ionic liquid was not used directly in the sensing process, its high temperature stability was exploited for the alloy formation, which could not be achieved with conventional volatile solvents. Finally, Gao *et al.*[161] suggested a disposable platform completely free of noble metals for the electrochemical detection of arsenic (As^{3+}) in water samples. The electrode consisted of a screen-printed carbon electrode modified with an iron oxide (Fe_3O_4)–RTIL composite film. The RTILs tested were [C_4dmim][NTf_2], [$N_{2,1,1,3}$][NTf_2], [C_4mim][FAP] and [C_4mim][NTf_2], with [C_4dmim][NTf_2] showing the best performance. Using SWASV, approximately linear calibration graphs were obtained in the range ~10–100 ppb, with the lowest ever LOD of 0.0008 ppb obtained in acetate buffer. The sensor was applied to detect As^{3+} in groundwater samples with reasonable accuracy.

These many reports have shown that RTILs can be successfully employed as electrode materials to detect heavy metal ions in many real-world water samples. In many cases, they show superior behavior to electrodes without RTILs, and by tuning the nature of the RTIL anion and cation, may result in more long-term stable sensors and even lower LODs for heavy metal analytes.

10.2.6 Other Analytical Targets

There are several reports of sensing in ionic liquids that do not fit neatly into one of the above sections, but appear to be interesting and worthwhile mentioning in this chapter to highlight the new advances in the field of ionic liquid-based sensing. On reviewing the articles, it appears that the detection of analytes takes place directly in the RTIL phase (*i.e.* dissolved in the RTIL at different concentrations), or at either film-modified glassy-carbon electrodes (GCEs) or screen-printed carbon electrodes (SPCEs). In the latter case, the detection of analytes often takes place in aqueous solutions using the film-modified electrodes. Therefore, this section is broken down into the three main sub-headings below.

10.2.6.1 Direct Detection in the RTIL

Villagran *et al.*[162] described the detection of trace chloride ions directly in the RTILs [C_4mim][BF_4], [C_4mim][NTf_2] and [C_4mim][PF_6] using cathodic stripping voltammetry at a silver disk electrode. Good analytical responses were observed, with LODs in the low ppm range in the three ionic liquids. The same group[163] later proposed a microfluidic device for the electrochemical determination of trace chloride in six different RTILs ([C_4mim][BF_4], [C_4mim][NTf_2], [C_4mpyrr][NTf_2], [C_4mpyrr][FAP], [C_4mim][OTf] and [$P_{14,6,6,6}$][NTf_2]). Using either square wave voltammetry or differential pulse voltammetry, LODs of approximately 5 ppm were calculated in the six RTILs. This presents a way to quantify the concentration of chloride impurities in RTILs, which can often be left over from their synthesis.

Lu *et al.*[31] used a task-specific ionic liquid (TSIL, *i.e.* an ionic liquid with a specific functional group) for the direct electrochemistry and detection of metal oxides. The TSIL was combined with bismuth oxide (Bi_2O_3) on a planar surface with working, reference and counter electrodes and used to simultaneously detect cadmium oxide (CdO), copper oxide (CuO) and lead oxide (PbO) (Figure 10.7). The TSIL consisted of a tetraalkylammonium cation with a carboxylic acid group on one carbon chain, along with a [NTf_2]$^-$ anion. The carboxylic acid group aided the solubilisation of the metal oxide into the solvent. Using anodic stripping voltammetry and a 120 s accumulation time, low detection limits of 0.28, 0.30 and 0.34 ng L^{-1} were obtained for CdO, CuO and PbO, respectively.

The inherent ability of some ILs to dissolve a range of materials has been investigated extensively with regards to biomass processing.[164] This opens up the route to use the IL as a digestion solvent and electrolyte all in one; small volumes of IL (relative to that of volatile solvents) can be used as there are no reflux or volatilisation issues, resulting in significant pre-concentration. Lau *et al.*[32] very recently investigated ILs for fresh chilli dissolution. CV of the solution resulted in quantification of both capsaicin (flavor indicator, and directly related to the spiciness of the chilli) and ascorbic acid (freshness indicator).

Shamsipur *et al.*[165] used square wave and differential pulse voltammetry to detect 2-furaldehyde (a chemical used in oil-refining processes) in three RTILs, [C_2mim][BF_4], [C_4mim][OTf] and [C_4mpyrr][NTf_2], on a glassy carbon electrode. Linear calibration graphs were observed in the approximate range 100–400 μg g^{-1}, with LODs of 1.4, 19.0 and 2.5 μg g^{-1} in [C_2mim][BF_4],

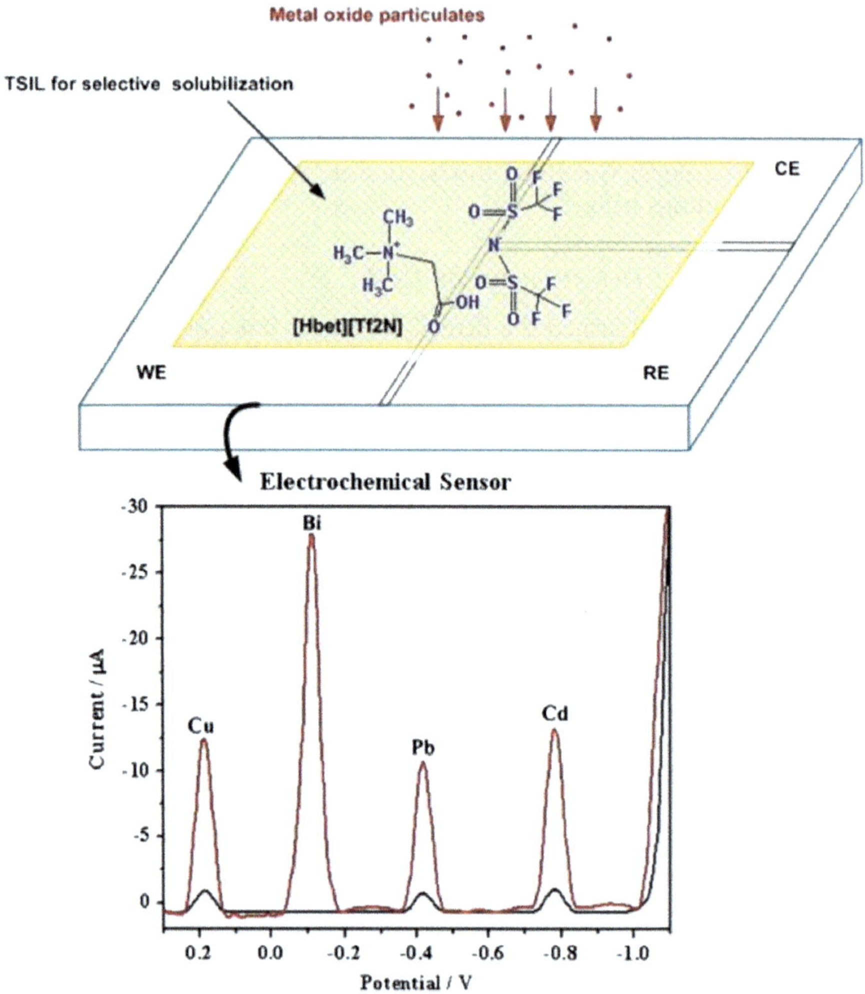

Figure 10.7 Top: Schematic diagram of an electrochemical sensor with a thin layer of task specific ionic liquid (TSIL) as a selective solubilisation medium. Bottom: Square wave voltammetric response of 8 ng L^{-1} PbO, CdO and CuO in the absence (black line) and presence (red line) of 10 ng L^{-1} Bi_2O_3.[31]
(Copyright Elsevier. Reproduced with permission.)

$[C_4mim][OTf]$ and $[C_4mpyrr][NTf_2]$, respectively, indicating that the nature of the RTIL anion and cation has an effect on the analytical response. Finally, Yang *et al.*[166] reported the detection of *N*-nitrosodiphenylamine (NDPhA), a powerful carcinogen, directly in the RTIL $[C_4mim][BF_4]$ on a porous gold working electrode with a large surface area. Using CV, peak currents were approximately linear *versus* concentration in the range 10–100 mM NDPhA, showing the analytical utility of RTILs for detecting these type of carcinogens.

10.2.6.2 Film-modified Glassy Carbon Electrodes

Dedzo and Detellier[167] used a GCE modified with a nanohybrid kaolinite material, containing kaolinite interlayer spaces in the ionic liquid 1-benzyl-3-(2-hydroxyethyl)imidazolium chloride, for the detection of iodide anions. The electrode exhibited excellent stability and had a detection limit of 1.5×10^{-7} M iodide. It was the first time such a material had been used for rigorous quantification of anions. Ma *et al.*[168] detected nitrobenzene in tap and lake water using a GCE electrode modified with macro-/meso-porous carbon materials. The materials were prepared by pyrolysis of a polymer ionic liquid, 1-allyl-3-ethylimidazolium tetrafluoroborate ([AEIm][BF_4]), wrapped around SiO_2 microspheres, followed by removal of the silica core. Good reproducibility and stability of the sensor was reported, with a linear response from 0.2 to 40 μM, and a LOD of 8 nM. Other nitroaromatic compounds were detected by Zhao *et al.*[169] using a film containing the RTIL [C_4mim][PF_6] with single-walled carbon nanotubes, deposited on a GCE. Six nitroaromatics were detected with LODs between 2 and 20 nM, and the sensor was used to determine the six compounds in lake water samples.

An interesting approach was employed by Wang *et al.*,[170] who used ionic liquid functionalised graphene composites to detect catechol and hydroquinone in phosphate buffer solutions using differential pulse voltammetry. The composites were prepared using the ionic liquid 1-butyl-3-methylimidazolium 2-amino-3-mercaptoproponic acid (L-cysteine) salt, [C_4mim][Cys], as both a reducing agent and functionalising agent. The electrode exhibited a good analytical response with LODs of 1 and 0.85 μM for catechol and hydroquinone, respectively. Norouzi *et al.*[171] reported the detection of ochratoxin A, a food contaminant, using a sensor based on a nanocomposite film of the RTIL [C_4mim][BF_4] with graphene nanosheets and gold nanoparticles. The sensor achieved a LOD of 0.22 nM, with a short response time (<7 s) and good long-term stability (60 days).

The antibiotic cefotaxime was detected by Yang *et al.*[172] using a GCE modified with a molecularly imprinted polymer, gold networks, the RTIL [C_4mim][BF_4], porous Pt nanoparticles and carboxyl graphene. The ionic liquid caused the gold to form a three-dimensional structure, and also played a role in immobilizing the porous Pt nanoparticles. A LOD of 0.1 nM was observed in BR buffer, and satisfactory results were achieved for the detection of cefotaxime in human serum samples. Xu *et al.*[173] reported the detection of the herbicide trifluralin on a GCE modified with acetylene black and the RTIL [C_4mim][PF_6]. In PBS (phosphate buffered saline), a detection limit of 10 nM and linear range of 80 nM to 12 μM was achieved, and the sensor was applied to determine trifluralin in soil samples. Finally, Zhao *et al.*[174] used a GCE modified with a film containing a molecularly imprinted polymer, the ionic liquid 1-(2-hydroxyethyl)-3-methylimidazolium bis(trifluoromethylsulfonyl)imide ([HeMIM][NTf_2]) and graphene oxide to detect the pesticide methyl parathion in PBS. A LOD of 6 nM was obtained and the sensor was applied to real samples such as cabbage and apple peel.

10.2.6.3 Film-Modified Screen-Printed Carbon Electrodes

Sulfide (S^{2-}) detection was reported by Chang *et al.*[175] using a screen-printed carbon electrode covered with a hydrophobic ionic liquid polymer film crosslinked between $[C_4Vim][PF_6]$ and $[C_9(Vim)_2][PF_6]_2$. A redox mediator, $Fe(CN)_6^{3-}$, was then exchanged into the polymeric film using CV. Using chronoamperometry, linear calibration graphs were obtained for sulfide at concentrations from 1 μM to 3 mM in phosphate buffer solution, with a LOD of 12.9 nM sulfide. The sensor was successfully used to determine sulfide content in hot spring water and ground water. Su *et al.*[176] used a SPCE modified with a functionalised ionic liquid ($[C_4mim]$[ferricyanide]) and covered with a layer of poly(3-(aminopropyl)trimethoxysilane) sol–gel to detect nitrite in acidic solutions. Good electrocatalytic activity was observed, with a linear range of 20–510 μM and LOD of 1.3 μM nitrite calculated using amperometry. Wu *et al.*[177] used a SPCE modified with a composite of MWCNTs and the RTIL $[C_4mim][NTf_2]$, followed by electrodeposition of luteolin into the film. The modified electrode was used to detect hydrazine, with two linear ranges observed, and a LOD for hydrazine of 6.6 nM in phosphate buffer solutions. The sensor was also used to determine hydrazine in spiked drinking water and river water with good results. Lastly, Wei *et al.*[178] employed a SPCE modified with a film of MWCNTs and the RTIL $[N_{2,1,1,3}][NTf_2]$ for the determination of tetrachlorobenzene (TeCB), an organic micropollutant, in aqueous solutions. Electrochemical impedance spectroscopy was employed, and the sensor showed a good sensitivity for TeCB, with a LOD of 0.05 μM reported.

From this section, ionic liquids clearly have many uses in sensor design for detecting a wide range of analytes—from simple ions, organic compounds, toxins, pesticides, carcinogens, pharmaceuticals *etc.* It has been shown that by tuning of the ionic liquid by adding functional groups may aid the solubilisation of more species in the IL, improving the range of analytes that can be detected. As with heavy metal determination (Section 10.2.5), RTILs are often incorporated into films and deposited onto electrodes (*e.g.* glassy carbon or screen-printed carbon) to detect analytes in aqueous solutions, since these are often the real-world samples that need to be analyzed. It is clear from the number of recent articles in this field that RTILs have much to offer in the sensing of many different analyte species.

10.3 Conclusions and Future Outlook

The chapter has discussed various aspects of ILs when applied in electrochemical sensing, quantification and electroanalysis. Early investigations, likely prompted by curiosity about the "new" electrolytes, have clearly resulted in significant discoveries and improvements in conventional techniques. The combined use of carbon nanomaterials and ILs has clearly benefited the application of both materials, resulting in stable composites, maximised surface areas and additional functionality introduced by the IL.

The inherent ability of ILs to result in (apparent)[179] electrocatalysis with respect to analytes has been beneficial. Generally, this stated electrocatalysis is an increase in current or slight shift in potential, due to the presence of the IL maximising surface area and pre-concentrating the analyte in its distinct phase at the electrode surface. Such changes clearly have positive results during sensing applications, as generally demonstrated throughout this chapter, and particularly demonstrated by the case of "carbon ionic liquid electrodes" that are used as replacements for classical "carbon paste electrodes". Electroanalysis is further enhanced by the incorporation of genuine electrocatalytic materials (*i.e.* gold nano-particles for glucose oxidation) where a combination with ILs can result in an additive or synergistic improvement in the system.

The inherent ability of ILs to stabilise certain biomolecules, and even enhance or alter their reactivity, will become of increasing relevance. Already it has been demonstrated that the incorporation of ILs can improve the thermal and chemical stability of such biomolecules; these are key issues that have prevented the wider application of biosensors.

The inherent ability of ILs to act as non-volatile electrolytes has prompted their application in areas such as gas sensors, pre-concentration and mini-aturised sensors. ILs remove volatility issues common at small volumes, and even allow such systems to be applied under extreme conditions, such as vacuum or elevated temperatures.

Many of these exciting developments have come about due to the inherent properties of readily available ILs. Even more exciting work is coming about as a direct result of improved physicochemical understanding and "task specific ionic liquids". Minor or major changes in the structure of the IL can introduce a further level of functionality to the sensor, increasing the selectivity, sensitivity, longevity, kinetics, thermodynamic, *etc.*

Key disadvantages are the major gaps that still exist in our knowledge regarding the synthesis, design and structure–property relationships of ILs; as such they can only be applied as "designer electrolytes" following con-siderable fundamental work. The expense, unfamiliarity of their synthesis, handling and application, accompanied with very few real demonstrations of their long-term application all go against the commercial adoption of ILs in sensors.

There is therefore a vast amount of work remaining to be done, but also a vast range of improvements, unique applications and new paradigms awaiting the application of ILs in electrochemical sensing.

References

1. M. C. Buzzeo, R. G. Evans and R. G. Compton, *Phys. Chem. Chem. Phys.*, 2004, **5**, 1106.
2. M. Armand, F. Endres, D. MacFarlane, H. Ohno and B. Scrosati, *Nat. Mater.*, 2009, **8**, 621.
3. D. S. Silvester and R. G. Compton, *Z. Phys. Chem.*, 2006, **220**, 1247.

4. P. Wasserscheid and W. Keim, *Angew. Chem., Int. Ed.*, 2000, **39**, 3772.
5. T. Welton, *Chem. Rev.*, 1999, **99**, 2071.
6. J. S. Wilkes, *J. Mol. Catal. A: Chem.*, 2004, **214**, 11.
7. J. Dupont, C. S. Consortia and J. Spencer, *J. Braz. Chem. Soc.*, 2000, **11**, 337.
8. M. J. Earle and K. R. Seddon, *Pure Appl. Chem.*, 2000, 72, 1391.
9. H. Zhao, *Chem. Eng. Commun.*, 2006, **193**, 1660.
10. C. Chiappe and D. Pieraccini, *J. Phys. Org. Chem.*, 2005, **18**, 275.
11. S. Chowdhury, R. S. Mohan and J. L. Scott, *Tetrahedron*, 2007, **63**, 2363.
12. G. A. Baker, S. N. Baker, S. Pandey and F. V. Bright, *Analyst*, 2005, **130**, 800.
13. S. Pandey, *Anal. Chim. Acta*, 2006, **556**, 38.
14. R. A. Sheldon, R. M. Lau, M. J. Sorgedrager, F. van Rantwijk and K. R. Seddon, *Green Chem.*, 2002, **4**, 147.
15. F. van Rantwijk and R. A. Sheldon, *Chem. Rev.*, 2007, **107**, 2757.
16. N. V. Plechkova and K. R. Seddon, *Chem. Soc. Rev.*, 2008, **37**, 123.
17. M. J. A. Shiddiky and A. A. J. Torriero, *Biosens. Bioelectron.*, 2011, **26**, 1775.
18. N. V. Shvedene, D. V. Chernyshov and I. V. Pletnev, *Russ. J. Gen. Chem.*, 2008, **52**, 80.
19. D. S. Silvester, *Analyst*, 2011, **136**, 4871.
20. W. Wei and A. Ivaska, *Anal. Chim. Acta*, 2008, **607**, 126.
21. F. Endres and S. Zein El Abedin, *Phys. Chem. Chem. Phys.*, 2006, **8**, 2101.
22. D. S. Silvester and D. W. M. Arrigan, *Electrochem. Commun.*, 2011, **13**, 477.
23. M. C. Buzzeo, C. Hardacre and R. G. Compton, *ChemPhysChem*, 2006, 7, 176.
24. E. I. Rogers, A. M. O'Mahony, L. Aldous and R. G. Compton, *ECS Trans.*, 2010, **33**, 473.
25. N. V. Ignat'ev, U. Welz-Biermann, A. Kucheryna, G. Bissky and H. Willner, *J. Fluorine Chem.*, 2005, **126**, 1150.
26. L. E. Barrosse-Antle, A. M. Bond, R. G. Compton, A. M. O'Mahony, E. I. Rogers and D. S. Silvester, *Chem. – Asian J.*, 2010, **5**, 202.
27. J. H. Davis, *Chem. Lett.*, 2004, **33**, 1072.
28. A. E. Visser, R. P. Swatloski, W. M. Reichert, R. Mayton, S. Sheff, A. Wierzbicki, J. H. Davis and R. D. Rogers, *Environ. Sci. Technol.*, 2002, **36**, 2523.
29. A. E. Visser, R. P. Swatloski, W. M. Reichert, R. Mayton, S. Sheff, A. Wierzbicki, J. H. Davis and R. D. Rogers, *Chem. Commun.*, 2001, 135.
30. J. L. McDonald, R. E. Sykora, P. Hixon, A. Mirjafari and J. H. Davis, *Environ. Chem. Lett.*, 2014, **12**, 201.
31. D. L. Lu, N. Shomali and A. Shen, *Electrochem. Commun.*, 2010, **12**, 1214.
32. B. B. Y. Lau, J. Panchompoo and L. Aldous, *New J. Chem.*, 2015, **39**, 860.
33. M. C. Buzzeo, C. Hardacre and R. G. Compton, *Anal. Chem.*, 2004, **76**, 4583.

34. K. Murugappan, J. Lee and D. S. Silvester, *Electrochem. Commun.*, 2011, **13**, 1435.
35. R. Wang, T. Okajima, F. Kitamura and T. Ohsaka, *Electroanalysis*, 2004, **16**, 66.
36. Z. Wang, P. L. Lin, G. A. Baker, J. Stetter and X. Q. Zeng, *Anal. Chem.*, 2011, **83**, 7066.
37. X. J. Huang, L. Aldous, A. M. O'Mahony, F. J. del Campo and R. G. Compton, *Anal. Chem.*, 2010, **82**, 5238.
38. L. H. J. Xiong, P. Goodrich, C. Hardacre and R. G. Compton, *Sens. Actuators, B*, 2013, **188**, 978.
39. X. Mu, Z. Wang, X. Zeng and A. J. Mason, *IEEE Sens.*, 2013, **13**, 3976.
40. C. G. Hu, X. Y. Bai, Y. K. Wang, W. Jin, X. Zhang and S. S. Hu, *Anal. Chem.*, 2012, **84**, 3745.
41. J. Lee, K. Murugappan, D. W. M. Arrigan and D. S. Silvester, *Electrochim. Acta*, 2013, **101**, 158.
42. X. Shen, X. Chen, J. H. Liu and X. J. Huang, *J. Mater. Chem.*, 2009, **19**, 7687.
43. S. Q. Xiong, Y. Wei, Z. Guo, X. Chen, J. Wang, J. H. Liu and X. J. Huang, *J. Phys. Chem. C*, 2011, **115**, 17471.
44. R. Toniolo, N. Dossi, A. Pizzariello, A. P. Doherty, S. Susmel and G. Bontempelli, *J. Electroanal. Chem.*, 2012, **670**, 23.
45. N. Baltes, F. Beyle, S. Freiner, F. Geier, M. Joos, K. Pinkwart and P. Rabenecker, *Talanta*, 2013, **116**, 474.
46. J. Gebicki, *Electroanalysis*, 2011, **23**, 1958.
47. J. Gebicki and A. Kloskowski, *Metrol. Meas. Syst.*, 2010, **17**, 637.
48. J. Gebicki, A. Kloskowski and W. Chrzanowski, *Sens. Actuators, B*, 2013, **177**, 1173.
49. N. Dossi, R. Toniolo, A. Pizzariello, E. Carrilho, E. Piccin, S. Battiston and G. Bontempelli, *Lab Chip*, 2012, **12**, 153.
50. R. Toniolo, N. Dossi, A. Pizzariello, A. Casagrande and G. Bontempelli, *Anal. Bioanal. Chem.*, 2013, **405**, 3571.
51. T. Carvalho, P. Vidinha, B. R. Vieira, R. W. C. Li and J. Gruber, *J. Mater. Chem. C*, 2014, **2**, 696.
52. X. B. Ji, C. E. Banks, D. S. Silvester, L. Aldous, C. Hardacre and R. G. Compton, *Electroanalysis*, 2007, **19**, 2194.
53. K. Murugappan, C. Kang and D. S. Silvester, *J. Phys. Chem. C*, 2014, **118**, 19232–19237.
54. B. A. Rosen, A. Salehi-Khojin and R. I. Masel, *2010 IEEE Sens.*, 2010, 365.
55. Z. Wang, X. Y. Mu, M. Guo, Y. Huang, A. J. Mason and X. Q. Zeng, *J. Electroanal. Chem.*, 2013, **160**, B83.
56. X. Y. Mu, Z. Wang, M. Guo, X. Q. Zeng and A. J. Mason, *2012 IEEE BioCAS*, 2012, 140.
57. L. H. Liu, C. Q. Duan and Z. N. Gao, *J. Serb. Chem. Soc.*, 2012, **77**, 483.
58. M. Nadherna, F. Opekar and J. Reiter, *Electrochim. Acta*, 2011, **56**, 5650.

59. M. Nadherna, F. Opekar, J. Reiter and K. Stulik, *Sens. Actuators, B*, 2012, **161**, 811.
60. S. R. Ng, C. X. Guo and C. M. Li, *Electroanalysis*, 2011, **23**, 442.
61. R. Toniolo, N. Dossi, A. Pizzariello, A. P. Doherty and G. Bontempelli, *Electroanalysis*, 2012, **24**, 865.
62. M. A. G. Zevenbergen, D. Wouters, V. A. T. Dam, S. H. Brongersma and M. Crego-Calama, *2011 IEEE Sens.*, 2011, 585.
63. E. S. Forzani, D. L. Lu, M. J. Leright, A. D. Aguilar, F. Tsow, R. A. Iglesias, Q. Zhang, J. Lu, J. H. Li and N. J. Tao, *J. Am. Chem. Soc.*, 2009, **131**, 1390.
64. A. D. Aguilar, E. S. Forzani, M. Leright, F. Tsow, A. Cagan, R. A. Iglesias, L. A. Nagahara, I. Amlani, R. Tsui and N. J. Tao, *Nano Lett.*, 2010, **10**, 380.
65. C. X. Guo, Z. S. Lu, Y. Lei and C. M. Li, *Electrochem. Commun.*, 2010, **12**, 1237.
66. S. J. Guo, D. Wen, Y. M. Zhai, S. J. Dong and E. K. Wang, *Biosens. Bioelectron.*, 2011, **26**, 3475.
67. C. H. Xiao, A. Rehman and X. Q. Zeng, *Anal. Chem.*, 2012, **84**, 1416.
68. L. Yu, Y. Huang, X. X. Jin, A. J. Mason and X. Q. Zeng, *Sens. Actuators, B*, 2009, **140**, 363.
69. Y. M. Wang, H. Y. Xiong, X. H. Zhang, Y. Ye and S. F. Wang, *J. Electroanal. Chem.*, 2012, **674**, 17.
70. V. V. Singh, M. Boopathi, K. Ganesan, B. Singh and R. Vijayaraghavan, *Electroanalysis*, 2010, **22**, 1357.
71. V. V. Singh, A. K. Nigam, M. Boopathi, P. Pandey, B. Singh and R. Vijayaraghavan, *Sens. Actuators, B*, 2012, **161**, 1000.
72. A. J. Bard, G. g. Inzelt and F. Scholz, *Electrochemical Dictionary*, Springer, Berlin, London, 2008.
73. N. Maleki, A. Safavi and F. Tajabadi, *Electroanalysis*, 2007, **19**, 2247.
74. P. Macikova, V. Halouzka, J. Hrbac, P. Bartak and J. Skopalova, *Sci. World J.*, 2012, 394756.
75. M. M. Musameh, R. T. Kachoosangi, L. Xiao, A. Russell and R. G. Compton, *Biosens. Bioelectron.*, 2008, **24**, 87.
76. L. Lu, X. R. Huang and Y. B. Qu, *J. Solid State Electrochem.*, 2012, **16**, 3299.
77. J. B. Zheng, Y. Zhang and P. P. Yang, *Talanta*, 2007, **73**, 920.
78. W. Sun, Q. Jiang, M. Y. Xi and K. Jiao, *Microchim. Acta*, 2009, **166**, 343.
79. L. H. Liu, C. Q. Duan and Z. N. Gao, *Croat. Chem. Acta*, 2010, **83**, 409.
80. Y. M. Zhang, C. Q. Duan and Z. N. Gao, *J. Serb. Chem. Soc.*, 2013, **78**, 281.
81. W. Sun, Y. Y. Duan, Y. Z. Li, T. R. Zhan and K. Jiao, *Electroanalysis*, 2009, **21**, 2667.
82. H. W. Gao, M. Y. Xi, L. Xu and W. Sun, *Microchim. Acta*, 2011, **174**, 115.
83. W. Sun, Y. H. Wang, Y. X. Lu, A. H. Hu, F. Shi and Z. F. Sun, *Sens. Actuators, B*, 2013, **188**, 564.

84. Y. Wang, H. Y. Xiong, X. H. Zhang and S. F. Wang, *Microchim. Acta*, 2010, **169**, 255.
85. Y. Zhang and J. B. Zheng, *Talanta*, 2008, **77**, 325.
86. Y. Y. She, Y. G. Tang, H. T. Liu and P. He, *Chem. Cent. J.*, 2010, **4**, 17.
87. X. Y. Sun, S. Hu, L. F. Li, J. Xiang and W. Sun, *J. Electroanal. Chem.*, 2011, **651**, 94.
88. A. Safavi and F. Farjami, *Electroanalysis*, 2011, **23**, 1842.
89. A. Safavi, S. Momeni and M. Tohidi, *Electroanalysis*, 2012, **24**, 1981.
90. A. Safavi and S. Momeni, *Electroanalysis*, 2010, **22**, 2848.
91. J. Vahedi, H. Karimi-Maleh, M. Baghayeri, A. L. Sanati, M. A. Khalilzadeh and M. Bahrami, *Ionics*, 2013, **19**, 1907.
92. P. Y. Chen and H. Y. Huang, *Electrochem. Commun.*, 2011, **13**, 1408.
93. X. L. Wang, C. C. Cheng, R. R. Dong and J. C. Hao, *J. Solid State Electrochem.*, 2012, **16**, 2815.
94. S. Hu, H. H. Zhu, S. Y. Liu, J. Xiang, W. Sun and L. Q. Zhang, *Microchim. Acta*, 2012, **178**, 211.
95. W. Sun, Y. H. Wang, S. X. Gong, Y. Cheng, F. Shi and Z. F. Sun, *Electrochim. Acta*, 2013, **109**, 298.
96. W. Sun, X. Z. Wang, H. H. Zhu, X. H. Sun, F. Shi, G. N. Li and Z. F. Sun, *Sens. Actuators, B*, 2013, **178**, 443.
97. M. Elyasi, M. A. Khalilzadeh and H. Karimi-Maleh, *Food Chem.*, 2013, **141**, 4311.
98. Z. R. Mo, Y. F. Zhang, F. Q. Zhao, F. Xiao, G. P. Guo and B. Z. Zeng, *Food Chem.*, 2010, **121**, 233.
99. R. T. Kachoosangi, G. G. Wildgoose and R. G. Compton, *Electroanalysis*, 2007, **19**, 1483.
100. M. Fouladgar and H. Karimi-Maleh, *Ionics*, 2013, **19**, 1163.
101. F. Shi, S. X. Gong, L. Xu, H. H. Zhu, Z. F. Sun and W. Sun, *Mater. Sci. Eng., C*, 2013, **33**, 4527.
102. C. M. Li, J. F. Zang, D. P. Zhan, W. Chen, C. Q. Sun, A. L. Teo, Y. T. Chua, V. S. Lee and S. M. Moochhala, *Electroanalysis*, 2006, **18**, 713.
103. F. Faridbod, M. R. Ganjali, B. Larijani and P. Norouzi, *Electrochim. Acta*, 2009, **55**, 234.
104. P. Norouzi, M. R. Ganjali, F. Faridbod, S. J. Shahtaheri and H. A. Zamani, *Int. J. Electrochem. Sci.*, 2012, 7, 2633.
105. H. T. Liu, Y. Liu and J. H. Li, *Phys. Chem. Chem. Phys.*, 2010, **12**, 1685.
106. M. Pumera, A. Ambrosi, A. Bonanni, E. L. K. Chng and H. L. Poh, *TrAc, Trends Anal. Chem.*, 2010, **29**, 954.
107. Y. R. Wang, P. Hu, Q. L. Liang, G. A. Luo and Y. M. Wang, *Chin. J. Anal. Chem.*, 2008, **36**, 1011.
108. L. Bai, D. Wen, J. Y. Yin, L. Deng, C. Z. Zhu and S. J. Dong, *Talanta*, 2012, **91**, 110.
109. F. H. Li, Z. H. Wang, C. S. Shan, J. F. Song, D. X. Han and L. Niu, *Biosens. Bioelectron.*, 2009, **24**, 1765.
110. B. Haghighi and R. Nikzad, *Electroanalysis*, 2009, **21**, 1862.

111. K. J. Fraser, S. Y. Yang, F. Cicoira, V. F. Curto, R. Byrne, F. Benito-Lopez, D. Khodagholy, R. M. Owens, G. G. Malliaras and D. Diamond, *Proc. SPIE*, 2011, **8118**, 81180C-1-12.
112. S. Y. Yang, F. Cicoira, R. Byrne, F. Benito-Lopez, D. Diamond, R. M. Owens and G. G. Malliaras, *Chem. Commun.*, 2010, **46**, 7972.
113. X. M. Wu, B. Zhao, P. Wu, H. Zhang and C. X. Cai, *J. Phys. Chem. B*, 2009, **113**, 13365.
114. Y. Liu, L. Liu and S. J. Dong, *Electroanalysis*, 2007, **19**, 55.
115. L. Liu, Y. Cheng, F. R. Sun, J. P. Yang and Y. Wu, *J. Solid State Electrochem.*, 2012, **16**, 1003.
116. R. F. Gao and J. B. Zheng, *Electrochem. Commun.*, 2009, **11**, 608.
117. Q. L. Sheng, R. X. Liu, J. B. Zheng, W. H. Lin and Y. Y. Li, *J. Solid State Electrochem.*, 2012, **16**, 739.
118. Z. L. Wei, Z. J. Li, X. L. Sun, Y. J. Fang and J. K. Liu, *Biosens. Bioelectron.*, 2010, **25**, 1434.
119. M. Tasviri, S. Ghasemi, H. Ghourchian and M. R. Gholami, *J. Solid State Electrochem.*, 2013, **17**, 183.
120. C. S. Shan, H. F. Yang, J. F. Song, D. X. Han, A. Ivaska and L. Niu, *Anal. Chem.*, 2009, **81**, 2378.
121. Q. Zhang, S. Y. Wu, L. Zhang, J. Lu, F. Verproot, Y. Liu, Z. Q. Xing, J. H. Li and X. M. Song, *Biosens. Bioelectron.*, 2011, **26**, 2632.
122. J. W. Li, J. J. Yu, F. Q. Zhao and B. Z. Zeng, *Anal. Chim. Acta*, 2007, **587**, 33.
123. H. Zhu, X. Q. Lu, M. X. Li, Y. H. Shao and Z. W. Zhu, *Talanta*, 2009, **79**, 1446.
124. H. Zhu, X. H. Liang, J. T. Chen, M. X. Li and Z. W. Zhu, *Talanta*, 2011, **85**, 1592.
125. A. Safavi, N. Maleki and E. Farjami, *Biosens. Bioelectron.*, 2009, **24**, 1655.
126. F. Xiao, F. Q. Zhao, D. P. Mei, Z. R. Mo and B. Z. Zeng, *Biosens. Bioelectron.*, 2009, **24**, 3481.
127. C. H. Chou, J. C. Chen, C. C. Tai, I. W. Sun and J. M. Zen, *Electroanalysis*, 2008, **20**, 771.
128. X. B. Lu, J. Q. Hu, X. Yao, Z. P. Wang and J. H. Li, *Biomacromolecules*, 2006, **7**, 975.
129. W. Sun, R. F. Gao and K. Jiao, *Electroanalysis*, 2007, **19**, 1368.
130. W. Sun, X. W. Qi, Y. Chen, S. Y. Liu and H. W. Gao, *Talanta*, 2011, **87**, 106.
131. B. Wang, Y. Li, X. J. Qin, G. Q. Zhan, M. Ma and C. Y. Li, *Mater. Sci. Eng., C*, 2012, **32**, 2280.
132. T. Wang, L. Wang, J. J. Tu, H. Y. Xiong and S. F. Wang, *Bioelectrochemistry*, 2013, **94**, 94.
133. Y. Liu, X. Q. Zou and S. J. Dong, *Electrochem. Commun.*, 2006, **8**, 1429.
134. P. Du, S. N. Liu, P. Wu and C. X. Cai, *Electrochim. Acta*, 2007, **52**, 6534.
135. Z. Dai, Y. Xiao, X. Z. Yu, Z. B. Mai, X. J. Zhao and X. Y. Zou, *Biosens. Bioelectron.*, 2009, **24**, 1629.
136. Q. Zhang, W. Wei and G. C. Zhao, *Electroanalysis*, 2008, **20**, 1002.

137. C. L. Xiang, Y. J. Zou, L. X. Sun and F. Xu, *Electrochem. Commun.*, 2008, **10**, 38.
138. J. S. Xu and G. C. Zhao, *Int. J. Electrochem. Sci.*, 2008, **3**, 519.
139. H. W. Gao, X. W. Qi, Y. Chen and W. Sun, *Anal. Chim. Acta*, 2011, **704**, 133.
140. Z. M. Liu, Y. Y. Cao, Z. J. Li, G. L. Shen and R. Q. Yu, *Sens. Lett.*, 2011, **9**, 563.
141. H. X. Fan, Y. Zhang, D. Wu, H. M. Ma, X. J. Li, Y. Li, H. Wang, H. Li, B. Du and Q. Wei, *Anal. Chim. Acta*, 2013, **770**, 62.
142. A. Salimi, B. Kavosi, F. Fathi and R. Hallaj, *Biosens. Bioelectron.*, 2013, **42**, 439.
143. B. Kavosi, A. Salimi, R. Hallaj and K. Amani, *Biosens. Bioelectron.*, 2014, **52**, 20.
144. R. A. Mundaca, M. Moreno-Guzman, M. Eguilaz, P. Yanez-Sedeno and J. M. Pingarron, *Talanta*, 2012, **99**, 697.
145. M. Moreno-Guzman, L. Agui, A. Gonzalez-Cortes, P. Yanez-Sedeno and J. M. Pingarron, *J. Solid State Electrochem.*, 2013, **17**, 1591.
146. S. C. Fernandes, S. K. Moccelini, C. W. Scheeren, P. Migowski, J. Dupont, M. Heller, G. A. Micke and I. C. Vieira, *Talanta*, 2009, **79**, 222.
147. K. D. Maguerroski, S. C. Fernandes, A. C. Franzoi and I. C. Vieira, *Enzyme Microb. Technol.*, 2009, **44**, 400.
148. J. B. Jia, L. Y. Cao, Z. H. Wang and T. X. Wang, *Electroanalysis*, 2008, **20**, 542.
149. J. Guo, Y. P. Luo, F. Ge, Y. L. Ding and J. J. Fei, *Microchim. Acta*, 2011, **172**, 387.
150. Y. H. Li, X. Y. Liu, X. D. Zeng, Y. Liu, X. T. Liu, W. Z. Wei and S. L. Luo, *Sens. Actuators, B*, 2009, **139**, 604.
151. J. F. Ping, J. Wu, Y. B. Ying, M. H. Wang, G. Liu and M. Zhang, *J. Agric. Food Chem.*, 2011, **59**, 4418.
152. Z. Q. Wang, G. Liu, L. N. Zhang and H. Wang, *Int. J. Electrochem. Sci.*, 2012, **7**, 12326.
153. H. Bagheri, A. Afkhami, H. Khoshsafar, M. Rezaei and A. Shirzadmehr, *Sens. Actuators, B*, 2013, **186**, 451.
154. A. Afkhami, H. Ghaedi, T. Madrakian and M. Rezaeivala, *Electrochim. Acta*, 2013, **89**, 377.
155. X. R. Zhai, L. Li, H. T. Gao, C. D. Si and C. Y. Yue, *Microchim. Acta*, 2012, **177**, 373.
156. A. Sbartai, P. Namour, N. Sarrut, J. Krejci, R. Kucerova, M. L. Hamlaoui and N. Jaffrezic-Renault, *J. Electroanal. Chem.*, 2012, **686**, 58.
157. X. H. Niu, H. L. Zhao and M. B. Lan, *Electrochim. Acta*, 2011, **56**, 9921.
158. A. Safavi and E. Farjami, *Anal. Chim. Acta*, 2011, **688**, 43.
159. N. Zhou, J. H. Li, H. Chen, C. Y. Liao and L. X. Chen, *Analyst*, 2013, **138**, 1091.
160. J. F. Huang and B. T. Lin, *Analyst*, 2009, **134**, 2306.
161. C. Gao, X. Y. Yu, S. Q. Xiong, J. H. Liu and X. J. Huang, *Anal. Chem.*, 2013, **85**, 2673.

162. C. Villagran, C. E. Banks, C. Hardacre and R. G. Compton, *Anal. Chem.*, 2004, **76**, 1998.
163. R. Ge, R. W. K. Allen, L. Aldous, M. R. Bown, N. Doy, C. Hardacre, J. M. MacInnes, G. McHale and M. I. Newton, *Anal. Chem.*, 2009, **81**, 1628.
164. M. M. Hossain and L. Aldous, *Aust. J. Chem.*, 2012, **65**, 1465.
165. M. Shamsipur, A. A. M. Beigi, M. Teymouri, Y. Ghorbani, M. Irandoust and A. Mehdizadeh, *Talanta*, 2010, **81**, 109.
166. H. L. Yang, G. H. Chu, Q. Chang, G. J. Zhou, J. H. Li and X. H. Xia, *Electroanalysis*, 2008, **20**, 2003.
167. G. K. Dedzo and C. Detellier, *Analyst*, 2013, **138**, 767.
168. J. Y. Ma, Y. S. Zhang, X. H. Zhang, G. B. Zhu, B. Liu and J. H. Chen, *Talanta*, 2012, **88**, 696.
169. F. Zhao, L. Liu, F. Xiao, J. Li, R. Yan, S. Fan and B. Zeng, *Electroanalysis*, 2007, **19**, 1387.
170. C. F. Wang, Y. J. Chen, K. L. Zhuo and J. J. Wang, *Chem. Commun.*, 2013, **49**, 3336.
171. P. Norouzi, B. Larijani and M. R. Ganjali, *Int. J. Electrochem. Sci.*, 2012, **7**, 7313.
172. G. M. Yang, F. Q. Zhao and B. Z. Zeng, *Biosens. Bioelectron.*, 2014, **53**, 447.
173. N. Xu, Y. Ding, H. Ai and J. Fei, *Microchim. Acta*, 2010, **170**, 165.
174. L. J. Zhao, F. Q. Zhao and B. Z. Zeng, *Sens. Actuators, B*, 2013, **176**, 818.
175. J. L. Chang, G. T. Wei, T. Y. Chen and J. M. Zen, *Electroanalysis*, 2013, **25**, 845.
176. W. Y. Su and S. H. Cheng, *Int. J. Electrochem. Sci.*, 2012, 7, 9058.
177. S. H. Wu, F. H. Nie, Q. Z. Chen and J. J. Sun, *Anal. Methods*, 2010, **2**, 1729.
178. Y. Wei, Z. G. Liu, X. Chen, J. H. Liu and X. J. Huang, *Anal. Methods*, 2013, **5**, 2440.
179. L. Aldous, A. Khan, M. M. Hossain and C. Zhao, in *Catalysis in Ionic Liquids: From Catalyst Synthesis to Application*, The Royal Society of Chemistry, 2014, p. 433.

Subject Index